Nutritional and Therapeutic Values of Fruits and Vegetables

THE AUTHOR

Dr. Kanaya Lal Bhat, Professor (rtd) Division of Vegetable Science and Floriculture, Sher-e – Kashmir University Jammu (J and K), received his Ph. D. in Vegetable Science from HPKV, Palampur in 1994. Before his superannuation, he was teaching to M. Sc and Ph. D students at SKUAST –Srinagar and SKUAST –Jammu for about twenty years and has guided a number of students in the discipline of Vegetable production. He has numerous research papers and has written extensively in scientific papers and for the general public. Dr, Bhat has participared in several national seminars and workshops on vegetable production. The author has to his credit three books *viz.*, Vegetables- Untapped Potential (2007), Physiological Disorders of Vegetable Crops (2009) and Brinjal (2011).

Nutritional and Therapeutic Values of Fruits and Vegetables

– Author –

Kanaya Lal Bhat

2023

Daya Publishing House®
A Division of
Astral International Pvt. Ltd.
New Delhi – 110 002

ISBN: 978-93-5461-377-7 (HB)

Publisher's Note:

Every possible effort has been made to ensure that the information contained in this book is accurate at the time of going to press, and the publisher and author cannot accept responsibility for any errors or omissions, however caused. No responsibility for loss or damage occasioned to any person acting, or refraining from action, as a result of the material in this publication can be accepted by the editor, the publisher or the author. The Publisher is not associated with any product or vendor mentioned in the book. The contents of this work are intended to further general scientific research, understanding and discussion only. Readers should consult with a specialist where appropriate.

Every effort has been made to trace the owners of copyright material used in this book, if any. The author and the publisher will be grateful for any omission brought to their notice for acknowledgement in the future editions of the book.

Published by : **Daya Publishing House®**
A Division of
Astral International Pvt. Ltd.
– ISO 9001:2015 Certified Company –
4736/23, Ansari Road, Darya Ganj
New Delhi-110 002
Ph. 011-43549197, 23278134
E-mail: info@astralint.com
Website: www.astralint.com

Preface

India is the 2nd largest producer of fruits and vegetables in the world. Today, horticulture alone accounts for 30 per cent of India's agricultural GDP from 8.5 per cent of the cropped area. It is playing a pivotal role in strengthening the country's nutritional security besides generating employment avenues. Nutritional knowledge of whatever we take, especially of fruits and vegetables is in high demand and result is a greater need for nutritional information, conveniently available and easily understood. Fruits and vegetables have been recognized as a good source of vitamins and minerals. They have been especially valuable for their ability to prevent vitamin C and vitamin A deficiencies. Fruits and **vegetables** give you high **nutrition** and low calories. They have enormous therapeutic value with their rich nutritional composition, high fibre and water content and boost **immunity**, improve stamina and protect us from several health problems. Fresh fruit and vegetables that are in season are the best, as if they are eaten raw and as fresh as possible, valuable nutrients are less likely to be lost. Fruits and vegetables should be included in your diet to maintain a healthy and balanced body, which then functions optionally thus leading a longer life span. They are **low in fat and high in fibre-may be associated with a reduced risk of some types of cancer** and cholesterol. The unnecessary pressure on our hospitals as many of today's illnesses are related to diet and could be avoided if people looked after themselves and ate more fruits and vegetables. Admittedly, a huge amount of money could be saved and thus spent where it is most needed. Therefore, make sure to include green, orange, and yellow fruits and vegetables - such as broccoli, carrots, cantaloupe, berries and citrus fruits. These fruits are rich in vitamin C and vitamin A, which may help protect against various types of cancer and other diseases. A **die**t high in fruits and vegetables will ensure that your body gets the small amounts of sugar it needs, while providing you with far more health benefits than refined sugar. Fruit, veggies and nuts contain powerful anti-ageing chemicals to keep you youthful. The organic plant based food viz., vegetables, fruits and grains are nutritionally superior and deliver bonafide health benefits as they give 20-40 per cent more antioxidants.

This book is intended to serve a nutritional dictionary, a quick reference book for dieticians and layman as well as for professionals. Its purpose is to provide basic information on nutritional status of various fruits and vegetables available round the year and can be used at home or while shopping form the market. Therefore, an attempt has been made in this book, ***Nutritional and Therapeutic Values of Fruits and Vegetables,*** to give as much information as is possible. I hope that this book will serve to increase public awareness of the exciting connection between nutrition, well being and longevity.

'Let us make fruits and vegetables our medicine'

Kanaya Lal Bhat

Contents

Chapter 1

Introduction

Let food be your medicine and medicine your food

– Ayurved

Fruits and vegetables give you high nutrition and low calories. They have enormous therapeutic value with their rich nutritional composition, high fibre and water content and boost immunity, improve stamina and protect us from several health problems. Research shows diets containing substantial amounts of varied vegetables and fruits reduce chances of cancer by 20 per cent and of stroke and cardio-vascular disease by 60 per cent. Consumption of natural fruits rich in flavonoids helps body to protect from lung and oral cavity cancers. Diets low in fat and high in fibre-containing grain products, fruits and vegetables may be associated with a reduced risk of some types of cancer.

Different coloured fruits and vegetables contain different minerals, nutrients and antioxidants and therefore it is recommended that we consume a wide variety of fruit and vegetables in order to receive the benefits from the various types. For example, dark green leafy vegetables such as watercress, cabbage or spinach contain certain carotenoids that protect, delay and may prevent the onset of degenerative age-related eye diseases such as cataracts or macular degeneration. They are also rich in vitamins C and E, which are both very powerful antioxidants. This means that eating dark green vegetables daily could help to protect the body from developing cancerous cells and from suffering heart disease. Fruits and vegetables are a natural source of energy and give the body many nutrients that may help protect you from chronic diseases. Red, orange and yellow coloured fruits and vegetables such as melon, tomatoes, carrots and apricots contain lots of vitamins A, C and E, which all help to fight certain types of cancer and act by neutralizing free radicals in the body.

As well as containing large amounts of vitamins A, C and E, fruits and vegetables are also rich in vitamins B and K plus minerals such as potassium, calcium, phosphorous, manganese and iron. As long as you are consuming lots of fruit and vegetables, it doesn't really matter whether they are raw, cooked, tinned, dried, fresh or frozen. But fresh fruit and vegetables that are in season are the best, as if they are eaten raw and as fresh as possible, valuable nutrients are less likely to be lost.

Nutritionists say that foods with high content of vitamin A (carrots), vitamin C (oranges, strawberries and broccoli) and vitamin E (peanuts, spinach, almonds, hazelnuts) optimize sperm count, whereas vegetables (selenium), tomato and watermelon (lycopene) help prevent oxidative damage to sperms. To avail maximum fruit nutrition benefits eat organic produce many 'wild' varieties of berries and 'tropical tree' fruits that still have not been treated with any kind of fertilizers or chemicals and can be readily purchased from the local market. Organic fruits tend to be smaller, they feature special flavour and richness in vitamin and minerals and are stuffed with numerous antioxidants.

There is unnecessary pressure on our hospitals as many of today's illnesses are related to diet and could be avoided if people looked after themselves and ate more sensibly and healthily. Admittedly, a huge amount of money could be saved and thus spent where it is most needed. Fruit and vegetables are packed full of goodness and often contain a number of essential vitamins and minerals that cannot be found in other types of foods or they may contain higher levels of these nutrients than other foods. They are made up of water, melons up to 94 per cent, which is also essential for the body and their skin and seeds contain plenty of fibre, which our body needs to help cleanse and rid itself of waste and toxins. Fibre is needed to keep bowel movements regular, lower cholesterol, prevents constipation, bowel cancer and other illnesses of the bowel and intestine such as diverticulosis.

Fruits and vegetables should be included in your diet to maintain a healthy and balanced body, which then functions optionally thus leading a longer life span. Fruits and vegetables of different colours contain diverse mixtures of phyto-nutrients (protective plant compounds). These can act as powerful anti-oxidants, protecting the body from harmful free radicals and helping to protect against certain chronic diseases such as cancer. Some fruits and vegetables are labeled as 'super foods' because they contain high concentrations of some phyto-nutrients, particularly antioxidants which appear to be beneficial to health. About 80 per cent of our food intake should be alkaline vegetables and alkaline fruits and no more than 20 per cent of the entire food intake should be acidic. Alkaline vegetables help maintain the ph balance of the body at 7.3. They can be included to form part of your diet and should be substituted for most of the acid food. The greener the vegetable, the more alkaline it is said to be. So when looking for alkaline vegetables, choose greens more. Vegetable juices make you feel active, lighter, your skin and hair feel better. Vegetable juices can be made from every vegetable, unlike fruit juices. They help you absorb all the nutrients from the vegetables and allow you to consume optional amount of vegetables in an efficient manner. A study conducted at the

Baylor College of Medicine (USA) has revealed that drinking at least one glass of low sodium vegetable juice daily may help overweight people with metabolic syndrome lose weight (Anonymous, 2009). Metabolic syndrome is defined by a cluster of risk factors, including excess body fat in the midsection, high blood pressure, high blood sugar and abnormal blood lipids.

There are a variety of fruits that you can choose from if you do not drink water that much. The best pick is a **watermelon** which has the maximum water content in it. Fresh fruits will give you the needed hydration to feel better and be healthier. Some more examples are: **apples, kiwi,** grape fruit, coconut, strawberries, and any type of berries. Eat plenty of foods rich in vitamin C such as blackcurrants, blueberries, broccoli, guava, kiwi, oranges, papaya, strawberries and sweet potatoes- all help to produce collagen that strengthens the capillaries that feed the skin. Similarly eat foods rich in vitamin E, such as almonds, avocado, hazelnuts, pine nuts, sunflower and corn oils. Zinc rich foods like pumpkin seeds help to repair skin damage. Foods rich in vitamin A help new skin to grow. Carrots are also high in beta- carotene which the body converts to vitamin A. Vitamin A is one of the most important nutrient for healthy skin. Skin with vitamin A deficiency appears dry and dull. Carrots are also rich in antioxidants to help prevent free radical skin damage, vitamin B6, C, K, biotin, thiamine, potassium and fibre. Sulphur rich foods such as onions and garlic help to keep the skin sooth. Eating a few dried apricots, rich in iron daily help improve skin tone. Blueberries are rich source of vitamin E, C, manganese and soluble fibre and in antioxidants that inhibit free radicals from damaging skin.

The green fresh vegetables provide nutrients that help the body to stay hydrated and not get too heavy with other foods. It is better to eat the vegetables raw. Cooking them will take out the natural moisture and hydration. Greens like spinach, amaranth, fenugreek leaves, argula not only have high iron content, but are horders of Vitamin K that helps maintain strong bones and healthy joints. A study shows that older adults with ample blood levels of K were less likely to develop osteoarthritis, compared to a low-in-K control group. However, Vitamin K also helps with blood clotting, so if you're taking blood thinners, check with your doc before boosting your K intake. A plant based diet with a variety of fruits, vegetables, nuts, grains and beans is the best organic way to fight cance. To prevent cancer, crucuferous vegetables *viz.* cabbages. Broccoli, Brussels sprout and cauliflower are important as they contain powerful anti-cancer molecules but are destroyed while boiling them, so steam briefly or stir-fry rapidly in a little olive oil.

While we know that fruits and vegetables are healthy, do we really know how and why they are so good for us? Why are we meant to eat 3-6 servings of fruits and vegetables each day? And how do they help in fighting diseases? A big reason behind their nutritional value is that they are full of antioxidants, which fight harmful chemicals in our body. Here we have the top miraculous antioxidant fruits and vegetables that will help in maximizing overall health

Green Leafy Vegetables

Packed with micronutrients, it is essential for building stamina and also to improve your RBC count. Green leafy veggies are rich in fibre and digest slowly while maintaining your blood-glucose level. We are blessed with foods that can help us maintain healthy hemoglobin levels without having to bust open the bank account. Beetroot, palak or spinach, green peas, rajma or kidney beans, cabbage, turnip, sweet potato, broccoli, lima beans, collards, black beans and cauliflower are some vegetables that are easily available in the market. Beetroot is the best natural remedy to boost blood count, it has the ability to regenerate iron content and activates red blood cells, supplying fresh oxygen to the blood. Raisins, prunes, dried figs, apricots, apples, grapes and watermelons not only get the red blood cells flowing but also improve the blood count. Citrus fruits like oranges, amla or Indian gooseberry, lime and grape fruit help to attract iron. They play a very important role in increasing blood count. Cherries have been found effective in promoting health by scavenging action against free radicals. Herbs like rosemary, thyme, oregano, basil and mint are rich in essential oils of the terpene. Watermelons, papayas and bananas (rich in potassium) increase blood flow by dilating arterioles (small blood vessels) and thus help in improving an erection. Save the greens

The next time you bring home a bunch of beets don't toss the tops away. These dark, leafy greens that are often overlooked are rich in iron, calcium, Vitamin A, K and C. They're loaded with vitamin K that plays a major role in blood clotting. An average male requires 120 micrograms of vitamin K while female adults require 90 micrograms – one cup of beet greens provides a whopping 152 micrograms of this vitamin. So these can be cooked just like spinach. They are slightly bitter as compared to the sweet bulb.

Raw Vegetables

Raw vegetable is a diet, that is uncooked, unprocessed and often organic foods. If between 80 per cent to100 per cent of an individual's food consumption is raw, she or he is considered a raw foodist. The enzymes in raw foods (amylases, proteases and lipases) aid digestion. Uncooked foods populate the digestive tract with beneficial gut flora or microorganisms that are key in promoting the early development of the gut's mucosal immune system. Avoid canned vegetables because they often contain a lot of sodium, which should be avoided if you have hypertension. Raw carrots are rich in fibre called pectin that help lower cholesterol. Other fruits that contain pectin include, apples, citrus fruits, straw berries, rasp berries and black berries.

Cooking is believed to kill the vitamins and minerals, but studies have found the opposite. While cooking may destroy some (but not all) vitamin C, the process boosts the uptake of disease-fighting nutrients – antioxidants. A 2008 study found that vegetables such as carrots, spinach, mushrooms, asparagus, cabbage and peppers supply more antioxidants when cooked than when eaten raw. This is because cooking breaks down vegetables' thick cell walls, making it easier for the body to absorb the nutrients they contain. Steaming is best, then gentle boiling. Frying preserves the least vitamins and minerals (Anonymous, 2013).

Our body possesses countless cells and tissues that have their own enzymes requirements and production to perform a whole range of tasks. The ones that are made availed from eating raw vegetables or juices work in conjunction with the already present ones. So one need not cut off raw vegetables completely. Striking the right balance is essential.

Benefits of Green Juices

Green juices are power-packed with nutrition. Green vegetables have the ability to transform sunshine into the food that all creatures consume. Greens produce chlorophyll, which oxygenate our body. This enables us to release stored toxins in the body. Hemoglobin (that has a similar molecular structure as that of your red blood cells) will get elevated, improving the blood circulation as a result and giving our body a boost of energy. A green juice will cleanse our digestive system, lungs, liver (which will rev up, if sluggish - causing us to retain weight) and uterus. So if you are a smoker, this is an excellent supplement to your diet.

Magnesium is essential for efficient calcium utilisation. The chlorophyll molecule has magnesium at its centre and its impact on the body's magnesium levels is highly significant. When green juices is include in daily diet, it helps absorb calcium better. All green plants have Vitamins A and C, which are important co-factors for calcium absorption. Chlorophyll foods also act as a form of 'stored sunshine', to regulate calcium and Vitamin D deficiency. So if you get no sunshine, then increase the greens in your diet. However, greater consumption of fruit juices is associated with a higher risk of type 2 diabetes, researchers led by Harvard School of Public Health (HSPH) found (Anonymous, 2013).

Researchers have found that nitrate-rich beetroot juice helps the body acclimatise more quickly and thoroughly at high altitude. Drinking beet juice can help you beat acute mountain sickness, caused by lower air pressures at high altitude which affect the ability of our bodies to take up oxygen, the findings showed. The study further showed that beet juice with high amounts of nitrate made the blood vessels relax and return to normal function, while beet juice with no nitrate (the placebo) did not have any effect (Anonymous, 2015).

Fruit juice or vegetable juice, both have properties to break down waste and flush them out. The juices, especially vegetable juices have phytochemicals that are great for those undergoing dialysis as they help prevent kidney failure. Fruit juices, especially berries are rich in antioxidants that avoid issues and improve kidney health. You can either have vegetable and fruits or squeeze them out for their juices.

Beet juice is a more concentrated source of betalains, but cooked beets will contain much more fibre. Traditionally, beet juice was used as a blood purifier and to cleanse the liver. It is also considered as a natural remedy for anaemia or iron deficiency. Beetroot, all juiced up, is a healthy way to get all nutrients that may be lost on cooking. It is also easier to digest and absorb nutrients in liquid form. Runners and athletes are often advised to drink beetroot juice that allows their muscles to use oxygen more effectively and boosts stamina. Pineapple juice is 500 per cent more effective at helping you to stop coughing than cough syrup is. Fresh pineapple contain a substance known as Bromelain, a specific type of enzyme that has anti-inflammatory characteristics which can combat infections and eradicate bacteria.

What is a Vegan Diet?

A vegan diet is one that consists of only plant-derived foods, as vegans do not use or consume any animals or animal products, including flesh, eggs, and milk. Like non-vegans, vegans eat soups, stews, stir-fries, salads, and casseroles. They may consume a wide variety of ethnic foods, as well as vegan versions of traditional favorites such as pizza, tacos, burritos, lasagna, burgers, barbeques, loaves, chili, pancakes, waffles, sandwiches, and desserts. Green leafy vegetable like spinach, fenugreek, pak choy, radish leaves, lettuce, *etc.* are healthy and are known to reduce the risk of heart diseases and cancer as well. That's because they are extremely low in fat, calories and high in dietary fibre. They also contain folic acid, magnesium, calcium, potassium, *etc.* These minerals are beneficial for the optimum functioning of the heart. Studies have shown that one daily serving of green leafy vegetables can lower the risk of heart diseases by 11 per cent. However, certain vegetables like carrots, spinach, cabbage, cauliflower, onion, garlic, sweet potatoes and pumpkin are also said to cause allergic reactions like skin rashe, abdominal pain and gastric problems. Similarly among fruits, bananas, strawberries, oranges and grapes are principal offenders. Citrus fruits like lemon is said to be cause a topic allergy (Meghna Maukherjee, 2012).

What is a Healthful Vegan Diet

A balanced vegan diet is made up of these four food groups: legumes, nuts, and seeds; grains; vegetables; and fruits. Because individual nutrient needs and energy requirements vary due to age, activity level, and one's state of health, this guide should only be considered a broad blueprint for a balanced vegan diet.

Legumes, Nuts, and Seeds (4+ servings per day)

The legume-nut-seed group includes beans, split peas, lentils, nuts, seeds, and soy products. These nutrient-dense foods are packed with protein, fibre, minerals, B vitamins, protective antioxidants, and essential fatty acids. Sample serving sizes from this group include: ½ cup cooked beans, 4 ounces of tofu or tempeh, 1 cup soy milk, 1 ounce of nuts or seeds, or 2 tablespoons of nut or seed butter. Seeds such as pumpkin, cucumber, melon and water- melon, are good sources of protein and fibre. These are valuable sources of vitamin E and B-complex. Pumpkin seeds are rich in **Panagamic acid**, an antioxidant. These four seeds are useful in stimulating 'anti stress' hormones and fighting fatigue. These also boost immunity, lower cholesterol, prevent arthritis and reduce food cravings. Pumpkin seeds are believed to positively influence prostrate health (Ishi Khosla, 2013).

Grains (4-6+ servings per day)

Whole grains provide B vitamins, fibre, minerals, protein, and antioxidants. They are preferable to refined grains because the refining process removes the healthiest nutrients. Also, intact whole grains–such as brown rice, oats, wheat berries, millet, and quinoa–are nutritionally superior to whole grain flours and puffed or flaked whole grains. A serving is one slice of bread, ½ cup cooked grain, or 1 ounce of ready-to-eat cereal. This group is fairly flexible with regard to servings per day. Vary your intake based on your individual energy needs.

Vegetables (4+ servings per day)

Eating a wide variety of colourful vegetables every day will ensure that you're getting an assortment of protective nutrients in your diet. A vegetable serving is ½ cup cooked, 1 cup raw, or ½ cup vegetable juice. For most vegetables, particularly calcium-rich leafy greens, it's nearly impossible to eat "too much."

Fruits (2+ servings per day)

Most fruits, especially citrus fruits and berries, are a great source of vitamin C; all fruits provide antioxidants and fibre. Choose whole fruits over fruit juices to get the most benefit, particularly from dietary fibre. A serving size is one medium piece, 1 cup sliced fruit, ¼ cup dried, or ½ cup of juice.

Eat Well and Feel Good

A healthy diet has shown experts that food could be the answer to combat diseases. Foods can be therapeutic and also make you feel good. Eat well not just for basic nutrition but to heal. After a lot of suffering, people realize it's better to eat well than pop in medicines. Food science introduces us to foods that heal rather than just provide basic vitamins. Increase your intake of yellow and orange vegetables to prevent infertility. Leafy vegetables are good for eyesight and detox the liver. If you want to avoid depression, just eat lots of seeds, nuts and watermelons for a better nervous system. Salads make the digestion stronger as they have enzymes. Eating fresh and dried fruit before a meal helps people feel fuller and eat slightly less during the main course, but being on fruit and vegetable heavy diet for months does not make difference on hunger and fullness.

Herbs Help Blood Pressure

Include herbs, green and herbal teas, and vegetables like celery, nettle (bichoo booti), lauki juice, cucumber, garlic, green coriander and parsley as they have diuretic effect, help in urine formation and control blood pressure. Include soya, nuts, fatty fish, mustard seeds, flaxseeds, fenugreek seeds (methi) in the diet.

How Healthy is a Vegan Diet?

According to the American Dietetic Association's 2009 Position Paper on Vegetarian Diets, vegan diets "are healthful, nutritionally adequate, and may provide health benefits in the prevention and treatment of certain diseases." A healthy vegan diet helps reduce your risk of heart disease, cancer, obesity, and diabetes. Eat your vegetables as they are, by trying to retain as much of their original colour and nutrition as possible. Cook them lightly, stir-fry, pan cook, flavour with as little masala that is palatable and learn to love the natural taste of healthy vegetables. Once you love these subtle flavours, you can enjoy then without trying to mask their taste in anyway.

Helps in Weight Management

Being low in **calories** and fat content and high in fibre, micronutrients and antioxidants with a high satiety value, they help in weight management. They work as fillers and can be treated as free foods with the exception of potatoes, sweet potatoes, jimikand arbi, which must be used in moderation

Keeps BP in Control

Owing to their high potassium and low sodium content, intake of fruits and vegetables is recommended for BP patients. These natural foods protect you from heart disease by preventing accumulation of cholesterol in arteries. Green leafy vegetables are a rich source of Omega-3 fats, the kind present in fish, which prevents heart disease.

Maintain Blood Glucose

The fibre in fruits and vegetables plays an important role in maintaining blood glucose levels. Since diabetics are more prone to oxidative cell damage leading to complications of kidney, nerves and eyes, the antioxidants in vegetables and fruits prevent these. Diabetics need to restrict their intake of mangoes, bananas and potatoes.

Osteoporosis Cause

Calcium deficiency is not the cause of this condition. It's a diet that's either very rich in acidic foods or high in protein. Proteins are broken down into amino acids which are acid yielding. And so, consuming large amounts of animal protein results in the body extracting calcium from the bones to neutralize the effect of acids. When the lost calcium is not restored, it causes osteoporosis over a period of time.

Sex Drive

There are fruits and vegetables that increase sex drive like asparagus-high in B-vitamin folate that increases the production of histamine, which in turn increases sex drive in men and women. Citrulline (watermelon) an amino-acid that relaxes blood vessels thereby increasing sex drive. Chilies contain capsaicin- a chemical which also increases blood flow and triggers the release of mood-enhancing endorphins that naturally pump up your libido. Similarly, avocado-loaded with potassium, boosts libido for both sexes and are rich in folic acid that provide energy and stamina- both of which are important once your libido is restored and pomegranate juice could also liven up libido. Almonds have a high quantity of a type of amino acid called arginine, which helps to relax blood vessels and improve blood circulation, therefore, for men who want a healthy sex life- start taking more almonds, oranges, green tea and pine apples are known to lower sperm acidity. Zinc which is present in bananas and almonds aids testosterone synthesis and increase sperm numbers, motility and volume. Nutritionists says foods with high content of vitamin A (carrot), vitamin C (oranges, strawberries, broccoli), vitamin E (peanuts, spinach, almonds, hazel nuts) optimum sperm health. Selenium rich foods such as garlic and lycopene. Foods such as tomato and watermelon help prevent damage to sperms (Anand Holla, 2012). Watermelons, papayas and bananas (rich in potassium) increase blood flow by dilating arterioles (small blood vessels) and thus help in improving an erection. Celery contains two pheromones called androstenone and androstenol that can boost your sex life. Chewing on the celery stalk releases these pheromones. Nuts like Brazli nuts are valuable source of selenium which has been seen to be beneficial for female fertility. Selenium is important for many biological functions such as immune response thyroid hormone production and acts as an antioxidant, helping to detoxify damaging chemicals in the body. This antioxidant plays key role in the development of healthy ovarian follicles. Ovarian follicles are responsible for the production of eggs. Goji, this amazing super food not only raises your spirits, but it also raises your libido. Goji berries raise testosterone levels, and, therefore, your sex drive goes up. Fennel contains an estrogen like substance **(estirol)** that turns out libido.

Nuts, milk and cheese contain zinc, which is good for the male sex hormone testosterone, which in turn is required for an erection. Also, garlic and onion contain allicin, which helps increase blood flow. Cherries have been found effective in promoting health by scavenging action against free radicals. Walnuts which are rich in omega 6 fatty acids and arginine help in the production of nitric oxide, thus causing relaxation of arterioles and increasing the blood flow.

The Scoop on some Important Nutrients

Like non-vegans, vegans need to be mindful of consuming all the nutrients they need in order to be healthy. Three nutrients that *everyone* needs to pay attention to are vitamin B_{12}, vitamin D, and Omega-3 fatty acids. Many common health problems can be prevented or alleviated with a healthy diet. Nutrients are important to lead a healthy life. Vitamins and minerals you get from your diet are just as important as carbohydrates, protein and fats; however, you only need them in small amounts. Vitamins and minerals usually function as co-enzymes, which mean they help chemical reactions in the body happen a lot faster. One of the most important vitamins for the body is Vitamin E. In order for the heart to stay healthy the body needs plenty of Vitamin E. Vitamin E is also an excellent option for helping limit bad heart conditions. The body is more prone to certain diseases if it doesn't have a proper amount of Vitamin E.

Vitamin B12 is necessary for proper red blood cell formation, neurological function, and DNA synthesis. It is manufactured by certain types of bacteria found in nature. Because plants vary widely in their levels of this bacteria (and most of us favour our food scrubbed squeaky clean), we cannot rely on plant foods to meet our B_{12} needs. We can ensure our dietary needs are met by consuming supplements and/or fortified foods. Our suggestion is to supplement with a vegan source of 2000 micrograms once a week or 10-100 micrograms a day (be advised that some B_{12} vitamins labeled as vegetarian are in a stomach base). Or, if you prefer not to use supplements, consume at least three servings of vitamin B_{12}-fortified food per day (each supplying at least 20 per cent of the Daily Value on the label), such as non-dairy milks, **breakfast** cereals, meal replacement bars, beverage mixes, and Red Star Vegetarian Support Formula Nutritional Yeast (read labels to ensure B_{12} content).

Vitamin B are vital, they help convert calories into energy and are components of enzymes that maintain normal skin function (including functioning of the oil-producing glands which keep skin moist and smooth). Poor intake of almost any B vitamin can cause dry or scaly skin. The best sources are bananas, whole grains and peanut butter *etc.*

Vitamin D, the "sunshine vitamin", is also a hormone; our skin manufactures it from the ultraviolet rays of the sun. It plays an important role in bone health and supports normal neuromuscular and immune function. Go od vitamin D status is linked to a lowered risk of osteoporosis, certain cancers, and other chronic diseases.

Vitamin D blood levels are an international public health concern. Getting enough of it is not as easy as we may think. The body's ability to produce vitamin D from sun exposure varies based on skin pigmentation, sunscreen, clothing, time of year, latitude, air pollution, and other factors, and the vitamin is found naturally in only a handful of foods. This is why all people–not just vegans–need to be mindful about vitamin D. The latest research suggest that getting even 100 per cent of the current Recommended Dietary Allowance (RDA) for vitamin D may be insufficient for many people. To ensure adequate vitamin D intake, take 1000-4000 International Units (IU) per day, depending upon your age and other individual needs.

Supplemental vitamin D comes in two forms: vegan D2 (ergocalciferol), usually synthetic or manufactured from yeast, and non-vegan D3 (cholecalciferol), manufactured from lanolin (from sheep's wool). A dose of sunlight along with almonds will benefit those who want a baby. Eat to beat infertility which is on the rise and have foods rich in iron like apples and radish. It's essential for menopausal women to take vitamins D and E, which are found in avocadoes, tomatoes, hazel nuts. Women can get vitamin D from butter and eggs too.

Omega-3 fatty acids: Omega-3 are polyunsaturated fatty acids, considered as essential nutrients that our body can't produce but needs to maintain good brain health and cognitive function. Omega-3 fatty acid is like an essential magic wand which helps to treat and prevent several ailments and diseases. The fat present in omega-3 fatty acid is essential for your health and proper functioning of various body organs. The two types of essential fatty acids are omega-3 fatty acids and omega-6 fatty acids. A balance between both is necessary for good functioning of the organs. Make sure to include enough of omega-3 and omega-6, both essential fatty acids in the daily diet plan. A proper balance of essential fats is important for optimal brain function, heart health, and infant/child development. Alpha-linolenic acid (ALA) is an omega-3 fatty acid that partly converts to DHA and EPA in the body. It is present in several plant foods, including flax products, hemp products, canola oil, walnuts, and leafy green vegetables. Aim to consume 2 to 4 grams of ALA per day. Beat the blues by eating right, as certain foods can help cut the chances of depression. Make sure you're having enough Omega 3 fats from mustard oil, wheat, lobia, flax seeds, methi seeds, soybean and green leafy vegetables. Even oily fish can be good to beat the blues. "Whole- grains are important, everything from oats to whole-wheat bread to barley are great sources of slow-release energy that will prevent your blood sugar from nose-diving. Seeds like flaxseeds, pumpkin seeds, soybeans, chickpeas and sunflower seeds *etc.* are again friends of sexual hormones. "Consuming one tablespoon of flaxseed every day helps to increase the testosterone level in the body. These seeds are rich in Omega 3 fatty acids, a drop in which affects hormone levels leading to a plummeting sex drive. Pumpkin seeds are also rich in zinc which is a mineral needed to produce testosterone." An imbalance of Omega-3 fats in the diet can also aggravate acne. Omega-3 fats are known to counter the effect of inflammatory chemicals that lead to breakouts. It wards off cancer and cardio-vascular diseases. It also improves your immune system and reduces joint pain. Omega 3 fatty acids, these can have a positive impact or even prevent serious degenerative illnesses rheumatoid arthritis and osteoporosis. One can get omega 3 fatty acids from various seeds like pumpkin seeds, flax seeds and walnuts. Research has shown that a diet with a high percentage of omega-3 fatty acids and a low percentage of omega-6 fatty acids has been linked with decreased inflammation and cortisol levels.

Omega-3s eicosapentaenic acid (EPA) and docosahexanoic acid (DHA) can boost heart health and lower triglycerides (TGs). So, taking fish oil supplemen**ts** can lower the elevated blood TGs. Having high levels of TGs is a risk factor for heart disease and DHA alone has also been shown to lower triglycerides.

Omega-3 fatty acid promotes a healthy heart by lowering the overall risk of death associated with heart disease. People with heart problem should include omega-3 fatty acids in their diet to lower their risk.

Omega-3 fatty acid also helps in boosting your mental health by reducing the level of depression and anxiety. Omega-3 fatty acids acts like an antidepressants to lower the level of depression. Besides, it also helps to reduce the depressive symptoms of various mental ailments like bipolar disorder. Omega-3 are polyunsaturated fatty acids, considered as essential nutrients that our body can't produce but needs to maintain good brain health and cognitive function.

A pregnant woman should take in sufficient amount of omega-3 fatty acids in their diet to boost the development of the fetus. Omega-3 fatty acids boost the health of pregnant women and the development of the fetus. Besides, it also helps in enhancing the visual and neurological development of the baby.

Women who consume good intake of omega-3 fatty acids have a minimum risk of developing osteoporosis and other bone loss ailments. Omega-3 fatty acids help to increase the level of calcium in the body. It further improves bone strength by proper absorption of calcium into the bones. Omega-3 fatty acid helps in giving you happy periods by reducing various menstrual pains like stomach cramps and back ache. Soy flour, it bursts with vitamins and minerals, and is also one of the best vegetarian sources of Omega-3 fatty acids. Soy protein is great for women post menopause and also for elderly women.

Be a Vegetarian, Live a Healthy Life

Today, an increasing number of people are turning towards a vegetarian diet. The reasons vary from religious, environmental, cruelty towards animals, to just the desire for a healthier lifestyle. A predominantly non-vegetarian **diet** poses many health risks. Here's why opting for a vegetarian diet achha hai.

The human body is anatomically and physiologically geared for a vegetarian diet. Some of the distinctions between vegetarians and carnivores are:

Natural Detox

Being vegetarian is a natural way to detox the body, because a vegetarian diet is more rich in fibre, vitamins, minerals and anti-oxidants, which help to cleanse the body's system. Meat and fish contain a lot of residue from toxic chemicals. It has been observed that the bodies of carnivorous animals contain 10 times more hydrochloric acid than that of herbivorous ones, but the human body should not have the same amount of hydrochloric acid. This establishes the fact that the human body is basically meant for a vegetarian diet. So digestion of vegetarian food is easier for our bodies. A vegetable diet contains dietary fibre (bottle gourd, pumpkin, spinach, cabbage) which flushes toxins out of the body. A diet containing only eggs, fish and mutton is a poor source of fibre. Brussel sprouts are packed with sulfur which is good for clearing your body of toxins.

Colourfully Appetizing

Good food is not restricted to the taste, its visual appeal is equally important. In fact, if the food served to us is a visual treat, we are bound to eat and enjoy it even more. Vegetables come in an array in colours, which makes them more appetizing and appealing to our eyes. Meat and fish usually come in boring shades of brown and beige. Therefore, using many coloured food items in your cooking is a great way to eat a variety of natural foods that will boost your health.

The best way to ensure that your diet is balanced is to colour code it. Pigmentation is what lends colour to fruits and vegetables, and also contains healthy nutrients. The deeper the colour, the higher the percentage of a specific nutrient the food item has. Purple coloured foods like plums and aubergines have phytochemicals called anthocyanins and resveratrol that fight ageing, reduce blood pressure and protect the body against liver diseases.

Orange-coloured foods such as oranges and carrots contain beta-carotene and beta cryptoxanthin that keep the skin and eyes healthy, and also protect against arthritis. Yellow foods such as bananas and sweet corn contain carotenoids that boost the immune system and keep the heart healthy. White foods, including garlic, onions, cauliflower and mushrooms, are rich in vitamin C, and antibiotic and antifungal compounds that fight cell damage and keep the fighter white blood cells healthy. Red foods like tomatoes and watermelons contain the powerful antioxidant lycopene which acts as an internal sunscreen and protects you from sunburn, prevents osteoporosis and heart diseases.

Greens such as peas, broccoli and spinach are packed with sulforaphane, isothiocyanate, and indoles which stimulate the liver to keep cancer producing chemicals in check

Good for Digestion

The most important difference between a vegetarian and a non vegetarian diet is that the former contains dietary fibres, whereas a non-vegetarian diet is lacking in fibre. This dietary fibres is very useful for the human body because people who have a diet rich in dietary fibre have low incidence of diseases like coronary heart diseases, cancer of intestinal tract, piles, obesity, diabetes, constipation, hiatus hernia, diverticulitis, irritable bowel syndrome, dental caries and gallstones. The food rich in this dietary fibre includes cereals and grains, legumes, fruits with seeds and citrus fruits. Complex carbohydrates in vegetarian foods are digested gradually providing steady source of glucose. Conversely, meats rich in fat and proteins are difficult to digest.

Boosts Cardio-vascular Health

Yes, it's true. A vegetarian diet boosts cardio-vascular health. Consumption of a vegetarian diet consisting of whole grains, legumes, vegetables, nuts, and fruits, and abstaining from meat and high-fat animal products, along with a regular exercise program, is consistently associated with lower blood cholesterol levels, lower blood pressure, less obesity, lower incidence of diabetes and consequently less heart disease. **Fruits and vegetables**, especially leafy greens and some legumes, are abundant in folic acid, which has been shown to help lower **homocysteine levels** (associated with heart disease). Many whole and unrefined plant foods are also important sources of minerals such as copper and magnesium, both of which can protect against cardiovascular disease. The many phytochemicals/flavonoids in fruits, vegetables, nuts and whole grains have properties that reduce the risk of heart disease. Various nuts are the source of heart healthy fatty acids (omega 3, MUFA, PUFA). On the other hand, a non-vegetarian diet (mainly red meat) is associated with an adverse impact on cholesterol levels, and increases the incidence of heart disease. Freshly made vegetable juices contain natural plant pigments and enzymes that act as cancer protection agents. These juices should be made fresh and consumed immediately, unstrained to retain the fibre. Researchers in the United States say their studies suggest that greater consumption of fruits and vegetables may reduce the risk of developing invasive bladder cancer in women. The study also suggested that women with the highest intake of vitamins A, C and E had the lowest risk of bladder cancer (Anonymous, 2013).

Reduces Risk of Cancer

A diet with adequate portions of green leafy vegetables and fruits ensures an intake of roughage or fibre. This has been shown to have beneficial effects on a number of cancers, notably cancers of the colon and rectum. A non-vegetarian diet, especially one rich in red meats and animal fats, has been shown to have a carcinogenic effect. Carcinogens are substances or agents that are directly involved in causing cancer.

Longevity

A vegetarian diet is associated with lower blood cholesterol levels, lower blood pressure, less obesity, less heart disease, less stroke, less diabetes, less cancer, and hence vegetarians have a chance of living a longer life. But based on statistics, it certainly appears that a vegetarian diet is far healthier than a non-vegetarian one. Studies have shown that people who eat mostly plant-based foods look younger than their real physical age. And if that is true, it sure makes sense to include plant-based, anti-ageing foods in our daily diet to get that ageless, young skin and delay the inevitable effects of natural ageing by improving skin elasticity and thereby staving off wrinkles.

Red meat requires maximum digestion effort for your body to process. Over-eating will lead to issues like perspiration and unwanted sweat. While vegetarians might not have such issues with their limited food intake, fenugreek and other strong flavoured plants cause similar results.

Stronger Bones

Gorging on meat can lead to protein over load. This can tax our kidneys, interfere with the absorption of calcium and prompt the body to extract existing calcium from the bones. Such calcium extraction is rare amongst the vegetarians. Our body is mostly alkaline. The human body's natural pH level stands at 7.45, making it alkaline. All fruits and vegetables, natural juices like coconut water are alkaline, which make them a better nutrition fit than say animal proteins, tea, coffee, sugar, alcohol, and packaged foods that have high acidic levels. An imbalance in pH levels, can lead to a disruption in cellular activity, causing health problems that range from heartburn to heart disease.

Carbohydrate deficiencies: A non-vegetarian diet is a poor source of carbohydrate. Carbohydrates deficiency can lead to ketosis- a condition where the body starts breaking fat (instead of carbs) as a source of energy.

Healthy Skin: Eating beet root, tomato, pumpkin and bitter gourd can clear off blemishes. And guava, apples and peaches, eaten along with their peel, promise a glowing complexion.

Weight Management: Avoiding meat is the simplest way to reduce fat intake. Instead, eating whole grains, legumes, vegetables, nuts and fruits, lowers cholesterol levels, blood pressure and obesity.

Easy on the Teeth: Our molars are more suitable for grinding grains and vegetables than tearing flesh. Digestion begins with the saliva, which can only digest complex carbohydrates present in plant foods.

Phyto nutrients: Diabetes, cancer, kidney disease, stroke and bone loss are partially preventable with a good intake of **phytonutrients**. As these are present only in vegetarian diet, the non- vegetarians are at a loss.

Fruit compared with animal food.

Energy: When you look at this table you see that one banana and a few strawberries give you as much as energy as an entrecote! You will not feel as fulfilled, but your body really can extract as much of energy out of fruit.

Proteins - Fat - Sugar: As the energy fruit contains consists of sugars the fruit is digested in 30 minutes. This chemical process has no toxic waste products that are difficult to remove from the body and it even stimulates the removing of toxic elements. The energy animal products contain consists of fat and proteins. Because of the high amount of proteins it is digested in 6-8 hours. This chemical process does produce toxic waste-products that have to be removed from your body. Protein is an important nutrient for the growth, development and repair of muscle and body tissues. Protein has a higher metabolic rate than fat, so an individual can burn more calories. It is also said to increase satiety, which prevents a person from overeating. Healthy sources of protein include lean chicken, fish, eggs and nuts.

Fibres: Another remarkable difference is that fruit contains a substantial amount of fibres while animal based products don't contain any. Fibres are very important. Your body does need dietary fibre, which might not carry nutritional value but acts as a cleaner for the digestive tract and intestines. It is essential to consume sufficient fibre to keep your body clean and light. Lignans are naturally occurring forms of the female hormone estrogen and are found in certain foods. Mucilage helps assures a healthy intestine and proper absorption of nutrients.

Water: The water percentage of fruit (80 per cent) is higher than that of meat (15 per cent) and comes more near the water percentage of the human body (80 per cent). Vegetarians sweat through their skin to get rid of excess heat and impurities, whereas carnivores pant or breathe rapidly and extend their tongues.

Vegetarians do not have long canines and claws, whereas meat-eating animals do for killing and tearing the flesh of prey.

The saliva of vegetarians contains the enzyme **ptyalin** to predigest starch whereas meat-eating animals do not have this enzyme.

Vegetarians secrete less hydrochloric acid during digestion in the stomach compared to flesh –eating animals, who secrete large quantities of it to digest meat and bones.

The jaws of vegetarians move up and down and also sideways for additional chewing and grinding, whereas those of carnivores open in an up and down motion only.

Vegetarian drink by suction through their teeth, whereas carnivores lap up liquid with their tongue.

Vegetarians naturally salivate on seeing fruits and vegetables, but do not do so on seeing live chickens, pigs or other animals. Carnivores, on the contrary, naturally salivate on seeing its animal prey, but do not do so on seeing fruits and vegetables.

Therefore meat- eating is incompatible with the human body, causing internal damage and disease and also result in much damage to the eco-system. Experts have warned that eating three eggs a week could significantly increase a man's chance of dying from prostate cancer. The researchers found that men who consume more than two and a half eggs on a weekly basis were up to 81 per cent more likely to be killed by the disease. They suggested the damage might be done by the large amounts of cholesterol or choline - a nutrient that help cells to function properly - that are found in eggs.

Did you know that our body digests fruits in one hour, vegetables, such as potatoes in three hours, but that the latest medical researches show that it takes 24 to 72 hours to fully digest meat? Also according to The Independent, red meat is linked to diseases like Alzheimer's, bowel cancer and high cholesterol.

Vegetarians tend to have faster digestion transit times than non-vegetarians.

The team at the Harvard School of Public Health in Boston examined the eating habits of 27, 000 men over a 14-year period. They found that a higher number of cancer deaths among those who admitted consuming lots of eggs.

In one of the largest studies conducted till date, it has been found that vegetarians live longer than meat eaters. After following more than 70, 000 men and women over six years, researchers have concluded that those who followed vegetarian lifestyles enjoyed 12 per cent lesser risk of death than non-vegetarians. The participants were divided into five dietary groups – non-vegetarians, semi-vegetarians, pescatarians, lacto-ovo-vegetarians and vegans. It was found that people who are vegetarian tended to be older, more highly educated and more likely to be married. They were also more likely to drink less alcohol, smoke less, exercise more and be thinner. According to researchers, all of these factors could play a role in their lower risk of death.

Interestingly, the study has found that the association between vegetarian diets and lower mortality was greater in men than in women. Men had a lower rate of cardiovascular disease and death from heart-related conditions.

Previously too, vegetarian diets have been linked to a lower likelihood of developing chronic diseases like heart disease or diabetes while red-meat consumption, because of its high levels of cholesterol and saturated fat, is likely to clog up arteries (Anonymous, 2013)

Freshly squeezed juices from fruits and vegetables are excellent sources of minerals and vitamins which catalyze chemical reactions occurring in the body. These enzymes also produce the energy needed for digestion, absorption, and conversion of food into body tissues. An increased intake of fruit and vegetable juices ensures that the body will efficiently absorb more minerals and vitamins.

Pure orange juice, for example, is very rich in Vitamin C, potassium, and folic acid. Other fruit juices that are rich in antioxidants, vitamins, and anthocynanins (water soluble pigments found in plants which act as antioxidants) include grape juice (purple), pomegranate juice, and cranberry juice. It is believed that juices that are made from fruits with rich colors such as grapes, pomegranates, cranberries, and blueberries are rich in vitamins and antioxidants.

Vegetables for Good Health

A wide variety fruits and vegetables contain minerals, vitamins and a range of phyto-chemicals which help to protect the body from damage, disease and memory impairment. Eating vegetables raw is at its best form because cooking it destroys some nutrients. Celery, onion, tomato, carrot, broccoli and garlic (Garlic is considered to be a wonder drug for the heart).

Fruits and vegetables give you high nutrition and low calories. They have enormous therapeutic value with their rich nutritional composition, high fibre and water content and boost immunity, improve stamina and protect us from several health problems. Research shows diets containing substantial amounts of varied vegetables and fruits reduce chances of cancer by 20 per cent and of stroke and cardio vascular disease by 60 per cent. Fruits and vegetables are lower in calories per gram compared to denser foods. To increase satiation, consume a cup of chopped water-rich fruits and vegetables, such as berries, citrus fruits, tomatoes, kiwi, celery, cucumbers, bell peppers, leafy greens and water chestnuts. If you don't like it plain, prepare a fruit or veggies salad by adding a few drops of vinegar and low-fat salad dressing to it. High-fibre foods can actually scrub your teeth clean. Fruits and vegetables that are crunchy and juicy like cucumber, watermelons, muskmelons apples, pears have water content that offsets the fructose. Vitamins, minerals, calcium, iron, potassium, zinc and other nutrients present in nuts like almonds, walnuts, cashews and peanuts are good for your teeth.

Fruits and Vegetables Recommended in Rheumatism, Anaphrodisiac, Hypnotics, Head-aches, Beauty and other Diseases

A. Fruits and Vegetables Recommended in Rheumatism

Fruits - gooseberry, cherry, red currant, lemon, apple, plum, grape.

Vegetables - paprika, onion, pumpkin, bean, parsnip, salad, celery, garlic, cabbage.

B. Fruits and Vegetables Recommended in Nervous System

Tranquilizers

Fruits: Cherry, mandarin, apple, plum.

Vegetables: salad, cabbage.

Anaphrodisiac (The opposite of aphrodisiac)

Vegetables: salad.

Hypnotics

Fruits: Apple, apricot.

Vegetables: onion, salad.

Headaches

Fruits: cherry, lemon.

Vegetables: cabbage, parsley.

C. Fruits and Vegetables Recommended in Beauty

Face Beauty

Fruits: plum, emon, strawberry.

Vegetables: parsley, cucumber, tomato.

For Dry Skin

Fruits: strawberry, quince, oranges.

Vegetables: cucumber, salad

Wrinkled Skin

Fruits: cherry, orange, almond.

Vegetables: cucumber.

Hands Beauty

Fruits: Lemon.

Vegetables: tomato.

D. Foods that Fight Acne

While fresh, leafy green vegetables like spinach, lettuce, artichokes and cabbages are good for health and aid in digestion, they are also beneficial in lessening acne.

Cucumbers are a source of vitamins A, C, E, water and amino acids, which help to fight against acne.

You may not like its strong odour but garlic will help you get rid of your zits. It has strong anti-bacterial properties that will improve your immunity.

Tomatoes are not just a very rich source of vitamin C; they also have bio-flavonoids that help to repair damaged or scarred skin. Consuming them regularly will reduce acne (Purvaja Swant, 2014).

E. Foods that Fight Wrinkles and Ageing

Leafy greens: Leafy greens such as spinach, kale, turnip greens, and romaine lettuce, are great sources of lutein and zeaxanthin. Studies have shown eating foods rich in these antioxidants can significantly reduce risk of AMD (age-related macular degeneration), as well as non-Hodgkin's lymphoma. Spinach has a very high ORAC (Oxygen Radical Absorbance Capacity) score. It is a measurement of a food's ability to destroy the free radicals that cause damage in your body. The higher the ORAC score, the better a food is for you.

Blueberries: These tiny berries are powerhouses of nutrition. They are loaded with antioxidants, which stops premature ageing and help prevent cancer. Wild blueberries also boost your health. A study suggests that anthocyanins in blueberries appear to combat oxidative stress, which is one of the main causes of ageing.

Yellow and orange root vegetables: Put plenty of beta-carotene on your plate. These super-antioxidants are good for your skin and eyes. Good choices include carrots, sweet potatoes, pumpkins, and squash.

Broccoli: Quercetin is a powerful antioxidant that is found in broccoli, along with other foods including cranberries, onions, and apples. It is a natural anti-inflammatory agent as well, fighting the number two cause of aging. Broccoli sprouts have 30 times more isothiocyanates (yet another antioxidant) than regular broccoli.

Aloe vera: Aloe vera's anti-ageing properties have been known for ages. Scientific evidence suggests that Aloe Vera juice can heal skin and reverse skin ageing. Skin ages because of UV light damage and loss of collagen in the skin. Collagen helps in keeping the skin firm and elastic. Ageing breaks down the collagen matrix in your skin which leads to wrinkles. Drinking aloe vera juice is known to reduce the wrinkle depth and improve skin elasticity due to increased collagen production.

Garlic: Allium is an antioxidant that packs a punch. Garlic, onions, and scallions are loaded with this free-radical fighter that is good for your skin and your immune system.

White Vegetables

Cauliflower: It contains sulfur compounds that are associated with fighting cancer, strengthening bone tissue and maintaining healthy blood vessels.

Mushroom: Mushrooms are low in calories, fat-free, cholesterol-free, gluten-free, with barely any sodium, and yet they carry a wealth of selenium, potassium, riboflavin, niacin and vitamin D. Mushrooms are also filling, so they can help you control your weight.

Garlic: It is believed to help in growing hair, cause acne to disappear and keep colds and flu at bay. Its antioxidant properties can help boost your immune system. Garlic is antiviral and antifungal. It has been used for centuries for its antibacterial properties. The chemical component of garlic, allicin, has been shown to prevent the growth of the *candida albicans* fungus.

Garlic can reduce cholesterol. Its powerful antioxidant properties prevent free radical damage to the arterial lining and stop the formation of scar tissue. Eat garlic to lower your blood pressure. It has the ability to decrease platelet stickiness, which will help in making your blood thinner. Garlic can help to regulate blood sugar levels, which is good for people with type 2 diabetes.

Potatoes: The white potato provides as much fibre as and more potassium than other commonly consumed vegetables or fruit. A medium skin-on baked potato weighs in at just 163 calories, a whopping 941 milligrams of potassium and 3.6 grams of fibre. Potatoes also provide vitamin C, vitamin B6 and magnesium in addition to small amounts of high quality protein

F. Fruits and Vegetables Recommended in other Diseases

Alcoholism

Vegetables: paprika, cabbage.

G. Fruits and Vegetables for Diabetics

Diabetics can have fruits, provided the sugar level of the patient is in control, but these fruits must be consumed in a limited quantity. To help diabetic patients, some fruits which will not affect blood sugar level, provided they're consumed in moderation are as follows (Renita Tisha Pinto, 2013).

Asparagus: Broccoli, Brussels sprouts, Cabbage, Cauliflower, Cucumber, Spinach, Tomato (lycopene – a powerful antioxidant).

High fibre vegetables such as peas, beans, broccoli and spinach/leafy vegetables should be included in one's diet. Also, pulses with husk and sprouts are a healthy option and should be part of the diet.

Fruits high in fibre such as papaya, apple, orange, pear and guava should be consumed. Mangoes, bananas, and grapes contain high sugar; therefore these fruits should be consumed lesser than the others. The fruits and vegetables recommended for diabetics are as blueberries, cranberries, raspberries, apples, watermelon, rock melon, red grape fruit, tomato (lycopene), asparagus (low calories), carrots, broccoli, red onion, Spinach (beta-carotene), Brussel sprouts (folic acid), cabbage (manganese), cauliflower, cucumber, nuts, beans

Kiwi: Many researchers have shown a positive correlation between kiwi consumption and lowering of blood sugar level

Black jamun: This is one of the best fruits for diabetics. It is known to improve blood sugar control. Seeds of these fruit can be powdered and consumed by patients to control diabetes.

White jamun: Diabetic patients can consume jamun fruit daily, to control their sugar levels. White jamuns are also high in fibre.

Starfruit: Similar to jamuns, star fruits are good for diabetics too as they help improve blood sugar control. But caution needs to be exercised in case the person has diabetes nephropathy.

Guava: Guava controls diabetes and it is good for constipation. Guavas are high in vitamin A and vitamin C and contain high amounts of dietary fibre. This fruit has a reasonably low GI.

Cherries: Their GI value is 20 (or even less in some varieties) which makes it a good healthy snack for diabetes patients at any time of the day.

Peaches: These tasty fruits are a great healthy treat, with a low GI and good for diabetics.

Berries: Numerous varieties of berries are available throughout the world and almost all seem to be a rich source of antioxidants. Diabetics can include a serving of different berries to keep their sugars in check. To name a few: Strawberries, blueberries and blackcurrants, raspberries, cranberries, chokeberries and blackberries, are good for diabetes patients.

Apples: Apples contain antioxidants, which help to reduce cholesterol levels, cleanse the digestive system, and boost the immune system. Apples also contain nutrients that help in the digestion of fats.

Pineapples: Good for diabetics, pineapples also benefit the body as they are rich in anti-viral, anti-inflammatory and anti-bacterial properties.

Pear: These delicious fruits are a good snacking option for diabetics as they are rich vitamins and fibre.

Papaya: They are good for diabetics because they are rich in vitamin and other minerals.

Figs: Their richness in fibre helps with insulin function in diabetes patients.

Oranges: These citrus fruits can be consumed on a daily basis by diabetics, as they are rich in vitamin C.

Watermelon: Although watermelons have a high GI value, their glycemic load is low, making them good fruits for diabetes patients. However, consume in moderation.

Grapefruit: It is a good option for diabetics as it slows down the blood sugar peak.

Pomegranate: These tiny red rubies help diabetic people improve their blood sugar statistics.

Cantaloupe: This fruit is high in the Glycemic Index, but has a good amount of fibre. Therefore, within moderation, it can easily be included in the diet of a diabetic.

Jackfruit: Jack fruit contains vitamin A, vitamin C, thiamin, riboflavin, niacin, calcium, potassium, iron, manganese and magnesium among many other nutrients. Good for diabetes as they improve insulin resistance.

Amla: A fruit loaded with vitamin C and fibre, Amla is a healthy addition to the diabetic diet.

Pulses are important in the diet as their effect on blood glucose is less than that of most other carbohydrate containing foods. Vegetables rich in fibre help lowering down the blood sugar levels and thus are healthy.

Good fats such as Omega-3 and mono-unsaturated fats (MUFA) should be consumed as they are good for the body. Natural sources for these are canola oil, flax seed oil, and nuts. These are also low in cholesterol and are trans fat free.

H. *i. Fruit and Vegetables that Battle Cancer*

One can keep cancer at an arm's length if they incorporate certain foods in their diet every day and exercise regularly. Here is a list of 12 foods that prevents cancer (Sobiya N. Moghul, 2013)

Peanuts: Rich in Vitamin E, peanuts reduces the risk of colon, lung, liver, and other cancers. Have a spoon of peanut butter on a slice of whole grain toast for a filling cancer-fighting snack.

Grapefruit: Oranges, grape fruits, broccoli all of them are high in Vitamin C. This element prevents the formation of cancer-causing nitrogen compounds. Grape fruits, Oranges, grape fruits, broccoli and other fruits reduce the risk of esophagus, bladder, breast cancer, cervical cancer, and stomach and colon cancer. So remember to fill your plates with veggies and fruits everyday.

Berries: Berries top the charts when it comes to cancer fighting foods. Raspberries, blueberries, and cranberries are very promising potential to help prevent cancer. These berries are packed with vitamins, minerals and anti-oxidants that have anti-cancer properties. Pterostilbene, an antioxidant found in berries is a storehouse of cancer fighting properties.

Sweet potatoes: Sweet potato is full of beta-carotene. Studies have revealed that consumption of high beta carotene foods reduce the risk of cancers especially. Colon, breast, stomach and lung cancers. In a study including premenopausal women who ate a lot of vegetables that included beta- carotene like sweet potatoes had lowered their breast cancer risk by half.

Walnuts: Contain nutrients which help body fight inflammation and halt the growth of breast cancer tumers. Therefore adding walnuts in your breakfast helps in reducing the rise of breast cancer.

Turmeric: It contains a compound known as Curcumin which helps in slowing down the spread of breast cancer due to its anti- oxidant and anti-inflammationatory properties.

Flaxseed: Omega- 3 fatty acids prevents in the growth of cancer by preventing the growth of cancer cells and interrupting the growth of cancerous tumor. Omega-3 fatty acids also help reduce inflammation. Along with fatty fish, one can also include a tbsp of flaxseed in your daily diet.

ii. Herbs and Spices for Cancer Prevention

Herbs and spices have been traditionally used for their flavour-enhancing and medicinal properties. The increasing prevalence of cancer in the world has prompted increased interest among researchers and the public in the potential of foods to reduce cancer risk Romana D'Souza, 2014).

Rosemary: Studies suggest that extracts of rosemary can delay chemically induced cancers. Carnosic and rosmarinic are active compounds in rosemary that inhibit the production of cancers like prostate cancer, lung cancer, and breast cancer. Rosemary can also reduce resistance to chemotherapy drugs.

Turmeric: Turmeric is a spice that contains the compound curcumin. Curcumin has powerful anti-inflammatory and anti-oxidant properties that make it a highly effective medicinal spice. It destroys the blood vessels that feed cancerous tumours, cutting off the blood supply supply to cancerous cells. Curcumin can also inhibit the development of metastasis. **Curcumin** (8-methy- n-vanillyl- 6-nonenamide), the active ingredient in turmeric, acts as both an anti-inflammatory and an antioxidant, and prevents cancer by interfering with aspects of cellular signaling. Use it liberally in all your chicken and fish curries. Turmeric contains the powerful polyphenol Curcumin that has been clinically proven to retard the growth of cancer cells causing prostrate cancer, melanoma, breast cancer, brain tumour, pancreatic cancer and leukemia amongst a host of others. Curcumin promotes 'Apoptosis'- (programmed cell death/cell suicide) that safely eliminates cancer breeding cells without posing a threat to the development of other healthy cells. In cases of conventional radiotherapy and chemotherapy, the surrounding cells too become a target in addition to the cancer cells. Therefore, the side-effects are imminent

Black Pepper: **Piperine,** an active chemical compound in black pepper, has powerful antioxidant properties. Studies have shown that pepper inhibits the growth of cancerous stem cells in breast tumours, while leaving healthy cells untouched.

Garlic: Garlic contains organosulfur compounds, which have immunity-strengthening and anticarcinogenic properties. Compelling evidence suggests that these compounds can lower the incidence of breast, colon, uterine and lung cancers. Garlic suppresses the formation of cancer-causing nitrosamine compounds.

Parsley: Apigenine, a natural oil in parsley, has the potential to prevent the formation of blood vessels that supply oxygen and nutrients to cancerous tumours.

Cumin: Thymoquinone, the primary component of black cumin seed oil, has been reported to reveal antioxidant and chemopreventive characteristics. Thymoquinone inhibits the production of the cells responsible for prostate cancer, leukemia and breast cancer. Yes, it aids digestion and probably that is why we like chewing a handful of cumin seeds at the end of every meal. However, its health benefits go beyond. A portent herb with anti-oxidant characteristics, cumin seeds contain a compound called 'Thymoquinone' that checks proliferation of cells responsible for prostate cancer. So, instead of loading your usual snack options with calories and oil, add this seasoning to your bread, fried beans or sauce and make the dish rich in flavour and high on health. You can rediscover the magic of cumin in your regular bowl of tadka dal and rice too!

Cayenne Pepper; Research has proven that cayenne peppers help with **weight loss** and lower blood pressure, but the active compound in cayenne peppers also kills cancer cells. Capsaicin, a primary component in cayenne peppers, stifles the growth of prostate cancer cells. It could also prevent the shrinking of cancerous tumours and the relapse of prostate cancer. A promising spice with anti-cancer properties, an overdose of chilli peppers however should be restrained. Capsaicin induces the process of apoptosis that destroys potential cancer cells and reduces the size of leukemia tumour cells considerably. It can be concluded that apart from setting our tongues on fire, chilli peppers can scare cancer pathogens off too.

Oregano: Carvacrol, a molecule present in oregano, may help counteract the spread of cancer cells. Marinating foods in oregano also reduces the production of heterocyclic amines (HCAs) – chemicals formed when meat is charred. HCAs are responsible for increasing cancer risk. Oregano confirms its worth as a potential agent against prostate cancer. Consisting of anti-microbial compounds, just one teaspoon of oregano has the power of two cups of red grapes! Phyto-chemical '**Quercetin'** present in oregano restricts growth of malignant cells in the body and acts like a drug against cancer-centric diseases.

Saffron; Crocetin is the primary anticarcinogenic element in saffron. It inhibits the development of cancer and also reduces the size of cancerous tumours. Aqueous saffron preparations have been reported to hinder chemically induced

skin carcinogenesis. A natural carotenoid dicarboxylic acid called 'Crocetin' is the primary cancer-fighting element that saffron contains. It not only inhibits the progression of the disease but also decreases the size of the tumour by half, guaranteeing a complete goodbye to cancer. Though it is the most expensive spice in the world for it is derived from around 250, 000 flower stigmas (saffron crocus) that make just about half a kilo, a few saffron threads come loaded with benefits you won't regret paying for. Saffron threads can be used in various ways

Cinnamo: Cinnamon blocks the formation of the blood vessels that supply oxygen and nutrients to cancerous cells. Cinnamon extracts suppress the growth of H. pylori, a bacterium that causes gastric cancer. It takes only a teaspoon of ground cinnamon a day to reduce cancer risk. A natural food preservative, cinnamon is a source of iron and calcium. Useful in reducing tumour growth, it blocks the formation of new vessels in the human body.

Mint: The healing power of mint is an apt example of the efficacy of phytochemicals in preventing and treating diseases in humans. The phytochemicals in mint destroy the blood vessels that supply cancerous tumours. They starve the cancer cells by blocking the supply of oxygen and nutrients.

Coriander: Studies suggest that coriander may be effective in preventing **colon cancer**. It reduces **cholesterol** and toxin levels in the colon, and offers protection against the harmful effects of the lipids in the colon that could cause **colon cancer**.

Dill: Dill has a variety of medicinal properties. Monoterpenes are protective compounds in dill that secrete enzymes with powerful antioxidant properties. These compounds neutralize carcinogens, specifically free radicals, thereby reducing cancer risk.

Ginger: Ginger contains the compounds ginerol and zingerone, which have antioxidant and anti-inflammatory properties that prevent precancerous tumours from forming a breeding ground for cancer cell growth. Ginger destroys cancer cells in two ways. Apoptosis is a process wherein cancer cells destroy themselves while leaving the surrounding healthy cells unaffected. In the second way, autophagy, compounds in ginger trick cancerous cells into consuming themselves. This humble spice boasts of medicinal qualities that help lowering cholesterol, boost metabolism and kill cancer cells. Easily added to vegetable dishes, fish preparations and salads, ginger enhances the flavour in cooking. Chew on fresh parsley if the odour bothers you.

Fennel: Fennel is packed with phyto-nutrients and antioxidants. Anethole, a compound in fennel, resists and restricts the adhesive and invasive action of cancerous cells. It suppresses the activity of enzymes that cause the multiplication of cancerous cells. Armed with phyto-nutrients and antioxidants, cancer cells have nothing but to accept defeat when the spice is fennel. 'Anethole', a major constituent of fennel resists and restricts the adhesive and invasive activities of cancer cells. It suppresses the enzymatic regulated activities behind cancer cell multiplication.

A tomato-fennel soup with garlic or fresh salads with fennel bulbs make for an ideal entre prior to an elaborate course meal. Roasted fennel with parmesan can be another star pick.

Other foods that too fight cancer arebeans and soy beans, grapes, green leafy vegetables and cruciferous vegetables.

iii. Anti-cancer Fruits and Vegetables

Blueberries, strawberries, goji berries, acai berries, avacado, citrus fruits, grapes, dragon fruit, avocado, pomegranate, kiwi, apple, mangosteen, noni, carrots, garlic, onion, leek and cruciferous vegetables

Fruits and vegetables consists of small quantities of photo-nutrients and these naturally occurring nutrients are as important as minerals and vitamins. Cruciferous the metabolism of carcinogens and stimulate the body's production of detoxification enzymes.

Tomato: According to sfgate.com, A study published in 2009 in the Journal of Clinical Oncology shows that tomatoes, which contain lycopene, help prevent prostate cancer and stops the cancer cell growth that causes breast cancer. Lycopene is a strong antioxidant that may also prevent other types of cancers, so Farquhar recommends three servings of tomatoes a week for men and women. Therefore, drink a glass of tomato juice daily to enjoy the benefits from it.

Spinach: According to a research, women who eats spinach have 40 per cent lower-risk of breast cancer. It is loaded with folic acid and anti-oxidants, which helps in flushing out the toxins from the body and also fights against cancer causing hormones.

Broccoli: Sulforaphane, a compound in broccoli contains cancer fighting properties that boosts the performance of enzymes and flushes out cancer causing chemicals. It is aslo rich in indoles which deactivate oestrogen and reduces the risk of breast cancer.

Pomegranates: They are rich in ellagic acid. According to a few studies, ellagic acid slows down cancer cell growth and deactivate cancer-causing compounds. So have this fruit in any form, a juice, as salad or milkshakes.

I. Foods that Seduce

Bananas: A banana is packed with vitamin A, B and C and also potassium. Vitamin B and potassium are known to increase sex hormone production in the body. Banana is also packed with Bromelain which increases the testosterone levels. High levels of sugar present in banana gives energy, which helps to last long.

Garlic: It has allicin which helps in blood circulation. So if there is enough blood going to your groin there surely wouldn't be any problem related to erection. Studies also show that garlic can stimulate the production of nitric oxide synthase which is responsible for erection.

Avocado: This is one fruit which has been associated to both female as well as male sexuality. The fruit is voluptuous and feminine in shape but as the fruit hangs in pairs from the tree, they are often said to resemble the male testicles. Avocado is a rich source of beta carotene, magnesium, vitamin E, potassium and proteins – all of these are good for your sexual appetite.

Fig: A fig when cut vertically seems to resemble the female sex organ. Since ancient times it has been associated with fertility. Figs contain vitamin A, vitamin B1, vitamin B2, calcium, iron, phosphorus, manganese, and potassium and these are known to decrease sexual weakness.

Asparagus: In the 19th century bridegrooms in France were served three courses of asparagus just a day before their marriage. Asparagus is a rich source of potassium, vitamin B6, vitamin A, C, thiamin and folic acid. It is said that folic acid boosts histamine production which help both male and females to reach orgasm. Folic acid also helps in reducing birth defects, so, asparagus is good for a pregnant woman. Asparagus is also known to increase blood circulation in the genitourinary system.

Basil: This sweet smelling herb is called 'kiss me Nicholas' in Italy. It is believed that basil increases sex drive and fertility. Basil has magnesium, iron, vitamin A, C and K. It relaxes blood vessels and prevents clotting in arteries, hence increases blood circulation. Basil is also known to cure all kinds of 'headaches' so, next time if you get this excuse it is time to use basil. The sweet scent is believed to make men lust after a woman wearing it. Ancient Greeks gave it to horses before breeding them.

Chilies: From bell peppers to the red chilies – all are considered **aphrodisiacs**. Chilies have capsaicin which increases blood circulation and heartbeats, raises body temperature and also produces sweat. All these symptoms occur while having sex so this might be another reason for calling it an aphrodisiac. Capsaicin also helps in releasing endorphins and stimulates the nerve endings which increases the pulse and makes the body sensitive.

Cloves: Used in aromatherapy to increase sexual desire. It improves blood flow and body temperature when eaten.

Coriander: In the tale, The Arabian Nights, a merchant who was childless for 40 years is cured by a concoction that includes coriander. Hippocrates made a wedding drink containing it to stimulate libido of the newlyweds.

i. Aphrodisiac Foods that Improve Sex Drive

Any healthy food is good for sex. However, there are certain fruits and vegetables that are particularly beneficial *viz.,* walnuts, strawberry, avocados, watermelons and almonds.

Leafy greens are essential for our sexual health because they help block absorption of some of the environmental contaminants thought to negatively impact our libido.

Strawberries and raspberries: The seeds of these fruits are loaded with zinc which is essential for sex for both, men and women. If women have high levels of zinc their bodies find it easier to prepare for sex. Experts say that strawberries are good for your sex life since they boost the adrenal gland, known to produce hormones that help women orgasm. In men, zinc controls the testosterone level which is responsible for producing sperm. It is important that men load up on zinc as their zinc levels reduces during intercourse. So, if you want to raise temperatures in the bedroom, tuck into some strawberries vanilla ice-cream with.

The colour red is known to help stoke the fire. A 2008 study found that men find women sexier if they are wearing red, as compared to cool colours such as blue or green. This is the reason that people give chocolate on Valentine's day, because it is full of libido- boosting methylxan thines. Strawberries are also an excellent source of folic acid, a B-vitamin that helps ward off birth defects in women.

Basil: Use this herb, which is known to improve blood flow and increase heart rate, to spice up your sex life. Whether you use it as a topping over soup or add it over food, basil is known to enhance desire

Celery: While the last thing you might feel like eating is celery, especially to get in the mood, it is a great stimulant. Try having some celery sticks with sour cream or a tangy dip or add it in a soup or salad. Celery has androsterone – a hormone known to excite men. Eating celery makes men release pheromones, known to attract women

Cinnamon: Experts say that cinnamon heats up the body, which in turn boosts your sex drive. Its anti-inflammatory properties also normalise blood sugar. Mix crushed cinnamon in a glass of milk before going to bed

Ginger: Ginger is known to increase circulation, body temperature, which in turn increases desire, and is believed to increase blood flow to sexual organs.

Aniseed: One of the most popular aphrodisiacs, aniseed is a favourite 'bedroom dish' (Anonymous, 2014)

Nutmeg: Valued it as an aphrodisiac by Chinese women, referred to as the "Viagra for Women" in Africa. Can produce hallucinations when used in quantity.

Walnuts: Walnuts improve the quality of sperm. It is known to improve the shape, movement and vitality of the sperm. Include walnuts in your diet to improve fer**tility.**

Avocados: The scientific reason why avocado make sense as an aphrodisiac is that they are rich in unsaturated fats and low in saturated fat, making them good for your heart and arteries. As anything that keeps the heart beating strong, help keep blood flowing all places. In fact men with underlying heart disease are twice as likely to suffer from erectile dysfunction. Folic acid and vitamin B6 are both necessary for a healthy sex drive. Folic acid pumps the body with energy, while vitamin B6 stabilises the hormones.

Watermelon: Watermelon improves your erection and increases your libido. They also contain citrulline which releases amino acids and arginine in the body. Arginine is responsible for vascular health.

Almonds: Almonds contain arginine which improves circulation and relaxes blood vessels. This amino acid found in almonds helps you maintain an erection. They have been purported to increase passion, act as a sexual stimulant and aid with fertility. Like aspergus, almonds are nutrient- dense and rich in serval trace minerals that are important for sexual health and reproduction, such as zinc, selenium and vitamin C. Zinc helps enhance libido and sexual desire.

Citrus: Any member of this tropical fruit family, is supper rich in anti Oxidants, vitamin C and folic acid- all of which are essential for mens reproductive health.

Figs: These funny shaped fruits have a long history of being a fertility booster and they make an excellent aphrodisiac, because they are packed with both soluble and insoluble fibre which is important for heart health. Plus high fibre foods help fill you up, not out, so it is easier to achieve that sexy bottom line or belly.

Peaches: Vitamin C present in peaches improves sperm count and the quality of the sperm. Peaches contain high levels of vitamin C that is great for reducing infertility.

Saffron: Saffron is a natural aphrodisiac and should be consumed to improve your sex drive and your performance in bed. Saffron can also boost stamina and energy. An extensive review of food aphrodisiacs done in 2011 found just a few threads can improve ED, but was not as effective as Viagra.

Fennel: Contains estirol, an estrogen-like substance. Ancient Egyptians used it to boost libido in women

Fenugreek: The seeds contain **saponins,** which play a role in increasing testosterone production. A 2011 study showed it raised libido in men.

Goji- This amazing superfood not only raises your spirits, but it also raises your libido! Goji berries raise testosterone levels, and, therefore, your sex drive goes up.

Drum Stick: It enhances sexual health.

ii. Foods to Avoid before Sex

Fruits and vegetables, which boost sex drive or raise sex quotient are given below. But did you also know that there are foods, which can be a total put off when consumed before sex.

Beans: Beans are a good source of fibre and proteins for our body. But they also contain a type of sugar that doesn't digest fast and causes bloating and gas. We can all agree that a gassy and bloating belly is not ideal for sex.

Garlic: Garlic adds extra flavour to our food, but it is really a good idea before sex. It not only has a strong scent which can be an instant turn off but also has starches that can cause bloating.

Peppermint: While you might think that chomping on a peppermint for fresh breath is a good idea before sex, studies show that the menthol present in peppermint can reduce testosterone levels and this can affect your sex drive.

iii. Foods for Harder Erection

Chilies, bananas, onions

a. High Sperm Count

Asparagus, bananas, walnut, pumpkin seed, garlic, Goji berries

b. Erectile Dysfunction

Peanuts, black beans watermelon, banana's and pistachio nuts

iv. Foods to Induce, Regulate your Periods

Many women suffer from irregular periods, where they get them every few weeks or not for months at a time. If you want to regulate your period and keep it from getting delayed, here are some natural emmenagogues, which are foods that Induce menstruation.

Parsley: Parsley leaves boiled in water have been used as an emmenagogue for centuries. Drinking parsley tea twice a day causes contractions in the uterus, which helps bring on your period.

Ginger: Ginger tea is also a powerful emmenagogue, but unlike parsley it can have some side effects, like acidity. For extremely delayed periods, a combination of parsley and ginger tea is recommended.

Celery: Eating celery stalks or drinking celery tea stimulates the blood flow to your pelvis and uterus, helping to induce your period

Cumin: Cumin seeds, also known as jeera in Hindi, belong to the same family as parsley and also have a similar effect.

Fennel seeds: Fennel seeds, also known as saunph in Hindi, can be boiled in water to make a fragrant tea that should be consumed every morning on an empty stomach in order to regulate your period and have a healthy flow.

Fenugreek seeds: Fenugreek, or methi, seeds are recommended by herbalists to induce your period. Breastfeeding mothers can also consume these seeds to improve their supply of milk.

Papayas: Eating a small-sized papaya every day for a week can help induce your period, since papayas are a fruit that create a lot of heat in your body. Fruits like mango and pineapple have lesser but still significant emmenagogic effects as well.

Almonds: Almonds induce your period, ensuring that you get it on time. They also have nutrients that help balance your hormones.

Grapes: Drinking a glass of fresh grape juice every morning can help regulate your period and prevent delays.

Aloe vera: Aloe vera juice is normally used to soothe an upset stomach, but it can be used to emmenagogue as well.

J. Foods that Contain Oxalates

The following are some examples of the most common sources of oxalates in fruits and vegetable. It is important to note that the leaves of a plant almost always contain higher oxalate levels than the roots, stems, and stalks.

Fruits: Blackberries, blueberries, raspberries, strawberries, currants, kiwifruit, concord (purple) grapes, figs, tangerines, and plums

Vegetables: Spinach, Swiss chard, beet greens, collards, okra, parsley, leeks and quinoa are among the most oxalate-dense vegetables

Celery, green beans, rutabagas, and summer squash would be considered Moderately Dense in oxalates.

Nuts and seeds: Almonds, cashews, and peanuts

Legumes: Soybeans, tofu and other soy products

K. i. Fat Fighting Fruits and Vegetables

Avocados: They are high in healthy fats and help keep our heart healthy.

Asparagus: Are high in nutrients and helps flush the body which removes unwnted fat deposits and toxins.

Almonds: Are full of high protein and fibre. Eating almonds throughout the day helps appease your hunger and add nutrients to your daily diet.

Spinach: It is low in calories but rich in fibre, iron and beta carotein, which helpfs lower cholesterol.

Broccoli: Great sources of vitamin A and C. It contains lots of fibre.

Walnuts: Eating walnuts before meals not only helps you from overeating, but it also give you a healthy dose of fibre, vitamin C and mega- 3 fatty acids to reduce bad cholesterol.

Apple: Rich in vitamins and fibre, helps to keep you feeling full.

Grape fruit: It contains **galacturonic acid,** which breaks down fat, while fibre helps keep you feeling full.

Pine apple: It is rich in fibre and vitamin C. It also contains **bromelain**, an enzyme that helps the body to process protein quickly. Even though the pineapple is very nutritious, it also naturally contains a lot of sugar so eat in moderation.

Vegetables that fight abdominal fat: Cruciferous vegetable like cabbage, cauliflower broccoli, brussel sprouts and kale

ii. Fat Burning Fruits and Vegetables

Celery, Bananas, Tomatoes (lycopene), Cinnamon, Hotpepper

Berries (Blue berries, raspberries, strawberries)

Nuts: Almonds, walnuts, cashews even peanuts.

iii. *Fruits and Vegetables for Weight Loss*

Apple, blueberries, cucumber, zucehii, tomato, green sweet peppers, asparagus, spinach, brinjal, cabbage, carrots and bean sprouts.

Apple: Apples are rich in antioxidants that aid in weight loss. If you consume an apple before eating your meal it tends to cut down on calories. Plus the antioxidants in apples may help prevent metabolic syndrome, a condition which is marked by excess belly fat or an apple shaped figure. Apples are said to contain four to five grams of fibre per serving, so they are sweet as well as crunchy and keeps you full for a long time. They also contain antioxidants which add essential vitamins and minerals in your body

Almonds: Did you know that these nuts are packed with the best nutrients and vitamins? Just a handful of almonds will give you the right boost of energy and will also keep hunger pangs at bay. Avoid the ones that are coated with chocolate or salt. Additionally, because of their high content of Zinc and vitamin B, they greatly help in reducing sugar cravings. Since the fibre content is high, you stay fuller for longer. They are highly satiating and provide bulk to food without really adding to the calorie content.

Avocado: Though they are known to be fatty, but you need not worry. Avocados are loaded with fibre and proteins and helps in weight loss. They contain 'good' mono-saturated fats that keeps you full for a long time.

Bell Peppers: Red, yellow or green - you can choose any, bell peppers contain vitamin C, and gives you the right amount of vitamins. You can have them raw, as a salad or in a bowl of soup.

Cinnamon: Instead of adding sugar to your beverages or meals, try adding a pinch of cinnamon for increased energy and a slower release of insulin in your system. Cinnamon has a sweet taste and is completely guilt-free.

Blueberries: Though all berries are good for you, but blueberries are best of the brunch. Rich in antioxidants, it helps you shed excess fat.

Quinoa: If you are craving carbs during your lunch or dinner hours, you can always go in for quinoa. Said to be an alternative for rice, it's rich in proteins and fibre and is a wonderful food that aids in weight loss.

iv. *Foods that cut Belly Fat*

Garlic: Consume garlic as a natural antibiotic and blood sugar regulator. By controlling blood sugar and insulin level, your body can maximize fat burning for energy, resulting in reduced belly fat. Garlic is also a thermogenic that boosts metabolism.

Mint: Add one tablespoon honey, a pinch of pepper, and some crushed mint leaves to a cup of hot water. Let it to steep for five minutes. Strain and drink the liquid to get a flat tummy. Mint soothes the abdomen while honey and pepper dissolve the fat and boost metabolism.

Cinnamon: This spice is loaded with antioxidants and has the ability to control your blood sugar. Thus, it directly affects the insulin level and controls your appetite. Once you stop craving for food, your body automatically uses up stored fat, especially from the mid-section. The best way to add cinnamon to your meal is by using it for flavouring your drinks. You can also add it to your lunch and dinner by cooking it along with the main course. Cinnamon works as a great fat burner. Take ½ tablespoon of cinnamon powder and steep it in water for five minutes. Add 1 tablespoon of honey after straining the cinnamon water. Drink this before breakfast and before going to bed.

Watermelon: Watermelon contains 82 per cent water, which helps your stomach not to crave for food. Watermelon is rich in vitamin C, which is beneficial for health. It is one of the healthiest foods to have if you are planning to go on a healthy diet. It is also a good mid-day snack.

Apple: Eating apple regularly can help in fighting many diseases and it can also help in reducing the fat from your belly. Apple helps your stomach to feel full because it contains potassium and many vitamins. So, eat an apple in breakfast to get the desired tummy size.

Bananas: Like apples, bananas are also rich in potassium and contain multivitamins. They are filling and help you to resist the cravings for fast food. Moreover, banana boosts the metabolic rate, thereby melting the belly fat.

Broccoli: This little green vegetable is rich in vitamin C and calcium, which increases the absorption of chemical elements in the body and boosts metabolism. The fibre and water content in broccoli is high, which means you feel full fast, won't indulge in overeating and are hydrated throughout the day. It's advisable to include broccoli in your meal at least four times a week. You can either steam or microwave the broccoli instead of boiling it for longer duration.

Avocado: While the fat content in avocado is slightly higher than that in other 'healthy foods', studies have found that the monosaturated fats in this fruit help in controlling cholesterol and blood sugar. They are also loaded with potassium, which helps in converting nutrients into energy. Avocados are also quite filling, as they have a higher fraction of fibre, vitamin B and amino acids. Try taking in avocados during breaks by substituting it with oily items.

L. Fat-Releasing Fruits and Vegetables

Apples: Besides keeping the doctor visits away, consumption of apple everyday also assists in reducing fat cells in the body. The skin of the apple possesses great magical properties towards fulfilling your goal of weight reduction. The presence of pectin limits the absorption of fat by the cells and also releases fat deposits through its water binding property (Pinto.2014).

Walnuts: Walnuts possess a healthy dose of omega-3 fat alpha-linolenic acid and mono-unsaturated fats. The presence of mono-unsaturated fat helps to burn a large amount of fat and boosts the metabolism at the same time. Just a handful of walnuts are needed to reduce weight in a healthy way. It is also one of the healthiest nuts available.

Beans: Beans are low-fat, low glycemic index and high - fibre and protein. This is the best food for protein intake for vegetarians. Besides, it is the best fat burning food as it provides great a metabolic environment for releasing and metabolizing fatty acids.

Ginger: Ginger possesses many magical properties. It helps to relieve digestive problems, reduces inflammation, increases blood flow and aids muscle recovery. If you are on a weight loss program, include ginger in your diet as it helps to boost calories and fats.

Hot peppers: Consuming hot peppers will help to speed up your metabolism by burning fat and calories at a faster rate, albeit for a short duration after meals. The presence of **capsaicin** helps to provide short-term stimulus to your body by releasing stress hormones. This process helps to boost your metabolism and hence burns calories and fats.

Fruits and Vegetable to Gain Weight Quickly

Potatoes: It is healthy food to have in your diet as it is high in protein, full of fibre and contains lot of vitamin C.

Nuts (mixed) : They contain 500 t0 600 calories per 100 grams. Nuts are calorie dense foods packed with nutrients like omega- 3 fatty acids, protein, vitamin C and fibre. Bananas are rich in potassium, carbohydrates and other important nutrients that give you energy and keep you healthy. One medium sized banana has about 120 calories.

Bananas: They are time tested remedy for weight gain. Traditionally they are used w

Fruits and Vegetables that Cleanse Liver

Diet and exercise are very important to stay healthy. Following are the fruits and vegetables that you can include in your daily diet to flush out toxins:

Grape Fruit

It is an excellent fruit to eat during a liver cleanse. It is not only high in anti-oxidants like vitamin A and C, but cleanse the liver due to its high level of flavonoids.

Benefits

Stimulates the liver

Improves insulin levels

Metabolizes fat better

Beets

They are rich in minerals like magnesium, calcium, iron, phosphorus, and high in vitamin A and C, folic acid (known to guard against certain cancers, like colon cancer. It is the betacyanin that guards you).

Benefits

Cleanse the liver plus tones it.

Cleanse the colon

Purifies the blood

Bitter Greens

Like endive, argula, mustard greens, radicchio, escarole, dandelion greens. Any thing you eat that is bitter will stimulate your liver and gallbladder. In salads, it is easy to add a bit of endive or arugula without getting overwhelmed by the pungent taste.

Benefits

High in anti-oxidants like vitamin C, betacarotene.

Excellent to stimulate digestion and breaks down food.

Apples

The foods that are sour and bitter are good for liver and gallbladder. Apples are sour and the skin has a slight bitter taste plus the extra pectin makes a clean sweep of your bowels

Benefits

High in anti-oxidants.

Its peel has cancer fighting properties, especially for liver, colon and breast cancer.

The fibre also helps you to reduce cholesterol levels that can contribute to gallbladder stone formation.

Lemons

Are a very alkaline type of fruit, even though it is considered a citrus. Adding 1 whole lemon to 1 quart water stimulates your liver and is especially helpful, if you have eaten a meal and you feel bloated. Lemons release enzymes that help excrete toxins from the body. Health experts usually advise people to drink a glass of warm water with lemon and honey in the morning as it helps cleanse the body of toxins. It also supplies vitamin C to the body and balances the acidity in food consumed.

Benefits

High in anti-oxidants.

Excellent to regulate bile flow.

Maintains good digestion

Ginger

This is considered one of the most potent disease-fighting spices, along with turmeric. It helps the metabolism, flushes out toxins and boosts liver function. You can chew on a small piece of ginger or even include it in your cooked food every day.

Parsley

While parsley is mostly used as a garnish, it has more medicinal values than you know. It keeps the kidney healthy and free of infection, and reduces bloating during menstruation in women.

Fruits and Vegetables that Flush out Toxic Waste from the Liver

Garlic (alicin)

Beet and carrots

Grean leafy vegetables (raw, cooked, juice)

Avocados (improves liver health)

Apples (pectin) and berries.

Cruciferous vegetables like broccoli, brussel sprouts, cabbage, cauliflower and cale *etc.*

Lime and lemon

Walnuts (aminoacids)

Turmeric

Cilantra, ginger

Fruits and Vegetables as Natural Diuretics

Diuretics flush out the water and toxins built–up through the urinary tract. Nature has naturally endowed certain foods like fruits, vegetable and spices with diuretic properties and play a vital role as an agent to alleviate body discomfort. A glass of cranberry juice aids in the removal of excess fluids. Two spoons of apple cider vinegar mixed in glass of water taken 3-4 times in a day, has tremendous flushing power. Similar potassium rich dandelion leaves tea act as powerful diuretic and also replaces the potassium lost through urine. It also has beneficial effects on people suffering from urinary tract infection. Potassium rich foods release the fluid built up in the system. Some of the foods rich in potassium include bananas, coconut water, tomatoes, apricot, strawberries and papayas. However, in case of kidney problem, excessive potassium can have negative consequences like severe muscular pain and irregular heartbeats (Mukerjee, 2014).

Fennel seeds have diuretic properties. It has cooling effect on the body when eaten with sugar and water.

Water melon and cucumber rich in sulphur and silicon stimulate the kidney to function more effectively.

Asparigine in asparagus boosts kidney performance and improves waste removal from the body.

Lettuce leaves aids in better metabolism and flushing of toxins.

Tomatoes are rich in vitamins A C, and aid in the metabolism and release of water from the kidney to flush out waste.

Horseradish, raw onion and radish are powerhouse of sulphur compounds which rave up your metabolism to flush out the toxins.

Coconut water is one of the nature's best tonics that naturally cools the body and maintains electrolyte balance in conditions like pre-menstrual syndrome.

Parsley is an excellent diuretic herb that increases the elimination of excessive water along with waste products like urea and other acidic metabolites.

Fruits and Vegetables that Damage Teeth

Dried fruits can be harmful for our teeth because of their concentrated sugar content and stickiness. Dried raisins, prunes, apricots, fig are gummy. So, parts of them easily adhere to teeth and the sugar in them encourage bacteria in the mouth that erodes tooth enamel. Dried fruits are packed with non-soluble cellulose fibre, which can bind and trap sugars on and around the tooth, making it worse than sweets. Chewing regular fruits, however, is less harmful, because even though they have sugar too, they aren't as sticky and the saliva helps in cleaning up

Fruits and Vegetables

High-fibre foods can actually scrub your teeth clean. Fruits and vegetables that are crunchy and juicy like cucumber, watermelons, muskmelons apples, pears have water content that offsets the fructose

Nuts

Vitamins, minerals, calcium, iron, potassium, zinc and other nutrients present in nuts like almonds, walnuts, cashews and peanuts are good for your teeth.

Fruits and Vegetables for Glowing Skin

These foods work on all types of skins, from oily to normal. So simply, add these foods to your diet and you will have glowing skin.

Colour your diet with wild. Berries bursting with antioxidants synergistically protect skin and delay signs of ageing. Black berries, blueberries, strawberries, plums, polyphenols in pomegranates, lycopene in tomato offer the scavenging effect.

Orange: it helps maintain great skin and vision**.** It is loaded with nutrients and is one of the richest sources of vitamin C, which is important in Collagen production. Orange juice is also believed to beat ageing effects. You can also rub orange juice on your face to clean clogged skin pores and are believed to prevent acne, pimples and fine lines. Rubbing its peel on your skin can prove beneficial. Like apple, **orange** too contains collagen that slows skin aging process. Rub the insides of orange on your skin to tighten the skin. Oranges can be dried and powdered and used as a natural scrub. Like lemon, oranges too help clear skin blemishes.

Apple: It has immense qualities that give your skin the required glow. It contains copper and vitamin C that are both skin friendly nutrients. The fruit helps retain essential nutrient. Potassium, which is an essential mineral for your skin is also present in apple.

Watermelon: Not only does the fruit keep your skin hydrated, it is rich source of lycopene, vitamin C and A that helps in slowing ageing signs from appearing. Vitamin A in the fruit helps decrease skin pores and oil secretions from the sebaceous glands. Applying watermelon slice, can work as a cleaner.

Orange: Orange is loaded with nutrients and is one of the richest sources of Vitamin C, which is important in collagen production. Orange juice is also believed to beat ageing effects. You can also rub orange juice on our face to clean clogged skin pores and are believed to prevent acne, pimples and fine lines. Not only that. You can also rub its peel on your skin for health benefits.

Bananas are rich in vitamin A, B and E and hence works as an anti-aging agent. A fresh mashed banana facial can do wonders for your skin.

Pomegranate: The fruit contains a high quantity of Punicic acid, Ellagic acid and anti-oxidants. Not only is rich in omage-3 fatty acid that helps in skin cells regeneration. It also increases red blood corpuscles (RBC) count and is useful for blood circulation. It contain vitamin A, C and E along with lot of important minerals. It is also a natural moisturizer and is perfect for dry and irritated skin.

Gralic purges the body and the blood from harmful bacteria. A clear system and good blood circulation improves the condition of the skin.

Beetroot purges our bodies of toxins; they clear the blood, liver and fight free radicals. Besides beetroots contains potassium, iron, magnesium, phosphorous, fibre and vitamins A, B, and C.

Fennel seeds has two strong properties:it is an antioxidant and detoxifier. It also tones the skinand improves blood circulation. Fennel also keeps the skin hydrated and moisturised.

Acne plaguing you into adulthood? Then chomp on some nuts to fight acne since they contains selenium and zinc. Pumpkin seeds and almonds help purify the blood and provide nutrients beneficial to the body.

Like apple, **orange** too contains collagen that slows skin aging process. Rub the insides of orange on your skin to tighten the skin. Oranges can be dried and powdered and used as a natural scrub. Like lemon, oranges too help clear skin blemishes.

Protect your skin from the sun with **tomatoes**. It also reduces acne and reduces the oiliness of your skin. Tomatoes also help cure sunburn.

Papaya is rich in antioxidants and contain a special enzyme called **papain** that can kill dead cells and cure skin impurities. It is a nutrientially rich fruit, that has some unique enzymes. It is believed to help in weight loss and even protect your heart. It helps keep your skin looking young and fresh. And if you have been dumping the black seeds, it is worthy to know that they help in flushing out toxins and removes dead skin cells. A glass of papaya milk or just applying the flesh of papaya on your skin can do wonders to your skin.

Almonds: Almonds are not only rich in Vitamin E but also they help you to have a fresh mind and fresh face after a hangover. Apart from boosting your Vitamin E levels almonds are also known as the best antioxidants for your skin. Also many believe that eating a handful of almonds on a regular basis also increases your memory power. So add a new task in your to-do-list of everyday: have a handful of almonds.

Strawberries: With vitamin C and other antioxidants, strawberries can work wonders for your skin! These anti-oxidants prevent cellular damage, slowing down skin ageing. Strawberries have salicylic acid that removes dead skin from the face and tightens pores because of its cleansing properties. You can use strawberries as a skin toner by mixing it with rose water. Both these ingredients have great astringent properties.

Raspberry oil: Helps your skin look younger due to its high ellagic acid content with antiaging properties.

Avocado: Sooths dry and itchy skin, also acts as an excellent nutrition for daily skin care.

Breath Freshening Fruits and Vegetables

Citrus fruits Oranges, sweet lime, lemons and kiwis (rich in Vitamin C) : These help in keeping the breath fresh. Vitamin C is known to fight the bacteria. It is also an antioxidant - thus it helps in reducing the toxins in the body, this includes those produced by the bacteria in the mouth. You may have these fruits as whole or even drink up or sip on their juices.

Spices and herbs: These are super amazing at keeping the breath free of any odour for a good period of time. Examples are cardamom (elaichi), fennel (saunf), spear mint, parsley, rosemary, eucalyptus, coriander (dhania daal), cinnamon (dal chini), cloves (laung) *etc.* From amongst these, the following are favorites with most people.e.g Cardamom and Eucalyptus both contain a substance called **cineole**. Cineole has antibacterial, antiseptic properties and thus attacks the stinker bacteria in the mouth. With these harbingers of foul smell gone for good, the mouth feels and smells fresh. Cardamom seeds can be kept as such in the mouth and biting into them intermittently to release that typical delicate flavour and aroma of cardamon which most of us tend to like.

Parsley and coriander: These are rich in the green pigment chlorophyll which because of it's alkaline properties cleanses the body in a generic sense. It attacks the bacteria of the mouth thus rendering them ineffective. Chlorophyll also possesses detoxifying ability.

Cinnamon (dal chini) : The essential oil present in cinnamon is antiseptic and kills the germs giving rise to bad breath.

Fennel seed (saunf) is an age old Indian remedy to keep the mouth fresh and the breath smelling sweet. Fennel is known to possess aromatic and medicinal properties.

Vegetables: Broccoli, carrots and cucumber are known to keep the mouth fresh by diminishing the bacterial content of the mouth. Chewing onto them floods the mouth with saliva.

Apples: When we crunch on to an apple, salivary secretion gets stimulated. This literally bathes the mouth and rinses out the odour generating bacteria in mouth thereby making the breath fresh.

Best Fibre Rich Fruits and Vegetables

Refined ingredients tend to get logged in the colon causing grief, flatulence and restricting bowel movement. Hence high fibre foods are highly recommended to keep your colon healthy. On an average, women need to consume 25 grams of fibre a day, while men need at least 38 grams.

i. Nuts

Nuts such as almonds, pistachios, groundnuts, and walnut are a good source of fibre. We can also mention that flax and sesame seeds also contain fibre that is good for your bowel movement. 28 grams of these nuts and seeds will increase your fibre count from 1-8 grams.

ii. Fruits

Berries are fibreous in content and a really good source; indulge in strawberries, raspberries, gooseberries and blueberries. One cup of berries will pump up your fibre content between four to eight grams depending on the fruit. You can simply eat fruits by themselves, or in a healthy dessert or snack. For high fibre fruits consume raisins, dried peaches, prunes, and figs, apples, pears, oranges and banana. These fruits will give you two to eight grams of fibre when you consume them whole.

iii. Vegetables

Spinach, turnips, potato and sweet potato, cauliflower, red cabbage, soy beans, pumpkin and lady fingers are rich in fibre. These veggies should be consumed in time of bowel and colon distress. You are assured of fibre anywhere between two grams to 8 grams. You cannot go wrong with these fibre rich foods.

Fruits and Vegetables that will Naturally Detox You

Cauliflower: This anti-oxidant rich cruciferous veggie aids your body's natural detoxification system and reduces inflammation.

Broccoli: A strong detoxifier, broccoli neutralizes and eliminates toxins while also delivering a healthy dose of vitamins.

Turnip Greens: A potent detoxifier, this cruciferous veggie also has been found to help prevent many types of cancer and is a great for reducing inflammation.

Grapefruit: This fibre-rich sweet and tangy fruit helps lower cholesterol, prevent kidney stones, and aids the digestive system.

Cucumber: Nutrient dense cucumbers, which are 95 per cent water, help flush out toxins and alkalize the body.

Sunflower Seeds: High in selenium and Vitamin E, sunflower seeds aid the liver's ability to detox a wide range of potentially harmful molecules. In addition, they help prevent cholesterol build up in the blood and arteries.

Hemp Seeds: These tiny nutritional powerhouses are an excellent source of omega 3 and 6 fatty acids, are an easily digestible plant based protein, have a strong anti-inflammatory effect, and also aid in elimination

Brussel sprouts: Are packed with sulfur which is good for clearing your body of toxins

Fruits and Vegetables for Osteoporosis

Nuts: Having a fistful of nuts daily provides one's body with the required minerals like calcium, magnesium, manganese and phosphorus. Almonds and Pistachios in particular are good sources of calcium.

Vegetables

Include brocolli, chinese cabbage, cauliflower, beetroot and okra in your daily diet to strengthen your bones.

Dates

Dates are particularly good sources of calcium with manganese. Copper and magnesium helps in improving the bone mineral density.

Fruits

Include fruits like oranges, guava, strawberries, and pineapples *etc.*, these fruits will provide vitamin C to the body which in turn will strengthen our bones. Also include banana and apple.

Pomegranates: Chemicals found in the exotic fruits can protect against Osteoarthritis, the most common form of arthritis, called phytochemicals, they help prevent damage to cartilage cells, which keeps bones healthy. Pomegranate seeds are also full of Punicic acid, which has powerful anti-oxidant properties and could reduce joint inflammation (Anonymous, 2017).

Dark Green Leafy Vegetables

Though dark green leafy vegetables are touted as one of the major means for improving one's calcium status. This one could be tricky because spinach has good amount of calcium but the presence of oxalate prevents the absorption of calcium.

Instead opt for collard greens, mustard greens, methi, amaranth, mustard green and turnip greens for getting enough calcium in the diet.

Fruits and Vegetables that Boost Immunity

Garlic: It contains the immune-stimulating compound allicin, which promotes the activity of white blood cells to destroy cold and flu viruses. It also stimulates other immune cells, which fight viral, fungal, and bacterial infections. Garlic kills with near 100 per cent effectiveness the human rhinovirus, which causes colds, common flu, and respiratory viruses.

Because allicin is released when you cut, chop, chew, or crush *raw* cloves, allow freshly chopped garlic to stand for 10 minutes and then cook it, sprinkle it over foods, drop it into soup, or swallow bits of garlic with some water like a pill. You can also drop a clove of garlic into some honey and swallow it immediately for a quick dose that tastes good!

Onions: Onions, like garlic, contain allicin. They also contain quercetin, a nutrient that breaks up mucus in your head and chest while boosting your immune system. Additionally, the pungency of onions increases your blood circulation and makes you sweat, which is helpful during cold weather to help prevent infections. Consuming raw onion within a few hours of the first symptoms of a cold or flu produces a strong immune effect. Chopping onions into your favorite soup or cooked recipe is a great way to enjoy them. Also, it may sound a little weird, but putting half an onion in your bedroom while you sleep can help absorb some of the circulating bacteria and potentially lessen the symptoms of your cold.

Ginger: Spicy, pungent, and delicious, ginger reduces fevers, soothes sore throats, and encourages coughing to remove mucus from the chest. Anti-inflammatory chemicals like shagaol and gingerol give ginger that spicy kick that stimulates blood circulation and opens your sinuses. Improved circulation means more oxygen is getting to your tissues to help remove toxins and viruses.

Research has indicated that ginger can help prevent and treat the flu. Ginger is also extremely helpful for stomachaches, nausea, and headaches.

If you're feeling a little sickly, a homemade ginger tea is one of the best things you can drink. Slice some fresh ginger root, place it into a pot with water, and bring to a boil. Then drop in a bit of lemon juice or cayenne, which makes the tea that much more effective at nourishing and purifying your system.

Cayenne: The cayenne family of hot peppers, contains capsicum – a rich source of vitamin C and bioflavonoids, which aid your immune system in fighting colds and flue. It does this by increasing the production of white blood cells, which cleanse your cells and tissues of toxins.

Cayenne pepperis also full of beta carotene and antioxidants that support your immune system and help build healthy mucus membrane tissue that defends against viruses and bacteria. Spicy cayenne peppers raise your body's temperature to make you sweat, increasing the activity of your immune system.

The fresher the pepper, the more effective it is. However, fresher also means spicier, so choose accordingly.

When you're sick, add organic cayenne powder to some warm water with lemon juice for an intense immune boost.

Squash: Squash is a good source of vitamin C and carotene. The six carotenoids (out of the 600 found in nature) found most commonly in human tissue – and supplied by squash and other gourds – decrease the risk of various cancers, protect the eyes and skin from the effects of ultraviolet light, and defend against heart disease.

One of them, alpha-carotene, helps slow down the aging process. Butternut squash is the strongest source of these nutrients, but you can also try acorn, Hubbard, delicata, calabaza, and spaghetti squash.

Kale: Like other leafy greens, kale offers up a good dose of vitamin E. This immunity-boosting antioxidant is known for increasing the production of B cells, those white blood cells that kill unwanted bacteria. Whether you eat kale raw in a salad, steam it, or lightly sauté it, you'll reap all of its wonderful benefits.

Citrus Fruits: Adding a bit of citrus to your diet goes a long way toward fending off your next cold or flu. Packed with vitamin C, oranges and grape fruits help increase your body's resistance to nasty invaders.

The best way to enjoy citrus fruits is to eat them whole. Otherwise, you can make fresh juice yourself (stay away from the premade stuff in cartons or in the freezer section at your supermarket).

Mushrooms: For centuries, people around the world have turned to mushrooms for a healthy immune system. Contemporary researchers now know why. Studies show that mushrooms increase the production and activity of white blood cells, making them more aggressive. This is a good thing when you have an infection.

Shiitake, maitake, chaga, and reishi mushrooms appear to pack the biggest immunity punch. Experts recommend eating a quarter ounce to an ounce a few times a day for maximum immune benefits (Anonymous, 2014).

Watermelon: It is packedwith a high dose of glutathione which helps boost our immune system.

Fruits and as Pain Killers

Painkillers can damage to your liver and kidneys if consume constantly, therefore, If you are looking for natural ways to relieve your pain, here are some foods that work as natural painkillers (Gupta, 2014).

Grapes: Eating just one cup of grapes every day can help keep back pain away. Grapes contain properties that increase the circulation in your lower back, thereby alleviating backaches.

Pineapples: Pineapple contains bromelain, a compound that improves your circulation, prevents cramps and thwarts inflammation. It gets rid of the inflammatory compounds that cause arthritis and also cures a bloated tummy and heaviness.

Garlic: Since ancient times, garlic has been used as a painkiller to treat various conditions. It is especially useful in treating joint pain, so you can try chopping up a clove of garlic and heating it in a little oil and applying the oil on your joints. You can also use it to treat a toothache by crushing a few pods of garlic, adding a pinch of salt and applying it to your tooth. Apply it on a skin rash or heat it with a tablespoon of olive oil and apply on the aching

Cloves: Cloves are also effective in treating toothaches. You can keep a clove in your mouth next to the tooth that hurts or dab some clove oil onto a piece of cotton and apply it to the tooth.

Turmeric: Turmeric contains curcumin, a chemical that has anti-inflammatory and pain relieving properties. Turmeric is effective against both arthritis and heartburn.

Cherries: Cherries contain anthocyanins, which are the compounds responsible for their bright red colour. These compounds are also natural painkillers and are especially effective against joint pain. According to research, these tiny, sweet fruits help in muscle repair. Have cherry juice or a bowl of this fruit post your work out.

Apple cider vinegar: Apple cider vinegar has alkaline-forming properties, so it can be used to combat heartburn. Mix one spoon of apple cider vinegar into a glass of water and sip it slowly for relief from acidity.

Blueberries: Eating blueberries can help ward off urinary tract infections, in addition to curing peptic ulcers, digestive problems and bladder infections. This is because blueberries contain antioxidants that kill the free radicals that harm your digestive lining.

Cranberries: Like blueberries, cranberries also cure ulcers and prevent urinary tract infections. The next time you are experiencing stomach pain for these reasons, drink cranberry juice instead of taking a painkiller. Not a store bought one mind you, but natural cranberry juice that hasn't been processed.

Peppermint: Peppermint is a natural remedy that can be used to treat anything from toothaches and joint pain to headaches and muscle pain. Peppermint is also effective in treating skin problems, joint pain and gastrointestinal problems like gas and bloating.

Grapes: A cup of grapes, eaten daily, can help relieve your backache. These fruits contain nutrients that increase blood circulation to the lower back, reducing the pain.

Cloves: This condiment helps fight toothache, and as a result, features in toothpastes. Dab a little clove oil on a cotton swab and put it on the aching tooth.

Ginger: Our ancestors have been using ginger to cure inflammation and pain for centuries and modern studies have only confirmed what they already knew. Ginger helps ease not only arthritic pain, but overall pain as well. According to Ayurveda, adding fresh or dried ginger to your diet helps prevent and cure muscular and joint pains. According to research, it also helps curb swelling and stiffness. joints.

Foods that Boost Metabolism

Metabolism plays a vital role in maintaining health and weight too. So take following foods regularly to help boost your metabolism.

Spinach: Leafy veggies like spinach help boost your metabolism too. Plus, they are high on proteins, which means less fats to digest.

Chilli: Chilli peppers contain chemicals that help increase the metabolism pace. Even mild pepper flakes can help you erase up to 100 calories a day.

Foods to keep your Eyes Healthy

Green leafy Vegetables like spinach, kale, corn, broccoli are rich in iron, folate, zeaxanthin and lutein. All these leafy vegetables are a good source of anti-oxidants, which make them perfect for eyes and reduce the risk of cataracts and other age related eye diseases.

Spinach: Loaded with vitamin C, beta carotene and large amounts of lutein and zeaxanthin, spinach contains nutrients that absorb 40 to 90 per cent of blue light intensity and as a result, act like sunscreen for your eyes. Have it as a side vegetable with dinner or saute it in an omelet.

Carrots: Rich in beta carotene, an antioxidant that helps reduce the risk of macular degeneration and cataracts. Beta carotene is effective for better functioning of retina. Carrots are the most important source of vitamin A, therefore, are recommended for improving your eyesight. Have them raw as a part of your veggie salad or with a curd dip.

Sweet potatoes: Containing beta carotene, sweet potatoes help promote healthy vision and also help the eyes to adjust to low levels of light at night, according to studies.

Seeds: Hemp seeds, China seeds and flax seeds can reduce the chances of age related vision loss and other eye problems. These super foods are rich in plant based Omega- 3, which strengthens the tissues of eyes

Almonds: Research suggests that almonds are filled with vitamin E, which slows macular degeneration. A handful of almonds are said to provide about half of your daily dose of vitamin E. Therefore, can prevent several diseases and improves your vision.

Walnuts: These make for a perfect substitute of fatty fish oil, which is very good for eyes as it has ample Omega -3. Much like fish oil or cod liver oil, walnuts are rich source of Omega -3 and gives strength to the eyes.

Berries: Berries like cranberries or blue berries, fresh or dried are rich sources of anti-oxidants that prevent loss of vision or weakness of eye sight.

Foods that make you Radiant

Blueberries: Blueberries has the highest antioxidant content of all fresh fruit and is rich in Vitamins A, B Complex, C and E as well as in anthocyanin, copper, selenium, zinc and iron say experts. It is one of the best sources of antioxidants in the world has been known to fight superbattles against disease, ageing, heart ailments and belly fat.

Pumpkin: Both the seeds and the flesh of this Superfood are bursting with benefits. The fruit itself is high in antioxidants, Vitamins A, C, K and E and is a rich source of magnesium, potassium and iron. Pumpkin seeds have been associated with treating osteoporosis, prostate problems, depression and preventing cancer.

Oranges: "What people do know is that oranges are a wonderful source of Vitamin C. What they might not know is that this simple fruit helps with asthma, kidney stones, high BP and arthritis as well as assists in the lowering of cholesterol and the prevention of diabetes, " says Pooja

Broccoli: Nutrient rich, this superfood helps those with blood pressure, diabetes, osteoporosis. It is believed to contribute to the control or prevention of cataracts, stomach and colon cancer, arthritis, heart disease, Alzheimer's, tumours, as well as ageing.

Walnuts: This nut, which looks like a brain, is actually good for your brain. Walnuts have also been known to do wonders for your heart, mood, bone health and weight loss as well as help with insomnia, gallstones, diabetes and cancer. Deepshikha states that nuts are calorie dense foods packed with protein, healthy fats and some essential vitamins and minerals and one must include a handful of these daily.

Spinach: Spinach is a nutrient-dense food, loaded with not just iron but a whole host of vitamins, minerals and phytonutrients. This vegetable protects your heart, promote gastrointestinal health and keep your brain young and active.

Tomatoes: This superfood derives its luscious red colour from lycopene, an antioxidant which does not occur naturally in the body and has to be derived from an external food source. Tomatoes are associated with prevention of cancer and heart disease as well as the lowering of cholesterol.

Bananas: A rich source of potassium and fibre, go bananas if you want to work on lowering your blood pressure, maintaining the health of your bones, having a healthy digestive tract and preventing the development of cancer, states Pooja.

Strawberries: The delicious strawberry doesn't just look and taste good. This is an antioxidant superfood and has been associated with cardiovascular benefits, blood sugar regulation (and is therefore good for diabetics) as well as with fighting cancer.

Fruits and Vegetables for Hair

Green Leafy Vegetables

Spinach: Greens like spinach are a rich source of vitamins, minerals and antioxidants. So add it to your meal or you can also enjoy it as a drink in the form of juice. Spinach contains iron, beta carotene, folate and vitamin C, which are known to keep hair follicles healthy and also ensure that scalp oils are circulated well. A good source of folate and iron, can go a long way in inducing hair growth. Folate helps building red blood cells, which then carry oxygen to the hair follicles. Have spinach regularly as a part of your salad.

Bell peppers: The colourful red, yellow and green bell peppers are great source of vitamin C, which is necessary for hair health. Vitamin C is needed to ensure that there is enough iron in red blood cells to carry oxygen to hair follicles. It is also used to form collagen, a structural fibre required for hair follicles to stay healthy. Vitamin C deficiency can lead to dry, splitting hair that breaks easily.

Lentils: Along with tofu, soybeans, starchy beans, and black-eyed peas, lentils are a great vegetarian source of iron-rich protein, which is necessary for cell growth, including hair cells.

Sweet potatoes: Rich in vitamin A and beta-carotene, sweet potatoes are great for hair growth. Some other beta-carotene-rich foods like carrots, kale, dark green lettuces, asparagus, and pumpkin can also work wonders on your hair (Anonymous, 2014). Sweet potatoes helps produce vitamin A, and produce oils that sustain your scalp. Inadequate vitamin A levels can result in an itchy scalp and dandruff problems. Other foods that help include carrots, cantaloupe, mangoes, pumpkin and apricots.

Walnuts: Walnuts have good amounts of omega-3 fatty acids, biotin and vitamin E that help protect your cells from DNA damage. And since your hair is rarely covered, walnuts are great protection. Very little biotin leads to hair loss. Walnuts also contain copper, which helps you keep your natural hair colour rich.

Blueberries: Blueberries contain vitamin C, which is an essential nutrient for hair. It helps circulation to the scalp and also supports the blood vessels that feed the follicles. When your body doesn't get enough vitamin C, it leads to hair breakage. Other options include kiwis, tomatoes and strawberries.

Carrots: This tasty root vegetable is full of beta carotene, which produces vitamin A. This vitamin helps in the production of certain oils that are nourishing for the hair. A deficiency of vitamin A can cause thinning of hair, thus much on raw carrots daily as a snack, as juice or cook them as a dish.

Some other high-biotin foods include peanuts, almonds and avocado.

i. Fruits and Vegetables that Affect the Blood Sugar Naturally

Diet is the most crucial part of the journey to regulate blood sugar, especially if you are a diabetic.

Onions and garlic, leafy greens, seeds (pumpkin), coconut, nuts (walnut, hazel nuts, peanuts or almond) and herbs like basil, cilantro, oregano, mint, lemons-balm are very good because they have no macronutrients, which is the protein, the fats and the carbs, but instead a lot of vitamins and minerals, so they have a great effect on the blood sugar; but gives great taste to food and have healthy vitamins and minerals and a lot of cleansing and healing properties.

High fibre vegetables such as peas, beans, broccoli and spinach/leafy vegetables should be included in one's diet. Vegetables for Diabetes are asparagus, broccoli, Brussels sprouts, cabbage, cauliflower, cucumber, spinach, tomato (lycopene – a powerful antioxidant).

Fruits high in fibre such as papaya, apple, orange, pear and guava should be consumed. Mangoes, bananas, and grapes contain high sugar; therefore these fruits should be consumed lesser than the others.

A diabetes diet should be high on fibre, must contain milk without cream, buttermilk, fresh seasonal fruits, green vegetables, etc. But remember to consume these components in moderation.

Pumpkin and its Seeds: Pumpkin and its seeds are used as a traditional remedy for diabetics especially in countries like Iran and Maxico, as they are great for regulation of blood sugar, as they are full of fibre and anti-oxidants. According to a study published in the Journal Molecules, powder and extracts can effectively reduce blood sugar levels in both animals and humans.

Nuts: Having nuts like almonds and peanuts and nu butter made with them can play a significant role in reducing blood sugar levels. In a study published in the Journal Molecule, it was found that consuming nut butter or almonds and peanuts throughout the day reduced post-meal blood sugar levels in people who have type 2 diabetes.

Broccoli: It contains a plant compound called Sulforaphane, which is produced when it is chewed or chopped. This compound holds properties capable of reducing blood sugar. In research published in Nutrients showed that broccoli rich in Sulforaphane has significant anti- diabetic effects. It lowers blood sugar, oxidative stress and enhances insulin sensitivity.

Okra: Okra or Bhindi is rich in blood sugar diminishing compounds such as flavonoids and polysaccharides. In countries like Turkey, seeds from Okra are as a treatment for diabetes. Okra is loaded with compound that have strong anti-diabotic impact and can successfully decrease blood sugar levels.

Beans and Lentils: They are full of nutrients that not just keep you happy but also actively reduce your blood sugar levels. They are high in resistant starch and soluble fibre, which slows down the digestion process and prevent the blood sugar levels from fluctuating. Adding chickpeas or black beans to your rice will help you better manage blood sugar when compared to having just rice.

Flax seeds Pack is a massive punch of healthy nutrients such as healthy fats and fibre. They also enable the body to control blood sugar levels. Research published in Clinical Nutrition Research found that people who have 30 grams of Flax seeds every day have better blood sugar levels when compared to those who consume plain Yoghurt.

ii. Fruits and Vegetables that Lower Blood Pressure

Eating more fruit and vegetables has been proven to help lower blood pressure.

Fruit and vegetables are full of vitamins, minerals and fibre to keep your body in good condition. They also contain potassium, which helps to balance out the negative effects of salt. This has a direct effect on your blood pressure, helping to lower it.

Apricot: Apricot is a very good source of potassium and vitamin A. It is very low in sodium content as well as saturated fat and cholesterol.

Avocado: Avocado is rich in assortment of vitamins and high in mono-unsaturated fat and potassium. It contains a unique fatty alcohol, called avocadene, which has a curative property for a number of ailments including high blood pressure. One cup (150 g of cubes) contains 727 mg of potassium (21 per cent of the daily recommended value for potassium) and only 10.5 mg of sodium. It is very low in cholesterol and it is a good source of dietary fibre.

Banana: Bananais a versatile fruit, is packed with potassium. A medium sized banana will provide 422 mg of potassium and 17 per cent of the daily recommended value for vitamin C. With 2.83 g of dietary fibre, this fruit will help you stay full for longer periods of time.

Cantaloupe: It is an excellent source of vitamin A and vitamin C. A cup of cubed cantaloupe (160 g) contains 494.5 mg of potassium 14.1 per cent daily recommended value for potassium.

Oranges and Lemons: Citrus fruits are best known for their high vitamin C content. Oranges are high in nutrition and low in calories. With a potassium content of 326 mg and no sodium, this is one of the best fruits **that lower blood pressure**. Limes, too, are a good source of potassium, calcium, phosphorus, vitamin A and folate. They contain 2.8 g of dietary fibre.

Grapefruit: This fruit has a distinctive, tangy taste. Select ripe grape fruits for best flavor and quality. The bioflavonoids present in grape fruit and other citrus fruits not only help lower blood pressure but also help lower cholesterol levels. Half a grape fruit (123 g) contains 166 mg of potassium and provides 5 per cent of daily recommended value for potassium.

Melons: Melon is a very good source of vitamin A, vitamin C, thiamin and potassium. One cup of frozen melon balls (173 g) 484 mg of potassium and provides 14 per cent of daily recommended value for potassium. It is also a good source of magnesium, folate and vitamin B6.

Prune. One cup of pitted prunes (174 g) contains 1274 mg of potassium and almost no sodium. Moreover, prune is a rich source of dietary fibre. A quarter cup of prunes supply 12.1 per cent of the daily value for fibre. The soluble fibre promotes a sense of satisfied fullness after a meal as it slows down the digestive process and thus helps with weight loss. So if you have high blood pressure and are overweight too, prunes may be the right fruit for you.

Spinach: A green leafy delight, **spinach** is low in calories, high in fibre, and packed with heart-healthy nutrients like potassium, folate, and magnesium – key ingredients for lowering and maintaining blood pressure levels.

Sunflower seeds are also a great source of magnesium. A quarter cup of these makes a nutritious snack – but be sure to buy them unsalted, since salted sunflower seeds are high in sodium, which you want to avoid.

Beans: Nutritious and versatile, beans (including black, white, navy, lima, pinto, and kidney) are chock-full of soluble fibre, magnesium, and potassium, all excellent ingredients for lowering blood pressure and improving overall heart health. Add beans to your favorite salads, soups, or wraps; as a bonus, they're pretty inexpensive. Since beans are high in fibre, they help you feel full, steady blood sugar, and even lower cholesterol. Moreover, they are said to burn body fat since they contain calcium. Make a bean and sprout salad or have them in your soup

Baked White Potato: Potatoes are rich in both magnesium and potassium, two vital nutrients for heart health. When potassium is low, the body retains extra sodium (and too much sodium raises blood pressure). On the other hand, when you eat a potassium-rich diet, the body becomes more efficient at getting rid of excess sodium. Like potassium, magnesium is also a key player in promoting healthy blood flow. Therefore, maintaining a healthy balance of both minerals can help keep high blood pressure at bay.

Sage: This herb boosts insulin secretion and activity, which helps to curb blood sugar in pre-diabetics and manage it in Type 2 diabetics. Besides, it also keep your liver functioning well, thereby improving your body's immunity. According to a German study, consuming sage on an empty stomach can reduce the blood glucose levels. This herb can also be added to your tea.

Flaxseeds: A good source of fibre and alpha-linolenic acid, which your body converts to omega-3, flaxseed is known to lower blood sugar and cholesterol, thereby reducing the risk of heart disease, heart attack, and other cardiovascular issues. Have them with oatmeal for breakfast or paneer.

Soybeans: Soybeans are another excellent source of potassium and magnesium.

In addition to these fruits, you can also eat raisins, dates, figs, Pineapple, apples, grapes, raspberries, kiwi, pomegranate, pear, cranberry, strawberry, tomato and molasses. They too contain a high amount of potassium. Dried fruits normally contain more potassium than fresh versions.

Dates: This brown and sticky fruit not only has a sweet taste but is rich in fibre, making it a diabetes-friendly snack. According to a study, they are richer in antioxidants as compared to a serving of grapes, oranges, broccoli and peppers. Have raw dates or stuff them with walnuts

iii. Spices and Herbs that Lower High blood Pressure

Oregano, basil, fennel, onion power, black pepper, parsley and tarragon

Fruits that are Good for High Blood Pressure

Leafy Greens: Foods high in potassium give you a better ratio of potassium to sodium. Improvements in this ratio can help with lowering blood pressure. Leafy greens like romaine lettuce, arugula, kale, turnip greens, collard greens, and spinach are high in potassium. Try to opt for fresh or frozen greens, as canned vegetables often have added sodium. Frozen vegetables, on the other hand, contain just as many nutrients as they do when fresh and are easy to store. Leafy greens like romaine lettuce, arugula, kale, turnip greens, collard greens, and spinach are high in potassium. Some easy ways to incorporate more potassium in the diet include eating more bananas, tomatoes, sweet potatoes, beans, lentils, dates and coconut water! Include more Omega 3 in your diet, include, walnuts, pecans, pine nuts, flax, and chia seeds.

Berries: Blueberries, raspberries, and strawberries especially blueberries, are rich in natural compounds called flavonoids. Consuming these compounds may prevent hypertension, and possibly help to reduce high blood pressure as well.

Potatoes: Potatoes are high in both potassium and magnesium, two minerals that can help to lower your blood pressure. They are also high in fibre, which is necessary for an overall healthy diet.

Beets: Researchers at the Queen Mary University of London found that patients with high blood pressure saw significant improvements in blood pressure from drinking beetroot juice. The study authors concluded that it was the nitrates in the juice that brought down the participants' blood pressure within just 24 hours.

Bananas: Bananas are a great way to add potassium to your diet. Adding foods that are rich in this mineral in our diet is better than taking supplements, and it's easy.

Celery: contains a compound called 3-n-butylphthalide, which has been proven to lower blood pressure. In animals, a small amount of this compound lowered blood pressure by 12-14 per cent, which is the equivalent of 4-6 stalks of celery. Celery lends itself really well as a base for juice and smoothies, so this amount would not be hard to incorporate!

Garlic: Eat a clove of garlic alongside each meal! Studies were conducted using dried garlic powder with 1.3 per cent alliin at 600-900mg (equivalent to 1.8-2.7g of fresh garlic per day), resulting in a drop of 11 mmHg in systolic blood pressure and 5 mmHg in diastolic blood pressure over a period of one to three months.

Beet: Beet juice contains l-arginine which helps relax blood vessels. A recent study proved that consuming one cup of beet juice per day was comparable to the effect of medication on patients with high blood pressure, which is approximately 9/5 mmHG.

Foods that can Effectively Remove Blockages in the Arteries and Maintain a Healthy Heart

Here are 20 best heart healthy foods to cleanse your arteries

Turmeric: Atherosclerosis is when the arteries harden which is caused by inflammation. Now turmeric or haldi has anti-inflammatory properties. This spice works in favour of reducing heart diseases. Curcumin that's present in turmeric reduces inflammation in the arteries and the deposits of fats or blockages.

Orange: Orange is a powerhouse of vitamin C that can fight common cold. But this fruit can also fight heart diseases, here's how. Oranges contain fibre pectin that lowers cholesterol; vitamin C strengthens the walls of the artery and sweeps out blockages. Orange juice can improve the functioning of the blood vessels.

Pomegranate: Fruits rich with antioxidants improve the walls of the arteries. Pomegranate is especially heart-friendly due to the presence of phytochemicals that are an antioxidant. Pomegranate ignites the production of nitric oxide that improves the flow of blood and expands the arteries.

Broccoli: Broccoli is a good source for proteins for vegetarians. Broccoli contains vitamin K which is beneficial for bone formation. Vitamin K also protects the arteries from damage. Besides vitamin K, broccoli is rich in fibre, which can lower cholesterol and high blood pressure.

Nuts: Almonds, peanuts, walnuts or hazelnuts are heart-friendly nuts. These nuts contain vitamin E that can protect the walls of the arteries. Nuts also contain fibres that also help to reduce cholesterol in the blood.

Olive oil: Olive oil is known to have several health benefits, but it is the best oil to prevent a build-up of cholesterol in the blood. Besides cholesterol, olive oil can also reduce high blood pressure. Olive oil comes under mono-unsaturated and polyunsaturated fats, which are both beneficial fats for the body.

Cinnamon: This spice should be brought in the limelight for the several health benefits it caters to overall health. Cinnamon is used as a spice as well as for dessert. This versatile spice lowers fat in the blood preventing blockages in the vessels.

Watermelon: Watermelon works great for skin care treatment, as a detox and a weight loss food. Amino acid found in watermelon helps reduce high blood pressure. But watermelon also contains nitric oxide which opens up the blood vessels to improve the flow of blood.

Spinach: This is another fierce ingredient to fight heart diseases. Spinach contains carotene that prevents cholesterol from getting clogged in the arteries. Spinach can reduce high blood pressure to due to the nutrients and minerals present in it.

Tomato: Like in pomegranate and tea, tomato too contains antioxidants that protects the arterial walls. In tomatoes, lycopene keeps the cholesterol levels low. If the tomato sauce for your pizza is prepared from fresh ingredients, then it is a healthy pizza.

Beans: Beans contain fibre and folic acid that prevents arteries from getting clogged. Beans are also a good source of carbohydrates.

Apples: Like oranges, apples too contain a fibre called pectin that absorbs cholesterol from the blood. Apples are a great way to reduce cholesterol as you can have it anytime, anywhere and you consume it the way you want. Slice them or have it as a dessert, apple is an all time favourite to improve health.

Grapefruit: The pink grape fruit is rich in antioxidant, lycopene which is also present in tomatoes. Pink grape fruit helps prevent damage to your arteries. Grapefruit is also a great fruit for diabetics.

Garlic: Garlic can keep vampires away, but it can prevent a build-up of cholesterol too. The compound thioallyls present in garlic helps fight blood clots.

Fruits and Vegetables that Lower Cholesterol

Beans, lentils and peas are good source of soluble fibre, high in protein and low in fat which helps lower cholesterol. Beans contain **Lecithin**, a nutrient that lowers cholesterol. White beans are also Cholesterol fighting super stars, as it helps in lowering cholesterol levels.

Apple, pears, orange, grape fruit and kidney beans contain soluble fibre which helps lower cholesterol. Apple rich in pectin (fibre), powerful antioxidant like quercetin, catechin, phloridzinand chlorogenic acid that help lower (LDL) bad cholesterol and helps increase HDL (good) levels.

Fruits like blue berries, cranberries and grapes protect LDL (bad) cholesterol from being damaged. Only damaged cholesterol forms sticky plaques that may lead to development of atherosclerosis and increases risk of heart diseases. Antioxidants in cranberries help slow down LDL cholesterol oxidation and help raise HDL (good) cholesterol levels. Blue berries contain antioxidant Pterostilbene that effectively helps lower cholesterol levels. Blue berries are also rich in soluble fibre and other powerful heart healthy vitamins. Antioxidants in cranberries help slow down LDL cholesterol oxidation and help raise HDL (good) cholesterol. Wild berries offer twice the anti-oxidants of regular blue berries.

Foods that are fortified with plant sterols and stanols substances naturally found in many plant foods, help block the absorption of cholesterol into the blood stream *e.g.* orange juice and yogurt drinks are often fortified with sterols and stanols.

Walnuts contain mono-unsaturated fats that helps reduce total cholesterol and LDL (bad) cholesterol levels. Omega- 3 fats and antioxidants in walnuts work reverse the arterial damage caused by saturated fats.

Eating garlic reduces LDL (bad) cholesterol and raises HDL (good) levels as it contains active substance called Allicn.

Avocados are rich in oleic acid, a mono-unsaturated fat that helps lower cholesterol.

Black beans are high in dietary fibre, which bind with cholesterol and helps excrete it from the body thus lowering cholesterol.

Pistachios, significantly lower LDL (bad cholesterol levels).

Any form of pomegranate whether the arils or the juice- could help control cholesterol by slowing its build up. Antioxidants in pomegranate help in reducing the cholesterol plaque build up and increases nitric oxide production that reduces arterial plaque. The active component in shitake mushrooms in eritadenine helps lower cholesterol levels. Shitake mushrooms also contain Lentinan, which not only lower cholesterol, but has anti-cancer properties and helps boost the immune system.

Pruns are wonderful sources of antioxidants and fibre, which is known to reduce LDL cholesterol.

Tar cherries-Anthocyanins, a type of antioxidant found in purple and dark red fruits and vegetables may help decrease risk of heart attack in women.

Plum: The fruit contains anthocyanins a.k.a. antioxidants-that help heart by lowering blood pressure and cholesterol. According to the study 3 or more servings of anthocyanin- rich fruit each week can lower your heart risk by 34 percent.

Foods that Boost Brain Power

Eating the right foods has proven to sharpen memory, boost intelligence, improve mood and emotional stability, increase mental clarity and keep the mind young. Eating right type of foods that contain abundant vitamins, minerals and anti-oxidants nurture the brain and protects it from oxidative stress (Thadani, 2017).

Beet root: Contains natural nitrates which help the blood to flow to the brain. This improves oxygenation, which in turn boosts information processing and cognition. The high quality of B6 in pistachios can increase the oxygen quantity in the blood and boost the haemoglobin in count.

Walnuts: Pack in a number of neuroprotective compounds including vitamin E, folate, melatonin and high concentration of omega-3 essential fatty acids. Even eating a handful daily will be beneficial.

Broccoli: Is a great source of vitamin K, which is known to enhance cognitive function and improve brain power. Studies have revealed that because broccoli is high in compounds called glucosinolates, it can slow the breakdown of the neurotransmitter acetylcholine, which we need for central nervous system to perform properly and to keep our brains and our memory sharp. The vitamin C content in broccoli can also help fight free radical damage to brain cells.

Pumpkin seeds: They are rich in various micro-nutrients that are critical for brain function including magnesium and zinc. Magnesium is essential for learning and memory whereas zinc prevents brain fog and boosts brain function.

Fruits and Vegetables that Prevent and Cure Kidney Stones Naturally

Kidneys perform the important task of detoxifying and filtering impurities from blood, removing waste products and regulating fluid balance. Hence it is very important to keep the kidneys healthy. Stones can be due to high intake of animal protein, sodium, refined sugars, fructose, grape fruit juice, apple juice and oxalates (found in spinach, nuts, strawberries and rhubarb).

Pomegranates: Pomegranates seeds or its juice is the easiest way of flushing stones and other toxins from the system.

Lemon juice and Olive oil: Mix 4 table spoons of olive oil with ¼ cup of lemon juice. Drink and follow with a glass of water. Repeat daily.

Lemonade: Drink lemonade as it prevents calcium formation and hence, reduces the risk of kidney stones.

Kidney beans: Soak the beans over night, boil in water or pressure cook. Use a muslin cloth to strain the water then allow it rest for about 7 to 8 hours to strain the water. Drink the strained water daily.

Citrus fruits: Fruits like oranges, grapes, banana and tomato help in breaking down the insoluable calcium.

Celery seeds: Pour 2 cups of water in a sauce pan and add one table spoon of celery seeds and allow to simmer. Let it cool and then strain and drink and drink once a day.

Basil: Chewing two to three basil leaves every morning can also help flush out the stones.

Apple cider vinegar: Mix apple cider vinegar and honey in 2: 1 ratio. Take this mixture with a cup of warm water twice a day.

Advantages and Disadvantages to Fresh, Canned, Frozen, Dried or Juiced Fruit to keeping mind

Fresh

Advantages: Fresh fruit is best if it can be consumed immediately. Popping fresh fruits in your greatest amount of nutrients, mouth straight from the field will give gustatory pleasure.

Disadvantages: As the time increases from the field to plate the amount of vitamins and the quality of the fruit will decrease.

Frozen and Canned

Advantages: Commercial products that are frozen or canned are usually processed immediately after harvesting. During processing, some nutrients are lost but the preservation helps to keep much of the nutrient composition. When fruit is frozen or canned, availability of the fruit is year round.

Disadvantages: Fruits that are frozen or canned are fully ripened and tend to be softer than fresh. Frozen and canned fruits may be available with or without added sugar or juice. People with diabetes need to check the labeling of the product so that the lower carbohydrate values may be selected and calculated to fit into the meal plan. The amount of added sugars can be quite high, especially in many canned products.

Dried Fruits

Advantages: The process of drying removes the water from the fruit and thus gives it a much longer shelf life.

Disadvantages: For a person with diabetes, the amount of carbohydrate in the fruit does not change as the water is removed. Measurement of the dried fruits is needed to calculate carbohydrate values. It is very easy to underestimate the amount consumed. Dried fruit may stick to teeth and contribute to cavities. Make sure teeth are brushed after eating.

Juice

Advantages: The most nutritious form of juice is the 100 per cent juice option. Sometimes frozen juice bars are wanted as a treat. The best choice would be 100 percent juice bars with fruit pieces. Make sure to take into consideration the amount of carbohydrates in each bar.

Disadvantages: Juice can raise your blood sugar rapidly. This may not be desired if blood sugar is not low. Most fruit juices do not contain fibre.

Phytochemicals (plant chemicals) in Fruit Are Healthy: People with diabetes should also include foods to help prevent cancer. According to the American Institute for Cancer research, some fruit contains phytochemicals that may help prevent cancer. A few examples are:

a. Carotenoids (found in red orange and green fruits such as apricots, oranges, cantaloupe and watermelon) may help to improve immune response and inhibit the growth of cancer cells.

b. Flavonids (found in apples and citrus fruits) may inhibit inflammation.

c. Polyphenols (found in grapes, berries, citrus fruits and apples) may help to prevent cancer formation.

Fruits are Particularly Good for Children: According to the NSW Ministry of Health in Australia: "Eating fruit and vegetables every day helps children and teenagers grow and develop, boosts their vitality and can reduce the risk of many chronic diseases - such as heart disease, high blood pressure, some forms of cancer and being overweight or obese. **Fruit should be a tasty**, nutritious part of everyone's diet including people with diabetes.

Fruits and Vegetables that are Hydrating

There are plently of hydrating fruits and vegetables that are easily available and work wonders for good health. Hydrating foods with high water content, will keep you hydrated and cool as the temperatures soar.

Watermelon: It has 92 per cent water content. They are the best summer food to binge on.

Broccoli: The dense green broccoli is not that heavy as the veggie has 91 per cent water content and has properties that help reduce cancer risk.

Spinach: It has 92 per cent water content and is very low in calories.

Carrots: Carotenoids rich carrots not only help sharpen your vision, they are hydrating too, with 87 per cent water content.

Zucchini: Has 95 per cent water content, which makes it a low caloric, that is rich in vitamin C, manganese and phytonutrients.

Tomato: Antioxidant rich tomato and these protect against cancer have 94 per cent water content making it a great summer food.

Raddishes: These have 95 per cent water content with a high content of riboflavin, fibre, magnesium and calcium which makes it a super healthy.

Celery: It is known for its low calories. They have 95 per cent water content and help keep you hydrated.

Cumcumber: They have cooling effect on body, have 96 per cent water content, therefore, good for summer

Capsicum (bell pepper) have 92 per cent water content therefore good for summer salads.

Lettuce: It is a hydrating vegetable with 95 per cent water content.

Apples have 84 per cent water content and help you keep hydrated and happy in the scorching summer heat.

Pears: This fruit too has 84 per cent water content and cuts your stroke risk by half.

Grape fruit: It is a citrus fruit with 91 per cent water content that can keep you well hydrated.

Grapes: Not only are grapes high in vitamins and antioxidants, they also have 81 per cent water which makes them cooling fruit.

Pineapple: Has a compound called bromelain which is an anti-inflammatory agent and contains 87 per cent water to keep you hydrated.

Cherries: Each cherry has 80 per cent water content and eating a full cup means that you will be well hydrated.

Berries Blue and black berries have 95 per cent water content which makes them a great fruit to eat in summer.

Straberries: full of vitamins and minerals, straw berries have 91 per cent water content.

Fruits and Vegetables that kill Cancer

Avocados: The secret weapon of the avocado is **glutathione** which strongly acts against free cancers radicals that invade the whole body and can prevent the body to absorb fat.

Broccoli: Broccoli is very useful in the fight against colon cancer. Green vegetables also release a special tool in the fight against cancer which converts **estrogen** and **estrogen** productive, which will reduce the risk of cancer.

Carrots: They are very powerful in the fight against breast cancer, bladder, lungs, tongue, stomach and prostate cancer.

Chili and jalapeno peppers: Doctors say that these types of peppers do wonders in the fight against cancer. Although you're probably thinking that your body burns once you eat only one pepper. However, they are very healthy.

Cauliflower, kale, Brussels sprouts and cabbage: This army of vegetables is extremely beneficial in the fight against some cancers, especially prostate cancer. They are also able to prevent additional spreading of the cancer.

Figs: This sweet fruit can help save your life. Japanese researchers have found that figs contain **benzaldehyde,** which shrinks the tumor.

Garlic: This plant is excellent for strengthening the immune system, preventing cancer spread; because a body that has strong immune system can fight against cancer more effectively. This food should be eaten by every sick person.

Grapefruit: It is known as an excellent food for body purification, thanks to its acidity. It is also excellent in the fight against breast cancer.

Red grape: It is one of the most effective fruit for fighting cancer. This juicy fruit kills cancer enzymes. This is another food that every cancer patient should consume.

Mushrooms: The lectin which can be found in mushrooms will surround the cancer cells and prevent its spread to other parts of the body. You can prepare them with garlic and you will see that you will feel better.

Lemon and orange: Lemon and orange stimulate lymphocytes that can fight cancer. Squeeze lemon or orange and kick start your day.

Nuts: They are very powerful, especially in the fight against prostate cancer.

Tomato: Tomatoes can fight against cancer cells and at the same time they can heal the damaged cells because they contain vitamin C

(http://healthydefinition. com/foods-that-kill-cancer/).

Fruits and Vegetables that are Alkaline

Kale: It is good for lowering cholesterol. It should be remembered that steamed kale is more efficient for reducing cholesterol than raw.

Broccoli: Incredibly powerful in inhibiting cancers, supporting the digestive system, the cardiovascular system and the detoxification processes in the body.

Avocado: It contains a wide range of nutrients that have serious anti-inflammatory, heart health, cardiovascular health, anti-cancer and anti-sugar benefits.

Celery: It quickly neutralizes acids and is high in potassium and sodium which makes it a great diuretic.

Spinach: It is rich in vitamin E, fibre, folate, zinc, magnesium, vitamin C, vitamin A and protein. It is a potent alkaliser.

Cucumber: It reduces the risk of cardiovascular disease as well as serval cancer types, including breast, uterine, ovarian and prostate cancers.

Alkaline Foods that can Prevent Obesity Naturally

Basically, the human body is alkaline in nature and it is our duty to maintain it by balancing the PH level. The junk and processed food and the modern life style, we often fail to balance the PH level in the body, which results in the distortion of the immune system, which leads to many health disorders like obesity and cancer.

Therefore, natural foods like fruits and vegetables work magically in balancing the PH level and thus help prevent obesity naturally.

Quinoa: It is rich in protein and contains twice the amount of fibre, as compared to other grains. Regular use of quinoa helps control bad cholesterol.

Melon: The water content of melon is very high and that keeps you hydrated. As you feel full, so you end up eating less and this keeps the weight in control.

Lemon: Regular use of lemon not only prevents constipation but also improves digestion and as a result reduces the chances of weight gain. It also helps fight cancer and controls blood pressure.

Grapes: They contain a high amount of polyphenols, an anti-oxidant that also reduces anxiety and hypertension. According to exports, hypertension contributes to obesity and this is an ideal medium to keep both in control.

Flax seed: Flax seed is considered a top alkaline food as it is rich in fibre, anti-oxidants and vitamin C. The daily use of tea spoon flax seed can control obesity and even heart related problems.

Carrot: It is high in vitamin K, C, A, B6, iron, potassium and fibre. Altogether they reduce the body fat and protect you from moving towards obesity.

Banana: If you really want to lose weight, start eating banana daily. The fruit balances the blood sugar levels, protects the heart and also improves digestion.

Avocado: It contains fibre and mono-unsaturated fatty acids that helps in absorption of nutrients from vegetables and fruits further controlling the levels of cholosyerols causing obesity.

Foods that keeps your Stomach Calm

Banana: It is easy to digest. It contains a substance called Pectin which increases bowel movements.

Funnel Seeds: The phytonutrients present in funnel seeds gives it an antioxidative property that helps effectively in reducing gas and bloating, the common signs of an upset stomach.

Papaya: If you suffer from constipation and upset stomach regularly, then you should incorporate papaya in your daily diet. There is a digestive enzyme called papain present that breaks up foods irritating the stomach.

Ginger: It has reveal medicinal properties. It is anti- inflammatory, that soothe the intestinal tract, while reducing bloating and heart burn.

Cucumin Seeds: They have a substance called thymol that increases the secretion of enzymes, which helps in digesting food easily while removing the feeling of bloating.

Fruits and Vegetables that Reduce Cholesterol

Olive oil and Olive products: Olive, rich in mono-unsaturated fatty acids and vitamin E, reduces 'bad' *i.e.* low-density lipoprotein (LDL) cholesterol and increases 'good' *i.e.* high-density lipoprotein (HDL) cholesterol. It contains henolics (plant substances) that help blood to clot easily in case of wounds and harmful clotting.

Almonds: Almonds, rich in unsaturated fats, raise healthy HDL cholesterol. Fats present in almonds helps to lower down the unhealthy LDL cholesterol by not letting it getting oxidized. Hence, saving your arteries from getting blocked, leading to normal blood circulation to your heart.

Legumes: Legumes are rich in high dietary fibre and protective nutrients. It contains minerals, vitamins and phytonutrients. They are low in fat density content and have a low glycemic index (GI). The dietary fibre knocks down the bad cholesterol level and reduces energy intake.

Avocado: This creamy textred fruit is rich in heart-healthy mono-unsaturated fats. It controls the bad LDL cholesterol and triglycerides and boosts the healthy HDL cholesterol.

Yoghurt and Dairy products: Fat-free Yoghurt includes calcium and minerals. It is rich in ***Lactobacillus, a*** micro-organism which reduces high blood cholesterol levels. Fat-free milk and cottage cheese are good for health and are very good essentials for a low-cholesterol diet.

Tomatoes: Tomatoes, richin **lycopene**, help to knock down the artery-clogging fat and lower the LDL cholesterol production in your body. Consumption of 25 milligrams a day is really beneficial in controlling high cholesterol levels.

Spinach: It is rich in beta-carotene. It contains a carotenoid named '***lutein***'. *Lutein* prevents buildup of cholesterol in the blood. You will be able to lower your cholesterol levels in no time with the intake of fresh spinach every day.

Oranges: Oranges are anti-oxidant rich citrus fruits. They are rich in Vitamin C. Vitamin C protect your heart and control your cholesterol level.

Onion and Garlic: Daily consumption of garlic, spring onions and other onions is helpful in lowering the bad cholesterol level and protect the heart.

Chapter 2

Fruits

A fruit is generally a fleshy seed associated part of a particular plant; **it is naturally and mostly edible and sweet in the raw state.** Fruits are the natural and staple food of man. They are excellent source of minerals, vitamins and enzymes. They are easily digested and exercise a cleaning affect on the blood and the digestive tract. Fruits and vegetables are lower in calories per gram compared to denser foods. They are alkalizing, energy boosters. The natural sugars in the fruits help to balance the pitta element of the body. These are natural cleansers and can act as an in-between snack for people who crave desserts or fried snacks. Fruits and vegetables are excellent sources of fibre. These are free of added colours, preservatives and additives when eaten raw and fresh.

To increase satiation, consume a cup of chopped water- rich fruits and vegetables, such as berries, citrus fruits, tomatoes, kiwi, celery, cucumber, bell peppers, leafy green vegetables and waterchest nuts. The research has shown that men and women whose skin has a yellow glow are thought to be particularly **attractive** and healthy - and yellow pigments called carotenoids, found in certain fruit and vegetables play a key role in giving the skin that hue. Scientists have found how micronutrients found in certain fruits and vegetables can have a rapid impact on your brain. They found that some foods like blueberries, cocoa and green leafy vegetables provide an instant boost to our mental sharpness and mood and help to ward off depression. "Certain micronutrients can have a big influence on moods", the Daily Star quoted Dr Emma Williams of the British Nutrition Foundation as saying. Fruits are good source of vitamins and minerals and are a readily available source. A fruit will boost you instantly. It is something which can quench your thirst and satisfy your hunger at the same time. Fruits are guardian angels for all the busy bees out there, who have no time to even have their breakfast. Since its handy they can be eaten even on the move. Fruits are meant to be nutritious and are our sole companion which makes them our best friends! Make more fruit friends here. When it comes to the uses of fruits, it is never-ending! Fruits are consumed as food by millions of people in the form of salads, soups, juices, jams and pickles.

In this study, which involved over 100, 000 women and men, researchers evaluated the effect of study participants' consumption of fruits; vegetables; the antioxidant vitamins A, C, and E; and carotenoids on the development of early ARMD or neovascular ARMD, a more severe form of the illness associated with vision loss. Food intake information was collected periodically for up to 18 years for women and 12 years for men.

While, surprisingly, intakes of vegetables, antioxidant vitamins and carotenoids were not strongly related to incidence of either form of ARMD, fruit intake was definitely protective against the severe form of this vision-destroying disease.

Three servings of fruit may sound like a lot to eat each day, but by simply tossing a banana into your morning smoothie or slicing it over your cereal, topping off a cup of yogurt or green salad with a half cup of berries, and snacking on an apricot, you've reached this goal.

Each fruit has a vital role to play in strumming up your mood. The key ingredients to look for are vitamin B and folate as they stimulate amino acids which produce serotonin that pumps up your gloomy mood. This is also your cure for seasonal inflammations. These goodies will also keep your anxiety, stress, anger and frustration under control. Improvement in your mood leads to a healthy body. Many fruits also contain fructose, but nature has provided the antidote, as these fruits are also packed with fibre which prevents your body from absorbing too much of it.

Colour Fruits are Healthy

Different colours indicate different nutrient profiles. Get a little of each colour in your diet, every day, to maximise the benefits.

Red fruits such as raspberries, guava and watermelon are rich in antioxidants, lycopene and anthocyanins. There is no set daily value for lycopene, but nutritionists recommend getting 2 to 30 mg per day.

Orange and yellow fruits are rich in beta-carotene, which your body converts to vitamin A, a nutrient that improves night vision and keeps your skin, teeth and bones healthy.

White fruits like apples, pears and bananas are high in dietary fibre and have antioxidant-rich flavonoids. These fruits protect you from high cholesterol and lower the risk of stroke.

Blue and purple fruits like blueberries and jamun contain anthocyanins, natural plant pigments with antioxidant properties that reduce the risk of cardiovascular diseases. They also contain flavonoids and ellagic acid, compounds that are known to destroy cancer cells.

In General

Keep a bowl of whole fruit on the table, counter, or in the refrigerator. Refrigerate cut-up fruit to store for later use. Buy fresh fruits in season when they may be less expensive and at their peak flavour. Fruits contain water-soluble vitamins and these are lost when fruits are exposed to air. So, it is to eat them whole or cut and eat them immediately. On an average, you should eat at least 2-3 servings of fruit a day (1 serving =100 gm). Fruits with higher sugar content, like mangoes, grapes, banana, sitaphal and chikoo are very high in sugar content. They needn't be curtailed completely for diabetics. The trick is to keep the serving size to not more than 50 gm at one time (versus 100 gm of other fruits). This helps prevent the sudden rise in blood sugar levels which is unwarranted in diabetics.

Fruits are also called negative calorie foods - another wonderful reason why those on weight loss mode must inculcate the habit of eating them as in-between fillers. Negative-calorie foods are foods that lead to the body using up more calories in digesting them versus calories they add to your diet. For example, an apple contains approx 90 kcal but to digest this apple (breakdown the sugar, fibre, extract the nutrients), your body expends 120 kcal! Thus, your body actually uses 30 kcal of its own energy reservoir to just digest the apple you ate. It's magical, isn't it? Grapefruit, orange, cranberries, strawberries and sweet limes are among other negative-calorie fruits (*Makhija,* 2014).

For the Best Nutritional Value

Fruits are alkalizing, energy boosters. The natural sugars in the fruits help to balance the pitta element of the body. These are natural cleaners and can act as an in-between snack for people who crave desserts or fried snacks. They are also excellent sources of fibre. These are free of added colours, preservatives and additives.

Make most of your choices - whole or cut-up fruit rather than juice, for the benefits dietary fibre provides. Select fruits with more potassium often, such as bananas, prunes and prune juice, dried peaches and apricots, cantaloupe, honeydew melon, and orange juice.

When it comes to the uses of fruits, it is never-ending! Fruits are consumed as food by millions of people in the form of salads, soups, juices, jams and pickles.

Fruits = taste + flavour+nutrition+satiety!!

The natural antioxidants in fruits and vegetables will help keep your body working at its best, so consuming a diet that meets your daily recommended amount of fruits and vegetables is one of the best ways to give your body a strong defense against disease. Fruits and vegetables are protective to health as they're helpful at reducing the risk of coronary heart disease, stroke and some cancers. They're also low in calories, which helps prevent obesity, a significant risk factor for type 2 diabetes, cancer and cardiovascular disease.

Fruits are good source of vitamins and minerals and are a readily available source. A fruit will boost you instantly. It is something which can quench your thirst and satisfy your hunger at the same time. Fruits give you all the essential nutrients and are rich in vitamins like C and A.

Eating more whole fruits, particularly blueberries, grapes and apples is significantly associated with a lower risk of type 2 diabetes, according to a new study.

Fruit Pleasure

All fruits, especially those rich in vitamin C can be vital in sexual gratifications. The body needs vitamin C to keep the sexual organs fine tuned. Tests reveal that consuming 500-1, 000 milligrams of vitamin C in a day increases the number and quality of sperm produced. Some fruits which fall in this category include citrus fruits, specially kiwi, blackcurrants and strawberries.

A dose of sunlight along with almonds will benefit those who want a baby. Eat to beat infertility which is on the rise and have foods rich in iron like apples and radish. A study done at Harvard revealed women with high iron intake in their diet were more fertile. It's essential for menopausal women to take vitamins D and E, which are found in avocadoes, tomatoes, hazel nuts. Women can get vitamin D from butter and eggs too.

Foods that Help you Distress

Certain food items in our diet can actually help reduce stress in our bodies. While apricots and greens like spinach replenish the body with magnesium, which is known to relax the muscles and help with headaches, nuts and dark vegetables like almonds, pistachios, broccoli, and okra are rich in vitamins B and E that help lower blood pressure. Citrus fruits, which are a storehouse for Vitamin C, also help in controlling blood pressure. Other food items like sweet potatoes provide insulin to the body, provide energy, and help deal with fatigue in a stressful situation. Below is a list of foods recommended for stressed-out people:

1. Cruciferous vegetables like cauliflower, cabbage, broccoli, *etc.*
2. Nuts like almonds, pistachios, *etc.*
3. Dark coloured vegetables like beetroot, okra, and peas
4. Beans or lentils– Asparagus- Artichokes- Sweet potatoes- Leeks- Watercress

Fruit Nutrition Facts

Why Fruits?

Fruits are nature's wonderful medicines packed with vitamins, minerals, enzymes, anti-oxidants and many phyto-nutrients (Plant derived micronutrients). They are absolute feast to our sight, not just because of their colour and flavour but help body keep fit and healthy! They are the natural and staple food of man. They are easily digested and exercise a cleansing effect on the blood and the digestive tract.

1. Fruits are low in calories and fat and are a source of simple sugars, fibre, and vitamins, which are essential for optimizing our health.
2. Fruits provide plenty of soluble dietary fibre, which helps to ward of cholesterol and fats from the body and to get relief from constipation as well.
3. Fruits contain many anti-oxidants like poly-phenolic flavonoids, vitamin-C, anthocyanins. These compounds, firstly, help body protect from oxidant stress, diseases, and cancers, and secondly, help body develop capacity to fight against these ailments by boosting our immunity level. Many fruits, when compared to vegetables and cereals, have very high anti-oxidant values which is something measured by their "Oxygen Radical Absorbent Capacity" or ORAC.
4. Anthocyanins are flavonoid category of poly-phenolic compounds found in some "blue fruits" like blue-black grapes, mulberries, acai berry, chokeberries, blueberries, blackberries, and in many vegetables featuring blue or deep purple colour. Eating fruits rich in blue pigments offers many health benefits. These compounds have potent anti-oxidant properties, remove free radicals from the body, and thus offer protection against cancers, aging, infections *etc.* These pigments tend to concentrate just underneath the skin.
5. Fruit's health benefiting properties are because of their richness in vitamins, minerals, micro-nutrients, anti-oxidants which helps body prevent or at least prolong the natural changes of aging by protecting and rejuvenating cells, tissues and organs in the human body. The overall benefits are manifold! Fruit nutrition benefits are infinite!
6. You are protecting yourself from minor problems like wrinkling of skin, hair fall, memory loss to major ailments like age related macular degeneration of retina in the eyes, Alzheimer's disease, colon cancers, weak bones (osteoporosis) and Breaking the list of fruit nutrition benefits never ends (Aninymous, 2013).
7. No matter which beauty product rules the market, a fact that nobody can deny, is that fruits are nature's most valuable beauty enhancers.
8. Purple fruits and vegetables are such in anthocyanins, possess anti-diabetaic, anti-cancer, anti-inflammatory and anti-microbial properties.

Tips to Eating Fruits

- ☆ Potassium leaches out into the water during cooking. So the best way to get potassium is through fruits.
- ☆ Keep the fruit out on the counter or in the front of the fridge. That way you'll be more likely to notice it and eat it.

- Choose colour and variety in fruits. Go for yellow, green, orange and red fruits.
- Fruits salads are an interesting way to eat fruits.
- Get some fruits at snack time too.
- Go for variety in fruits. Try some new fruits.
- Cooked vegetables can prove to be more nutritious than raw ones, For example, a cooked tomato has more of the anti-oxidant lycopene than raw tomatoes.
- Dates are an excellent source of natural fibre that are essential for the body as a good bowel movement.
- Fruit juice can be unhealthy if consumed in excess. 230 millilitres of apple juice has around 115 calories- more than coca cola.
- Sucking a piece of clove after meals helps in reducing acidity problem.

To Get the Most from your Fruit and Vegetables

- Don't buy fruit and vegetable dishes that come with sauces. They often contain a lot of fat, salt and sugar.
- Dried, frozen and tinned products can be just as good as fresh, but watch out for added salt, sugar or fats.
- Vary the types of fruit and vegetables you eat. Each has different health benefits and it will keep your meals interesting. By eating a wide range of fruit and vegetables, you will ensure that your body is getting all the nutrients it needs.
- Don't add sugar to fruit or salt to vegetables when you cook or serve them.
- Try to eat fresh fruit and vegetables as soon as possible. They will lose their nutrients over time, so if you want to store your ingredients for a while, it is best to freeze them or buy frozen packets
- Avoid leaving vegetables open to the air, light or heat if they have been cut. Always cover and chill them, but don't soak them because the vitamins and minerals can dissolve away.
- Vegetables keep more of their vitamins and minerals if you lightly steam or bake them, instead of boiling or frying them.
- If you boil vegetables, use as little water as possible to help keep the vitamins and minerals in them.
- Experiment with other ways of cooking vegetables, such as roasting or grilling them, for new tastes and flavours.
- Stir-fries are great for getting lots of vegetables into one meal. So are freshly-made soups.
- **Spinach:** According to experts spinach should never be warmed. Its consumption must be immediately after preparation. The nitrates in

spinach turn into nitrites when spinach is warmed. So, reheated spinach is carcinogenic.

- ☆ **Celery:** In soups, we often use celery. This vegetable also contains nitrates than turn into nitrites when warmed. If you prepared soup and you plan to reheat it, be sure that there is no celery in the soup. The same may happen to carrots, so don't reheat soups that contain carrots.
- ☆ **Beets:** Beats as well as spinach and celery contain nitrates that are useful for the body. Yet, the nitrites are dangerous and carcinogenic, so don't reheat this vegetable ever again.
- ☆ **Potatoes:** Experts have proven that potatoes are healthy for our overall health. Yet, reheating and consuming potatoes after 24 hours of preparation destroys the nutritional and healing qualities of these vegetables. Reheated potatoes will be harmful for your body rather than healthy.
- ☆ **Mushrooms:** Reheated mushrooms can also affect your body and overall health negatively. Mushrooms should be eaten immediately after preparation. You can eat the remaining mushrooms the following day if you don't warm them.

Why Eat more Fruits and Veggies?

- ☆ **Colour and Texture.** Fruits and veggies add **colour,** texture and *appeal* to your plate.
- ☆ **Convenience.** Fruits and veggies are nutritious in any form – **fresh, frozen, canned, dried and 100 per cent juice,** so they're ready when you are!
- ☆ **Fibre.** Fruits and veggies provide **fibr**e that helps fill you up and keeps your digestive system happy.
- ☆ **Low in Calories.** Fruits and veggies are naturally **low in calories.**
- ☆ **May Reduce Disease Risk.** Eating plenty of fruits and veggies may help reduce the risk of many diseases, including heart disease, high blood pressure, and some cancers.
- ☆ **Vitamins and Minerals.** Fruits and veggies are rich in **vitamins and minerals** that help you feel healthy and energized.
- ☆ **Variety.** Fruits and veggies are available in an almost infinite variety there's always something new to try!
- ☆ **Quick, Natural Snack.** Fruits and veggies are nature's treat and easy to grab for a snack.
- ☆ **Fun to Eat!** Some crunch, some squirt, some you peel some you don't, and some grow right in your own backyard!
- ☆ **Fruits and Veggies** are Nutritious and Delicious!

☆ There are many white fruits and vegetables viz coconut, banana, white gourd, garlic, potato, cashew nut, white pumpkin, cauliflower and onions are nutritious and good for health.

When to Eat Fruits?

The best time to eat fruits is first thing in the morning after a glass of water. Eating fruits right after a meal is not a great idea, as it may not be digested properly. The nutrients may not be absorbed properly either.

You need to leave a gap of at least 30 minutes between a meal and a fruit snack. Ideally, one should eat fruits an hour before the meal or two hours after, if they have diabetes or any other digestive problem like acidity. This is because, sometimes, diabetes is accompanied with digestive problems.

Can you mix fruits with other foods? You can mix fruit with yogurt or salt as long as you do not have any digestive problems like indigestion or acidity. You can also mix fruits like pineapple, oranges, melons or pomegranate with salads if you like. There is no harm in mixing berries and dry fruits with cereals either (Sumitra Nair, 2013).

Dried Fruits

In between your meals you snack on a packet of dried fruits instead of a processed fruits – you are unaware that even dried fruits can be harmful for our teeth because of their concentrated sugar content and stickiness. Dried raisins, prunes, apricots, anjeer are gummy. So, parts of them easily adhere to teeth and the sugar in them encourage bacteria in the mouth that erodes tooth enamel. Dried fruits are packed with non-soluble cellulose fibre, which can bind and trap sugars on and around the tooth, making it worse than sweets. Chewing regular fruits, however, is less harmful, because even though they have sugar too, they aren't as sticky and the saliva helps in cleaning up.

Chemical Constituents of Fruits and Nuts

Fruit	*Antioxidants*	*Chemical Compounds*
Apple		Quercetin, Epicatechin, Procyanidin B2, Elastin and Collagen
Apricot	Antioxidants	Lutein, Zeaxanthin, β-cryptoxanthin, Carotenes, Laetrile (B_{17})
Avacoda	Monosaturated fatty acid	Oleic acid and Palmitoleic acid, ω-3 –fatty acids, Palmitoleic acid
	Poly saturated fatty acid	Omega-6-poly- unsaturated fatty acid, Tannin (poly phenolic compound).
	Flesh	Cryptoxanthin, Lutein, Xeaxanthin, β and α carotenes

Fruit	*Antioxidants*	*Chemical Compounds*
Almond	Unsaturated fatty acids Amino acid	Oleic, Palmitoleic acids, Alpha-tocophenol, Arginine (helps relax blood vessels and improve blood circulation) and Folic acid
Banana		Bromelain (increases libido), TNF, Vitamin B6), Dopamine Trypothan (Converts into serotonin in the body)
Blueberry	Poly phenolic anthocyanidin Flanovoids	Amyl acetate, Pterostibene (remedy for colon and liver cancers), Proanthocyanidins Chlorogenic acid, Tannins, Myricetin, Quercetin, Kaempferol, Carotene β, Lutein, Zeaxanthin.
Black berries	Phenolic flavonoid phytochemical	Xylitol <-a calorie sugar substituteAnthocyanins, Ellagic acid tannins, Quercetin, Gallic acid, Cyanidins, Pelargonidins, Catechins, Kaempferol and Salicylic acid
Brazil nuts	Mono-unsaturated fatty acid	Oleic acid, Palmitoleic acid and Selanium, Copper, Mg. Mn.
Cranberries	Anti-oxidant compounds	Oligomeric Proanthocyanidins (OPC'S), Flavonoids, Cyanidin, Peonidin and Quercetin.
Cashewnut		Oleic acid and Palmitoleic acid
Citrus Fruits	Bitterness Aromatic principae	Narangi and Limonoids (Flavonoid compounds) Limonene and other essential Oils
Cranberry		Proanthocyanidins (prevents plaque formation on teeth)
Chestnut		Oleic acid and Palmitoleic acid, Vit. C, Folate
Coconut	Saturated fatty acids	Lauric acid, * Cytokinins**
Cherry	Poly phenolic flavonoid compounds	Anthiocyanin glycosides, Melatonin***, Phenols, Quercetin
Cherimoya	Annonaceous	Asimicin, Bullatacinare#
Cocoa (powder)		Catechin, Epicatechin
Dates	Anthocyanidin flavonoids	Cyanidin, Peonidin and Quercetin, β carotene, Tannins
Drum stick		Terigospermin. Moringin
Fig	Poly phenolic flavonoid antioxidant	Chlorogenic acid, Carotenes, Luteins, Tannins, ω-3-fatty acids, ω-6-fatty acids. Lycopene, Cryptoxanthin
Fennel		Estirol (estrogen like substance)
Grapes	Polyphenolic Phytochemical Compound	Resveratrol, Anthocyanins, Catechins, zeaxanthin Pterostilbene (lowers cholesterol levels), polyose
Grape fruit	Antioxidant	Naringenin, Lycopene, β-carotene, Xanthin, Lutein, Pectin
Guava		Lycopene, Vit. A, Vitamin C, Potassium.

Fruit	*Antioxidants*	*Chemical Compounds*
Jackfruit		Carotene β, Xanthin, Lutein, Cryproxanthin-β, Vitamin A, Vit. B6, Folate
Jamun		Jambosine and Glycoside Jambolin or antimellin
Kiwifruit		Omega-3-fatty acid, Serotonin, Manganese, Potassium, Vit. C and K
Kumquat	Essential Oils	Limonene, Pinene, α-bergamotene, Cryophyllene, α-humulene, α-muurolene, Vitamin C.
Lemon Lime	Phytochemical	Hesperetin and Naringenin, Ascorbic acid. Hesperidine
Liquat	Antooxidants	Chlorogenic acid, Neo-chlorogenic acid, Hydroxybenzoic acid, Feruloylquinic Acid, Protocatechuic acid, Epicatechin, Coumari acid, Ferulic acid
Litchi	Polyphenol	Oligonol+ Vit. C, Copper.
Mango	Flavonoids	β-carotene, α-carotene, β- cryptoxanthin, Vit. A and C, Vit. B6
	Unripe fruits	Anacardic acid
	Antioxidants	Glutamine acid (An important protein for concentration and memory)
Mulbery	Berries (Phyto-chemicals)	Anthocyanins, Resveratrol, Zeaxanthin
Orange	Phytochemicals	Hesperetin and Narigenin- Flavonoids.
Olive	Pungent/bitter taste	Oleuropein aglycone- a poly phenol
		Oleuropein- anti-oxidant, anti- inflammatory, anti ageing, anti-cancer, anti-viral and anti-microbial and
		Oleocanthal- a phytonutrient (Pain killer)
Papaya	Proteolytic enzyme	Papain
	Alkaloid	Carpaine++, β-carotene, Lutein, Zeaxanthin, Cryptoxanthins Potassium.
Peaches	Phenolic anti oxidants	β-carotene, lutein, Zeaxanthin, β-cryptoxanthin, K, Fluride, Fe.
Pears		Non Soluble Polysaccharide (NSP), cOPPER
Persimmon	Polyphenolic antioxidants	Catechins## and Gallocatechins, Betulinic acid, Zeaxanthin, Lycopene, Vitamin C.
Pineapple	Proteolytic enzyme	Bromelain, Folates, Copper, Manganese, Vit. C
Plums		Sorbitol, Isatin, β-carotene, Vit. A and C, Carotene, Oxalic acid
Pomegranate	Polyphenolic oxidant compound	Punicalagin and Ellagic acid
Peanuts	Polyphenolic antioxidants	P-coumaric acid^, Resveratrol^^, Vit. C, Folates, Niacin.

Fruit	*Antioxidants*	*Chemical Compounds*
Peacannuts	Mono unsaturated fatty acid	Oleic acid
	Ployphenolic antioxidate	Ellagic acid
Pine nuts	Mono-unsaturated fatty acids	Oleic acid, (ω-6) Pinolenic acid^^^, Vitamin E.
Pistachio		Oleic acid, carotenes, Vitamin E, Copper, Iron.
Raspberry	Phenolic flavonoids phytochemicals	Xylitol<, Anthocyanins, Ellagic acid (a tannin), Quercetin, Gallic acid, Cyanidins, Pelargonidins, Catechins, Kaemferol, Salicylic acid, Vitamin acid.
Strawberry	Phenolic flavonoids	Anthocyanins, Ellagic acid, Vitamin C, Manganese
Sita phal	Seeds/Fruits	Anonine, michelabbine, oxoushinsuie, corydine, isocorydine, alanine, β-alanine
Walnut+++	Mono-unsaturated fatty acid	Oleic acid and Omega -3-fatty acid like linoleic acid, α-linolenic acid (ALA) and Arachidonic acid
	Phytochemical	Melatonin, Ellagic acid, Vit. E, Carotenes. Jugladic acid

<: Alow calorie sugar substitute present in the fruit fibres, absorbs more slowly than sugar and does not contribute to high blood sugar levels. !: Helps lower blood sugar levels and controls blood glucose levels in type 11 diabetes mellitus condition. *: Helps rise HDL cholesterol level in blood. **: Anti ageing anti cercirogenic and anti thrombotic effects. ***: Can cross the blood brain barrier easily and produce soothing effects on the brain neurons, calming down nervous system irritability which helps relieve neurosis, insomnia and headache condition. #: Are powerful cytotoxins and have been found to have anti cancer, anti malerial and anti helminthes properties. +: Antioxidant, anti-influenza. It improves blood flow in organs, reduces weight and protect skin from harmful UV Rays. ++: Which could be dangerous when eaten in high doses (unripe fruit, seeds, latex, leaves). ##: Have anti infective, anti inflammatory and anti hemorrhagic properties. ^: Is believed to reduce the risk of stomach cancer by reducing the formation of carcinogenic nitrosamines. ^^: It reduces stroke risk by alteration of molecular mechanism in blood vessels and by increasing production of vasodilator hormone, nitric oxide. ^^^: Causes the triggering of hunger suppressant enzymes cholecystokinin and glucagon like peptide. In addition Pinolenic acid may have LDL lowering properties by enhancing hepatic LDL uptake. +++: Walnut contains 13541 µmol TE (Trolex equivalents) of oxidant radical absorbant capacity (ORAC).

Based on a study in 2010, reported by the Dietary Guidelines of America, you need 1.5 - 2 cups of fruits daily, but the amount depends on the number of calories per day recommended for your healthy weight. Also, it is recommended that you eat five servings of fruits and veggies a day, but it depends on your age and gender: Women between the ages of 19-50 should eat 7-8 servings, while men should eat 8-10 servings. And for those above 50, men and women should consume seven servings of fruits and vegetables a day (Trina Remedios, 2013).

Fruits help fight cancer, obesity and heart disease. Fruits provide nutritional vitamins. Certain fruits contain a good amount of natural sugar and fluids, which also keep us hydrated. Fruits are rich in fibre, and are great for bowel movement and weight loss. It is also advised that you consume fruits whole, instead of skinned or juiced. Always opt for fresh fruits as their nutritional value is higher and try different seasonal fruits from time to time.

Eating many of the numerous coloured fruits and vegetables, can supply your body with a wide range of vitamins and minerals like fibre, folate, potassium and vitamins A and C. Some examples include green spinach, orange, sweet potato, black beans, yellow corn, purple plums, red watermelon and white onion.

Orange, red and dark green vegetables like pumpkin, carrots, cantaloupe, kale and yellow corn are rich source of natural alpha- carotene, where as potatoes, carrots and pumpkins, spinach and leafy vegetables are rich in beta- carotene. The red- orange colour in carrots shows the presence of beta-carotene which helps boost your immune system and provides your body with antioxidants.

Green vegetables, such as spinach and kale contain lutein and zeaxanthin, which may help protect against age related diseases. Green foods are particularly good for the circulation of system. Green foods include minerals and vitamins B.

Red foods are specially for reducing internal damage. Red foods include phytochemicals so take watermelon, tomatoes, strawberry.

Yellow foods are specially for reducing the risk of developing cataracts. So take Yellow Corn, Oranges, Pineapple.

The fibre and water content of fruits and vegetables help to maintain a healthy digestive tract and weight control (as they are lower in calories than most packed foods).

Diets that are high in insoluble fibre may offer the best protection against **Diverticulosis.**

Fruits and vegetables are high in cellulose–a type of insoluble fibre.

Studies have reported that high fruit and vegetable intake can reduce the risk of a stroke by up to 25 percent.

Unadorned fruits and vegetables are naturally low-in calories, and may be an important way to prevent and treat obesity. One study concluded that diets that are high in fibre are associated with lower body weight.

The higher intake of fruits and vegetables seemed to increase the ventilation function of the lungs. All fruits are rich in vitamins and plant derived nutrients known as phyto-nutrients like beta-carotenes, flavonoids and anthocyanins. Vitamin A and C commonly found in most of the fruits have powerful antioxidant properties.

Another helpful benefit of fruits and fruit juices is their ability to promote detoxification in the human body. **Fruits** help to cleanse the body, especially those with high acid levels. Tomatoes, pineapples, and citruses such as oranges, red grape fruits, and lemons are known for their detoxifying properties. While these fruits promote cleansing, they still provide the body with a high boost of Vitamin C!

Nutrition is the supply of 'food' required by organisms and cells to stay alive. The benefits of good nutrition are many; they range from helping you maintain a healthy weight and help in enhancing body functions all over the body.

Vitamin A: Vitamin-A is also required for maintaining healthy mucus membranes and skin. Consumption of natural fruits rich in carotenes helps protect body from lung and oral cavity cancers. Vitamin A is a powerful antioxidant and is essential for vision. It is available incarrot, melon, mangoes, apricotetc. Carotenoids give fruits and vegetables their bright orange, red, or yellow colours, and are a source of dietary vitamin A. Findings suggest that consuming carotenoid-rich foods may help prevent or delay the onset of ALS (amyotrophic lateral sclerosis). It's important to have adequate levels of vitamin A for female sex hormone production and regular reproductive cycles. Eat apricots, cabbage, carrot, chillies, greens, grape fruit, lettuce, mango, peppers, spinach, sweet potato, tomato, and watermelon, that are all rich in the vitamin.

Vitamin C: Vitamin C is a powerful natural antioxidant. Regular consumption of foods rich in vitamin C helps body develop resistance against infectious agents and scavenge harmful, pro-inflammatory free radicals from the body. Further, the vitamin is required for collagen synthesis and wound healing, anti-viral, anti-cancer activity and helps prevent from neuro-degenerative diseases, arthritis, and cold/fever *etc.* by removing oxidant free-radicals from the body. Collagen is the main structural protein in the body required for maintaining the integrity of blood vessels, skin, organs, and bones. Vitamin-C is one of powerful natural anti-oxidant which has many essential roles like, anti-viral, anti-cancer activity. Vitamin C helps strengthen your immune system and protects you from infections like cold and cough, which sap your energy. Also, when you are outdoors, you are more susceptible to bacterial infections which hamper your health. Thus ensure that you consume foods and fruits rich in vitamin C. It is known to be a great source of antioxidant. Apart from giving a strong immune system, it also helps in giving a radiant skin, heal blemishes and keeps you away from cold and cough. The best sources of Vitamin C are known to be blackcurrants, blueberries, broccoli, guava, kiwi fruits, oranges, papaya, strawberries and sweet potatoes.

Eat fruits rich in vitamin C: It fights against cortisol buildup in the body, fruits like oranges, berries and other vitamin C rich fruits and vegetables eliminate cortisol. Vitamin C is known to increase your immunity because it is anti oxidant in nature. It also helps to build the skin with its collagen boosting property. It adds extra glow to the skin. In many places, food supply could be affected because of extreme winter, hence the intake of iron can take a back seat. Iron gets absorbed to its maximum in Vitamin C; hence it also helps maintain the iron content in your blood. Vitamin C can be sources from all kinds of citric fruits like oranges, sweet lime, lemon as well as dates and vegetables like amaranth.

Vitamin C is necessary for the formation and maintenance of tissues, ligaments, blood vessels and cartilage in the body. It is also used by the body to produce hormones like ATP, dopamine, peptide hormones, and tyrosine. A powerful antioxidant, it lowers oxidative stress and protects the body against health conditions caused by free radicals, ranging from skin ageing to cancer.- Vitamin C helps synthesise sex and fertility hormones like androgen, estrogen and progesterone and it also helps you get aroused. Vitamin C is an immunity booster and keeps your joints smooth and pain-free. Vitamin C is found in citrus fruits, strawberries, kiwi, cantaloupe, and sweet and chilli peppers.

The daily requirement for Vitamin C is 90 mg for men and 75 mg for women. One orange contains about 70 mg of Vitamin C.

Unfortunately, Vitamin C is water soluble, so your body doesn't store it. You have to consume it regularly, in order to ensure that you meet your nutritional requirements.

Guava contains	377mg of Vitamin C
Yellow bell pepper	155mg (1/2 cup)
Red bell pepper	190mg/CUP
Papaya	238mg (½ of large papaya)
Chillies	108mg (1/2 cup)
Cauliflower	128mg (A small head of cauliflower)
Broccoli	82mg (1 cup)
Kale	80mg (1 cup of chopped)
Kiwis	122mg (1/2 cup of exotic fruit)
Mango	122mg (1mango)
Pineapple	80 mg

Vitamin B: Bananas, sweet potato and beans. Cure your crankiness and satiate your food carvings by increasing vitamin B in your diet.

Vitamin E is a powerful lipid soluble antioxidant, required for maintaining the integrity of cell membrane of mucus membranes and skin by protecting it from harmful oxygen free radicals. Present in avocados, soaked almonds. Getting enough of vitamins C and E may also help to perk up a flagging memory and to keep it sharp. Vitamins C and E are anti-oxidants, which help to ease the stress put an brain cells by free radicals. This vitamin helps in building up the skin, because of its moisturising properties. Our skin tends to become flaky during this season; hence the intake of this vitamin is crucial. Vegetables like spinach, broccoli, *etc.* and nuts as well as tamarind is an excellent source of this vitamin. Vitamin E.

It is also called the 'sex vitamin, ' vitamin E not just pumps your nether regions with blood and oxygen, it also regulates your sex hormones so that your libido gets a boost. Its anti-ageing properties keep you looking sexy and feeling energetic. Get your vitamin E from walnuts, sweet potato, spinach, asparagus, chickpeas, chestnuts, broccoli and tomatoes.

Vitamin K is essential for many coagulant factors in the blood as well as in bone metabolism. Vitamin K has potential role in the increase of bone mass by promoting osteotrophic activity in the bone. It also has established role in Alzheimer's disease patients by limiting neuronal damage in the brain. Spinach, basil, kale, cabbage, mustard, parsley, broccoli, asparagus, celery, lady finger, cucumber, lettuce, carrots, chilli powder, olive oil, cloves and dried fruits are vitamin K rich.

Vitamin B-6 or pyridoxine is required for GABA hormone production in the brain. It also controls homocystiene levels in the blood, which may otherwise be harmful to blood vessels resulting in CAD and stroke. Research suggests that increasing the intake of certain B vitamins, especially B6 and B12, can help enhance memory abilities. Vitamin B6, helps you get smooth skin, riboflavin helps cure angular cheilitis, commonly known as cracks on the lips or corner of the mouth while as thiamine helps in curing scaly and flaky skin. The intake of Vitamin B6 is very important for women, especially those who are pre-menopausal. Vitamin B6 regulates the sex hormones and ups your libido by controlling raised levels of prolactin and regulating the production of oestrogen, testosterone, red blood cells, serotonin, and dopamine. Bananas, avocados, baked potatoes and tomatoes will give you healthy dose of Vitamin B6.

Vitamin B3 or niacin, is key for orgasms as it helps blood flow easier and faster. It also helps the adrenal glands produce sex hormones. Vitamin B3 is available in brown rice, whole wheat bread, peanuts, and sun-dried tomatoes.

Carotene: Consumption of natural fruits rich in carotene is known to protect from lung and oral cavity cancers. Alpha carotene reduces the risk of lung cancer and boosts the immune system. Beta- carotene functions as a powerful antioxidant protecting the protein, fat and DNA in cells from free radicals damage.

Folic acid is essential for DNA synthesis and cell division.

Resveratrol: This antioxidant has anti-ageing and disease preventing properties. It is very good for heart health, because it helps to reduce inflammation and bad cholesterol. Besides, it also helps to prevent cancer and Alzheimer's.

Zeaxanthin, an important dietary carotenoid, selectively absorbed into the retinal macula lutea in the eye where it is thought to provide antioxidant and protective light-filtering functions; thus it protects eyes from "Age related macular degeneration" (ARMD) disease in the elderly. Zeaxanthin has photo-filtering effects on UV rays and thus protects eyes from age related muscular disease in the elderly populations. Yellow-coloured fruits and vegetables are the best sources of lutein and zeaxanthin. It is highest in orange and yellow coloured pumpkin, yellow corn, yellow and orange bell pepper, carrots, yams, musk melons, peaches, grapes, oranges, mangoes and dried apricots, spinach, collards, kale and sweet corn.

Lutein: Both lutein and zeaxanthin, the anto-oxidants help keep our eyes healthy. It is the only anti-oxidant present in the retina and lens of the eye and it is called the eye vitamin. It protects the eyes from strong anti-oxidant damage by ultraviolet rays. Lutein can be found in the back of the eye where the retina is located. When lutein levels in the eyes decrease with age, cataract formation takes over. Therefore, eat plenty of lutein and other other anti-oxidant rich foods like spinach, broccoli, lettuce, dark leafy green, green peas, tomatoes, corn, watermelon, fenugreek, parsley, mint, green coriander and kiwi fruit, avocados which are rich in lutein.

Anthocyanins: This antioxidant possesses great anti-inflammatory and anti-carcinogenic property. It helps in lowering the risk of cardiovascular diseases, obesity and diabetes. Anthocyanins are red, purple or blue pigments found in many fruits and vegetables, especially concentrated in their skin, known to have powerful anti-oxidant properties. They are another class of polyphenolic anti-oxidants present abundantly in the red grapes. These phyto-chemicals have been found to have anti-allergic, anti-inflammatory, anti-microbial, as well as anti-cancer activities. Anthocyanin is a naturally occurring bioactive compound in berries and other fruits and vegetables including brinjals, blackberries, blackcurrants, are most likely to reduce the risk of developing Parkinson's disease (Anonymous, 2012). Anthocyanins which give colour to most red, purple and blue fruits and vegetables, can possibly slow the growth colon cancer cells.

Lycopene is the most powerful flavonoid antioxidant, studies have shown that it protects skin damage from UV rays, and offers protection against prostate cancer. Carrots, yams, sweet potatoes, squash, tomatoes, apricots, beets *etc.* contain Vitamin A and Lycopene, which slow down the growth of aggressive cancers.

Omega-3-fatty acids: Omega 3 fatty acids maintains the high-density lipoprotein (HDL) cholesterol level in the body, which aids the body to remain healthy. A lot of people experience joint pains during the winter Omega 3 significantly reduces stiffness and joint pain which are the symptoms of rheumatoid arthritis. In fact, it also helps induce bone strength by increasing the calcium levels in the body. Consumption of foods rich in ω-3 fatty acids may reduce the risk of coronary heart disease, stroke, and help prevent development of ADHD, autism, and other developmental differences in children. For omega-6 use safflower, sunflower and corn oils. Omega-3 fatty acids help prevent toxins from entering the cell membranes, aids nutrients in entering and helps in excretion of waste products. Omega -3 also helps prevent inflammation which can damage skin. Lack of Omega 3 may be linked to pre mature birth or low weight babies. Fish is the best source of Omega 3 but there are vegetarian sources of Omega 3 like flax seeds and walnuts. Omega 3 is responsible for regulating hormones and a healthy uterus is essential for improving fertility. Omega -3 also improves the blood supply to reproductive organs (Trina Remedios, 2013).

Folate helps prevent birth defects such as spina bifida. Fruits and vegetables such as oranges, spinach, broccoli, and dried beans are good sources of folate.

Antioxidants are essential to optimizing health by helping to combat the free radicals that can damage cellular structures as well as DNA.

Melatonin can cross the blood-brain barrier easily and produces soothing effects on the brain neurons, calming down nervous system irritability which helps relieve neurosis, insomnia and headache conditions.

Pectin, which by acting as bulk laxative helps to protect the colon mucous membrane by decreasing exposure time to toxic substances in the colon as well as binding to cancer causing chemicals in the colon. Pectin has also been shown to reduce blood cholesterol levels by decreasing its re-absorption in the colon by binding bile acids resulting in its excretion from the body.

Tannins have shown to lower blood pressure and stimulate the immune system, while anthocyanins help protect blood vessels and reduce inflammation.

Fluoride is a component of bones and teeth and is important for prevention of dental caries.

Calcium is an important mineral that is an essential constituent of bone and teeth, and required by the body for muscle contraction, blood clotting, and nerve impulse conduction. Calcium helps to strengthen bones and teeth, but you will be amazed to hear that calcium also helps to control hunger pangs. Consumption of dairy products and other calcium rich products have low fat mass in their body and have more control over appetite. Hence, consume more of calcium rich products, if you want to shift your body to a fat burning mode. Lack of calcium in the body can pose a risk of osteoporosis.

Calcium-Rich Foods for your Bones

Leafy greens: Many vegetables, especially the leafy green veggies, are rich sources of calcium. Opt from spinach, turnip, kale, romaine lettuce, celery, broccoli, cabbage, asparagus and mushrooms or toss all of them into a salad.

Beans: For another rich source of calcium, try kidney beans, white beans or baked beans.

Herbs and spices: Herbs like basil, thyme, cinnamon, mint and spices like garlic not only enhance the flavour of your **food** but also provide calcium to your body.

Oranges: Citrus fruits like oranges are not only rich in calcium but also contain vitamin D that is essential to absorb the calcium in the body. Have it as a fruit or squeeze out its juice and have it as a breakfast.

Potassium: The third-most abundant mineral in the body, Potassium is an essential electrolyte which plays a crucial role in body's hydration and athletic performance. Potassium is also needed to lower blood pressure. Other health benefits of potassium include stroke prevention, alleviation from heart and kidney disorders, relief from anxiety and stress, as well as improved muscle strength, metabolic rate, water balance, electrolytic functions and nervous system.

Potassium is a heart-healthy mineral, it is an important component of cell and body fluids helps controlling heart rate and blood pressure; thus counters the bad influences of sodium, thus offers protection against stroke and coronary heart diseases. Yet another mineral that has a significant effect on your PMS mood swings. Fresh fruit salads, with oranges and apples are good options for an early evening snacks. Potassium is an important mineral that controls the balance of fluids in the body and helps to lower blood pressure. Potassium, is good for your body because it helps regulate your blood pressure. Potassium found in bananas also benefits your bones. By helping to suppress calcium excretion, bananas reduce the risk of developing osteoporosis and brittle bones as you age.

How does Potassium Help Reduce Blood Pressure?

Potassium is a very important mineral for the proper functioning of all cells of our body. Along with sodium, calcium and magnesium, potassium helps maintain the electrolyte balance of the body. Too much salt or sodium causes water retention in the body. This increases the blood volume and puts pressure on the artery walls resulting in high blood pressure. Again, low potassium levels and high sodium levels, makes the heart and blood vessels to work harder and therefore increase pressure on the walls. So, keeping the right sodium–potassium balance is important for proper functioning of the body. Increasing the dietary consumption of potassium can help lower blood pressure. Since our diet is normally high in sodium, it is important that we increase the intake of potassium. And if you have high blood pressure, you need to decrease the intake of sodium and increase the intake of potassium to get better effects. Potassium can lessen the effects of sodium on blood pressure. And fruits are a great source of potassium.

Magnesium is found in green leafy vegetables and banana. It is impotant for immunity, bone and heart health. It is often described as an anti-stress mineral, plays an important role in learning and memory. It is also vital for the production and transfer of nerve impulses, in muscal contraction and relaxation, in protein synthesis and in many biochemical reactions, which aid in improving memory. It relaxes muscles and steadies nerves. It also helps build strong bones, stabilizes heart beat and regulates blood pressure and blood sugar levels. Magnesium contributes to the structural development of bones and it is required for energy production. Increase consumption of green leafy vegetables (spinach, legumes), seeds, nuts and fruits will ensure adequate leaves for optimum functioning. The mineral helps in the production of sex hormones like androgen and oestrogen that affect your libido. Studies have shown that taking supplements of this mineral can up your libido. If anxiety is stopping you from getting an orgasm, this mineral could help as it has calming and relaxing properties. Magnesium is found in rice, wheat, oat bran, dried coriander, dill, sage, basil, pumpkin seeds, watermelon seeds, cocoa powder, flax seeds, sesame seeds, sunflower seeds, almonds and cashews

Manganese is used by the body as a co-factor for the antioxidant enzyme, *superoxide dismutase*. It is essential for bone strengthening and has cardiac-protective role as well. Adding magnesium (Spinach, beans, cashews) to your diet reduces PMS symptoms drastically. Foods that are rich in magnesium help regulate serotonin, which in simple terms is the mood balancer of the brain. Serotonin, a brain chemical, plays a vital role in mood fluctuations and pain sensitivity. Thus keeping it under control, significantly influences the onset of PMS. Increasing your magnesium intake will help cure symptoms like breast tenderness and abdominal bloating. Magnesium is essential for bone growth.

Copper: It is required in the bone metabolism as well as in production of white and red blood cells. Copper is a co-factor for many vital enzymes, including cytochrome c-oxidase and superoxide dismutase (other minerals function as co-factors for this enzyme are manganese and zinc. It is helpful in collagen formation that plays a crucial role in tissue's health. It is also important for proper growth of body and regulates heart rhythm.-

Copper Rich Fruits and Vegetables

Mushrooms, kidney beans, cashews, avocados, dried prunes and sesame seeds are rich in copper.

Iron is essential for red blood cell production. It is required for various body functions. Lack of iron in our diet can cause various health issues such as loss of stamina, breathlessness, low levels of energy and even dizziness. You can get iron from meat, beans, nuts and some vegetables such as broccoli and spinach. Iron is required for red blood cell formation as well for cellular oxidation. Iron, being a component of hemoglobin inside the red blood cells, determines the oxygen carrying capacity of the blood. Iron rich foods are as under!

Spinach: This green leafy vegetable is loaded with iron. In 100 grams of spinach, there is 2.7 mg of iron.

Lentils: Lentils are not only loaded with protein but also enriched with iron. A cooked cup of lentils provide.

Mushrooms: Certain varieties of mushrooms contain high amount of iron. Oyster mushrooms contain up to twice as much iron as button mushrooms.

Kidney beans: They are nutritious and contain iron. A cup of kidney beans contain 4 milligrams of iron and is also loaded with protein.

Chromium, the trace mineral that helps tissue cells respond appropriately to insulin levels in the blood; thus helps facilitate insulin action and control sugar levels in diabetes.

Iodine: Human body needs Iodine to make thyroid hormones and the absence of iodine can pose problems like fatigue, depression, high cholesterol levels and swelling of thyroid glands. Iodine rich fruits and vegetables (Thyroid diet) Crane berry, dried prunes, straw berry, navy beans, potato and yogurt.

Cobalt: It is essential part of vitamin B12 and is also called as cobalamin. According to health experts, the human body needs a very little of cobalt for proper functioning but that does not mean it should be neglected- nuts, broccoli and oats are rich in cobalt.

Zinc: It is a trace element that is found in cells throughout our body. It is a nutrient which is essential for healthy immune system, good for brain function and decreases the risk of certain diseases. As our body is not capable of producing zinc, so we need to consume zinc rich foods to supply our body with this essential element. Pomegranate, guava, apricot and peach are rich source of zinc. In green leafy vegetables, green peas, spinach and bhindi are the ones rich in zinc. Zinc is required for immunity, cell biology, smell and taste and carbohydrate metabolism.

Benefits of Zinc, Especially for Children

It strengths the immune system

Boosts learning and memory

Decreases the severity of common cold

Helps body produce protein and DNA

Speeds up the wound healing process.

Prevents age related mucular degeneration

Most seeds are a rich source of Zinc, *e.g.* seeds like sesame, hemp, mustard, Pumpkin and chia seeds.

Legumes: All kinds of legumes like lentils, beans have an abundant amount of Zinc. However, absorption of Zinc found in legumes is less because of the presence of the anti-nutrient **phylates.**

Nutes, like almonds are an excellent source of Zinc. Use roasted nuts like cashews, peanuts, walnuts as snacks for your teenage children.

Folate, has many functions in the body such as developing proteins along with B-12 and vitamn C, in cell biology, prevention of B12 related anemia. Folate is found richly in aspergus, mango, green veafy vegetables and oranges.

What are some of the Good Things in Fruits and Vegetables?

- **Vitamins**
- **Minerals**
- **Flavonoids** - plant chemicals that act like antioxidants
- **Saponins** - plant chemicals that have a bitter taste
- **Phenols** - organic compounds in foods
- **Carotenoids** - vitamin A-like compounds
- **Isothiocyanates** - sulfur-containing compounds.
- Several types of **dietary fibre**

Functions of Antioxidants and Phytonutrients

- They are scavengers of free radicals
- Anti-cancer properties

- ☆ Anti-ageing properties
- ☆ Anti-allergic properties
- ☆ Natural immune boosters.

Vegetables and Fruits

The orange foods help to prevent cancer. Carrots (Gajar), Pumpkin (Kaddu), Sweet Potatoes (Shakar kand), Apricots (Khubani), Cantaloupe (Kharbuja) and other orange foods are particularly good for our eyes. The orange foods contains vitamin A.

White and green foods contain a variety of minerals. Include Garlic, Onions in your meals which are anti-bacterial and anti fungal.

Red and Blue foods are protecting against heart disease by improving circulation of blood. Add foods in your meals like Apple, Beets (Chukandar), black grapes. Overall, blue colour foods help in boosting overall health, by preventing diseases and promoting longevity. They are also good memory boosting foods. The best blue foods are blueberries, blackberries, figs, plums, raisins and purple cabbage. Food is blue in colour because of its richness in health-boosting phytochemicals known as anthocyanins and resveratrol.

APPLE

Apple (*Malus domestica*) a day keeps the doctor away is cliched, but its health benefits are undisputable. Apple's antioxidant property prevents cell and tissue damage. Studies by nutritionists have shown that apples contain abundant amounts of **elastin** and **collagen** that help keep the skin young. Applying a mixture of mashed apple, honey, rose water and oatmeal can act as a great exfoliating mask on your skin. An apple a day, will surely keep the doctor away because they contain **guercetin**, a photochemical containing ant-inflammatory properties. It also helps in prevention of blood clots. Eat apples for breakfast with your cereal or eat them as a snack when you're hungry instead to pigging out on deep-fried chips. The apple is a highly nutritive food. It contains minerals and vitamins in abundance. The food value of the apple is chiefly constituted by its contents of sugar which ranges from 9 to 51 per cent. Of this, fruit sugar constitutes 60 per cent and glucose 25 per cent and cane sugar only 15 per cent. Apples are a source of both soluble and insoluble fibre. Soluble fibre such as pectin actually helps to prevent cholesterol buildup in the lining of blood vessel walls, thus reducing the incident of atherosclerosis and heart disease. The insoluble fibre in apples provides bulk in the intestinal tract, holding water to cleanse and move food quickly through the digestive system. Although an apple has a low vitamin C content, it has antioxidants and flavonoid which enhances the activity of vitamin C thereby helping to lower the risks of colon cancer, heart attack and stroke. Apple improves bone health, treats asthama, prevents lung, breast, colon, lever cancer. It is also best fruit for diabetic patients.

It is a good idea to eat apples with their skin. Almost half of the vitamin C content is just underneath the skin. Eating the skin also increases insoluble fibre content. Most of an apple's fragrance cells are also concentrated in the skin and as they ripen, the skin cells develop more aroma and flavour. The skin of apple should not be discarded when taking it in raw form as the skin and the flesh just below it contain more vitamin C than the inner flesh. The vitamin content decreases gradually towards the center of the fruit. The skin also contains five times more vitamin A than the flesh.

They have been used as a beauty aid for decades. A cup of apple juice added to your bath will cleanse and smothen your skin. Apple juice applied to your scalp helps prevent dandruff. A final rinse of your hair with apple juice after shampooing brings about an added shine. Apples are also a good conditioner and toner, and help fight acne.

Peeling apples before eating can cause loss of many vital nutrients. Laboratory analysis has shown that 100 gm of unpeeled apple contains about 16 mg of vitamin C. The same quantity of peeled fruits contain only 8 mg this vitamin. Nearly half of an apple's vitamin C content lies beneath its skin. It contains as calcium, potassium, phosphorus, folate and iron. The peel is a significant source of both insoluble and fat-soluble fibre called pectin. It accounts for around 50 per cent of total fibre of the apple. Peels are rich in flavonoids and phenolic acids. It has triterpenoids. These compounds can inhibit or kill cancer cells.

Health Benefits of Apple

Apples combine certain nutrients in a way that sets them apart from all other fruits and makes them a food of choice for achieving several health goals. Here's what apples can do for you when it comes to your health. Delicious and crunchy apple is one of the popular fruit that contain an impressive list of essential nutrients, which are required for normal growth and development and overall nutritional well-being (Sobiya, 2013).

Unique Support for Heart Health

Apples are a rich source of phytonutrients known as **quercetin**, **catechin**, **phloridzin** and **chlorogenic acid**. These unique phytonutrients help protect the cardiovascular system from oxygen-related damage. In addition, the high fibre content (pectin) in apples helps keep your LDL ("bad") cholesterol levels low and under control. People who eat two apples per day may lower their cholesterol by as much as 16 percent.

Alzheimer's Prevention

A study on mice at Cornell University found that the quercetin in apples may protect brain cells from the kind of free radical damage that may lead to Alzheimer's disease.

Lung Cancer Prevention

According to a study of 10, 000 people, those who ate the most apples had a 50 per cent lower risk of developing lung cancer. Researchers believe this is due to the high levels of the flavonoids quercetin and naringin in apples.

Breast Cancer Prevention

A Cornell University study found that eating one apple per day can reduce the risk of breast cancer by 17 percent. By eating three apples per day can reduce the risk by 39 per cent and those who eat six apples per day can reduce their risk by 44 percent.

Colon Cancer Prevention

One study found that rats fed an extract from apple skins had a 43 per cent lower risk of colon cancer. Other research shows that the pectin in apples reduces the risk of colon cancer and helps maintain a healthy digestive tract.

Liver Cancer Prevention

Research found that rats fed an extract from apple skins had a 57 per cent lower risk of liver cancer.

Diabetes Management

The pectin in apples supplies galacturonic acid to the body which lowers the body's need for insulin and may help in the management of diabetes.

Weight Loss

A Brazilian study found that women who ate three apples or pears per day lost more weight while dieting than women who did not eat fruit while dieting.

Bone Protection

French researchers found that a flavanoid called phloridzin that is found only in apples may protect post-menopausal women from osteoporosis and may also increase bone density. Boron, another ingredient in apples, also strengthens bones.

Asthma Help

One recent study shows that **children** with **asthma** who drank apple juice on a daily basis suffered from less wheezing than children who drank apple juice only once per month. Another study showed that children born to women who eat a lot of apples during **pregnancy** have lower rates of asthma than children whose mothers ate few apples.

Apples are low in calories; 100 g of fresh fruit slices provide only 50 calories. The fruits contain no saturated fats or cholesterol; but rich in dietary fibre, which helps, prevent absorption of dietary LDL cholesterol in the gut. The dietary fibre also help protect the mucous membrane of the colon from exposure to toxic substances by binding to cancer causing chemicals in the colon.

It contains good quantities of vitamin-C and beta-carotene. Vitamin C is a powerful natural antioxidant. Consumption of foods rich in vitamin C helps body develop resistance against infectious agents and scavenge harmful, pro-inflammatory free radicals from the body.

They are rich in antioxidant phyto-nutrients flavonoids and polyphenols. The important flavonoids in apples are **quercetin, epicatechin,** and **procyanidin B2**.

Apples are also good in **tartaric acid** that gives tart flavour to them. These compounds help body protect from deleterious effects of free radicals.

Apples are good source of B-complex vitamins such as riboflavin, thiamin, and pyridoxine (vitamin B-6). Together these vitamins help as co-factors for enzymes in metabolism as well as in various synthetic functions inside the body.

Apple also contains small amount of minerals like potassium, phosphorus, and calcium.

Apple is a fair source of fibre.

Apple contains Pectin and Malic acid.

Iron contained in the apple helps in formation of blood.

Raw apples are good for constipation.

Cooked or baked apples are good for diarrhea.

Apples have been found useful in acute and chronic dysentery among children. Ripe and sweet apples should be crushed into pulp and given to the child several times a day.

Apples are of special value to heart patients. They are rich in potassium and phosphorus but low in sodium. It is also useful for patients of high blood pressure.

Apple is also said to be beneficial to gout patients caused by increase of uric acid in blood.

Apples, boiled to a jelly, make a very good liniment for rheumatic pains. They should be rubbed freely on the affected area.

Sweet apples are valuable in dry hacking cough.

Apples are useful in kidney stones.

The apple peel water is an excellent medicine for the inflamed eyes as an eye wash. The over-ripe apples are useful as a poultice for sore eyes. The pulp is applied over the closed eyes.

Tooth-decay can be prevented by regular consumption of apples as they possess a mouth cleansing property.

The apple is the best fruit to tone up a weak and run-down patient. It removes deficiencies of vital organs and makes the body stout and strong. It tones up the body and the brain as it contains more phosphorus and iron than any other fruit or vegetable. Its regular consumption with milk promotes health and youthfulness and helps build healthy and bright skin.

Here are some serving tips

1. Eat apple fruit as they are with skin to get maximum health benefits.
2. Sliced apple turns brown (enzymatic brownish discolouration) on exposure to air due to conversion in iron form from *ferrous oxide* to *ferric oxide*. If you have to serve them sliced, rinse slices in water added with few drops of fresh **lemon.**
3. Cloudy apple juice is a good drink with dinner.
4. Apple fruit is also used in the preparation of fruit jam, pie, and fruit salad

Here are things you may not know you can do with an apple:

1. **Ease your allergies**. Apples are abundant in the plant compound quercetin. The compound seems to slow down the secretion of histamine,

a chemical your body releases during an allergic reaction. A 2000 study in the journal Thorax also found that men who ate five or more apples a week enjoyed better lung function.

2. **Mitigate a migraine**. Some studies have found that the smell of green apples can reduce the brain-thumping symptoms of a migraine, and can shorten a migraine episode. Researchers are also trying to figure out if it could help alleviate joint pain. If you feel a headache coming on, you can cut a green apple in half and smell it. You be the judge if it works or not! Just don't sniff artificial apple fragrances–they could cause a headache!
3. **Ripen other apples, and other fruit**. Apples, like peaches, pears, and bananas, release off ethylene gas, which speeds ripening. If you want to ripen certain fruits fast, put them in a paper bag with a ripened apple. On the flipside, if you don't want your fruit to ripen too quickly, make sure you don't store these fruits close to each other in the fridge.
4. **Clean your hair (with apple cider vinegar).** To combat oily hair, put a teaspoon of apple cider vinegar in a pint of water, and rinse it through your hair while taking a bath or shower. The vinegar solution will take out soap residue that can cling to and weigh down oily hair. The vinegar scent will go away when your hair dries.

APRICOT

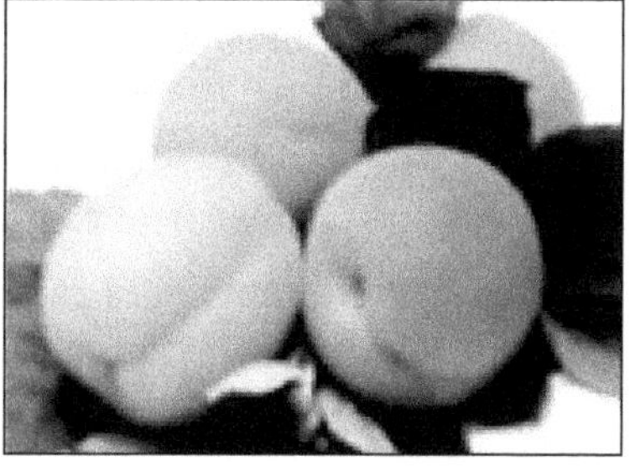

Apricot (*Prunus armeniaca*) can be easily categorized as one of the most beneficial and healthy fruits. The shape of the stone fruit, apricot, is quite similar to that of a peach and has a velvety skin, which is golden orange in colour. It is slightly smaller in size when compared to a peach and has a seed nut in it. Though it is not appropriate for making juice, the fruit can definitely be blended with other fruit juices. The taste is very fruity and feels like a cross between peach and plum. Apricot is widely used in jams, making pastry and also distilling into brandy and liquor. The fruit, when taken raw, is a bit acidic in nature. However, the acidity decreases as it ripens and the sugar content increases. Read on to find nutritional value and benefits of eating apricot. The mineral content in apricots is high, which makes them helpful in overcoming disorders like anemia, tuberculosis, asthma, bronchitis, etc. The high copper and iron content in apricots increases the hemoglobin in the blood. The vitamin A content in apricots is high and helps in reducing skin disorders like pimples and zits. Vitamin A also aids in improving eye-sight. A certain substance called lycopene is present in apricots, which helps in the prevention of cancer and also protects the body from high cholesterol, thereby preventing heart disease. Eating apricots helps in removing gallstones and also destroys intestinal worms. It also aids in digestion, since it reacts as an alkaline, breaking down food particles. The fruit is considered to be a laxative and consuming it regularly will help you get rid of constipation. It is very helpful in getting rid of the waste in the body, due to its high fibre content.

Nutrients in apricots can help protect the heart and eyes, as well as provide the disease-fighting effects of fibre. The high beta-carotene content of apricots makes them important heart health foods. An apricot in its raw state is somewhat acidic but the acidity decreases as it ripens and its sugar content increases. When it ripens, the vitamin A within also doubles.

Apricots contain nutrients such as vitamin A that promote good vision. Vitamin A, a powerful antioxidant, quenches free radical damage to cells and tissues. Free radical damage can injure the eyes' lenses.

The degenerative effect of free radicals, or oxidative **stress,** may lead to cataracts or damage the blood supply to the eyes and cause macular degeneration. Researchers who studied over 50, 000 registered **nurses** found that **women** who had the highest vitamin A intake reduced their risk of developing cataracts by nearly 40 per cent. Apricots are a good source of fibre, which has a wealth of benefits. These include preventing constipation and digestive conditions such as diverticulosis. A healthy, whole-foods diet should include apricots as a delicious way to add to your **fibre** intake of the body.

The apricot has natural sugars that are easily digestible, vitamins A and C, Riboflavin and Niacin, and also minerals like calcium, phosphorus, iron and traces of sodium, sulphur, manganese, cobalt and bromine. Interestingly, when the fruit ripens, its vitamin A content doubles. When the fruit is dried not only does it double in calories, but its calcium, phosphorus and iron content also increases.

Apricot is a juicy fruit rich with nutrients and healing properties.

Apricot contains beta carotene, fibres, vegetable fats, vegetable hormones, vitamin B complex, vitamin B-17 (in the seed), vitamin C, iron, essential oils, essential fatty acids, potassium, calcium, magnesium, mineral salts, sodium, organic acids, proteins, starch, sulfur, phosphor, carbohydrates, cellulose and sugar. All these components make the apricot very effective food in the prevention from bacterial infection, restoration of harmed tissues, teeth and bone development and improving vision. It is very rich source of beta carotene which makes it very beneficial for the heart, because it reduces the bad cholesterol.

Apricots are rich with fibre which makes them helpful for the digestive system. Thanks to the cellulose and pectin it acts like a laxative and eating a single apricot before having a meal will ease the digestion. Because of the high levels of iron it is highly recommendable for the people who suffer from anemia, since iron increases hemoglobin levels.

Nutritional and Medicinal Benefits

Apricots are recommended for treating oral cavity, heart diseases, blood vessels, constipation and strengthening of the immune system. It is also helpful against depression.

Fresh apricot juice can be useful against fever and skin diseases. It speeds up the waste removal from the body. Apricots are very beneficial for the skin which makes them very effective remedy for eczema and burns.

Dried apricots have the same health benefits therefore they are highly recommendable for eating as a substitution for other snacks.

Apricot kernel has amazing powers, especially when it comes to the fight against all types of cancers. The apricot kernel contains a lot of vitamin B-17. The molecule of vitamin B-17 contains two units of sugar, cyanide and benzaldehyde, which are released by the human organism only in the areas affected by malignant cells. The malignant cells require sugar in order to survive and grow.

Malignant cells are rich with the enzyme called beta-glucosidase which creates the toxic compound named Hydrogen Cyanide when combined with cyanide and benzaldehyde. This is how the cancer is destroyed. The healthy cells don't contain high levels of beta-glucosidase, which is the reason why they aren't poisoned. If you consume them for this purpose, you shouldn't eat other forms of sweet.

The apricot has highly health-building virtues. The fresh fruit is rich in easily-digestible natural sugars, vitamins A and C, riboflavin (B2) and niacin (B3).

It is also an excellent source of minerals like calcium, phosphorus, iron and traces of sodium, sulphur, manganese, cobalt and bromine. Apricots are often dried, cooked into pastry or eaten as jam. The calories in apricots multiply many times over when dried, but the amount of calcium, phosphorus and iron also increased significantly.

The beta-carotene and lycopene in this golden fruit helps protect the LDL cholesterol from oxidation, which in turn helps prevent heart disease. The apricot seed is a nut that is rich in protein and fat like any other nuts. It also has an extremely high content of vitamin B17 which is known as Laetrile. Daily consumption of this seed is claimed to be highly effective in preventing cancer. Cancer patients on Laetrile Cancer Therapy have reported that their tumors have shrunk with high doses of vitamin B17 (healthy diet.com/apricot).

These bitter seeds may be chopped up or ground and swallowed with a teaspoon of honey.

Studies show that the fruit, kernel, oil and flowers of the apricot have been used in medicine and medical treatment from ancient days.

Its high iron content makes it beneficial for people suffering from anemia.

Its cellulose and pectin content make apricot a gentle laxative and therefore, useful to treat constipation.

It aids digestion, and its high Vitamin A content improves eyesight.

The beta-carotene and lycopene in this golden fruit help protect LDL or good cholesterol from oxidation, which in turn helps prevent heart disease.

Due to their high fibre to volume ratio, dried apricots are sometimes used to relieve constipation or induce diarrhea. Effects can be felt after eating as few as three.

Health Benefits

The fruit, kernel (inner softer part of the seed), oil and flowers of the apricot have always been used in medicine and medical treatment from ancient days.

The kernel yields an, oil that is similar to that of the almond and is widely used for their sedative, anti-spasmodic relief to strained muscles. It is also useful for healing of wounds, expelling worms and as a general health tonic.

Anemia: The high content of iron in apricot makes it an excellent food for anemia sufferers. The small but essential amount of copper in the fruit makes the iron available to the body. Liberal consumption of apricot can increase the production of hemoglobin in the body. This is ideal for women after their menstrual cycle, especially those with heavy flow.

Constipation: The cellulose and pectin content in apricot is a gentle laxative and are effective in the treatment of constipation. The insoluble cellulose acts as a roughage which helps the bowel movement. The pectin absorbs and retains water, thereby increasing bulk to stools, aiding in smooth bowel movement.

Digestion: Take an apricot before meal to aid digestion, as it has an alkaline reaction in the digestive system.

Eyes/Vision: The high amount of vitamin A (especially when dried) is essential to maintain or improve eyesight. Insufficiency of this vitamin can cause night blindness and impair sight.

Fever: Blend some honey and apricots with some mineral water and drink to cool down fevers. It quenches the thirst and effectively eliminates the waste products from the body.

Skin Problem: Juice fresh apricot leaves and apply on scabies, eczema, sun-burn or skin itchiness, for that cool, soothing feeling.

Nutrients in apricots can help protect the heart and eyes, as well as provide the disease-fighting effects of fibre. The high beta-carotene content of apricots makes them important heart health foods. Beta-carotene helps protect LDL cholesterol from oxidation, which may help prevent heart disease.

Apricots contain nutrients such as vitamin Athat promote good vision. Vitamin A, a powerful antioxidant, quenches free radical damage to cells and tissues. Free radical damage can injure the eyes' lenses.

The degenerative effect of free radicals, or oxidative stress, may lead to cataracts or damage the blood supply to the eyes and cause macular degeneration.

Apricots are a good source of fibre, which has a wealth of benefits including preventing constipation and digestive conditions such as diverticulosis.

Nutrients in apricots can help protect the heart and eyes, as well as provide the disease-fighting effects of fibre. The high beta-carotene content of apricots makes them important heart health foods.

An apricot in its raw state is somewhat acidic but the acidity decreases as it ripens and its sugar content increases. When it ripens, the vitamin A within also doubles.

Consumption Tips

Apricots are usually picked when they are still firm. An unripe apricot is often yellow and hard. When ripe and soft, its colour turn a consistent golden-orange hue. At this time, handle the fruit with care as it is easily bruised.

Stored in the fridge, these fruits can last for three or four days. When overripe, the fruit turns soft and mushy.

Caution

Fresh apricots contain a small amount of oxalates. Individuals with a history of calcium oxalate-containing kidney stones should not consume too much of this fruit.

Whereas dried apricots contain sulfur-containing compounds such as sulfur dioxide. These compounds may cause adverse reactions in people who suffer from asthma.

Apricot seeds are best known for the controversy surrounding their ability to purportedly cure cancer. The seeds, or kernels, are found inside pits of apricots. Aside from the controversial substance in them called amygydalin, the kernels have mono-unsaturated fats, protein and carbohydrates.

The good: This food is very low in Saturated Fat, Cholesterol and Sodium. It is also a good source of Dietary Fibre and Potassium, and a very good source of Vitamin A and Vitamin C. But very high in sugar.

BANANA

Banana (*Musa acuminata colla*) is a tropically grown and is a very beneficial fruit. This is one fruit that's abundantly available in India all through the year. We know it's a good source of iron, magnesium and potassium and helps reduce menstrual cramps. Though banana is a very nutritious food, it's significance is often underestimated as it is easily available, can grow in plenty, and are considerably cheap compared to other fruits, but this does not change the nutrition facts. Calcium can be considered as the main nutrition that a banana provides and as we know calcium helps in strengthening the bones and that is just a tip of the iceberg, banana has ample benefits both in terms of nutrition and energy. The fact that banana is very high in calories, but at the same time low in terms of fats is very less known and this makes it a very good option for those who are very much concerned about their weight and fitness level. It is also easy to digest making it a good food for small kids because they do not have trouble digesting it. Banana also is rich in anti-oxidants and it is a well known fact that anti-oxidants help in increasing immunity, increase energy, reduce high blood pressure, *etc.* Not only this, one of the several banana nutrition are minerals. Banana is loaded with **trytophan** that increases **serotonin** levels in the body. Apart from this, the fruit has many important acids, enzymes and physiologically important chemical compounds. The banana develops its particular aroma on ripening because of the presence of a chemical compound called **Amylacetate.** Iron and potassium in banana are wholly available. Its intake help children to retain many mineral nutrients. The fruit is a fair source of B vitamins and calcium. Banana is a very energy rich fruit, *i.e.* two fruits can provide fuel for body for two hours. That is why banana is favoured fruit of sports persons. Besides energy, bananas are one of the best sources of potassium, an essential mineral for maintaining normal blood pressure and heart function. An average banana contains a whopping 467mg of potassium and only 1 mg of sodium. So a banana a day helps in preventing high blood pressure and protect against hardening of arteries (Parmar, 20012). Bananas are a particularly good energy source and these are best eaten a few hours before you intend to exert yourself. Along with being great sources of carbs, they are also effective because they trigger the release of 'dopamine' – a chemical that builds your concentration and focus. Banana is the happiest fruit (it is actually a berry) and contains bromelain an enzyme that triggerrs testosterone production, its potassium and vitamin B elevate energy level.

The effect of banana on skin too is not something that can be ignored. Bananas are rich in vitamin A, B and E and hence works as an anti-aging agent. A fresh mashed banana facial can do wonders for your skin.

Full ripe banana with dark patches on yellow skin produces a substance called TNF (tumor necrosis factor) which has the ability to compact abnormal cells. The more darker patches it has the higher will be its immunity enhancement quality: Hence, the riper the banana the better the anti-cancer quality.

When they are in unripe stage, the resistant starch content in banana's is higher. Resistant starch is a type of starch that resists digestion. Even though we can not digest it, resistant starch is very important for us. Since it ends up in the colon, where fermentation takes place, promoting the growth of bacteria. This, in turn, can protect us from the risk of developing colon cancer. As well as promoting good bacteria, green banana's also promote the absorption of calcium and other essential nutrients. This is achieved when good bacteria triggers the release of certain key digestive enzymes in the body. These help our bodies to absorb nutrients better, from the food we eat. However, they are good for people with 2 type diabetes. Since green banana's haver a lower sugar content (theopenews.com/should-babanas-).

Best muscle building foods: Bananas have three types of sugars. Fructose, sucrose and glucose, these sugars are prime for pre/post training. Bananas are fat and cholesterol free, extremely portable, and nutrient dense.

The fully ripe banana produces a substance called TNF which has anti cancer properties. The degree of anti-cancer effect corresponds to the degree of ripeness of the fruit. Bananas increase the number of white blood cells, enhance the immunity of the body and produce TNF. Eating one to two bananas a day increases the body's immunity to diseases like cold, flu and others. The bananas with dark spots are eight times more effective in enhancing the property of white blood cells than the green one (Charnjit Parmar, 2012).

The fruit **banana** can be called the perfect food item for a human body. **Banana nutrition facts** tell us that it has all the components that are needed to make health, sound and playful. It helps our body to fight against **diseases** and provides useful **minerals**, like **potassium**. Banana makes your body to **generate probiotics**, reduces the side effect of salt and helps to **retain calcium** in the body. This tropical fruit can do wonders to our body nutrition system with its **minerals**, **vitamins** and **micro nutrition**.

This is one fruit that's abundantly available in India all through the year. We know it's a good source of iron, magnesium and potassium and helps reduce menstrual cramps. The effect of banana on skin too is not something that can be ignored. Banana is rich in vitamin A, B and E and hence works as an anti-aging agent. A fresh mashed banana facial can do wonders to your skin.

Boasting three naturally-occurring sugars - sucrose, fructose, and glucose - bananas are an obvious choice for a quick energy boost.

Bananas also contain fibre, which will help sustain the glucose levels in your blood, thus giving you a steady source of energy over a longer period of time.

Bananas are a great sex food. Bananas are ideal to give you endurance in the bedroom as they are a rich source of vitamin B that converts carbohydrates into energy. It also helps your body to produce sex hormones such as **testosterone.** Having a banana a few hours before getting into action can keep you pepped up! Sexual pleasure and stimulus aside, here's a thought not many have had - why not nibble on quick energy foods so that you can last longer in bed? After all, you don't want your energy dipping just when your partner starts reciprocating. Nibble on this storehouse of B vitamins and potassium slowly and it is sure to give you that extra kick of energy.

Bananas are known to contain bromelain enzyme which is known to give a boost to your libido. Bananas contain the **bromelain** enzyme, that increases libido and reverse impotence in men. Also, they are potent sources of potassium and B vitamins like riboflavin, which increase the body's overall energy levels. A great way to incorporate bananas into your sex fantasy is to place cut banana pieces on your lady love's body and slowly eat them.

Researchers have found that eating three **bananas** cuts the risk of a stroke.

In the latest research, scientists analysed data from eleven different studies - dating back to the mid-Sixties - and pooled the results to get an overall outcome. They found a daily potassium intake of around 1, 600 mg, less than half the UK recommended daily amount for an adult of 3, 500mg, were enough to lower stroke risk by more than a fifth.

The average banana contains around 500 milligrams of potassium, which helps to lower blood pressure and controls the balance of fluids in the body.

Too little potassium can lead to an irregular heartbeat, irritability, nausea and diarrhoea.

If consumers ate more potassium-rich foods and also reduced their salt intake, the annual global death toll from strokes could be cut by more than a million a year.

They said that having one banana for breakfast, one for lunch and one in the evening would provide enough potassium to reduce the chances of suffering a blood clot on the brain by around 21 per cent.

The findings suggest that thousands of strokes could be prevented by the consumption of other potassium-rich foods such as spinach, nuts, milk, fish and lentils.

Although some previous studies have suggested bananas could be important for controlling blood pressure and preventing strokes, results have not always been consistent.

But if consumers ate more potassium-rich foods and also reduced their salt intake, the annual global death toll from strokes could be cut by more than a million a year.

The nutrition that can be found in an average banana has many useful benefits to our body. As the fruit contains high vitamin C, this vitamin can protect a human body from infections. Not only this, the vitamin is useful for the combination of connective type of tissues. Vitamin C helps to retain iron in our body and enhances the blood formation system. Banana does not help us by Vitamin C only; it supplies another good mineral, which is Potassium.

Potassium makes our body strong, and helps the growth of body muscles. Not only this, banana provides potassium, which reduce high stoke diseases, *i.e.* hypertension. The mineral combination of Banana is really helpful for body system. It contains only potassium and no sodium. This combination of nutrition can reduce high blood pressure as well. Banana frees us from the risks such as formation of red blood, as it contains Vitamin B6. Vitamin B6 also keeps the center of nerve functioning well.

Potassium, is good for your body because it helps regulate your blood pressure. Potassium found in bananas also benefits your bones. By helping to suppress calcium excretion, bananas reduce the risk of developing osteoporosis and brittle bones as you age.

Banana Nutrition Facts

Potassium: It is good for your body because it helps regulate your blood pressure. Potassium found in bananas also benefits your bones. By helping to suppress calcium excretion, bananas reduce the risk of developing osteoporosis and brittle bones as you age. A diet rich in potassium is said to reduce the risk of hypertension and stroke. As bananas are free from sodium and very rich in potassium, they can be included in the diet to reduce the risk of high blood pressure.

Fibre: Bananas are also an excellent source of fibre. To aid in maintaining a healthy digestive system, bananas help food move smoothly through the digestive tract. This also helps in bananas relieve the symptoms of heartburn.

Vitamins and Mineral: With high levels of Vitamin B6, bananas are essential to producing healthy blood. They contain good levels of Vitamin C, magnesium and manganese which enhances your immunity.

Ulcer beater: Besides fighting cancer, bananas also help to protect against stomach ulcers.

Reduces stress: The best bonus benefit of bananas is that they affect your mood and stress levels. Bananas are a good source of trypothan, a chemical when released in the body converts into serotonin. With proper serotonin levels, bananas help improve your mood, reducing stress.

Vitamin C which is found in bananas, helps the body to defend and heal against infections. This vitamin also proves valuable in the synthesis of the connective tissue, absorption of iron and the formation of blood.

Not only are bananas rich in vitamin C, they also contain potassium. Potassium is a mineral that helps in the building of muscles and protein synthesis. This is done as potassium stimulates nerve impulses for muscle contraction.

Bananas contain three natural sugars, sucrose, fructose and glucose, along with fibre. A banana thus gives an instant and substantial boost of energy.

A banana contains vitamin B6, which helps in the synthesis of antibodies in the immune system, apart from red blood cell formation, protein metabolism and functioning of the central nervous system.

No other fruit contains more digestible carbohydrates than bananas. This is advantageous, because the body burns off calories from carbohydrate more easily and quickly than calories from fat or protein.

Sweet banana, can provide you with the necessary energy to work all day long. This fruit gives our body immediate energy supply, because it has a sufficient amount of sugar and glucose. We know that banana contains carbohydrates, but it is a digestible carbohydrate. This type of micronutrient can help in reducing the amount of calories from our body. Eating a banana will help us in balancing the body fluid and getting dehydrated.

Now we have a clear idea about all the nutrition's that a banana can supply to our body. All the minerals and vitamins that banana contains help us to build our body properly and helps our system to fight against the harmful diseases.

Banana benefits our body by treating constipation. The fibre and minerals in a banana help the bowel move in a healthy way. Furthermore, when we eat food, the extra water needs to be absorbed to digest the foods properly. Banana provides pectin, which absorbs the water and helps in reducing constipation. Moreover, as we have discussed earlier, the potassium helps to prevent heart related diseases and bananas are a good source of potassium.

Banana can be used in making remedial foods. It will reduce the acids in the stomach, which in turn results in avoidance of ulcers. Regular eating of bananas does not only ensure the body growth, but also enhance the body immune. This provides you with a clear picture ofbanana nutrition facts.

Rich in potassium and vitamins A, C and E, bananas are ideal for skin and hair (they help restore dull and damaged hair). While **vitamin** A restores natural oils of the skin, vitamin Erepairs damaged skin and lightens age spots. Vitamin C, on the other hand, prevents cell oxidation and wrinkle

You don't need to eat bananas for the potassium! (Although it is present in bananas, potassium is the predominant nutrient among most all fruits and vegetables.)

Bananas are high in sugar, so they should not be eaten if you have blood sugar problems.

Don't eat bananas on an empty stomach; combining them with a bit of protein will help to normalize the insulin response caused by the sugar in the banana.

Green-tipped bananas are better for your health than over-ripe bananas.

Banana is naturally radioactive and in fact babana equivalent dose of radiation is used in measuring radioactivity.

They keep the heart healthy, are good for the gut and keep the bones strong.

Health Benefits

Banana fruit is rich in calories, but very low in fats. The fruit contains good amounts of health benefiting anti-oxidants, minerals, and vitamins.

Banana pulp is composed of soft, easily digestible flesh with simple sugars like fructose and sucrose that when eaten replenishes energy and revitalizes the body instantly; thus, for these qualities, bananas are being used by athletes to get instant energy and as supplement food in the treatment plan for underweight children.

The fruit contains good amount of soluble dietary fibre that helps normal bowel movements; thereby reducing constipation problems.

It contains many health promoting flavonoid poly-phenolic antioxidants such as lutein, zeaxanthin, beta and alpha carotenes in small amounts. These compounds help act as protective scavengers against oxygen-derived free radicals and reactive oxygen species (ROS) that play a role in aging and various disease processes.

Banana is a very good source of vitamin-B6 (pyridoxine), provides about 28 per cent of daily-recommended allowance. Pyridoxine is an important B-complex vitamin that has beneficial role in the treatment of neuritis, anemia, and decreasing homocystine (one of the causative factor for coronary artery disease (CHD) and stroke episodes) levels in the body.

The fruit is also good source of vitamin-C. Consumption of foods rich in vitamin-C helps body develop resistance against infectious agents and scavenge harmful oxygen free radicals.

Fresh bananas provide adequate levels of minerals like copper, magnesium, and manganese.

Fresh banana is a very rich source of potassium. 100 g fruit provides 358 mg potassium. Potassium is an important component of cell and body fluids that helps control heart rate and blood pressure, countering bad effects of sodium.

Once ripened, bananas are very fragile and start decaying in short time. In the field, bananas are usually harvested while they are green as it is easy to transport when the fruits are raw and firm.

Ready to eat bananas should be quite firm, bright yellow in colour and emanate rich fragrance, and the skin should be peeled off easily. Ripened, fresh bananas are nutritionally enriched and sweeter in taste.

Avoid mushy or damaged bananas, as they are un-appealing.

They keep the heart healthy, are good for the gut and keep the bones strong.

Now we have a clear idea about all the nutrition's that a banana can supply to our body. All the minerals and vitamins that banana contains help us to build our body properly and helps our system to fight against the harmful diseases.

Bananas comes with nature gifted protective outer layer of skin, therefore are less likely contaminated by germs and dust.

Just remove the peel and enjoy!

Banana fruit sections are a great addition to the fruit salads.

Fresh "banana-milkshake" with sugar syrup is a delicious drink.

Bananas have also been used in the making of fruit jams.

Banana fritters can be served with ice cream as well.

Banana chips are a snack produced from dehydrated or fried banana or plantain slices.

Mash ripe banana fruits and add to cakes, muffins, bread pudding *etc.*

Plantain is raw unripe banana that is used as vegetable in recipes.

Caution

Banana fruits are sometimes known to cause skin and systemic allergic reactions. In "oral allergy syndrome" which causes itching and swelling in the mouth or throat within hours after ingestion and is related to birch tree and other pollen allergies.

The other type of reaction is related to latex allergies and causes urticaria and potentially serious gastrointestinal symptoms like nausea, vomiting, and diarrhea.

CHERRY

Cherries (*Prunus cerasus*) are certainly one of today's most popular dessert fruits, but they have been recognized for their medicinal purposes since the 1400's. One cup of sweet cherries has just 90 calories and is a good source of fibre and vitamin C. Cherries are packed with antioxidant value and vitamins. They are slightly high on natural sugar content, but they're still very low-carb and are an excellent source of important fibre. Cherries are amazing with feta and greens. With a glycemic index of just 22, one should go for a daily serving of cherries, when they are in season locally. They are sweet, sour and astringent fruits. It cleanses the system and is good for dental problems. It is useful for lack of sleep. Cherries rank among the top 20 foods with the highest concentration of antioxidants. In fact, the standard one-cup serving of cherries has the capacity to carry 4, 873 antioxidants! Antioxidants are substances found in foods that may protect cells from damage caused by unstable molecules, known as free radicals. Cherries are especially rich in a phytochemical called anthocyanin. They also contain melatonin, phenols and quercetin. Melatonin is a harmone that is naturally present in the human body and aids sleep. Like many fruits, raw sweet cherries provide carbohydrates, potassium and vitamin C. They also are a source of cholesterol-lowering phytosterols.

Researchers tracked the participant's sleep habits and after drinking the cherry juice, they found significant improvements in sleep behaviour, most notably longer sleep time, less daytime napping and increased overall sleep efficiency. They attribute the sleep benefits to the melatonin content of the red super fruit - a powerful antioxidant critical for sleep-wake cycle regulation. Each serving of the tart cherry juice concentrate was estimated to contain the equivalent of 90-100 tart cherries, providing a significant level of melatonin in the juice and ultimately in the bodies of the participants (Anonymous, 2012).

Carbohydrates and Sugar: One cup of raw sweet cherries contains 87 calories, most of which are from the 22.1 g carbohydrates per serving. One cup of cherries contains 17.7 g sugar, most in a combination of the 7.4 g fructose and 9.1 g glucose per cup. The remaining sugars are a combination of less than 1 g each of sucrose, maltose and galactose. One cup of cherries contains no starch and 2.9 g fibre, which is just over 10 per cent of the DRI.

Protein-Cherries contain very little protein–one cup contains only 1.46 g. This protein is made up of small amounts of 18 different amino acids, but like all plant foods, cherries do not contain all of the essential amino acids and are not a complete protein.

Fat-Cherries also contain very little fat, with one cup containing only.3 g. This fat is a combination of trace amounts each of saturated, polyunsaturated and mono-unsaturated fats. Cherries contain no cholesterol and, in fact, contain **phytosterols**, which can help create lower cholesterol levels by competing for absorption with cholesterol.

Minerals-Cherries contain significant amounts of some minerals, while supplying very little of others. One cup of cherries has 15.3 per cent of the DRI for potassium and 9.2 per cent of the DRI for copper, but less than 5 per cent of calcium, magnesium, manganese, phosphorus, selenium but have no sodium.

Water-Soluble Vitamins- Cherries also supply varied amounts of water-soluble vitamins. One cup provides 12.9 per cent of the DRI for vitamin C for women and 10.8 per cent for men. Vitamin C acts as an antioxidant and is important for iron absorption. One cup also supplies approximately 5 per cent of pantothenic acid and B6, but contains less than 4 per cent of all other water-soluble vitamins, all of which are B vitamins.

Fat-Soluble Vitamins-Cherries supply very little fat-soluble vitamins. One cup provides 3.2 per cent of the DRI for vitamin K and less than 1 per cent of vitamins D, E and A.

Lutein-Cherries also contain lutein, a pigment in many plants that works as an antioxidant in the body. Lutein is thought to promote eye and cardiovascular health

Cherries and Melatonin

Melatonin is a naturally occurring hormone produced by the pineal gland in the brain. It plays a key role in regulating the body's internal clock and helps determine when we fall asleep and when we wake up. Eating a handful of cherries just before bed is a great way to naturally regulate your sleep cycle. According to a study Cherries grown in sun-rich Central Otago are natural source of melatonin at levels 30 times higher than their northern hemisphere counterparts.

Cherries, Arthritis and Pain Relief: Great news for arthritis sufferers! A bowl full of cherries may help alleviate pain and inflammation associated with arthritis and gout, the most severe form of arthritis. A gout attack occurs when excessive amounts of uric acid (waste product found in the blood) accumulate in the joints, and cause inflammation and pain.

A rich source of antioxidants: Cherries are a rich source of antioxidants and they prevent or repair the damage that is done to your body's cells by free radicals. They are also known to be potent anticancer agents. Apart from this, they relieve arthritis pain which is on the rise during the rainy season. Cherries also relieves stress and is known to prevent premature ageing.

Cherries and Heart Health

Cardiovascular disease- One of the many health benefits of cherries is that they contain powerful antioxidants called **anthocyanins**, which may reduce a person's risk of developing cardiovascular disease.

Just like red wine, anthocyanins give cherries their deep red colour and also protect cells from damage during an interaction with oxygen. This important process also serves to protect the heart and surrounding tissue, inhibit plaque formation and reduce inflammation.

Cherries and Brain Health

Cherries are one of the few foods that contain **melatonin**. In addition to helping regulate sleep patterns, melatonin is an important antioxidant that helps maintain optimum brain functioning and may deter the onset of age-related chronic diseases like Alzheimer's. Research also suggests that the anthocyanins found in cherries further protect neural cells and promote brain health.

Health Benefits

Cherries are one of the very low calorie fruits; yet are rich source of nutrients, vitamins and minerals. Both sweet as well as tart Cherries are packed with numerous health benefiting compounds that are essential for well being.

Cherries are pigment rich fruits. These pigments are in fact polyphenolic flavonoid compounds known as anthocyanin glycosides. Anthocyanins are red, purple or blue pigments found in many fruits and vegetables, especially concentrated in their skin, known to have powerful anti-oxidant properties.

Anthocyanins in the cherries are found to act like anti-inflammatory agents by blocking the actions of **cycloxygenase-1** and 2 enzymes. Thus consumption of cherries has potential health effects against chronic painful episodes such as gout arthritis, fibromyalgia (painful muscle condition) and sports injuries.

Studies also suggest that tart cherries are help body to fight against cancers, aging and neurological diseases and pre-diabetes.

Cherry fruits are very rich in stable anti-oxidant melatonin. Melatonin can cross the blood-brain barrier easily and produces soothing effects on the brain neurons, calming down nervous system irritability which helps relieve neurosis, insomnia and headache conditions.

Cherries are good source of minerals such as potassium, iron, zinc, copper and manganese.

The fruits, especially tart cherries are exceptionally rich in many health promoting flavonoid poly phenolic anti-oxidants such as lutein, zeaxanthin and beta carotene.

Cherries' anti-inflammatory functions, effective in reducing heart disease risk factors by scavenging action against free radicals.

COCONUT

Coconut (*Cocos nucifera*) is highly nutritious and rich in fibre, vitamins, and minerals. It is classified as a "functional food" because it provides many health benefits beyond its nutritional content. Coconut oil is of special interest because it possesses healing properties far beyond that of any other dietary oil and is extensively used in traditional medicine among Asian and Pacific populations. Pacific Islanders consider coconut oil to be the cure for all illness. The coconut palm is so highly valued by them as both a source of food and medicine that it is called "The Tree of Life." For thousands of years coconut products have held a respected and valuable place in local folk medicine. Coconut is used to treat a wide variety of health problems including the following: abscesses, asthma, baldness, bronchitis, bruises, burns, colds, constipation, cough, dropsy, dysentery, earache, fever, flu, gingivitis, gonorrhea, irregular or painful menstruation, jaundice, kidney stones, lice, malnutrition, nausea, rash, scabies, scurvy, skin infections, sore throat, swelling, syphilis, toothache, tuberculosis, tumors, typhoid, ulcers, upset stomach, weakness, and wounds. Young coconuts are good for most people, but are particularly good for carb and mixed types. Mature coconuts, on the other hand, are best for protein types.

Health Benefits

Coconuts provide vitamins, minerals and fibre that prevent diseases like gall bladder stones, and liver diseases, inflammation and skin diseases, to name a few. If you want proteins, coconut milk is a good source. It also provides iron, selenium, sodium, magnesium, phosphorus and potassium; vitamin C, E, B1, B3, B5 and B6. Surprisingly coconut milk is far richer in calcium, than regular milk and supports the immune system, helps in weight loss and is great for digestion less.

Coconuts can add flavour, variety and–best of all–healthy nutrients to your diet.

Coconuts are rich in **lauric acid**, which is known for being antiviral, antibacterial and antifungal, and boosts the immune system.

Fresh coconut juice is one of the highest sources of electrolytes known to man, and can be used to prevent dehydration, for instance in cases of diarrhea or strenuous exercise, instead of a sports drink.

Help you lose weight, or maintain your already good weight.

Reduce the risk of heart disease. Lower your cholesterol, improve conditions in those with diabetes and chronic fatigue, improve Crohn's, IBS, and other digestive disorders

Prevent other disease and routine illness with its powerful antibacterial, antiviral and antifungal agents.

Increase metabolism and promotes healthy thyroid function, boost your daily energy, rejuvenate your skin and prevent wrinkles. Modern medical science is now confirming the use of coconut in treating many of the above conditions and may provide a wide range of health benefits. Some of these are summarized below:

Kills viruses that cause influenza, herpes, measles, hepatitis C, SARS, AIDS, and other illnesses.

Kills bacteria that cause ulcers, throat infections, urinary tract infections, disease and cavities, pneumonia, and gonorrhea, and other diseases.

Kills fungi and yeasts that cause candidiasis, ringworm, athlete's foot, thrush, diaper rash, and other infections.

Expels or kills tapeworms, lice, giardia, and other parasites.

Provides a nutritional source of quick energy.

Boosts energy and endurance, enhancing physical and athletic performance.

Improves digestion and absorption of other nutrients including vitamins, minerals, and amino acids.

Improves insulin secretion and utilization of blood glucose.

Relieves stress on pancreas and enzyme systems of the body.

Reduces symptoms associated with pancreatitis.

Helps relieve symptoms and reduce health risks associated with diabetes.

Reduces problems associated with mal-absorption syndrome and cystic fibrosis.

Improves calcium and magnesium absorption and supports the development of strong bones and teeth.

Helps protect against osteoporosis.

Helps relieve symptoms associated with gallbladder disease.

Relieves symptoms associated with Crohn's disease, ulcerative colitis, and stomach ulcers.

Improves digestion and bowel function.

Relieves pain and irritation caused by hemorrhoids.

Reduces inflammation.

Supports tissue healing and repair.

Supports and aids immune system function.

Helps protect the body from breast, colon, and other cancers.

Is heart healthy; improves cholesterol ratio reducing risk of heart disease.

Protects arteries from injury that causes atherosclerosis and thus protects against heart disease.

Helps prevent periodontal disease and tooth decay.

Functions as a protective antioxidant.

Helps to protect the body from harmful free radicals that promote premature aging and degenerative disease.

Does not deplete the body's antioxidant reserves like other oils do.

Improves utilization of essential fatty acids and protects them from oxidation.

Helps relieve symptoms associated with chronic fatigue syndrome.

Relieves symptoms associated with benign prostatic hyperplasia (prostate enlargement).

Reduces epileptic seizures.

Helps protect against kidney disease and bladder infections.

Dissolves kidney stones.

Helps prevent liver disease.

Is lower in calories than all other fats.

Supports thyroid function.

Promotes loss of excess weight by increasing metabolic rate.

Is utilized by the body to produce energy in preference to being stored as body fat like other dietary fats.

Helps prevent obesity and overweight problems.

Applied topically helps to form a chemical barrier on the skin to ward of infection.

Reduces symptoms associated the psoriasis, eczema, and dermatitis.

Supports the natural chemical balance of the skin.

Softens skin and helps relieve dryness and flaking.

Prevents wrinkles, sagging skin, and age spots.

Promotes healthy looking hair and complexion.

Provides protection from damaging effects of ultraviolet radiation from the sun.

Helps control dandruff.

Does not form harmful by-products when heated to normal cooking temperature like other vegetable oils do.

Has no harmful or discomforting side effects.

Is completely non-toxic to humans.

Coconut water has many health benefits and one of them is losing weight quickly. Thus 'coconut' is packed with anti-aging properties and has the right number of calories to keep you going. The water inside the green coconuts is ones to be had. One cup of this water has around 46 calories compared to 550 calories that coconut milk has. The water contains no fat and consists mostly of natural sugars and minerals. Coconut water is very rich in magnesium and potassium, as well as being very high in fibre.

It is so hydrating because of the electrolyte composition, which is very similar to human plasma. Staying hydrated will help you feel full for longer, which in turn will reduce hunger and sugar cravings. This will effectively lead to you eating less, which means your total calorie consumption will go down. Drinking this water will known to flush out toxins which will help you to lose weight faster and also reduce those cravings and hunger pangs.

So, the next time you feel thirsty or feel a hunger/craving pang, go get this nature's pure drink and sip it. It is sweet, oily fruit. It is considered as an auspicious fruit and is used in almost every religious ritual

Coconut water contains organic compounds that have been known to help.

A Few Benefits as follows

The coconut is known for being a light moisturiser that removes the excess oil from the ***skin.*** You can also add it in your bath too. And, can be used as a conditioner or shampoo for oily hair.

Regular consumption of the coconut water helps in blood circulation and prevents from excessive heat burn problems and also controls acidity.

It is best for acne and blackheads and also a natural toning and cleansing mask for oily skin.

Coconut water is a rich source of natural fibre, natural calcium, magnesium, and natural potassium.

Coconut water is anti-viral, anti-fungal.

Coconut can be used as medicine. It helps in the problems of kidney, reduces the risk of cancer, and diabetes. It also helps in digestion.

Coconut water is a rich source of natural fibre, natural calcium, magnesium, and natural potassium- making it a healthy electrolyte drink.

Coconut Oil

While coconut possesses many health benefits due to its fibre and nutritional content, it's the oil that makes it a truly remarkable food and medicine. Once mistakenly believed to be unhealthy because of its high saturated fat content, it is now known that the fat in coconut oil is a unique and different from most all other fats and possesses many health giving properties. Coconut oil has been described as "the healthiest oil on earth."What makes coconut oil so good? All fats and oils are composed of molecules called fatty acids. There are two methods of classifying fatty- acids. The first, is based on saturation (saturated, mono- unsaturated, and polyunsaturated fats). Another system of classification is based on molecular size or length of the carbon chain within each fatty acid. Fatty acids consist of long chains of carbon atoms with hydrogen atoms attached. In this system you have short-chain fatty acids (SCFA), medium-chain fatty acids (MCFA), and long-chain fatty acids (LCFA). Coconut oil is composed predominately of medium chain fatty acids (MCFA), also known as medium-chain triglycerides (MCT). The saturated fatty acids in coconut oil are predominately medium-chain fatty acids. Both the saturated and unsaturated fat found in meat, milk, eggs, and plants (including most all vegetable oils) are composed of LCFA. MCFA are very different from LCFA. They do not have a negative effect on cholesterol and help to protect against heart disease. MCFA help to lower the risk of both atherosclerosis and heart disease. It is primarily due to the MCFA in coconut oil that makes it so special and so beneficial. There are only a very few good dietary sources of MCFA. By far the best sources are from coconut and palm kernel oils.

Benefits of Coconut Oil

Apart from being good for the skin and hair of a person, coconut oil has been found to be beneficial in case of the following ailments, stress, heart diseases, high cholesterol levels, too much weight, kidney problems, poor digestion, low metabolism, high blood pressure, low immunity, dental problems, diabetes, low bone density, HIV, cancer, premature aging, pancreatitis.

DATE

Mature Dates

Dates (*Phoenix dactylifera*) are even rich in several **vitamins** and minerals. These natural products contain calcium, sulphur, iron, potassium, phosphorous, manganese, copper and magnesium which are advantageous for health. It is said that consumption of one date daily is necessary for a balanced and healthy diet. You can have dates just like that. Or you can add it to those mouth-watering desserts. Or you can use dates to make chutney with either tomatoes or tamarind. You can store it in the refrigerator. It is delicious to have it with snacks. Fresh date is made of soft, easily digestible flesh with simple sugar like fructose and dextrose that when eaten replenishes energy and revitalizes the body instantly. Precisely that's why it is used to break fast. The fruit is rich in dietary fibre, which prevents dietary **LDL cholesterol** absorption. It is also a good bulk laxative. Dates are a good source of Vitamin-A which is known to have antioxidant properties and is essential for vision. Vitamin A also required maintaining healthy mucus membranes and skin. Consumption of natural fruits rich in vitamin A is known to help protect from lung and oral cavity cancers. The fruit is very rich in antioxidant flavonoids such as beta carotene, lutein, and zeaxanthin. These antioxidants have the ability to help protect cells and other structures in the body from oxygen free radicals. Dates are an excellent source of iron.

Wonderfully delicious, dates are one of the most popular fruits with a fabulous list of essential nutrients. Dates are even rich in several vitamins and minerals. These natural products contain calcium, sulphur, iron, potassium, phosphorous, manganese, copper and magnesium which are advantageous for health. It is said that consumption of one date daily is necessary for a balanced and healthy diet. You can have dates just like that. Or you can add it to those mouth-watering desserts. Or you can use dates to make chutney with either tomatoes or tamarind. You can store it in the refrigerator. It is delicious to have it with snacks. Fresh date is made of soft, easily digestible flesh with simple sugar like fructose and dextrose that when eaten replenishes energy and revitalizes the body instantly. Precisely that's why it is used to break fast. The fruit is rich in dietary fibre, which prevents dietary LDL cholesterol absorption. It is also a good bulk laxative. Dates are a good source of Vitamin-A which is known to have antioxidant properties and is essential for vision. Vitamin A also required maintaining healthy mucus membranes and skin. Consumption of natural fruits rich in vitamin A is known to help protect from lung and oral cavity cancers. The fruit is very rich in antioxidant flavonoids such as beta carotene, lutein, and zeaxanthin. These antioxidants have the ability to help protect cells and other structures in the body from oxygen free radicals. Dates are an excellent source of iron.

Health Benefits

The natural sugar found in Dates is the perfect alternative to regular sugar. It is easy to digest and reduce hunger. By its composition dates are real natural multivitamin table, which can be used by children and adults. Anaemia, high blood pressure, cholesterol, cancer there is almost no disease that is not treatable with Dates.

Excellent source of iron: Anaemic patients should eat it more, because they are an excellent source of iron. 3oz of dates contains about 0.35oz of iron, which is about 11 per cent of the recommended daily value. Iron, as part of the haemoglobin in the red blood cells, determines the capacity of oxygen in the blood. Children that are going through puberty and pregnant women have the highest iron requirements.

Vitamins for the eyes: Considering that contain lutein and zeaxanthin, dates are called "vitamins for the eyes". Lutein and zeaxanthin are especially important for the net and the macula of the eye, because they support the epithelial cells for good vision and can prevent damage to the macula, which occurs in old age. So if you want to preserve good vision, you should eat more Dates.

Stops diarrhea: Dates contain potassium, which is useful in stopping diarrhea. In addition, dates help to gut flora as quickly renew. Regular intake of dates makes it easier intestines to create "friendly" bacteria.

Cure for constipation: Just as stopping diarrhea, dates can help food to digest faster and easier and expel from the body. It is necessary to take a couple of dates and in the evening put them in clean water. During the night dates will relist their juice, which is an excellent laxative and stimulate the sluggish bowels. (3.3oz 100g) of dates (10-21 pieces) contains 0.3oz (0.90g) grams of fibre.

Facilitates childbirth: Women who ate before birth dates, gave births more easily and more quickly, than those women who did not use this fruit.

Regulates body weight: Due to the high share of nutrients, dates in humans causes a feeling of satiety and facilitate weight loss. If you take on an empty stomach, it will regulate bowel movements and give necessary sugar. Dates do not have cholesterol, but have a lot of sugar, so with excessive intake you can gain weight.

Lowers cholesterol: Dates have the ability to reduce the level of cholesterol in the body. So this is an excellent food for people with high cholesterol, which causes fatty deposits in the arteries that lead to the formation of blood clots and other heart diseases.

Strengthening heart: People with weak heart can also use this amazing fruit. During the night leave the dates in water. In the morning, remove the seeds from them, grind them in a blender along with the water in which they stood during the night. Take this medicine several times a day in order to strengthen the heart.

Reduce blood pressure: Dates are very suitable food for people who suffer from high blood pressure. They contain very little sodium and rich in potassium. Standard servings of five or six dates gives about 2.8oz (80mg) of magnesium, an essential mineral that contributes to the spread of blood vessels

Protects against stroke: One of the key benefits of this wonderful fruit is its ability to regulate a healthy nervous system, thanks to the rich content of potassium. Studies have shown that a higher intake of potassium 14oz (400 mg) can reduce the risk of stroke by 40 percent.

Brain food: Dates contain phosphorus, and it plays an important role as food for the brain. For this reason it is recommended in diet for people who are engaged in intellectual work.

Energy stimulant: Dates are large energy booster, containing natural sugars such as glucose, sucrose and fructose. The maximum energy output you can get from Dates is created if you mix it with milk.

Increases sexual stamina: A handful of dates soaked in goat milk and let stand overnight. In the morning, they blended with milk, add honey and cardamom (kind of spice) and drink. This syrup can drink both men and women, because it strengthens the body and increases energy levels.

Baby food: Sugar from Dates is a type of sugar that is most easily absorbed and digested, so it is suitable for new –born, dates should be chewed or soaked in water before giving the child, to make them easier to it.

Strengthen tooth enamel: Dates contain fluoride, which is an essential mineral in the fight against tooth decay, because it prevents further tooth decay. Dates can be given to little baby's (healthybeauty365.com/health–beauty/15-amazing-benefits-and-reasons-to-eat-this-fruit).

Dates are also good for treatment of the liver and cleanse the body of toxic substances. Taking juice of this fruit helps in the treatment of sore throat, different types of fever and colds. Dates are an excellent remedy against alcoholic stupor too.

Fresh date is made of soft, easily digestible flesh with simple sugar like fructose and dextrose that when eaten replenishes energy and revitalizes the body instantly. Precisely that's why it is used to break fast.

The fruit is rich in dietary fibre, which prevents dietary LDL cholesterol absorption. It is also a good bulk laxative.

Dates are a good source of Vitamin-A which is known to have antioxidant properties and is essential for vision.

Dates very rich in antioxidant flavonoids such as beta- carotene, lutein, and zeaxanthin. These antioxidants have the ability to help protect cells and other structures in the body from oxygen free radicals. Dates are an excellent source of iron.

Delicious dates are good source of essential nutrients *viz.* vitamins, and minerals, required for normal growth, development and overall well-being.

Fresh dates are made of soft, easily digestible flesh with simple sugars like fructose and dextrose that when eaten replenishes energy and revitalizes the body instantly.

The fibre content helps to protect the colon mucous membrane by decreasing exposure time and as well as binding to cancer causing chemicals in the colon.

They contain many health benefitting flavonoid polyphenolic antioxidants known as tannins.

Zeaxanthin, an important dietary carotenoid selectively absorbed into the retinal macula lutea where it is thought to provide antioxidant and protective light-filtering functions; thus it offers protection against age related macular degeneration, especially in elderly populations.

Dates are an excellent source of iron, contains 0.90 mg/100 g of fruits. Iron, being a component of hemoglobin inside the red blood cells, determines the oxygen carrying capacity of the blood.

They are good in potassium. Potassium is an important component of cell and body fluids that help controlling heart rate and blood pressure; thus offers protection against stroke and coronary heart diseases.

They are also rich in minerals like calcium, manganese, copper, and magnesium.

Further, the fruit has adequate levels of B-complex group of vitamins as well as vitamin K. It contains very good amounts of pyridoxine (vitamin B-6), niacin, pantothenic acid, and riboflavin. These vitamins are acting as cofactors help body metabolize carbohydrates, protein, and fats. Vitamin K is essential for many coagulant factors in the blood as well as in bone metabolism.

FIG

Ripe Fig Fruits

Fig (*Ficus carica*) is naturally rich in many health benefiting phyto-nutrients, anti-oxidants and vitamins. Dried figs in fact are concentrated source of minerals and vitamins. The fully ripe fig has bell or pear shape with succulent flesh. Figs are very good in preventing constipation. Figs promote bowel function as every three grams of this fruit is loaded with five grams of fibre. It also prevents stomach ache and indigestion. For people who suffer from weight related issues figs can be their answer. The fibre in the figs helps in reducing weight when taken regularly. A type of fibre that is found in fig is pectin. Pectin is very good in absorbing cholesterol from the digestive system and carrying them out of the body. So, a regular diet of figs can help you cut down on your cholesterol. Figs are also rich in phenol, omega-3 fatty acids and omega-6 fatty acids. These compounds are natural heart boosters and so figs can contribute towards reducing the risk of coronary heart diseases. Though figs are seasonal fruits, they are available throughout the year in the dried form. They taste more yummy in its dried form and are very nutritious as they contain vitamins A, B, and minerals like phosphorus, calcium, iron and manganese.

It is also believed that the fibre in figs absorb cancer causing substances, thus reducing the risk of various types of cancer. Figs are especially good in preventing colon and post-menopausal breast cancer. People suffering from diabetes should adopt a diet of high fibre. For this figs are especially good. This fruit is rich in potassium that controls which helps in controlling the blood sugar. Fig leaves are also known to have diabetic properties. The leaves effectively reduce the amount of insulin that is required by diabetic patients through injections. Hypertension is caused because of low levels of sodium and high levels of potassium in the body. Since figs are high in potassium and low in sodium it can avoid hypertension. Figs also have a high concentration of calcium and so it strengthens the bones.

However, a high calcium diet can lead to urinary calcium loss and this is avoided by the potassium content of the figs.

Age results in macular degeneration which leads to vision loss. A diet of figs prevents this. Sore throats can be effectively healed with figs as it also has high mucilage content. Figs are also known to boost the health of the liver. Figs are also high in alkaline and so they keep the body balanced because of the acidic diet by regulating the pH of the body. Figs can also heal various respiratory diseases like asthma and whooping cough. They are also good for earache, boils, venereal diseases and boils. Figs are excellent sources of amino acids that increase libido. They can also improve sexual stamina. **Figs** are perfect fertile booster foods which are full of soluble and insoluble fibre. This juicy taste fruit is a treat for your senses. Good for heart health, figs are a great way to rev up your sexual stamina. Figs are excellent sources of amino acids that increase libido. They can also improve sexual stamina.

Health Benefits

This fruit is very nutritious and has many health benefits (Anonymous, 2013).

Pectin, a soluble fibre found in figs is beneficial for the digestive system. This soluble fibre helps in clearing out the accumulating cholesterol in the body.

Hypertension may arise when there's more intake of sodium and low potassium levels in the body. Figs however, are low in sodium and high in potassium and hence, helps in prevention of hypertension.

Dried figs are said to contain omega-3 and omega-6 fatty acids along with phenol. These are useful in preventing coronary heart diseases.

Figs are also a source of calcium, which is good for strengthening of the bones.

Because figs are rich in potassium, it helps in regulating blood sugar. This is beneficial for people who are diabetics. Also, people whose diet is sodium may suffer from urinary calcium loss. The high potassium content in figs, helps in prevention of this.

Eating figs are said to be an effective cure for treating sore throat because of their high mucilage content.

Traditionally, figs have also been used for treating sexual weakness. The remedy requires one to soak about two to three figs in milk overnight and eat the same next morning.

Because figs contain iron, they are useful in treating anemia.

Mash about fresh figs and them on your face and leave for 10 to 15 minutes. This is useful for treating acne and pimples.

It is said that eating figs can relieve fatigue and boost brain power too.

As one grows older, one tends to suffer from macular degeneration i.e. weakness of the vision. Eating figs can help prevent this.

Fig fruit is low in calories. 100 g fresh fruits provide only 74 calories. However they contain health benefiting soluble dietary fibre, minerals, vitamins and pigment anti-oxidants that contribute immensely for optimum health and wellness.

Dried figs are excellent source of vitamins and anti-oxidants and minerals, like **calcium**, copper, potassium, manganese, **iron**, selenium and zinc. 100 g of dried figs contain 640 mg of potassium, 162 mg of calcium, and 2.03 mg of iron. In fact dried fruits are concentrated sources of energy.

Fresh figs, especially black mission, are good in poly-phenolic flavonoid anti-oxidants such as carotenes, lutein, tannins, chlorgenic acidetc. Their anti-oxidant value is comparable to that of apples at 3200 umol/100 g.

Figs contain adequate levels of some of anti-oxidant vitamins such as vitamin A, E, and K. Altogether these phyto-chemical compounds in fig fruit help scavenge harmful oxygen derived free radicals from the body and thereby protect us from cancers, diabetes, degenerative diseases and infections.

The **chlorogenic acid** in these berries help lower blood sugar levels and control blood glucose levels in type-II diabetes mellitus (Adult onset) condition.

Fresh as well as dried figs contain good levels of B-complex group of vitamins such as niacin, pyridoxine, folates and pantothenic acid. These vitamins function as co-factors for metabolism of carbohydrates, proteins and fats.

If you suffer from osteoporosis, include figs in your diet. Its high calcium content promotes bone health.

Caution

Fig leaves and unripened fruit produce white latex which can penetrate the skin causing burning discomfort. Fig latex contains several compounds like furocoumarins, 5-methoxypsoralen (5-MOP) *etc.* which can elicit cell-mediated allergic reactions. If left untreated, there may occur severe allergic eruptions all over the exposed parts. Eating fig fruit may also elicit allergic reactions ranging from vomiting, diarrhea, and itching of skin and mucus membranes in some sensitized individuals. It is therefore people with history of allergy to figs may be advised to avoid eating them.

GUAVA

Ripe Guava Fruit on Tree

Guava (*Psidium guajava*) is another tropical fruit rich in nutrition. With its unique flavour, taste, and health-promoting qualities, the fruit easily fits in new functional foods category, often called "super fruits."A new study led by an Indian origin researcher has found that guavas are the 'ultimate superfood' with the highest concentration of antioxidants that protect against cell damage which age`s skin and can cause cancer. A series of tests conducted on Indian fruits, including Himalayan apples and pomegranates, bananas from the south and grapes from Maharashtra, found that the guava, the poor man's fruit in India, has the highest concentration of antioxidants as compared to all the other fruits. According to scientists from India's National Institute in Hyderabad, the Indian plum, the custard apple and India's beloved mangoes, come after guavas in antioxidant richness. The study found that while there is a presence of antioxidant concentrations of just under 500 milligrams per 100 grams in guavas, 330mg in plums and 135mg in pomegranates, apples have a quarter of the antioxidants in guavas and bananas merely have a tiny fraction with 30 mg per 100 grams. Watermelons and pineapples were found to offer the least protection for the body's fight against free radicals, which can cause cell damage, whereas mangoes, despite a high fructose content, have 170 mg of antioxidants, which is more than three times that of papaya, and grapes were found to be three times more beneficial to the body than oranges.

Guava is a rich source of antioxidants, a rich source of fibre. It's a poor man's fruit because they're quite cheap. A guava a day keeps a doctor away,. The fruit is one of the richest sources of vitamin C. It contains 4 to 10 times more vitamin C than do some citrus fruits. The guava contains very little vitamin A or carotene. However, it is fairly rich in most other mineral nutrients.

Guava is one of the best fruits you can eat if you are trying to lose weight. A medium-sized guava fruit has 8 g of dietary fibre that will help support your weight loss diet. Fibre not only keeps your appetite under control, but also keeps your blood sugar stable by slowing the rate that glucose is absorbed and metabolized. Keeping your blood sugar stable is vital to achieving your weight-loss goals. When you raise your blood-sugar levels, your body releases insulin, which promotes fat storage. You want to prevent insulin release as much as possible if you are trying to lose weight. Its fruit is tasty appetizing, quenches thirst, enhances semen formation and strengthens the heart. It is anthelmintic rich and anti vomiting. It is purgative and cures anthelmin cough. Immature fruits are highly astrigent and not edible. The fruit is best when it is just ripe- that is when the skin has acquired a yellow tinge. Fruits are often plucked right off the tree and eaten just as they are.

Guava leaves have been proven to be effective for increasing platelets for patients with dengue fever, but can also prevent hair loss. Experts say that the leaves can avoid and overcome hair loss if used as a daily hair care. The main reason for this is because they are packed with vitamin B which is the most important vitamin that protects the health and hair fertility.

Health Benefits

The magnesium in guava helps relax the muscles and nerves. So after a hard days work a guava punch is what you need to relax your muscles, combat stress and give your system an energy.

Guavas are beneficial in regulating blood pressure. It is said that one guava contains almost a similar amount of potassium that's present in bananas. Potassium reverses the effects of sodium, thereby regulating the balance of blood pressure. Also, it reduces cholesterol levels in the blood by preventing it from thickening.

Consumption of guavas helps slow down the absorption of sugar in the blood. It is rich in fibre and is helpful for diabetic individuals. Also, studies have shown that a diet that is high in fibre (5.4 gm per 100 gm of fruit) is linked to a lower risk of developing Type 2 diabetes. Individuals suffering from constipation problems, too, can benefit from the high fibre content.

Did you know that guavas contain four times more Vitamin C than oranges? Vitamin C contains antioxidant properties that protect cells from the damage of free radicals and is useful in lowering the risk of cancer.

Even though guavas don't contain iodine, they are still beneficial in promoting healthy thyroid function, because it contains copper, which aids the production and absorption of hormones.

Guavas are a good source of manganese that acts as an enzyme activator utilizing **nutrients** like thiamine, biotin and ascorbic acid.

If you want to optimize your brain function, turn to this fruit. Guavas are rich in the B group of vitamins. Niacin, better known as Vitamin B_3, promotes blood circulation, thereby stimulating brain function. Vitamin $B_{6,}$ that is pyridoxine, helps in brain and nerve function.

Women with fertility problems can eat guavas as they contain a good amounts of folate, which contain fertility-promoting properties.

Eye problems can be kept at bay as guavas contain an abundance of Vitamin A that helps in improving vision.

Guava is good for the skin, too. Because of its Vitamin E content, astringent properties and antioxidants, the skin is nourished. Skin ailments like scurvy can be dealt with due to the high Vitamin C content in guavas.

Pink guavas are said to contain twice the amount of **lycopene** present in tomatoes. Lycopene protects the skin from being damaged by UV rays and also works against prostate cancer.

Even the leaves of guava have medicinal properties. The juice of the leaves is said to provide relief from cold and cough by reducing the formation of mucus, disinfecting the respiratory tract and preventing bacterial activity in the throat due to its astringent properties.

Guavas are low in calories and fats but contain several vital vitamins, minerals, and antioxidant poly-phenolic and flavonoid compounds that play pivotal role in prevention of cancers, anti-aging, immune-booster *etc.*

The fruit is very rich source of soluble dietary fibre, which makes it a good bulk laxative. The fibre content helps protect the colon mucous membrane by decreasing exposure time to toxins as well as binding to cancer causing chemicals in the colon. Due to the rich fibre content and low polycaemic index, guava prevents the development of diabetes.

Fresh guava-fruit is an excellent source of antioxidant **vitamin-C;** provides more than **three times the DRI** (daily-recommended intake). Outer thick rind contains exceptionally higher levels of vitamin C than central pulp.

Regular consumption of fruits rich in vitamin C helps body develop resistance against infectious agents and scavenge cancer causing harmful free radicals from the body. Further, the vitamin is required for collagen synthesis in the body. Collagen is the main structural protein in the body required for maintaining the integrity of blood vessels, skin, organs, and bones.

The fruit is very good source of **Vitamin-A** and flavonoids like beta-carotene, **lycopene**, lutein and cryptoxanthin. The compounds are known to have antioxidant properties and are essential for optimum health. Further, vitamin-A is also required for maintaining healthy mucus membranes and skin. Consumption of natural fruits rich in carotene is known to protect from lung and oral cavity cancers. Vitamin A helps improve the vision. It also helps slow down the appearance of cataract and macular degeneration.

Guava seeds are excellent laxative and helps in the formation of healthy bowl movements.

Lycopene in pink guavas prevents skin damage from UV rays and offers protection from prostate cancer.

Fresh fruit is a very rich source of potassium; contains more potassium than banana per 100 g of fruit weight.

Its leaves have a powerful anti-inflammatory and anti-bacterial ability, which helps fight infection and kills germs. Guava leaves work as a fantastic home remedy for toothache, swollen gums and even mouth ulcers,

Since it contains 80 per cent water, it helps to keep you hydrated.

Fig is also a moderate source of B-complex vitamins such as pantothenic acid, niacin, vitamin-B6 (pyridoxine), vitamin E and K, and minerals like magnesium, copper, and manganese.

CITRUS FRUITS

Citrus fruits, as such, have long been valued for their wholesome nutritious and antioxidant properties. Citrus fruits like oranges, tangerines, lemons and grape fruit contain anti- inflammatory flavonoids, which help reduce tumour inflammation. It is scientifically established that citrus fruits, especially oranges, by virtue of their richness in vitamins and minerals, have many proven health benefits. Moreover, it is now beginning to be appreciated that the other biologically active, non-nutrient compounds found in citrus fruits such as phyto-chemical antioxidants, soluble and insoluble dietary fibre have been found to be helpful in reduction in the risk for cancers, many chronic diseases like arthritis, and from obesity and coronary heart diseases. These fruits should be consumed in their natural form as salads and fruit chaat *etc.*

ORANGES

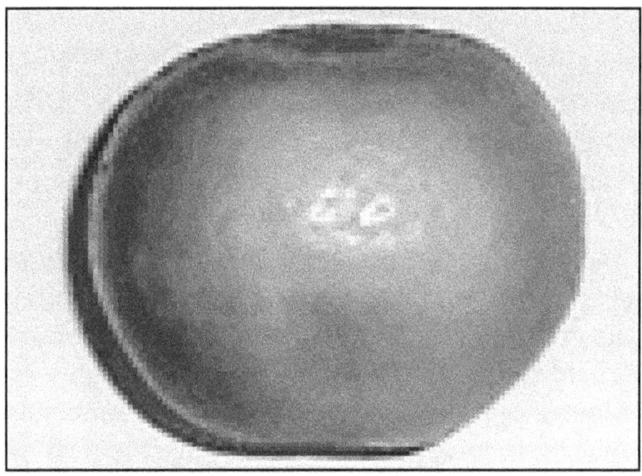

Nagpur Orange

Oranges (*Citrus reticulata*) are delicious and juicy orange fruit, contains an impressive list of essential nutrients, vitamins, minerals for normal growth and development and overall well-being. If you want to avoid all the extra calorie, it's better to have oranges just like that. So, now you can carry an orange in your bag. Rich in Beta-carotene, oranges can protect our cells. Orange peels are very good for skin too. The fibres in oranges are good for the digestive system and relieving Fibres also reduce cholesterol which is helpful in keeping you in good health and fibres are also good in keeping the blood sugar levels in control.

Oranges are also rich in calcium, associated with a healthy bone structure and teeth. Vitamin C stimulates the production of white cells in our body, thus improving the immune system. So, eating oranges regularly will keep you away from viral infections. Regular consumption of oranges significantly reduce the risk of developing kidney stones. They're mainly available in two categories - sweet and bitter, with the former being the type most commonly consumed. Generally an orange should have smoothly textured skin and be firm and heavy for its size. These will have higher juice content than those that are either spongy or lighter in weight. **Oranges** are extremely high in antioxidants. They help brighten the skin. Orange, a citrus fruit, contains vitamin C, that helps prevent conditions like skin inflammation. One thing known about vitamin C is that, it helps improve the absorption of Vitamin E and iron in the body, and this in turn gives you a radiant complexion. You might have noticed, there are many skin care products that have orange extracts in them. They are an excellent source of vitamin C. One orange supplies 116.2 per cent of the daily value for vitamin C. Good intake of vitamin C is associated with a reduced risk of colon cancer as it helps to get of free radicals that cause damage to our DNA. Vitamin C, which is also vital for the proper function of a healthy immune system, is good for preventing colds and preventing recurrent ear infections. One glass of orange juice a day will give you a good dose of vitamin D. Anti-oxidants in oranges help protect skin from free radical damage known to cause signs of aging. An orange a day can help you look young even at 50! Oranges, being rich in Vitamins B6, help support the production of hemoglobin and also help keep blood pressure under check due to the presence of magnesium.

According to a study by US and Canadian researchers, a class of compounds found in citrus fruit peels called **Polymethoxylated Flavones** (PMFs) have the potential to lower cholesterol more effectively than some prescription drugs without side effects. Fibre in oranges help by keeping blood sugar levels under control thereby making oranges a healthy snack for people with diabetes. The natural fruit sugar in oranges, fructose, can help keep blood sugar levels from rising too high after eating. Consuming foods rich in beta-cryptoxanthin, an orange-red carotenoid found in highest amounts in oranges, corn and pumpkin may significantly lower one's risk of developing lung cancer. Oranges like most citrus fruits, produce more juice when warmer - juice them when they are at room temperature. Rolling the orange under the palm of your hand on a flat surface will also help to extract more juice. Vitamin C gets destroyed fast when exposed to air, so eat an orange quickly once cut up. Oranges are rich in vitamin C that improves skin texture. Like apple, they too contains collagen that slows skin aging process. Rub the insides of orange on your skin to tighten the skin. Oranges can be dried and powdered and used as a natural scrub. Like lemon, oranges too help clear skin blemishes (Anonymous, 2013).

Tangerines may be smaller than oranges, but they are several times higher in vitamins, minerals, and antioxidants such as vitamins C and A, calcium, magnesium and selenium, and beta-carotene and luteins. Tangerines contain anti-viral and anti-cancer properties and are particularly beneficial for colds, fever, flu, arthritis, epstein barr, shingles, constipation, age related macular degeneration, bone repair, neurodegenerative diseases, heart disease, and breast, colon, lung, stomach, ovarian, and prostate cancer. Tangerines are an excellent food to help boost the immune system and keep it functioning at optimum levels. Their high fibre content also helps to lower cholesterol and protect the heart and cardiovascular system from illness and disease. Tangerines are high in pectin which can significantly improve digestion and elimination. Tangerines are also a great weight-loss food that can help to keep you full, satisfied, and full of energy. They are also a perfect food to eat after working out as they can deeply hydrate and replenish the body on a cellular level. Tangerines are typically easy to peel and are sweeter and juicier than oranges which makes them easy to eat out of hand and enjoyable for adults and children alike. Tangerine peels can also be added to water and/or tea to provide additional anti-bacterial and medicinal benefits.

Health Benefits

Oranges have lots of health benefits.

Boosts your immunity: Oranges are great source to meet your daily requirement of Vitamin C. This vital nutrient helps improve your immunity, keeping you free from diseases and infections.

Good for your skin: As we grow older, our skin goes for a rapid change. Oranges are packed with antioxidants and Vitamin C which slows down the process and makes you look younger than your age, therefore, grab that orange and enjoy the happiness of having glowing skin.

Great for your eyes: Oranges are rich in nutrients like Vitamin A, Vitamin C and potassium which are great for your eyes. So, if you want your vision to be fine, eat an orange every day.

Keeps you free from stomach ulcers: Oranges are a good source of fibre which helps keep your stomach healthy. A diet rich in fibre will ensure that you are not affected with ailments like stomach ulcers and constipation. So, oranges should figure in your diet.

Nutrients in oranges are plentiful and diverse. The fruit is low in calories, contains no saturated fats or cholesterol, but is rich in dietary fibre, **pectin**, which is very effective in persons with excess body weight. Pectin, by its action as bulk laxative, helps to protect the mucous membrane of the colon by decreasing its exposure time to toxic substances as well as by binding to cancer causing chemicals in the colon. Pectin has also been shown to reduce blood cholesterol levels by decreasing its re-absorption in the colon by binding to **bile acids** in the colon.

Oranges, like other citrus fruits, is an excellent source of **vitamin C.** Vitamin C is a powerful natural antioxidant. Consumption of foods rich in vitamin C helps body develop resistance against infectious agents and scavenge harmful, pro-inflammatory free radicals from the blood. One orange supplies 116.2 per cent of the daily value for vitamin C. Vitamin C, which is also vital for the proper function of a healthy immune system, is good for preventing colds and preventing recurrent ear infections.

Orange fruit contains a variety of phytochemicals. **Hesperetin** and **Narigenin** are flavonoids found in citrus fruits. Naringenin is found to have a bio-active effect on human health as antioxidant, free radical scavenger, anti-inflammatory, and immune system modulator.

Oranges also contain very good levels of vitamin A, and other flavonoid antioxidants such as alpha and beta-carotenes, beta-cryptoxanthin, zea-xanthin and lutein. These compounds are known to have antioxidant properties.

It is also a very good source of B-complex vitamins such as thiamin, pyridoxine, and folates. These vitamins are essential in the sense that body requires them from external sources to replenish.

Orange fruit also contains a very good amount of minerals like potassium and calcium. Potassium is an important component of cell and body fluids that helps control heart rate and blood pressure through countering sodium actions.

Anti-oxidants in oranges help protect skin from free radical damage known to cause signs of aging. An orange a day can help you look young even at 50!

Oranges, being rich in Vitamins B6, help support the production of hemoglobin and also help keep blood pressure under check due to the presence of magnesium.

According to a study by US and Canadian researchers, a class of compounds found in citrus fruit peels called Polymethoxylated Flavones (PMFs) have the potential to lower cholesterol more effectively than some prescription drugs without side effects.

Fibre in oranges help by keeping blood sugar levels under control thereby making oranges a healthy snack for people with diabetes. The natural fruit sugar in oranges, fructose, can help keep blood sugar levels from rising too high after eating.

Consuming foods rich in beta-cryptoxanthin, an orange-red carotenoid found in highest amounts in oranges, corn and pumpkin may significantly lower one's risk of developing lung cancer.

Oranges like most citrus fruits, produce more juice when warmer - juice them when they are at room temperature. Rolling the orange under the palm of your hand on a flat surface will also help to extract more juice. Vitamin C gets destroyed fast when exposed to air, so eat an orange quickly once cut up.

Caution

Insecticide sprays are widely applied over orange crops. Therefore, it is recommended to wash the fruits in cold running water before use. Organic orange fruits are devoid of these chemicals and are best suited for zest preparation.

GRAPE FRUIT

Grape fruit (*Citrus paradise*) Refreshing and delicious grape fruit is rich in phytonutrients such as vitamin A, beta-carotene, and lycopene. It is revered as fruit of "paradise" for its unique health-promoting as well as disease-healing properties, especially among health-conscious, fitness freaks. Grape fruit is called so because it grows in clusters like grapes. In Latin, grape fruits are called citrus paradisi meaning the citrus from paradise. One of the most popular weight loss diets is the Grapefruit diet. This diet involves drinking of at least three glasses of grape fruit juice everyday, as it helps burn calories. The grape fruit looks similar to an orange but is larger in size and belongs to the citrus family of fruits. However, the flesh of the grape fruit is usually red or pink in colour. Grape fruits taste sour and bitter but have a lot of health benefits. Though grape fruits are sour and acidic in taste, they are known to produce an alkaline reaction after digestion. This helps treat acidity which can affect the digestive system. Grapefruit is rich in vitamin C which helps maintain the elasticity of the arteries. It also strengthens them. Pectin, a compound, is known to help reduce the accumulation of arterial deposits. Consuming grape fruit on a regular basis will help you boost your immune system. Grape fruits are highly beneficial for diabetics as they help in bringing down blood sugar levels in the body.

Eating grape fruit before bedtime gives good sleep thus, reducing symptoms of insomnia. Pregnant women can reduce water retention in their bodies and swelling in their legs by consuming grape fruit. Grape fruits contain fat-burning enzymes that can promote weight loss. That's because they prevents sugars and starch getting stored in the body. Drinking fresh grape fruit juice can help relieve a sore throat.

Grapefruit contains anti-oxidant properties that helps prevent cancer.

Health Benefits

Delicious, grape fruit is very low in calories, consists of just 42 calories per 100 g. Nonetheless, it is rich in dietary insoluble fibre **pectin**, which by acting as bulk laxative helps to protect the colon mucous membrane by decreasing exposure time to toxic substances in the colon as well as binding to cancer causing chemicals in the colon.

Pectin has also been shown to reduce blood cholesterol levels by decreasing re-absorption of cholesterol binding bile acids in the colon.

The fruit contains very good levels of vitamin-A and flavonoid antioxidants such as **naringenin**, beta-carotene, xanthin and lutein.

Vitamin A also required maintaining healthy mucus membranes and skin. Consumption of natural fruits rich in vitamin-A and flavonoids helps to protect from lung and oral cavity cancers.

It is a good source of antioxidant vitamin-C; Vitamin-C is a powerful natural anti-oxidant and helps body develop resistance against infectious agents and scavenge harmful free radicals; also is required for the maintenance of normal connective tissue as well for wound healing. It also facilitates dietary iron absorption from the intestine.

Fresh fruit is very rich in potassium. Potassium is an important component of cell and body fluids, helps controlling heart rate and blood pressure through countering sodium effects.

Red varieties of grape fruits are especially rich in the most powerful flavonoid antioxidant, **lycopene**.

It contains moderate levels of B-complex group of vitamins such as folates, riboflavin, pyridoxine, and thiamin in addition to some resourceful minerals such as iron, calcium, copper, and phosphorus.

LEMON

Citrus Lemon Fruit

Lemon (*Citrus lemon*) juice is an important ingredient in most Indian recipes. This is also a fruit of all seasons and almost always finds place on our kitchen shelf or refrigerator. With its vitamin C content, its juice will keep your skin beautiful. A glass of warm water with a tsp of honey and a dash of lemon juice on an empty stomach every morning is a great skin cleanser. With its astringent properties, it can be used to lighten the skin tone and also diminish acne scars. Rub the inside of a lemon peel on your elbow remove dark spots. Mix lemon and honey and use it as a natural bleach on your skin. Lemons are an excellent source of vitamin C. Having a lemon squeezed in warm water first thing every morning will cleanse your digestive system. It also increases the levels of immunity thus cutting the chances of infections. Because it has anti-bacterial and anti-viral properties, you can battle flu and colds by downing a glass of lemon juice. It is also known to purify blood, hence getting rid of toxins from the body. Lemon is also used in aromatherapy. Since it smells so 'clean' it induces the feeling of freshness and purges the feelings of inadequacy thus building confidence. The essential oil is known to increase the level of concentration. To get rid of a tan apply lemon on hands and feet. Section hair and rub a lemon on hair and voila, you get natural-looking highlights like you just returned from a day at the beach. Lemon juice has more benefits that you'd ever imagine. Lemons have five per cent of citric acid, which gives it its unique taste. Rich in vitamin C, it also has vitamin B, calcium, phosphorus, magnesium, proteins and carbohydrates. Citrus fruits, are valued for their wholesome nutritious and antioxidant properties. Citrus fruits, especially lemons and oranges, by virtue of their richness in vitamins and minerals, have many proven health benefits. Other biologically active, non-nutrient compounds found in citrus fruits such as phytochemical antioxidants, soluble and insoluble dietary fibre have been found to be helpful in reduction in the risk for cancers, many chronic diseases like arthritis, and from obesity and coronary heart diseases.

Lemon juice is an important ingredient in most Indian recipes. This is also a fruit of all seasons and almost always finds place on your kitchen shelf or refrigerator. With its vitamin C content, its juice will keep your skin beautiful. A glass of warm water with a tsp of honey and a dash of lemon juice on an empty stomach every morning is a great skin cleanser. With its astringent properties, it can be used to lighten the skin tone and also diminish acne scars. Mix lemon and honey and use it as a natural bleach on your skin (Anonymous, 2013). Lemons are believed to stimulate and purify the liver. It also helps digestive acids with digestion and elimination.

Lemons are known for their 'skin and hair' cleansing properties. Dandruff can be prevented by rubbing lemon on the scalp. Adding two teaspoons of lemon juice in your bath not only deodourises the body but also keeps it fresh. Lemon rinds, when rubbed on scars, acne or dark spots, helps heal them. Rub lemon slices on rough areas of heels and elbows to soften them. Lemons are an excellent source **of vitamin C**. It also increases the levels of **immunity** thus cutting the chances of infections. Because it has antibacterial and anti-viral properties, you can battle flu and colds by downing a glass of lemon juice. It is also known to purify blood, hence getting rid of toxins from the body. Since it smells so 'clean' it induces the feeling of freshness and purges the feelings of inadequacy thus building confidence. The essential oil is known to increase the level of concentration. To get rid of a tan apply lemon on hands and feet. Section hair and rub a lemon on hair and voila, you get natural-looking highlights like you just returned from a day at the beach. It is really soothing. Daily consumption of lemon water has a number of health benefits. Citrus fruits like lemon are high in vitamin C and ascorbic acid. Vitamin C can help fight colds and the ascorbic acid helps iron absorption which also plays a role in immune function.

Good for stomach: Lemon is really helpful when it comes to digestion problems. Due to the digestive qualities of lemon juice, symptoms of indigestion such as heartburn, bloating and belching are relieved. Traditionally grandmoms always recommended a glass of lemon juice when somebody suffered from fever or stomach upset.

Glowing skin: Lemon, being a natural antiseptic medicine, can participate to cure problems related to skin. Lemon is a **vitamin C** rich citrus fruit that rejuvenates skin and results in a glowing skin. Lemon also acts as an anti-aging remedy.

Lose the flab: In India, many people drink lemon water with a dash of honey early in the morning to lose weight.

Cures throat infections: Lemon is an excellent fruit that aids in fighting problems related to throat infections, sore throat as it has an antibacterial property.

Controls high blood pressure: Lemon water works wonders for people having heart problem owing to its high potassium content. It controls high blood pressure, dizziness, nausea as well as provides relaxation to mind and body.

Health Benefits

Lemons are packed with numerous health benefiting nutrients. The fruit is low in calories, 29 calories per 100 g, one of the lowest among citrus group.

It contains no saturated fats or cholesterol, but is rich in dietary fibre.

Its acidic taste is due to **citric acid**. Citric acid is present up to 8 per cent in its juice. Citric acid is a natural preservative, aids digestion. Studies found that citric acid help dissolve kidney stones.

Lemons, like other citrus fruits, are excellent source of **ascorbic acid** Ascorbic acid or vitamin-C is a powerful water soluble natural anti-oxidant. This vitamin is helpful in preventing scurvy. Besides, consumption of foods rich in vitamin-C helps body develop resistance against infectious agents and scavenge harmful, pro-inflammatory free radicals from the blood.

Lemons like oranges contain a variety of phytochemicals. **Hesperetin** and **naringenin** are flavonoid glycosides commonly found in citrus fruits. Naringenin is found to have a bio-active effect on human health as antioxidant, free radical scavenger, anti-inflammatory, and immune system modulator. This substance has also been shown to reduce oxidant injury to DNA in the cells in-vitro studies.

They also a good source of B-complex vitamins such as pantothenic acid, pyridoxine, and folates. These vitamins are essential in the sense that body requires them from external sources to replenish.

They contain healthy amount of minerals like iron, copper, potassium, and calcium. Potassium in an important component of cell and body fluids helps control heart rate and blood pressure.

Suffer from indigestion? Mix a few drops of lemon juice with warm water and sip on it. This is useful for treating nausea, heartburn, diarrhea, bloating and burping.

Since lemon is a natural antiseptic, it is great to cure skin problems. It clears your skin and also acts as an anti-ageing agent by eliminating wrinkles and blackheads.

Have a toothache? Apply fresh lemon juice where it hurts. If you have bleeding gums, applying lemon juice can curb the bleeding and stop bad breath.

A sore throat can be cured by gargling with lemon juice and water regularly.

Nimbu paani is high in potassium, which controls high blood pressure, dizziness, nausea and reduces stress. It also cures respiratory disorders like breathing problems and asthma.

Lemon water

It improves the digestion.

Boosts immunity and energy.

Keeps the body hydrate.

Gives you rejuvenated and healthy skin.

Reduces inflammation and can aid in the weight loss.

Alkalizes the body and has cleansing powers.

Has anti-bacterial and anti-viral powers.

Reduces phlegm and mucus and can refresh your breath.

Can boost the power of your brain.

Effective against cancer as it neutralizes the acids in the body because cancer cells need acid environment to survive, develop and spread.

Caution

Lemon juice is very low in PH, about 2.0. Its sour taste sometimes causes burning sensation if encounters with mouth, tongue, and lip ulcers. In addition, if taken large amounts may exacerbate acid-peptic disease and stomach ulcer conditions.

PAPAYA

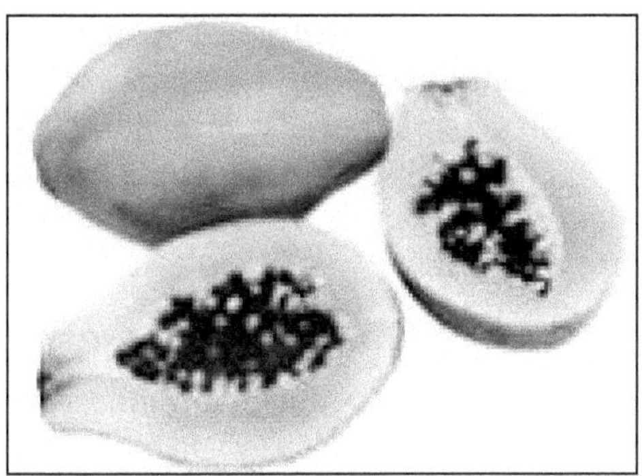

Papaya (*Carica papaya*). The fruit is one of the favourite of fruit lovers for its nutritional, digestive, and medicinal properties. It is a very wholesome fruit. As a source of vitamin A it is unrivalled by any other fruit except perhaps the mango. It is a cheap fruit, available in all seasons, throughout the country. With its deep yellow colour, the ripe papaya fruit contains large quantities of a substance called carotene. This pigment is similar to that found in carrots, beetroot, dark green leafy vegetables like drumstick leaves, palak, curry leaves *etc.* Carotene in food is converted into vitamin A in our body. Therefore, regular consumption of papaya will ensure a good supply of vitamin A and C, which are both essential for good health. Thus eating fruit will help prevent blindness caused by vitamin A deficiency. Papaya has more carotene compared to other fruits such as apples, guavas, sitaphal, plantains *etc.* Hence it gives the maximum nutritional returns for the money.

It is cultivated for its fruits and latex **papain**, an enzyme that is used in food industry. Sensual, earthy and almost wildly exotic in an olfactory sense, the humble papaya is one of earth's finest gifts to mankind. Here are a few reasons why you should include this fruit in your diet: Low in calories, high on nutrition, it's an excellent choice for those on a diet. The peel of the papaya helps in healing skin wounds and injuries. The fruit has anti-inflammatory and anti-cancerous properties. People affected with osteoporosis, arthritis and oedema should consume it regularly. The fruit is good for hair growth and is known to check the growth of dandruff. Papaya is one of the most trusted detox solutions. If taken in moderation, the fruit can help get rid of toxins from the body. Those suffering from constipation and indigestion can rely on the papaya for relief. The benefits of this fruit on skin have perhaps been talked about since the time of our ancestors Papaya is rich in **antioxidants** and contain a special enzyme called **papain** that can kill dead cells and cure skin impurities. Papain is present in all parts of the tree and fruit. It is excellent aid to digestion. It is an enzyme which helps to digest the protein in food. Hence it is used in various medicinal preparations. A glass of papaya milk or just applying the flesh of papaya on your skin can do wonders to your skin. **Papaya** is considered a wonderful wholesome fruit as it meets with some of our most essential vitamins, minerals and protein needs. Both ripe and raw papaya can be used to rejuvenate the skin. Ripe papaya is excellent for exfoliation of the skin and brings about newer looking skin and generally **suits** all skin types.

Papayas are good cleansers. It has Enzyme Papain which removes blemishes and rejuvenates an undernourished skin. Mash papayas and mix some honey to make a pack which you can smear all over your hands, neck and face. Wash after 15-20 minutes for a blemish free skin. Rich in vitamin A and enzymes, papayas act as a great exfoliator. Mashed papaya, when applied to the face, helps get rid of dead skin and gives your face that glow.

Health Benefits

Papaya is a very diverse fruit that supplements different aspects of health. This fruitful fruit improves immunity, builds energy and has anti-inflammatory benefits.

The humble **papaya** is one of earth's finest gifts to mankind. Here are a few reasons why you should include this fruit in your diet:

Low in calories, high on nutrition, it's an excellent choice for those on a **diet.**

The peel of the papaya helps in healing skin wounds and injuries.

The fruit has anti-inflammatory and anti-cancerous properties.

People affected with osteoporosis, arthritis and oedema should consume it regularly.

The fruit is good for hair growth and is known to check the growth of dandruff.

Papaya is one of the most trusted detox solutions. If taken in moderation, the fruit can help get rid of toxins from the body.

Those suffering from **constipation** and indigestion can rely on the papaya for relief.

The fruit is very low in calories (just 39 cal/100 g) and contains no cholesterol; but is a rich source of phyto-nutrients, minerals, and vitamins.

Papayas contain soft, easily digestible pulp/flesh with good amount of soluble dietary fibre that helps to have normal bowel movements; thereby reducing constipation.

Fresh, ripe fruit is one of the fruit with highest **vitamin-C** content.

It is also an excellent source of **Vitamin-A**and flavonoids like beta -carotenes, lutein, zeaxanthin and cryptoxanthins.

Papaya fruit is also rich in many essential B-complex vitamins such as Folic acid, pyridoxine riboflavin, and thiamin. These vitamins are essential in the sense that body requires them from external sources to replenish and play vital role in metabolism.

Fresh papaya also contains good amount of potassium and calcium. Potassium is an important component of cell and body fluids and helps controlling heart rate and blood pressure countering effects of sodium.

Papaya has been proven natural remedy for many ailments. In traditional medicine, papaya seeds are anti-inflammatory, anti-parasitic, and analgesic, and they are used to treat stomachache and ringworm infections.

Pieces of papaya laid on wounds and surgical incisions are reported to speed up their healing.

Papaya is best muscle building food, as it Contains papain which breaks dietary protein into easily absorbable compounds.

Beauty Uses

Since the fruit is a good source of Vitamin A and Papain which is a kind of protein, it helps in removing dead skin cells, along with breaking down the inactive proteins.

If you apply a finely grounded paste of raw papaya on your face, it will help reduce pimples and blemishes!

Mashed papaya can be used for treating the sore and cracked heels.

The peel (skin) of the papaya can not only be used on the face, but it should also be used for whitening the skin on your legs.

Using papaya on your face regularly will reduce the signs of ageing.

If you have rough and dry skin, mash a papaya with honey and apply this mixture for hydration of the skin.

Papaya also helps in controlling dandruff. Frequent use of its paste on your hair will improve its texture (Purvaja Sawant, 2013).

Green Papaya has many Nutritional Benefits

Since enzyme levels decline as the fruit ripens, raw papaya is picked when it is still green to retain all of its natural enzymatic qualities.

Healing problematic skin conditions and injuries: Fresh papayas possess dead cell dissolving ability that gives you a perfectly glowing skin. It is protective against skin infections and wounds too. This nutritional benefit of papaya is used by all beauty product companies (Sinha, 2014).

Cure for menstrual pain: The nutritional benefits of papaya are more useful for women, as papaya leaves also works as a cure for menstrual pain. You can take papaya leaf, tamarind and salt along with water that is helpful in frequent pain in women menstruation cycle.

Controls bowel movements: Papaya and its seeds possess anti-amoebic and anti-parasitic characters which controls the bowel movements. It cures diseases like indigestion, constipation, acid reflux, heart burn, irritable bowel syndrome, stomach ulcers and gastric problems also.

Protects from heart diseases: It controls flow of blood maintain the proper blood pressure. It regulates harmful sodium effects inside the body. So it protects you from heart diseases. This is why nutritional benefits of papaya do wonders for heart patients.

Burns calories and extra fat: Papayas contains Vitamin C, E and A, folate, it also gives only 39 calories per a 100 gram. Presence of antioxidants burns your calorie down and extra fat deposits. So you can have a healthy breakfast with papaya. Papaya's nutrition benefits are amazing, you can eat it as in salads, juice. The cocktails of papaya have lot of health benefits.

The budding leaves of papaya tree cure dengue fever and helps in removing excess toxins from the body. The intake of its juice from crushed leaves can helpin rising the platelet-count (Anonymous, 2015).

Caution

Papayas contain white milk like latex substance, which can cause irritation to skin and provoke allergic reaction in some sensitized persons. Ripe papaya fruit can be safely used by pregnant women. Unripe green papaya should be avoided in pregnant women as it contains lot of papain, a proteolytic enzyme that used commercially to tenderize meat. Unripe papaya seeds (in very small quantities), latex, and leaves also contain **carpaine,** an alkaloid which could be dangerous when eaten in high doses. Unripe papaya, however, used safely as cooked vegetable. **Carpaine** in large quantities, is said to lower the pulse rate and depress the nervous system. Fortunately, the fleshy part of the fruit is completely free from this toxic substance. Hence, the delicious fruit can be safely eaten, once the seeds are removed.

GRAPE

Grape (*Vitis vinifera*) Grapes, "queen of the fruits" are storehouse of numerous health promoting phyto-nutrients such as poly-phenolic antioxidants, vitamins and minerals. So, include them in your regular diet, be it in the form of fresh fruits, juice or in salads! Grapes contains nutrients, antioxidants, minerals and vitamins

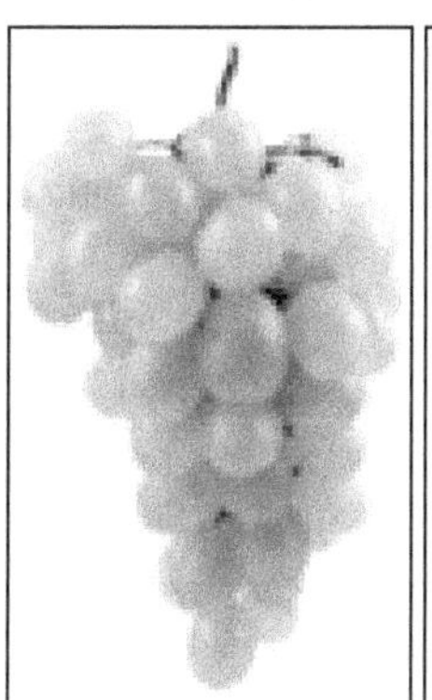

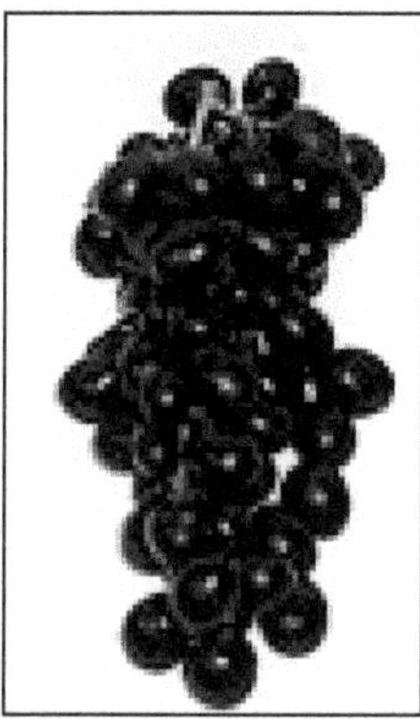

Grapes Green, Red and Black Varieties

Health Benefits

Power Up Your Weight Loss

Protect Your Heart

Mop Up Brain Damaging Plaques

Improve Brain Power

Protect Your Skin from Cancer

Protect Your Body from Radiation

Turn On Your Longevity Gene

Fight Diabetes

Turn Down Inflammation

Supports Muscle Recovery

Studies say that **grapes** help relieve migraine. Either eat them naturally or as ripe grape juice as soon as you begin your day.

Grapes are also great for brain health and delay the onset of neurodegenerative **diseases.**

Applying grape seed extract on your skin protects it from ultraviolet radiation by acting as a sunscreen. It also helps heal sunburned skin.

Grapes are also good for your eyes because they are rich in Lutein and Zeaxanthin - known for maintaining good eye sight.

A compound called **Pterostilbene,** present in grapes, helps lower cholesterol levels.

Asthmatics should consume grapes often because of the high assimilation power of the fruit, which helps to increase the level of moisture present in your lungs.

Grapes help in treating constipation since they are rich in sugar, organic acid and polyose (Anonymous, 2014).

Grapes are rich in polyphenolic phytochemical compound **resveratrol.** Resveratrol is one of powerful anti-oxidant, which has been found to play protective function against cancers of colon and prostate, coronary heart disease (CHD), degenerative nerve disease, Alzheimer's disease and viral/fungal infections.

Resveratrol reduces stroke risk by altering the molecular mechanisms in the blood vessels. It does so firstly by reducing susceptibility of blood vessels damage through decreased activity of angiotensin (a systemic hormone causing blood vessel constriction that would otherwise elevate blood pressure) and secondly, through increased production of the vasodilator substance, nitric oxide (a beneficial compound that causes relaxation of blood vessels).

Anthocyanins are another class of polyphenolic anti-oxidants present abundantly in the red grapes. These phyto-chemicals have been found to have anti-allergic, anti-inflammatory, anti-microbial, as well as anti-cancer activities.

Catechins, a type of flavonoid tannin group of anti-oxidants found in white/green varieties has also shown to have these health protecting functions.

The berries are very low in calories. 100 g fresh grapes just provide 69 calories but zero cholesterol levels.

Grapes are rich source of micronutrient minerals like copper, iron and manganese.

They are also good source of **vitamin-C**, vitamin A, vitamin K, carotenes, B-complex vitamins such as pyridoxine, riboflavin, and thiamin.

PEARS

Pears (*Prunus communis*): Asian varieties feature crispy texture and firm consistency that do not change after harvesting or storage makes them ready-to-eat. Whereas, European types generally becomes soft and juicy when allowed to ripen. In structure, pear fruit has bell or "pyriform" shape; around 5-6 inches long and weigh about 200 gm. Fresh fruit is firm in consistency with mild '**apple**' flavour. Externally, its skin is very thin and depending upon the cultivar type, the fruit may have green, red-orange or yellow-orange colours. Internally, it's off white colour pulp is crunchy and juicy. However in fully matured ones, the pulp turns to grainy texture with gritty sensation while cutting with knife. Center of the fruit is more or less similar to apple with centrally located tiny inedible seeds. **Pears come in different colours - green, red, yellow and brown.**

Health Benefits

It is an excellent source of dietary fibre that is good for human heart.

Low glycerin index and high fibre content, helps in contributing the sugar levels in the blood.

Folic acid present in pear helps in keeping the body safe from birth defects.

The cooling effect of pear is highly effective in fever.

It is recommended during hot days, as it helps in overcoming shortness of breath due to heat.

This multifunctional fruit can be a savior from summer heat and also provides strength to the foetus.

Packed with phytonutrients known as flavonols, pears provide antioxidant protection against several diseases.

The skin of a pear comprises even more phenolic phytonutrients as its flesh. These include antioxidants and anti-inflammatory flavonoids, apart from anti-cancer phytonutrients such as cinnamic acids. Also, rich with fibre, pear skins should always be consumed.

Some researchers say that flavonoids, which have three groups – flavonols, flavan-3-ols, and anthocyaninsin – are abundantly found in pears. This is especially beneficial for those who have type 2 diabetes (Zeenia Baria, 2013).

Pears contain phytonutrients, which are known to have anti-inflammatory benefits. This results in a decreased risk of common chronic diseases that start with chronic inflammation and oxidative stress like type 2 diabetes and even heart disease.

The low acidic nature of pears makes it easy to digest.

Pears are also known as being a hypoallergenic food, which means that the chances of one getting an allergy from a pear are very rare.

The dietary fibre in pears makes them a good source of immune-supportive vitamin C and bone-building vitamin K.

Pear fruit is packed with health benefiting nutrients such as dietary fibre, anti-oxidants, minerals and vitamins which are necessary for optimum health. They provide about 3.1 g of dietary fibre per 100g. Regular eating of pears may offer protection against colon cancer. Most of the fibre is **non soluble polysaccharide** (NSP), making them a good bulk laxative. Also, the gritty fibre content binds to cancer causing toxins and chemicals in the colon, protecting its mucous membrane from contact with these compounds.

Pear fruit is one of the very low calorie fruits, just provides 58 cal per 100g. Just a few sections a day in the diet can bring significant reduction in weight and blood LDL cholesterol levels.

It contains good quantities of vitamin C. Fresh fruits provide about 7 per cent of RDA per 100 g.

Low glycerin index and high fibre content, helps in contributing the sugar levels in the blood.

Folic acid present in pear helps in keeping the body safe from birth defects.

The cooling effect of pear is highly effective in fever.

They are rich in antioxidant flavonoids phyto-nutrients such as beta carotene, lutein and zea-xanthins. These compounds along with vitamin C and A help body protect from harmful free radicals.

The fruit is a good source of minerals such as **copper**, iron, potassium, manganese and magnesium as well as B-complex vitamins such as folates, riboflavin and pyridoxine (vitamin B-6).

Pears are among the least allergenic of all fruits and are therefore recommended by health practitioners as safe alternative in the preparation of food products for allergy sufferers.

Pears have suggested being useful in treating colitis, chronic gallbladder disorders, arthritis and gout.

It is recommended during hot days, as it helps in overcoming shortness of breath due to heat. This mutifuctional fruit can be a savoir from summer heat and also provides strength to the feotus.

PEACHES

Peaches (*Prunus persica*): This fruit is best for dry skin. A combination of peach paste and yogurt, when applied to the skin and rinsed off with lukewarm water not only moisturizes the skin but also leaves it supple and soft. Gorging on these yummy fruits is not enough, applying them on the skin and hair helps a lot as well. So, the next time you lunge at that favourite fruit of yours, look beyond its culinary value, for you'd be surprised at its cosmetic benefits.

Health Benefits

Peaches are low in calories and contain no saturated fats. Nonetheless, they are packed with numerous health promoting compounds, minerals and vitamins.

The total measured anti-oxidant strength of 100 g peach fruit is 1814 TE (Trolex equivalents).

Fresh peaches are a very good source of antioxidant vitamin C.

Fresh peaches are also moderate source of **vitamin A** and **β-carotene**. β-carotene is a pro-vitamin, which converts into vitamin A in the body. Vitamin A is essential for vision. It is also required for maintaining healthy mucus membranes and skin.

They are rich in many vital minerals such as potassium, fluoride and iron. Iron is required for red blood cell formation. Fluoride is a component of bones and teeth and is important for prevention of dental caries. Potassium is an important component of cell and body fluids that help regulate heart rate and blood pressure.

Peaches contain many health promoting flavonoid poly phenolic antioxidants such as lutein, zeaxanthin and β-cryptoxanthin.

PLUM

Plum (*Prunus domestica*) is a soft round smooth-skinned sweet **fruit,** it is also known as a stone fruit tree in the genus Prunus, and the ripe fruit is usually dusty-white outer layer that gives them the appearance of glaze and easily erased. Each fruit is about the size of medium sized tomato, measuring about 5-6 cm in diameter and weigh about 70 g. It has central depression at the stem side. Internally, the pulp is juicy and vary widely from creamy yellow, crimson, light blue or light green in colour depending upon the cultivar type. The main benefit contained in **Plum** fruit is much beneficial **antioxidants** for the body, some **studies** also reported that the **plums** have anti-cancer substances. There is centrally placed single, smooth but hard stone like seed. Seeds are inedible. It has sweet and tartly taste pulp with pleasant aroma. Some common cultivars of plums are: cherry plum, damson, blackthorn plum. Plums are a good source of iron and potassium. They are low in fat, sodium and cholesterol free. They are also good source of vitamin A and vitamin E. A typical 66 gram plum contains about 36 calories, 0.5 grams of protein and about 9 grams of carbohydrates. Plums are also strong anti-oxidants, which means that they are able to fight off free radicals. Although pruns are better known for their laxative effects, plums are just as effective and both forms of the fruit are great sources of dietary fibre. The taste of the plum fruit ranges from sweet to tart; the skin itself may be particularly tart. It is juicy and can be eaten fresh or used in jam-making or other recipes.

Health Benefits

It protect your heart

Keep your bowels regular

Lower blood sugar, boost bone health and Improve your memory

The, fleshy, succulent plums are low in calories and contain no saturated fats; but contain numerous health promoting compounds, minerals and vitamins.

Plum fruits contain, dietary fibre such as, **sorbitol,** and **isatin** are known to help regulate the functioning of the digestive system and thereby used in constipation conditions.

Fresh plums are an excellent source of **vitamin C**, which is also a powerful natural antioxidant.

Fresh plums, especially yellow Mirabelle type, are very good source of **vitamin A** and **beta- carotene**. Vitamin A is essential for vision. It is also required for maintaining healthy mucus membranes and skin.

The fruit is also good in health promoting flavonoid poly phenolic antioxidants such as lutein, cryptoxanthin and zeaxanthin in significant amounts. **zeaxanthin,** an important dietary carotenoid selectively absorbed into the retinal macula lutea where it is thought to provide antioxidant and protective light-filtering functions.

Plums are rich in minerals like potassium, fluoride and iron. Iron is required for red blood cell formation.

Rich in B-complex group of vitamins such as niacin, vitamin B-6 and pantothenic acid. These vitamins are acting as cofactors help body metabolize carbohydrates, proteins and fats. They also provide about 5 per cent RDA levels of vitamin K. Vitamin K is essential for many clotting factors function in the blood as well as in bone metabolism and helps reduce Alzheimer's disease in the elderly.

Caution

Plums contain oxalic acid, a naturally occurring substance found in some fruits and vegetables which may crystallize as oxalate stones in the urinary tract in some people. Therefore, people with known oxalate urinary tract stones are advised to avoid eating plums. Adequate intake of water is therefore advised to maintain normal urine output even if these individuals want to eat them.

LITCHI

Litchi Bunch

Litchi (*Litchi chinensis*) Its outer surface is covered with rough leathery rind or peel featuring pink colour. The peel can be easily removable in the ripe fruits. Inside; consists of edible portion or aril that is white, translucent, sweet and juicy. The fruit has sweet, fragrant flavour and delicious to savor. The pulp has single, glossy brown nut-like seed, 2 cm long, and 1–1.5 cm in diameter. The seeds, like in **sapodilla,** are not poisonous but should not be eaten. Fresh lychees are readily available in the markets from July to October, about 120-140 days after flowering. The litchi fruit contains vitamins and minerals that promote a healthy diet. Studies have shown that litchi fruit can fight the growth of cancer cells as it contains flavonoids, flavones, quercitin and kaemferol in its pulp, which reduce the proliferation of cancer cells especially breast cancer. It is also a great source of vitamin C that helps the body to fight heart disease apart from cancer. Vitamin C is also good for bones, skin and tissue. Due to high vitamin C content, the fruit benefits those suffering from cold, fever and sore throat. Litchi also helps the body to digest food properly for the best nutrition and is said to have the ability to relieve pain too (Manu Vipin, 2012). Litchi contains a high amount of soluble fibre which protects from colon cancer. The polyphenols present in litchi promote heart health, normalize blood pressure and heart rate thereby protecting against strokes and coronary heart diseases. The fruit is a great source of vitamin C, complex vitamins, folates, phosphorus, calcium, magnesium and protein. It also has oligonol, apolyphenol which has anti-influenza virus actions. It also helps to improve blood circulation, reduces weight and protects the skin from UVA rays. It decreases deep fat and raises sideline blood circulation (Parmer, 2013).

Health Benefits

Litche fruits are low in calories, contains no saturated fats or cholesterol, but rich in dietary fibre which, can be very important for individuals who are concerned about their excess body weight.

Lychee fruit contains **oligonol**, a low molecular weight polyphenol. **Oligonol** has been found to have several anti-oxidant, anti-influenza virus actions. In addition, it helps improve blood flow in organs, reduce weight, and protect skin from harmful UV rays (Takuya Sakurai (Kyorin University, Japan), Biosci. Biotechnol. Biochem., 72 (2), 463-476, 2008).

Litchi, like citrus fruits, is an excellent source of vitamin C. Consumption of fruits rich in vitamin C helps body develop resistance against infectious agents and scavenge harmful, pro-inflammatory free radicals.

It is a very good source of B-complex vitamins such as thiamin, niacin and folates. These vitamins are essential since they function by acting as co-factors to help body metabolize carbohydrates, protein and fats.

Litchi also contains a very good amount of minerals like potassium and copper. Potassium is an important component of cell and body fluids help control heart rate and blood pressure; thus offers protection against stroke and coronary heart diseases. Copper is required in the production of red blood cells.

MANGO

Mango Fruits

Mango (*Mangifera indica*) fruit is one of the most popular, nutritionally rich fruit with unique flavour, fragrance, taste, and heath promoting qualities making it a common ingredient in new functional foods often called "super fruits". Mango is one of the delicious tropical seasonal fruit and believed to be originated in the sub-Himalayan plains of Indian subcontinent. **Mangoes**, rightly called the king of fruits for not just its taste but also for health benefits. The soft pulpy fruit has an amazing effect on skin too. Rich in vitamin-A and antioxidants, it fights skin aging, regenerates skin cells and restores the elasticity of skin.

The mango is the **national fruit** of India, **Pakistan** and the Philippines. It is also the **national tree** of Bangladesh. In Hinduism, the perfectly ripe mango is often held by Lord **Ganesha** as a symbol of attainment, regarding the devotees potential perfection. Mango blossoms are also used in the worship of the goddess **Saraswati.** No Telugu/Kannada new year's day called Ugadi passes without eating ugadi pacchadi made with mango pieces as one of the ingredients. In Tamil Brahmin homes mango is an ingredient in making vadai paruppu on Sri Rama Navami day (Lord Ram's Birth Day) and also in preparation of pachchadi on Tamil new year's day.

The Jain goddess Ambikais traditionally represented as sitting under a mango tree. Mango leaves are used to decorate archways and doors in Indian houses and during weddings and celebrations like **Ganesh Chaturthi.** Mango **motifs** and **paisleys** are widely used in different Indian embroidery styles, and are found in **Kashmiri** shawls, **Kanchipuram silk sarees**, *etc.* Paisleys are also common to Iranian art, because of its pre-Islamic **Zoroastrian** past.

In Tamil Nadu, the mango is considered, along with banana and jackfruit, as one of the three royal fruits (Occupying first place in terms of sweetness and flavour. Ma-pala-vazhai.

As the saying goes, 'an apple a day keeps the doctor away'. This old adage could well be used for mangoes because, like apples, they're simply brimming with goodness. Mangoes are high in energy, low in fat, and are a great source of calcium and vitamins essential for good health. Mangoes have a high content of beta carotene and rank among the top providers of beta carotene. Beta-carotene is a powerful anti-oxidant that can help protect the body against disease and also fight the signs of ageing by assisting with the growth and repair of cells, tissues and skin.

Not only do they taste great, but **mangoes** are also loaded with several qualities that are excellent for your health. They are rich in powerful antioxidants that are known to neutralize free radicals that cause damage to cells and lead to health problems like heart disease, premature aging and cancer among other things.

However they are also a great source of nutrients in the concentrated form. In order to reduce the consumption of calories, mangoes are a very good choice, " said an expert. Beta carotene is also known to reduce the risk of certain forms of cancer. To get the right amount of nutrition not too many mangoes need to be consumed. A single fruit is capable of providing of almost a day's supply of Vitamin C to the body. It also helps in making bones stronger since it contains minerals, calcium and magnesium as well as B vitamins. They are also rich in a carotenoid called lycopene, which is an effective antioxidant. In order to lose weight, mangoes can form an important part of the diet however it has to be coupled with other **fruits** and a complete meal, only after consultation with a nutritionist. Not only losing weight, daily consumption of mangoes make the skin complexion brighter and makes skin softer. A 200g serve of ripe mango (the equivalent of less than one mango) provides you with up to three times your recommended daily intake of Vitamin A and Vitamin C. Vitamin C, an antioxidant important in protecting the body from infection, is required in the formation of collagen, a protein that gives structure to bones, cartilage, muscle and blood vessels. Vitamin C also aids the absorption of iron, necessary for transporting oxygen from the lungs to cells all around the body. It also promotes healthy immune function. Mangoes also provide more beta-carotene than any other fruit and are a rich source of fibre and potassium.

In some countries mangoes are used to treat a variety of diseases and conditions. It is believed that immature mango, in which the seed has not fully formed, can be combined with salt and honey to treat diarrhea, dysentery, piles, morning sickness, chronic dyspepsia, **indigestion** and constipation. Treatments for heat stroke, bilious disorders and scurvy have also been recorded. It's wonderful when a fruit that tastes so delicious has so many wonderful nutrients in it. They are high in Vitamin C, Vitamin B6, Potassium, Copper and Vitamin A. Mangoes also contain several important phytochemicals including: Cryptoxanthin, Lutein, Gallic Acid and Anacardic acid. They are an incredibly healthy snack. You can eat the entire fruit for just over a hundred calories. High in fibre, virtually fat-free, and mangoes contain numerous vitamins. Mangoes also contain beta-carotene which may help slow the aging process, reduce the risk of certain forms of **cancer,** improve lung function, and reduce complications associated with **diabetes.**

One cup of mangos is just 100 calories, so it's a sweet treat that won't weigh you down. One serving of mangos provides 100 per cent of your daily vitamin C and 35 per cent of your vitamin A, both important antioxidant nutrients. Vitamin A is important for vision and bone growth. Mangos provide 12 per cent of your daily dietary fibre. Mangos contain over 20 different vitamins and minerals. Each serving of mango is fat free, sodium free and cholesterol free. If your skin frequently feels parched, just eat a mango! Mangoes are bursting with vitamins and minerals, especially vitamin A. Vitamin A normalizes the production and life cycle of skin cells. Mangoes also improve your skin if you suffer from acne. In skin with acne, there is an overproduction of cells in the stratum corneum, which is the outermost layer of the skin and is composed of biologically 'dead' cells. These excess dead cells combine with sebum (the skin's own natural oil) to form comedones - the pore plugs that are the defining element of acne. Taken as a dietary supplement, vitamin A helps to prevent overproduction of skin cells in the stratum corneum. You can add 3 to 4 slices of **mango** to every meal as a refreshing and hydrating naturally sweet treat. Mango fruit is rich in pre-biotic dietary fibre, vitamins, minerals, and has antioxidant compounds. It is beneficial for digestion too.

According to a new research, mangoes have been found to protect against colon, breast and prostate **cancers**. Fresh mango is a rich source of **potassium,** which is an important component of cell and body fluids that helps to control **heart** rate and **blood pressure.** Vitamin E, which is abundantly present in mangoes, helps to regulate **sex hormones** and boosts sex drive. Mango helps to clear clogged pores that cause acne. Just slice a mango into thin pieces and keep them on your face for 10 to 15 minutes and then take bath or wash your face and see the results. It's a known fact that mangoes are rich in Iron. People who suffer from anaemia can take mangoes regularly along with their dinner. It is especially good for **women** after menopause. High level of soluble dietary fibre, Pectin and **Vitamin C** present in mangoes helps to lower serum cholesterol levels. A recent study shows that not only the flesh of a mango but the leaves can fight diabetes.

Mango, rightly called the king of fruits for not just its taste but also for health benefits.

Health Benefits

With its high iron content, mangoes are excellent for pregnant women and those who suffer from anemia.

Constantly complaining about clogged pores? Place mango slices on your skin and then wash off after 10 minutes.

If you suffer from indigestion problems, nothing will help you as much as a mango. They're known to give relief from acidity and aid proper digestion since they contain digestive enzymes that help break down proteins.

Rich in potassium, mangoes reduce high blood pressure. They also contain pectin, a soluble dietary fibre that is known to lower blood cholesterol levels.

Trying to put on weight? Include mangoes in your diet. Since it is rich in calories as well as carbohydrates, it could be the perfect fruit to have.

Some studies say that eating mangoes reduces the risk of kidney stone formation.

In Chinese medicine, mangoes are considered sweet and sour with a cooling energy. They are useful for those suffering from anemia, bleeding gums, cough, fever, nausea and even sea sickness.

This fruit is rich in **glutamine acid**- an important protein for concentration and memory. Instead of snacking on unhealthy chips and cookies, why not feast on slices of mangoes instead.

Though they are traditionally not considered as aphrodisiacs, mangoes contain Vitamin E which helps boost one's sex life. The vitamin works to regulate the body's sex hormones.

Mango fruit is rich in pre-biotic dietary fibre, vitamins, minerals, and poly-phenolic flavonoid antioxidant compounds. It is beneficial for digestion too.

Mango fruit has been found to protect against colon, breast, leukemia and prostate cancers. Studies suggest that *polyphenolic anti-oxidant* compounds in mango are known to offer protection against breast and colon cancers.

It is an excellent source of **Vitamin-A** and flavonoids like **beta-carotene, alpha-carotene,** and **beta-cryptoxanthin**. Together; these compounds are known to have antioxidant properties and are essential for vision. Vitamin A is also required for maintaining healthy mucus membranes and skin.

Fresh mango is a very rich source of potassium which is an important component of cell and body fluids that helps controlling heart rate and blood pressure.

It is also a very good source of vitamin-B6 (pyridoxine), **vitamin-C** and vitamin-E. Consumption of foods rich in vitamin C helps body develop resistance against infectious agents and scavenge harmful oxygen free radicals.

Copper is a co-factor for many vital enzymes, including cytochrome c-oxidase and superoxide dismutase (other minerals function as co-factors for this enzyme are manganese and zinc). Copper is also required for the production of red blood cells.

Mango peels are also rich in phytonutrients, such as the pigment antioxidants like carotenoids and polyphenols.

Fresh mango is a rich source of **potassium**, which is an important component of cell and body fluids that helps to control **heart** rate and **blood pressure.**

Vitamin E, which is abundantly present in mangoes, helps to regulate **sex hormones** and boosts sex drive.

Mango helps to clear clogged pores that cause acne. Just slice a mango into thin pieces and keep them on your face for 10 to 15 minutes and then take bath or wash your face and see the results.

It's a known fact that mangoes are rich in Iron. People who suffer from anemia can take mangoes regularly along with their dinner. It is especially good for **women** after menopause.

High level of soluble dietary fibre, Pectin and **Vitamin C** present in mangoes helps to lower serum cholesterol levels.

A recent study shows that not only the flesh of a mango but the leaves can fight diabetes.

The soft pulpy fruit has an amazing effect on skin too. Rich in vitamin-A and rich antioxidants, it fights against skin aging, regenerates skin cells and restores the elasticity of skin

Health tip: Before going to bed put some 10 or 15 mango leaves in warm water and close it with lid. The next day morning **filter** the water and drink it in empty stomach. Do this regularly.

Caution: Mango latex allergy especially with raw, unripe mangoes is common in some sensitized individuals. Immediate reactions may include itchiness at the angle of the mouth, lips and tip of the tongue. In some people, the reactions can be severe, with manifestations like swelling of the lips, ulceration at the mouth angles, respiratory difficulty, vomiting, and diarrhea. This reaction develops because of **anacardic acid** present in raw, unripe mangoes. Cross-allergic reactions with other anacardiaceae family fruits like "**cashew apples"** are quite common. It is quite rare with fully ripen fruits; however, people with known case of mango fruit allergy may have to avoid them.

LOQUAT

Loquats in Bunch

Loquat (*Eriobotry japonica*): Succulent, tangy and sweet, wonderfully delicious loquat fruits are rich in vitamins, minerals, and anti-oxidants. This unique fruit is originated in the mountainous, evergreen rain forest of South-eastern China, from where it spread all across the world including India. Loquat is a good source of Vitamin A, which is crucial for the visual and dental health of an individual. Extract from loquat leaves is an important ingredient for lung ailments and has been used by Chinese, since ancient times. Loquats are appropriate for maintaining optimum health, as they are low in saturated fat and cholesterol. People with a deficiency of Vitamin A should have loquat in large number, because of its rich content. Loquats are rich in fibre, making them suitable for those who wish to lose weight. The loquat leaf is said to alleviate coughing and nausea. It even dissolves phlegm and is an expectorant. Loquat paste helps in soothing the digestive and respiratory systems of a person. The loquat leaf is known to shorten the recovery time from respiratory illness. Loquat contains malic acid, tartaric acid, citric acid, vitamins A, B and C, and B17. Vitamin B17 is known as amygdaline, laetrile or the anti-cancer vitamin, as it helps prevent cancer. Loquat is very low in saturated fat, very low in sodium but has no cholesterol or sugar, but very high in vitamin A, and high in vitamin 6. It is high in dietary fibre, manganese and potassium. It has same properties as that of a mango but comparatively low in sugar and calories.

Health Benefits

Delicious, loquats are very low in calories; provide just 47 cal per 100 g, however, rich in insoluble dietary fibre, pectin. Pectin retains moisture in the colon and thus functions as bulk laxative and by this way, it helps to protect the colon mucous membrane by decreasing exposure time to toxic substances as well as binding to cancer causing chemicals in the colon.

Pectin has also been shown to reduce blood cholesterol levels by decreasing its re-absorption in the colon by binding bile acids resulting in its excretion from the body.

Loquat fruit is an excellent source of vitamin-A (provides about 1528 IU per 100g), and phenolic flavonoid antioxidants such as chlorogenic acid, neo-chlorogenic acid, hydroxybenzoic acid, feruloylquinic acid, protocatechuic acid, epicatechin, coumaric acids and ferulic acid. Ripen fruits have more chlorogenic acid concentrations.

Vitamin A maintains integrity of mucus membranes and skin. Lab studies have shown that consumption of natural fruits rich in vitamin-A and flavonoids helps to protect from lung and oral cavity cancers.

Fresh fruit is very rich in potassium and some B-complex vitamins such as folates, vitamin B6 and niacin and contain small amounts of vitamin-C. Potassium is an important component of cell and body fluids, helps controlling heart rate and blood pressure.

It is also a good source of iron, copper, calcium, manganese, and other minerals. Manganese is used by the body as a co-factor for the antioxidant enzyme, superoxide dismutase. Copper is required in the production of red blood cells. Iron is required for as a cofactor in cellular oxidation as well for red blood cell formation.

Caution

The seeds of loquat fruit contain many toxic alkaloids like cyanogenic glycosides which when consumed can cause serious life threatening symptoms like vomiting, breathlessness, and death. Therefore, especially children are advised to avoid chewing seeds and should be supervised by adults while eating loquat fruits.

POMEGRANATE

Pomegranate Fruit

Pomegranate (*Punica granatum*) is among the most popular, nutritionally rich fruit with unique flavour, taste, and heath promoting characteristics. Along with berries, and some tropical exotics such as **mango,** it too has novel qualities of functional foods often called as "super fruits."The impressive health values of this fruit had already been known since ancient days. The size of a ripe pomegranate can be as small as an orange or as big as a grape fruit, approximately 7-12 cm in diameter, depending on its variety. It has a rounded hexagonal shape, with thick yellowish to reddish outer layer. Inside a pomegranate is about 700-800 tightly packed seed casings called **arils** that are deep red in colour when nicely ripe. The taste of the juice differs depending on the variety and its state of ripeness. But basically, it can be sweet, sour or tangy. The potent anti-oxidant in pomegranate juice is beneficial in preventing the hardening of arteries. This food is very low in Saturated Fat, Cholesterol and Sodium. It is also a good source of Dietary Fibre and Folate, and a very good source of Vitamin C and Vitamin K but a large portion of the calories in this food come from sugars. **Pomegranates** are a wonderfully hydrating source of antioxidants. They may be the world's most prolific source of polyphenols. The unique combination of elements in pomegranates increases the protective abilities of sunscreens, which can help prevent sun damage. Add pomegranates to your morning cereal or yogurt, fresh green or fruit salads for a refreshing snack or breakfast.

Pomegranates have had a reputation as a super food for some time, with its high levels of antioxidants thought to reduce the risk of heart disease and cancer It has high levels of iron and vitamin A, C and E. Iron helps maintain a good supply of oxygen to the body, vitamin E can reduce aging in the brain and might help prevent Alzheimer's while vitamin C boosts skin, teeth and bones. Pomegranate polyphenolics, tannins and anthocyanins could also have beneficial effects.

Polyphenols help the body rid itself of cancer causing agents. With pomegranates thought to be particularly good at warding off prostate cancer and possibly breast cancer. Tannins- have been shown to lower blood pressure and stimulate the immune system, while anthocyanins help protect blood vessels and reduce inflammation. It also has effective anti-bacterial properties and can ease the symptoms of **osteoarthritis**. Pomegranate juice increases sexual urges, strengthen bones and muscles, and increases testosterone which in turn help improve a person's mood and memory and even relieve stress such as pre-match nerves. In men higher levels of **testosterone** increases facial hair, deepens the voice and stimulate the sex drive. While the men have higher levels of the hormone, **testoterone** is also found in women, produced in the advenal glands and ovaries. (Annamous, 2012). The dried bark of both root and stem, has long been used in the treatment of tapeworms. Extracted of different parts of the tree exhibit anti- biotic activity. The seeds of pomegranates yield a drying oil which contain punicic acid forming up to 72 per cent of the fatty acids. This oil possesses antibacterial properties. Pomegranate, is highly recommended for preventing breast cancer. It contains polyphenol- an ellagic acid with anti-oxidant properties that prevent cancer growth. Include this delicious fruit in your diet and discover effective health benefits.

Health Benefits

Pomegranates are an excellent weight loss food and also benefit the body by boosting the immune system, improving circulation, and offering protection from cancer and Alzheimer's disease. Pomegranates are packed with antioxidants and particularly one called **Punicalagin** which has been shown to effectively reduce the risks of heart disease by scavenging harmful free radicals from the body. Punicalagin also has potent anti-microbial properties making pomegranates fantastic in warding off bacterial and viral infections. Pomegranates act like a natural aspirin in the body and help to prevent blood clots. Pomegranates are also the perfect "brain food" as they help to increase cognitive function and memory recall. Pomegranates are also great for joints and may help to prevent cartilage deterioration making them essential for the prevention of osteoarthritis. Pomegranates contain powerful anti-inflammatory compounds which makes them a highly beneficial food for those with autoimmune disorders such as fibromyalgia, COPD, bursitis, Lyme disease, rheumatoid arthritis, Chronic Fatigue Syndrome, and lupus. Consuming pomegranates or their juice daily has been shown to effectively protect against diabetes, lymphoma, urinary tract infections, and breast, colon, lung, and prostate cancer. Pomegranate juice has also been shown to keep PSA levels stable in men thereby reducing the need for further treatments such as hormone therapy or chemotherapy. Pomegranates have also been shown to help lower LDL (bad) cholesterol and raise HDL (good) cholesterol as well as lowering systolic blood pressure for those who need it. Pomegranate juice is excellent for dental health and has been shown to naturally prevent dental plaque and gum disease. Pomegranate seed oil is an excellent source of essential fatty acids and can be taken internally or applied topically to the skin to help improve skin elasticity, skin tone, and skin conditions such as eczema, psoriasis, and sunburn. It is also excellent for revitalizing hair and protecting it from damage.

An average pomegranate contains around 600 arils, which are loaded with nutritional benefits. Here are some of the key nutrients found in pomegranate seeds, and how you can include them in your diet.

Fibre: One of the key health benefits of pomegranate seeds is weight loss, because like most fruits they are low in calories and rich in fibre. A 100 gram serving of these seeds has only 83 calories and 4 grams of dietary fibre. This fibre helps manage your weight in addition to improving your digestion and preventing chronic illnesses.

Vitamins: Pomegranate seeds contain high amounts of Vitamins E, C and K. Vitamin C boosts your immune system, heals wounds, improves the absorption of iron and keeps your gums healthy. Vitamin K strengthens your bones and keeps your skinyoung, while Vitamin E lowers your risk of developing heart disease and cancer, by protecting you from environmental toxins.

Minerals: Pomegranate seeds are packed with potassium, which keeps your heart and muscles healthy. The red seeds also contain plenty of iron, which improves your circulation and protects your immunity. They also contain small amounts of calcium, zinc and magnesium, which help keep your bones and teeth healthy.

According to the study, the superfruit contains antioxidants, called polyphenols, that keep arteries free of fat. Polyphenols also help arteries expand and contract to maintain blood flow and keep them from hardening, which is the leading cause of heart attacks. The leaf, fruit rind, seeds, dried bark, stem, root and the fruit itself have a host of benefits. Often called Nature's Power Fruit because of its usefulness, pomegranates are great for your heart since they are an excellent antioxidant.

Fights breast cancer: Studies show that pomegranate juice destroys breast cancer cells and leaves the healthy cells alone.

Stabilizes PSA levels: In a study of men who had undergone treatment for prostate cancer, eight ounces of pomegranate juice every day kept their PSA levels stable thereby reducing the need for more treatment such as chemotherapy or hormone therapy.

Protects your arteries: Pomegranates are known to help prevent plaque from building up in your arteries and reverse previous plaque buildup.

Lowers cholesterol: This fruit is known to lower LDL (bad cholesterol) and raise HDL (good cholesterol).

Free radicals: We may not be aware but pomegranates are a rich source of antioxidants. Therefore, it helps to protect your body's cells from free radicals, which cause premature aging. Free radicals are formed due to exposure to the sun and harmful toxins from the environment.

Natural blood thinners: There are two kinds of blood clots: The first kind speeds skin recovery from topical injuries like cuts and bruises. Here it is important that the blood clots immediately to avoid lose of blood. The second kind of blood clot is internal and dangerous; examples would include blood clots in the heart and arteries, also urinary retention. Here you don't want the blood to clot as the effects are lethal.

To smoothen out the blood you need pomegranate seeds, as the antioxidant properties help it act like a 'thinner for paints'. The seeds prevent your blood platelets from coagulating and forming clots.

Prevention of atherosclerosis: With age, and bad lifestyle habits, the artery walls harden with cholesterol and other substances, causing blockages. The antioxidant properties of a pomegranate prevent low-density lipoprotein or bad cholesterol from oxidizing. This essentially means that pomegranates prevent the hardening of the artery walls with excess fat, leaving your arteries fat free and pumping with antioxidants.

Pomegranates act like an oxygen mask: In simple words, pomegranate juice pumps the level of oxygen in your blood. The antioxidants reduce cholesterol, fight free radicals and prevents blood clots. This eventually helps the blood to flow freely in your body in turn improving the oxygen levels in your blood.

Arthritis prevention: Pomegranate health benefits run bone deep; it can reduce the damage on the cartilage for those hit with arthritis. This fruit has the ability to lessen the inflammation and fights the enzymes that destroy the cartilage.

Fight erectile dysfunction: This could well be on Ripley's Believe it or Not! Pomegranate can also cure this embarrassing problem. But mind you, it is not a wonder drug; pomegranate juice can improve erectile dysfunction only moderately. Research is inconclusive, but this theory has found some supporters.

Fights prostrate cancer and heart diseases: Again, this is inconclusive and not binding, but two separate studies claim that pomegranate juice helps fight prostrate cancer. In one lab experiment, the juices "slowed the growth of the cultured cancer cells and promoted cell death". In the second experiment, pomegranate juice improved the condition of the blood, hence improving the health of individuals down with cardiovascular diseases (Tina, 2014).

Other benefits include lowering blood pressure, protecting your teeth (drinking pomegranate juice is a natural way to prevent dental plaque) and even prevent cartilage deterioration

Pomegranates have very high content of **punicalagins**, a potent anti-oxidant component found to be responsible for its superior health benefits.

The capacity of anti-oxidant in this fruit is two or three times higher than that of red wine and green tea.

The level of anti-oxidant is even higher than those of other fruits known to have high-levels of anti-oxidant, including blueberries, cranberries and oranges. This was attributed to the very high polyphenol content in the fruit.

They are also a good source of vitamin B (riboflavin, thiamin and niacin), vitamin C, calcium and phosphorus. These combination and other minerals in pomegranates cause a powerful synergy that prevents and reverses many diseases.

Drinking pomegranate juice frequently is extremely beneficial in fighting the hardening of arteries (atherosclerosis). It reduces the oxidation of bad LDL cholesterol which contributes to artery clogging and hardening.

Not only does the juice significantly reduce the blood vessel damage, it is found to actually **reverse** the progression of this disease.

Pomegranates contain a powerful agent against cancer, particularly prostate cancer.

The dried bark of both the root and stem, is used in the treatment of tapeworms. The extracts of different parts of tree exhibit antibiotic activity.

The seeds of pomegranates yield a drying oil which contains **punicic acid** forming up to 72 per cent of the fatty acids. This oil possesses antibacterial properties.

This is highly recommended for preventing breast cancer. It contains polyphenol- an ellagic acid with anti-oxidant properties that prevent cancer growth. Include this delicious fruit in your diet and discover effective health benefits.

Some studies say that pomegranate juice is known to kill breast cancer cells while leaving healthy cells alone.

The fruit is also known to prevent plaque from building up in the arteries as well reversing previous plaque buildup.

Pomegranate can significantly lower LDL (bad cholesterol) and increase HDL (good cholesterol).

It is beneficial for those suffering from high blood pressure because it is known to lower blood pressure levels.

Pomegranate juice is good for your teeth because it prevents dental plaque. The fruit also prevents cartilage deterioration (Anonymous, 2013).

Pomegranate juice helps lower cholesterol levels and prevent hypertension because it is packed with vitamins, antioxidants and minerals. Some studies have also suggested that having pomegranate juice daily can slow down the progress of Prostate cancer.

Here are some common ailments that are known to react positively with the use of pomegranate or its juice:

Anemia: Add a teaspoon of ground cinnamon with a little honey to a cup of pomegranate juice. Especially beneficial for women after monthly loss of blood due to menstruation.

Anal Itch: Itching in the anal region is often caused by parasites in the intestines that go to the anal area to lay their eggs. Roast some pomegranate skin until it is brown and brittle. Then crush it to a fine powder form and mix with a little olive oil. Apply this concoction to the anus to kill the worms.

Anti-aging: We all know that anti-oxidant is highly effective in helping to protect the skin from free radical damage known to cause signs of aging.

Asthma: The high content of ascorbic acid (vitamin C) in this fruit is a powerful anti-inflammatory agent. It can greatly reduce wheezing in young children with asthma.

Atherosclerosis: The highly cleansing power of this miracle juice scrubs away the old build-up of arterial deposits, reducing the risks of heart diseases and stroke.

Bleeding Piles: Pound the (clean) skin of one fruit from the sour variety. Boil the pound pulp in about two cups of water. Sweeten with honey and drink twice a day until healed.

Cancer prevention: The high anti-oxidant content protects cells from damages by free radicals. Regularly drinking juices high in anti-oxidants keep cancer at bay.

Cholesterol: Drinking juices high in anti-oxidant has been proven to fight the oxidative stress that is the main culprit in oxidizing the LDLs in the blood.

Dysentery: Drinking fresh pomegranate juice is an excellent remedy to soothe the pain and inflammation caused by severe diarrhea with blood and mucus in stools.

Immune booster: The anti-oxidant nutrients in pomegranates are critical in building up your immune system. Drink juice high in anti-oxidant when you feel a cold coming.

Loss of Appetite: If you can't eat, at least you can drink! Pomegranate juice can help increase your appetite.

Morning Sickness/nausea: Mix and drink an equal amount of honey with pomegranate juice for relief.

Sore throat: The anti-inflammatory agent in pomegranate juice significantly reduces the soreness and redness in the throat.

Pomegranate is touted as a wonder fruit since it shows promise in improving depression and bone mineral density.

Other healing effects of pomegranate juice:

Reduces diarrhea

Controls weight

Prevents cartilage deterioration

Lowers blood pressure

Slows down Alzheimer's disease

Caution

If you are on any medication, consult with your doctor before you start consuming pomegranate juice regularly. There is some concern that the juice may affect the metabolism of some prescribed medications. A word of caution: Pomegranate juice may react to a heart patient's medication.

PINEAPPLE

Pineapple (*Ananas comosus*): It is both juicy and fleshy with the stem serving as the fibrous core. The rough, tough, scaly rind may be dark green, yellow, orange-yellow or reddish when the fruit is ripe. Juicy flesh ranges from creamy white to yellow in colour and has mix of sweet and tart taste with rich flavour. Each fruit measures in size up to 30cm long and weigh 1/2 to 4kg or more. A great skin emollient, pineapples help rejuvenate and cleanse the skin. Rubbing a slice of pineapple on spots like knees, elbows and heels helps soften the skin. It can also act as a loofah or sponge while taking a shower. Though only 60 per cent of it is edible, pineapple is a good source of thiamine (vitamin B1) compared to other fruits and is very rich in vitamin-C. Fresh pineapple juice contains an enzyme called Bromelin which aids digestion. Pineapple is highly rich in vitamin C and so acts as a great antioxidant, which prevents colds and flu. It is also a good source of thiamine, B12, folate, copper and dietary fibres. A few slices of pineapple a day in your meal can work wonders for your body and it is a good choice for weight watchers due to its low calorie content (100gm pine apple= 50Kcall). Besides helping you lose weight, pineapple has an important nutrient that reduces arthritic pain. Bromelain is a magical component which is present in the fruit that has many health benefits. Although it is present in the entire fruit, the stem of a pinappple has the most amount of bromelain. Bromelain extracts aid in digestion and act as an anti-inflammatory agent thus in turn it helps with arthritis pain. Being a proteolytic enzyme, bromelin helps to tenderise your meats by breaking down the protein in it, but remember bromelain can coagulate the milk and prevent the jelly to set hence remember if you want to make a pineapple milkshake you need to cook it to destroy the bromelain content (Trina Remedios, 2013).

Health Benefits

Fresh pineapple is storehouse of many health promoting compounds, minerals and vitamins that are essential for optimum health.

The fruit is low in calories, contains no saturated fats or cholesterol; but rich source of soluble and insoluble dietary fibre like pectin.

Pineapple fruit contains a proteolytic enzyme **bromelain** that digests food by breaking down protein. Bromelain also has anti-inflammatory, anti-clotting and anti-cancer properties. Consumption of pineapple regularly helps fight against arthritis, indigestion and worm infestation. Bromelain (enzyme) is known as a treatment for inflammation and swelling of the nose and ear.

It also contains good amount Vitamin A and beta-carotene levels. These compounds are known to have antioxidant properties.

In addition, this fruit is rich in B-complex group of vitamins like **folates**, thiamin, pyridoxine, riboflavin and minerals like **copper**, **manganese** and potassium.

Best muscle building foods: Pineapples have a protein-digesting enzyme named bromelein in abundance. It also decreases muscle inflammation, thus it can be a great post-workout meal.

Juice of ripe fruit is considered to be a diuretic (helps to pass urine easily) while juice from the unripe fruit acts as purgative.

Coughs and Colds-A single serving of pineapple has more than 130 per cent of the daily requirement of vitamin-C for human beings, making it one of the richest and most delicious sources of ascorbic acid. Vitamin C has strong immune system boosting power.

Pineapples also contain a relatively rare proteolytic enzyme called bromelain, which is also connected with the reduction of phlegm and mucus build up in the respiratory tracts and sinus cavities. So, it prevents the illnesses that cause phlegm and mucus build-up. Moreover, it treats them by loosening those materials and helping you eliminate them from your body if you've already contracted an illness or infection.

Bone Health: Pineapples have an impressive amount of manganese, which is another trace mineral that is essential in the strengthening of bones, as well as their growth and repair. Manganese is the most prominent mineral in pineapple, and more than 70 per cent of your daily requirement of this essential mineral can be delivered by a single serving.

Digestion: This fruit is abundant in fibre, but its even more important property is that it contains both soluble and insoluble fibre. Fibre can bulk up stool, which promotes the passage of food through the digestive tract at a normal rate, and also stimulates the release of gastric and digestive juices to help food dissolve.

Also, it bulks up loose stool, which helps with diarrhea and IBS. Fibre also strips the blood vessels clean of excess cholesterol and eliminates it from the body, thereby boosting cardiovascular health. This means that eating a healthy amount of pineapples can protect you from a vast amount of health conditions, including diarrhea, irritable bowel syndrome, atherosclerosis, constipation, blood clotting, as well as blood pressure.

Arthritis Relief: Another extremely popular property of the pineapple is its ability to reduce the inflammation of joints and muscles, particularly those associated with arthritis. The bromelain in pineapples is primarily associated with breaking down complex proteins. Furthermore, bromelain also has serious anti-inflammatory effects, and has been positively correlated with reducing the signs and symptoms of arthritis.

Immune System: Pineapples are extremely beneficial in boosting our immune system, mostly due to the high content of vitamin C. Vitamin C is mainly associated with reducing illness and boosting the immune system by stimulating the activity of white blood cells and acting as an antioxidant to defend against the harmful effects of free radicals.

Free radicals are dangerous byproducts of cellular metabolism that can damage various organ systems and disrupt function, as well as cause healthy cells to mutate into cancerous ones. The vitamin C defends against this.

Tissue and Cellular Health: Another critical benefit of vitamin C is its essential role in creating collagen. Namely, collagen is the essential protein base of blood vessel walls, skin, organs, and bones. Thus, high vitamin C content defends the body against illness and infections, and helps you heal wounds and injuries to the body quickly.

Cancer Prevention: Pineapple has directly been related to preventing cancers of the mouth, throat, and breast. Vitamin C has an antioxidant potential in the battle against cancer, but as stated above, pineapples are also rich in various other antioxidants as well, such as vitamin A, beta carotene, bromelain, various flavonoid compounds, and high levels of manganese.

Manganese is vital co-factor of superoxide dismutase, an extremely potent free radical scavenger that has been associated with a number of different cancers.

Loaded with Vitamin C - Did you know that one cup of pineapple chunks contains 131 per cent of the daily value of Vitamin C, which helps protect the body against viruses and builds thereby building a strong skin defence mechanism against infection. So, instead of restricting yourself to the regular orange or grape fruit, enjoy the juicy flavour of a tropical pineapple to get your daily requirement of this important nutrient.

It helps you have strong bones - Single cup of approximately 165 grams of pineapple contains about 76 per cent of the recommended daily value of manganese which is essential for maintaining strong bones and healthy connective tissues. Thus, in addition to brushing your teeth and flossing, you should add a healthy dose of pineapple to your dental routine.

It is a good digestive aid - To add to its benefit, pineapples also contain bromelain, an enzyme that aids indigestion, controls coughs, and loosens mucus. In order to maximize its health benefits on the digestive system, its best to eat pineapple in between meals as a snack. Bromelain is also very effective at reducing inflammation from infections and injuries thus it helps reduce swelling, bruising, healing time, and pain after physical injuries and surgery.

Idle for weight watchers - With its natural sweet and filling qualities, it is a good option for those looking for a sweet snack with the good taste coming in from the natural sugar it contains. At the same time weight watchers should be happy that they consume Pineapples which are 87 per cent water and relatively few calories as compared to high-sugar or high-fat foods.

Improves your eye sight - Along with carrots that keep your eyes bright and healthy eating three or more servings of fruit a day too may lower risk of vision loss in older adults by as much as 36 percent. By adding more pineapple to your diet you ensure getting the antioxidants you need to slow or halt the development of ARMD.

With age the breasts become saggy. However, pineapple will become your favorite fruit after learning its effect on women's breasts. Women's breasts sag because the skin becomes stretched. The nutrients in pineapples give the skin elasticity, so the breasts return to a higher location. Pineapples prevent the sagging of the breasts. So, adding pineapples to your diet will keep your breasts elevated (chosehealthylife.com).

The antioxidants in pineapple are excellent for the entire body. Pineapples are even effective against cancer.

Pineapple can also help with pain in the muscles and joints caused by arthritis. The ingredient in pineapple that helps with this medical issue is bromelain. Bromelain helps with inflammations in the body and breaks down complex proteins.

Pineapples also contain vitamin C and ascorbic acid. Eating one serving of pineapple will provide you with the required daily amount of vitamin C. Pineapples also increase the activity of the white blood cells which boosts the immune system.

Other Pineapple Benefits

Pineapples contain soluble and insoluble fibres, so they improve the process of digestion. Manganese in pineapples protects the health of the bones. Vitamin C and bromelain in pineapples also protect you from cough and colds.

Regular consumption of pineapples will improve your blood circulation, your eyes and your blood pressure. Pineapples are also helpful with the menstrual cycle. However, pregnant woman should not consume too much pineapple, since it can lead to miscarriage.

To conclude, pineapples protect from cancer, fight arthritis and digestive problems, and are useful for many other medical conditions. Due to these effects of pineapples, every person should make pineapples part of their nutrition. So, buy pineapples and make them part of your daily diet.

Caution

Pineapple fruit contains a proteolytic enzyme bromelain that may cause excessive uterine bleeding if consumed in large quantities during pregnancy. Consuming excess amount of pineapple can lead to irritation and soreness in mouth due to high acid content hence portion control is important.

JACKFRUIT

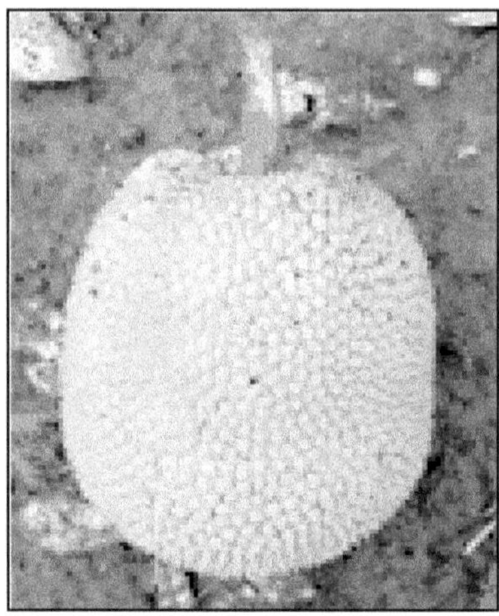

Jackfruit (*Artocarpus heterophyllus*) tree is believed to be indigenous to the Southwestern rain forests of India. It is widely cultivated in the tropical regions of Indian subcontinent, Thailand, Malaysia, Indonesia and Brazil for its fruit, seeds, and wood. The tree grows best under tropical humid and rainy climates but rarely survives cold and frosty conditions. Jackfruit is a huge tree that grows to as high as 30 meters, higher than the mango tree. During the season, each tree bears as many as 250 large fruits, supposed to be the largest tree-borne fruits in the world. The fruit varies widely in size, weigh from 3 to 30 kg, and has oblong or round shape measuring 10 cm to 60 cm in length, 25 to 75 cm in diameter. The unripe fruits are green in colour; When ripen, might turn to light brown colour and gives strong sweet, fruity smell. The fruit is very low in saturated fat and sodium. There is no sugar or cholesterol but is high in vitamin C. It is a good source of fibre, vitamin B2, copper, magnesium and potassium. The aroma of ripe fruit is so strong that either you love it or can not stand it. Research studies have revealed that about 16 esters, half a dozen alcohols, volatile vapours and many other compounds combine to present its typical aroma. It also provides appreciable amounts of carotene, thiamine, riboflavin, niacin, vitamin C. In addition it has crude fibre, calcium, iron, potassium and sodium. The seeds of Jack fruit are also nutritious. They provide 7grams of protein, but contain a powerful enzyme inhibitor- particularly carbonic anhydrase. This accounts for the occasional abdominal (stomach upsets) that jackfruit seeds cause.

Health Benefits

Improved immunity. Thanks to the vitamin C included, it will keep you safe from bacteria and infections by increasing and assisting the function of white blood cells.

Cancer protection. Jackfruit includes phytonutrients such as lignas, saponins and isoflavones. These properties of jackfruits are vital in the fight against cancer due of their anti-cancer and anti-aging compounds.

Assist digestion. Thanks to its anti-ulcer characteristics it keeps you safe from digestive issues. Jackfruit is helpful against constipation.

Increase energy. The fructose and sucrose included in jackfruit are providing the body with plenty of energy without distributing the sugar level in the system.

Reduce your BP. Thanks to its large amount of potassium your blood will be controlled which will automatically reduce the chances of heart attack and stroke. Being rich in potassium, jackfruit has been found to be helpful in the lowering of blood pressure.

Asthma control. The root of this fruit has been found to be beneficial for those suffering from asthma. In some studies was proved that if you boil the jackfruit root and its extract you will reduce your asthma issues. Try it if you are suffering from asthma.

Preventing anemia. Jackfruit can be of great help with your anemia issues. The women are more vulnerable than man. The jackfruit stimulates and improves of the blood and heals the anemia.

Thyroid will not cause any more problems. Thyroid requires in order to work well. The jackfruit has plenty of copper in its structure. It will support the function of the thyroid and the hormones will be produced without any problem. The medications will not be needed anymore.

Firmer bones. You already know that the bones are requiring calcium. Jackfruit is rich source of magnesium, which is very helpful in absorbing calcium. Your bones will be stronger and more resistant to osteoporosis and arthritis. If you have some low density issues you can use the help of jackfruit.

Prevents aging and helps your skin. This fruit is rich in antioxidants. They are helpful in decreasing the aging, while the water included in jackfruit will make your skin moisturizer. It is also helpful against crinkle. Boasting of anti-ageing properties, the fruit can help slow down the degeneration of cells and make the skin look young and supple.

Improves vision. This fruit is also rich in vitamin A which is helpful in keeping your eyes strong and powerful. It protects them from UV rays.

Prevent Colon cancer and piles. Thanks to the antioxidants you will lower the risk of colon cancer. The fibres included in jackfruit will prevent constipation and the symptoms of piles.

Helpful for pregnant women. The vitamin B (niacin) included in this fruit is very helpful for the pregnant and the women who are breastfeeding, monitoring the hormones and boosting the immune system.

The jackfruit seeds can be useful too. Triturate the jackfruit seeds and mix them with honey and milk in order to get a face mask. Put it on your face and leave it to work for 30 minutes. Wash it with water. Use this face mask regularly and you will have great results-face cleaner of wrinkles.

More benefits. Consume dried seeds like snacks. You can use unripe-jackfruit for many different recipes. They are great source of proteins, minerals and starch.

The fruit is made of soft, easily digestible flesh (bulbs) with simple sugars like fructose and sucrose that when eaten replenishes energy and revitalizes the body instantly.

Jackfruit is rich in dietary fibre, which makes it a good bulk laxative. The fibre content helps to protect the colon mucous membrane by decreasing exposure time and as well as binding to cancer causing chemicals in the colon.

Fresh fruit has small amounts of vitamin-A and flavonoid pigments such as carotene-β, xanthin, lutein and cryproxanthin-β. Together, these compounds play vital roles in antioxidant and vision functions.

It is one of the rare fruits thatis rich in B-complex group of vitamins. It contains very good amounts of vitamin B-6 (pyridoxine), niacin, riboflavin, and folic acid.

Fresh fruit is a good source of potassium, magnesium, manganese, and iron.

The extract of Jackfruit root is believed to help cure fever as well as diarrhea.

Jackfruit contains phytonutrients, with health benefits ranging from anti-cancer to **antihypertensive.**

Jackfruit proves to be a very good source of vitamin C, which is known for its high **antioxidant** properties.

The fruit contains **isoflavones, antioxidants**, and **phytonutrients**, all of which are credited for their cancer-fighting properties.

Jackfruit is known to contain anti-ulcer properties and is also good for those suffering from indigestion.

Jackfruit serves as a good supply of proteins, carbohydrates and vitamins, for the human body.

It is believed that the fruit can help prevent and treat tension and nervousness.

Since it contains few calories and a very small amount of fat, jackfruit is good for those trying to lose weight.

If you are suffering from constipation, regular consumption of the fruit will surely prove beneficial.

The root of jackfruit is said to be good for the treatment of a number of skin problems.

Jackfruit is known to be a good source of vitamin C and anti-oxidants that help strengthening of the immune system and enhancing the functioning of the white blood cells in the body.

Being a good source of potassium, jackfruit is good for regulating blood pressure and thus, reducing one's risk of getting strokes and heart diseases. Potassium helps to maintain the electrolyte balance in the body.

Due to the high fibre content present in jackfruit, it is beneficial in reducing constipation and aiding digestion. The fibre content of jackfruit is said to do away with the carcinogenic chemicals in the colon.

Besides vitamin C, phytonutrients like saponins, isoflavones and lignans present in jackfruit is said to slow down degeneration of cells in the body due to formation of free radicals that can cause cancer.

To maintain healthy skin and vision, vitamin A present and anti-oxidants present in jackfruit are beneficial.

Natural sugars like sucrose, fructose present in jackfruit, make it a good source for energy and makes it easily digestible as well.

Jackfruit also contains other minerals like manganese, iron, vitamin B6, niacin, folic acid, *etc.* whose presence in small quantities is required for the optimum functioning of the body.

The seeds of the jackfruit are quite nutritious and a good source of **protein.**

Jack fruits: Jackfruit seeds are indeed very rich in protein and nutritious. In general, the seeds are gathered from the ripe fruit, sun-dried and stored for use in rainy season in many parts of South Indian states. Different variety of recipes prepared in Southern India where they are eaten either by roasting as a snack or added to curries in place of lentils.

QUINCE

Quince (*Cydonia oblonga*) belonging to the same family as apples and pears, is regarded as their distant relative. The fruit is native to the warm-temperate areas of southwest Asia, falling in the Caucasus region. Quince is very much like pear in shape and gets a golden-yellowish outer-layer, when mature. The Asian variety is softer and much juicier. Apart from being eaten raw, quinces are commonly made into preserves and jellies as well. When you hear the word "quince, " you probably think of jams and jellies, or maybe desserts. Unlike apples, pears and other pome fruits, quinces are rarely eaten raw because the raw fruit has an unpleasant and astringent taste. When cooked, quinces get soft and sweet. They can be poached, baked or braised. Quinces can be round, oval or somewhat pear shaped. Their appearance resembles a golden apple or pear. Choose those that are firm with a pale yellow skin. The yellow skin is often somewhat mottled with brown spots that don't affect the flavor or quality. Quinces that are shriveled, soft, or brown all over are no longer fresh. In Medieval times, Europeans thought quinces aided the digestion and prepared them frequently along with meats. The English called the combination chardeqynce meaning flesh of quince.

The good: This food is very low in Saturated Fat, Cholesterol and Sodium. It is also a good source of Dietary Fibre and Copper, and a very good source of Vitamin C.

The nutritional value and health benefits of the quince makes it ideal for:

Maintaining optimum health

Weight loss

Avoid including quince in your diet if you're interested in:

Weight gain

Health and Nutrition Benefits of Eating Quince

Being rich in dietary fibre, quince is good for those people who are trying to lose weight and maintain a healthy body.

Quince boasts of antioxidant properties, which helps the body fight against free radicals and reduces the risk of cancer.

Researchers have revealed that quinces might be rich in various anti-viral properties.

Consumption of quince has been found to be beneficial for people suffering from gastric ulcer.

Quince juice is known to have tonic, antiseptic, analeptic, astringent and diuretic properties.

It is believed that eating quince is good for maintaining the optimum health of an individual.

Regular consumption of quince not only aids in digestion, but also helps lower cholesterol.

The presence of potassium in quince helps the body keep high blood pressure in check.

The vitamin C present in quince helps reduce the risk of heart disease in individuals.

If consumed on a regular basis, quince proves beneficial for those afflicted with tuberculosis, hepatic insufficiency, diarrhea and dysentery.

Those suffering from liver diseases and eye diseases would surely benefit from regular quince consumption.

Being rich in antioxidants, quince is believed to be helpful in relieving stress and attaining calm.

Quince is low in saturated fat, cholesterol and sodium, while having lots of vitamin C, dietary fibre and copper.

Quince juice is good for those suffering from anemia, cardiovascular diseases, respiratory illnesses, diseases of the gastrointestinal tract and even asthma.

The juice as well as pulp of boiled or baked quince fruit serves as a good anti-emetic remedy.

Caution

The seeds contain nitriles (RCN), which are common in seeds of the rose family. In the stomach, enzymes or stomach acid or both cause some of the nitriles to be hydrolyzed and produce HCN (hydrogen cyanide), which is a volatile gas. The seeds are only likely to be toxic if a large quantity is eaten.

BERRIES

All berries are rich sources of antioxidants and phytochemicals which play a role in reducing oxidation and cancer cell formation in the body. All fruits have incredible healing and energizing properties, but wild blueberries are the most phenomenal of them all. Wild blueberries are loaded with antioxidants, vitamins, and minerals. These sweet and satisfying berries are excellent liver and blood cleansers, and are vital to the health of eyes and the digestive tract. Wild blueberries also have the ability to repair tissues and cells and keep the immune system strong. They are a power house food not to be missed.

Strawberries, Cranberries, Blueberries, *etc.* contain antioxidant phytonutrients that get rid of free radicals in the blood and guard your collagen, thereby boosting skin repair. Collagen makes your skin supple, smooth and plump. "Antioxidants that are present in berries such as blueberries, and cranberries and in strawberry, can help keep your skin vibrant and glowing." The canned cranberry juice works for your skin. Eat one to two cups of blueberries a day for your daily 'dose' of anthocyanin, an anti-ageing antioxidant that increases the potency of Vitamin C. Berries atleast a week could cut their risk of developing the disease (parkinson`s) by quarter compared with those who never eat Anonymous, 2012). Flavonoids can protect neurons against disease of brain suchas Parkinson's. Cranberries have five times the antioxidant content of broccoli, which means they may protect against cancer, stroke and heart disease. But limit the cran -berries to about 8 tbsp per pint of juice. Berries have abundant antioxidant properties that support brain function and blood circulation. Raspberries are said to contain the most amount of ellagic acid that improves liver functions, regulating cholesterol levels and eliminating toxins. Since berries are sweet, they are the healthiest choices to satisfy your sugar cravings.

STRAWBERRY

Strawberry (*Fragaria xananassa*). Delicious and nutrition-rich red coloured strawberries are among the most popular berries. These berries have the taste that varies from cultivar to cultivar and ranges from quite sweet to acidic. The berry features red pulp with tiny yellow colour seeds piercing through its surface from inside. Top end has small, green leafy cap and stem that is adorning its crown. Each berry features conical shape, weighs about 25 g and measures about 3 cm in diameter. Strawberries have higher levels of vitamin C, fibre, folate and potassium than most other fruits like bananas, apples and even oranges. Apart from the obvious health benefits, it has been shown that eating one serving (about 8-10 strawberries) a day can significantly decrease blood pressure, which may reduce the risk of heart disease. Other studies showed additional nutrition benefits: Strawberries are found to reduce risk of cancer, enhance memory function and rheumatoid arthritis. Strawberries also contain **Ellagic acid** which is also found in raspberries, blackberries, cranberries, grapes, cherries, walnuts, pecans and Brazil nuts, acts as a scavenger to "bind" cancer causing chemicals, making them inactive. It inhibits the ability of other chemicals to cause mutation in bacteria. Strawberries are the only known fruit to carry their seeds outside, with an average of 200 seeds per fruit. The leaves and roots of strawberries are claimed for their medicinal benefits in terms of easing diarrhea, digestive upsets and gout. You'll find strawberries leaves in blended herbal teas. The fruit juice is also used to treat sun-burn, skin blemishes and discolored teeth. Nutrients found in strawberries include vitamin A, C and B6, fibre, potassium, folate and various antioxidants and flavonoids. Your immune system needs these fighters to protect you against diseases related to heart and cancers.

Strawberries are known to be brilliant as it benefits the skin in more than one way. This accessory fruit is known to be rich in coenzyme Q10, a substance that helps fight free radical damage, and lessens the pace of aging, increases both the adeptness and rate of cellular energy production. Quite high in **alpha-hydroxyl acid**, this fruit helps get rid of dead skin cells and helps the body build collagen. Another great thing about strawberries, it's highly concentrated with vitamin C, an essential antioxidant that helps prevent skin blemishes, fights the aging process and the harmful effects of the sun.

Compared to fruits like apples, oranges or bananas, they score highest in most of the nutrients. For example, comparing the Vitamin C level in 100g of strawberries and 100g of oranges, strawberries exceed oranges by 10mg. Best of all, they're low in calories and absolutely fat-free!. The good thing is they are very low in Saturated Fat, Cholesterol and Sodium. It is also a good source of Folate and Potassium, and a very good source of Dietary Fibre, Vitamin C and Manganese.whereas, alarge portion of the calories in this food come from sugars. One serving of strawberries - about eight medium strawberries - is high in fibre and an excellent source of vitamin C. The star player in strawberries is vitamin C, an antioxidant that helps repair the body's tissues, boosts immunity, and fights excess free radical damage and with powerful pain-reducing properties, according to research. Some studies suggest vitamin C may help people experience less pain after breaking a bone or having orthopedic surgery. This red, wet drippy fruit makes for an excellent foreplay fruit. Drip the juices over your partner's body and you will do just fine. Strawberries are an excellent source of Vitamin C, a powerful antioxidant. These pretty red berries are low in calories, but have health-promoting phyto-nutrients, minerals and vitamins that are essential for optimum health. They have high amounts of phenolic flavonoid phyto-chemicals called **anthocyanin** and ellagic acid and studies claim they may help against ageing, inflammation and neurological diseases.

Best eaten: Tastes best raw with fresh cream. Nutritional facts per cup Vitamin C: 149 per cent Manganese: 29 per cent Vitamin K: 4 per cent Iron: 3 per cent Calcium 2 per cent

Health Benefits

Strawberry is low in calories and fats but rich source of many health promoting phyto-nutrients, minerals and vitamins that are essential for optimum health.

Strawberries have significantly high amounts of phenolic flavonoid phyto-chemicals called **anthocyanins** and **ellagic acid**. Consumption of straw berries have potential health benefits against cancer, aging, inflammation and neurological diseases.

Fresh berries are an excellent source of **vitamin-C** which is also a powerful natural antioxidant. Consumption of fruits rich in vitamin C helps body develop resistance against infectious agents, counter inflammation and scavenge harmful free radicals.

The fruit is rich in B-complex group of vitamins. It contains very good amounts of vitamin B-6, niacin, riboflavin, pantothenic acid and folic acid. These vitamins are acting as co-factors help body metabolize carbohydrates, proteins and fats.

Strawberries contain vitamin A, vitamin E and many health promoting flavonoid poly phenolic antioxidants such as lutein, zeaxanthin, beta carotene in small amounts.

They contain good amount of minerals like potassium, manganese, fluorine, copper, iron and iodine. Potassium in an important component of cell and body fluids that helps controlling heart rate and blood pressure. **Manganese** is used by the body as a co-factor for the antioxidant enzyme superoxide dismutase. Copper is required in the production of red blood cells. Iron is required for red blood cell formation. Fluoride is a component of bones and teeth and is important for prevention of dental caries.

Fruits like strawberries that are rich in vitamin C help absorb iron, which is necessary for healthy hair and their growth. Eating just a few strawberries a day will help you get stronger strands (Purvaja Sawant, 2013).

Caution

Strawberries may cause serious allergic reactions in some sensitized individuals. Some of the most common symptoms of strawberry allergy include swelling and redness of mouth, lips and tongue, eczema, hives, skin rash, headache, runny nose, itchy eyes, wheezing, gastrointestinal disturbances, depression, hyperactivity and insomnia. Individuals who suspect allergy to these fruits may want to avoid

RASPBERRY

Raspberry Plant

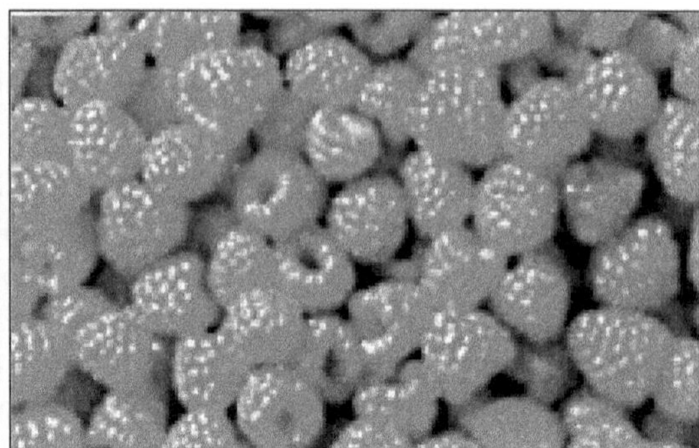

Raspberries

Raspberries (*Rubus idaeus*) :Are of various colours *viz.* black, *purple and even golden.* They are wonderfully delicious and rich red coloured raspberry among the most popular berries grown all over the world. They are rich source of many health promoting plant-derived nutrients, minerals, and vitamins that are essential for optimum health. Raspberries are an excellent source of fibre, manganese and vitamin C. They are a very good source of vitamin K and a good source of magnesium, folate, **omega-3 fatty acids**, copper, vitamin E and potassium. In addition, they contain significant amounts of the anti-cancer phytochemical **ellagic** acid. Rasberries as a are very low in Saturated Fat, Cholesterol and Sodium. They are also a good source of Vitamin K and Magnesium, and a very good source of Dietary Fibre, Vitamin Cand manganese and a large portion of the calories in this food come from sugars. The leaves of raspberry plant are also commonly used for their medicinal teas and also make delicious herbal teas. The leaves are also reputed to be effective in regulating the menstrual cycle in women. Raspberries rank in the top 10 antioxidant-high fruits and vegetables! They possess almost 50 per cent higher antioxidant activity than strawberries. **Antioxidants** are believed to help prevent and repair oxidative stress, a process that damages body cells and has been linked to the development of diseases such as cancer, heart disease and Alzheimer's disease. Raspberries are low in calories, but are a very rich source of dietary fibre. The large amount of Vitamin C present in them makes these berries very good antioxidants. High levels of a compound known as ellagic acid, a polyphenol prized for its anti-carcinogenic properties is present in them.

Best eaten: Can be eaten raw, in fruit salads, as a puree or jam. Make a smoothie in the blender with milk or yogurt. Nutritional facts per cup Vitamin C: 54 per cent Manganese: 41 per cent Vitamin K: 12 per cent Calcium: 3 per cent Iron: 5 per cent

Health Benefits

Rich in fibre

High in vitamin C

Antioxidant for bone and skin health

Anti cancer activity

Heart disease prevention

Protection against Macular degeneration

Raspberries are low in calories and saturated fats but are rich source of dietary fibre.

Raspberries have high levels of phenolic flavonoid phytochemicals such asanthocyanins, ellagic acid **(a tannin), quercetin, gallic acid, cyanidins, pelargonidins, catechins, kaempferol and salicylic acid.**

Xylitol, a low-calorie sugar substitute extracted from raspberries. Xylitol contains just 9.6 calories, as compared to one teaspoon of sugar, which has 15 calories. Xylitol absorbs more slowly than sugar and does not contribute to high blood sugar levels; can be helpful in diabetics.

Fresh raspberries are excellent source of **vitamin-C**, which is also a powerful natural anti-oxidant and may contain potential anti cancer agents.

Raspberry contains anti-oxidant vitamins like vitamin A, and vitamin E. In addition to the above mentioned antioxidants, is also rich in many other health promoting flavonoid poly phenolic antioxidants such as *lutein, zeaxanthin,* and *beta carotene* in small amounts.

Raspberry has an **ORAC value** (oxygen radical absorbance capacity) of about 4900 per 100 grams, crediting it among the top-ranked ORAC fruits.

Raspberries contain good amount of minerals like potassium, manganese, copper, iron and magnesium.

They are rich in B-complex group of vitamins and vitamin K. Contain very good amounts of vitamin B-6, niacin, riboflavin, and folic acid.

Raspberries have also been linked to a positive promotion in cardiovascular health and cancer.

Raspberries provide important anti-inflammatory, including anthocyanins (the pigments in red, **purple** and blue fruits), which are believed to help reduce cardiovascular disease and diabetes, help to improve eyesight and memory functions.

Ellagic acids, which may slow down the growth of some cancer cells, is found in raspberries (including the seeds!), in fact raspberries are a leading source of the acid. Research has also shown that ellagic acid from raspberries is easily absorbed by the body. The seeds of raspberries also serve as beneficial "roughage" in the digestive tract.

Raspberries, especially the seeds, may become more and more important in the booming cosmeceuticals market (skin care products with health benefits). The oil in the seeds of the raspberries is rich in vitamin E, omega-3 fatty acids and has a natural SPF (sun protection factor) of 25 to 50.

Raspberry Leaf Tea has been praised by many herbalists and midwives for it's medicinal benefits in women's reproductive health. For centuries, leaves of the red raspberry plants have been used to help during pregnancy, in childbirth, with breastfeeding, with irregular menstruation, infertility, and other conditions in connection with women's health.

Raspberry leaves - This is beneficial for women facing fertility problems, as raspberry leaves encourage the normal growth of the embryo. These can also bring disturbed hormones back to normal (Anonymous, 2014).

Caution

Raspberry may cause serious allergic reactions in some sensitized individuals. Some of the most common symptoms include swelling and redness of mouth, lips and tongue, eczema, hives, skin rash, headache, runny nose, itchy eyes, wheezing, gastrointestinal disturbances, depression, hyperactivity and insomnia. Individuals who suspect allergy to these fruits may want to avoid.

KIWI FRUIT OR CHINESE GOOSE BERRY

Kiwi Fruits

Chinese Gooseberry (*Actinidia chinensis*): It is exceptionally high in vitamin C, in fact it contains even more vitamin C than an orange. It also contains high amounts of vitamins E, A, and K as well as flavonoids, antioxidants, and minerals such as magnesium, potassium, and iron. Kiwi is particularly beneficial for the respiratory system and has been shown to help shorten the duration of colds as well as to help prevent asthma, wheezing, and coughing. Kiwi fruit contains anti-inflammatory properties which is good for those who suffer with autoimmune disorders such as Lupus, Fibromyalgia, CFS, and Lyme disease. Kiwi seeds are an excellent source of **omega-3 fatty acids** which are essential for cognitive function and can help prevent the development of ADHD and autism in children. Kiwi contains enzymes similar to those in papaya and pineapple which makes them useful in aiding in digestion and elimination. Kiwi fruit has also been shown to help protect DNA from mutating which is an incredible form of protection against illnesses and diseases such as atherosclerosis, heart disease, osteoarthritis, asthma, rheumatoid arthritis, and cancer. Kiwi fruit is also known to help remove excess sodium buildup in the body which can help reduce bloating, swelling, and water retention. Kiwi is good for promoting eye health and for preventing age-related macular degeneration. It is also highly beneficial for those who have weak or sensitive immune systems and are useful at keeping ear, nose, and throat infections at bay. Kiwi is also great for diabetics by helping to keep their blood sugar levels under control and for cardiovascular health as it has been shown to help lower triglycerides or blood fat in the body. Kiwi contains certain compounds that act as a blood thinner, similar to the way aspirin works which helps prevent blood clot formation inside the blood vessels and can protect the body from stroke and heart attacks.

The fruit has a soft texture and a sweet but unique flavour. Kiwi fruit or Chinese gooseberry is one of the delicious fruits with full of promising health promoting phyto-chemicals, vitamins and minerals. Named after a flightless bird of the same name, the kiwi is also known by other names like 'macaque peach' and Chinese gooseberries'. And, did you know that it is the national fruit of China? Kiwi fruit, with its tangy-sweet refreshing flavour, is ideal for daily consumption. Cut, scoop out and relish the delicious flesh that bursts with a tangy, juicy taste. It can be a part of every food course and can be used in fruit salads, desserts, smoothies, jams, chutneys or curries.

Each fruit measures about the size of large hen's egg and weighs up to 125 g. Internally the flesh is soft, juicy, emerald green in colour with rows of tiny, black, edible seeds. Fruit texture is similar to strawberry or sapodilla and the flavour resembles a blend of strawberry, apple and pineapple fruits. Similarly, consuming blue berries help protect brain from Alzheimer's disease and dementia. Kiwi fruit stands number one in nutrient content compared to 27 other fruits. It is packed with twice the amount of Vitamin C – compared to oranges (per 100 mg) – and has twice the amount of nutrients – compared to apples (per 100 mg). Kiwi fruit is an excellent source of Vitamin E, fibre, potassium, folic acid, carotenoids, antioxidants and trace minerals (Anonymous, 2013).

Vitamin E

It's quite unusual for a low fat food - let alone fruit - to be a good source of vitamin E, which is well-known for its heart health and antioxidant properties. However, kiwifruit is a great low-fat, natural source of vitamin E.

Folic Acid

Kiwifruit are a natural source of folic acid which is needed to prevent cardiovascular disease, help brain and cognitive development and prevent neural defects in babies - both before and during pregnancy.

Potassium

Kiwifruit has about the same levels of potassium as bananas but only half the calories - making kiwifruit an excellent low-sodium, high-potassium fruit which can be beneficial in the maintenance of blood pressure and for heart health.

High in vitamin C, dietary fibre, potassium and magnesium

Low in saturated fat, sodium and cholesterol

Health Benefits

Perhaps, the biggest plus point of this fruit is that it is an abundant source of vitamin C. In fact, they contain twice the amount of vitamin C than that of oranges. This advantage is useful when it comes to boosting the immune system. Also, the abundance of vitamin C helps in fight against the damage of free radicals in the body, reducing the risk of developing cancer. Also, this reduces the signs of ageing through reducing wrinkles, and dark spots.

It is said that kiwis promote respiratory health as well. This helps in reducing instances of coughing, shortness of breath, wheezing, *etc.*

Being a good source for dietary fibre, kiwis are great for reducing problems like constipation and improving digestion.

Kiwi fruit also contains a high level of **serotonin**, which is known as a happy hormone and reduces stress.

Who would have thought that kiwis are good for impotence? **Arginine**, an amino acid is useful for treating **impotence** in men.

Kiwis are great for cardiovascular health. Pectin found in kiwis are known to lower cholesterol. Also, deficiency of magnesium is associated with heart diseases and hypertension. **Cartenoids**, phenolic compounds, antioxidants found in kiwis help prevent oxidation of good cholesterol (HDLs).

A naturally occurring sugar, **inositol** plays a factor in regulating diabetes.

Kiwis are great for people who want to lose weight as it contains healthy fats, healthy sugars and is low in calories.

If you want to improve eyesight, then you can turn to this fruit. The combination of vitamins A, C and E aid in maintaining ahealthy eyesight and preventing the macular degeneration of the eyes.

Kiwifruit is a very rich source of soluble dietary fibre, which makes it a good bulk laxative. The fibre content helps to protect the colon mucous membrane by decreasing exposure time to toxins as well as binding to cancer causing chemicals in the colon.

Kiwi fruit contains very good levels of vitamin-A, vitamin-E, **vitamin-K** and flavonoid anti-oxidants such as beta-carotene, lutein and xanthin. Vitamin K has potential role in the increase of bone mass by promoting osteotrophic activity in the bone. It also has established role in **Alzheimer's** disease patients by limiting neuronal damage in the brain.

Kiwi-fruit functions as blood thinner similar to aspirin; helps to prevent clot formation inside the blood vessels and protects from stroke and heart attacks.

Kiwi seeds are an excellent source of omega-3 fatty acids. Consumption of foods rich in ω-3 fatty acids may reduce the risk of coronary heart disease, stroke, and help prevent development of ADHD, autism, and other developmental differences in children.

Freshfruit is a very rich source of heart healthy electrolyte "potassium". Potassium is an important component of cell and body fluids that helps controlling heart rate and blood pressure by countering effects of sodium.

It also contains good amounts of minerals like manganese, iron and magnesium. Manganese is used in the body as a co-factor for the powerful antioxidant enzyme, *superoxide dismutase*. Magnesium is an important bone-strengthening mineral like calcium.

Other Health Benefits

Consumption of kiwi fruit makes metabolism stronger as well as improves the nerve function.

Prevents wheezing and coughing, particularly in children.

Helps to prevent colon cancer

Boosts the immune system of the body so helps to fight cold and flu

Provides vitamins and **antioxidants** in a healthy amount

Prevents **asthma.**

Helps people with impotence.

Rich source of minerals such as magnesium, potassium and copper.

Also acts as a natural blood thinner.

Assists people suffering from depression.

Great for pregnant women as it contains folate which is necessary for cell development.

An anti mutagenic component present in kiwifruit helps to prevent the mutations of DNA that may start the cancer process.

CRANBERRY

Cranberries (*Vaccinium oxycoccus*) are rich in Vitamin C, but their acidic nature and the unusual nature of their proanthocyanidins make them extremely beneficial for health. The entire fruit, and not just the juice, shows presence of phytonutrients that is good for overall health. Cranberriesare an excellent source of vitamin C, A, and beta carotene. They are packed with antioxidants andt an ideal anti-aging and memory enhancing food. Cranberries have amazing anti-inflammatory and anti-cancer properties and are a vital food and supplement for anyone struggling with any chronic illness or disease. They are known to significantly boost the immune system and have a natural antibiotic effect in the body. Cranberries contain one of nature's most potent vasodilators which opens up congested bronchial tubes and pathways making it essential for healing any respiratory condition. Cranberries are very high in **tannic acids** which gives them there powerful ability to protect and heal urinary tract, bladder, and kidney infections. These tannic acids are made up of compounds called **proanthocyanidins** which essentially coats the infection forming bacteria, such as E.coli and H. Pylori, with a slick cover and prevents them from sticking to the walls of the urinary tract and digestive tract. Since the bacteria are unable to attach themselves to anything they are flushed out of the system and unable to cause any infection or harm. This anti-adhesion ability also help to prevent stomach ulcers, gum disease, and cavities. This ability also helps to prevent cardiovascular disease by stopping cholesterol plaque formation in the heart and blood vessels and by lowering LDL (bad) cholesterol and increasing HDL (good) cholesterol levels in the blood. Cranberry juice has also been shown to increase the desirable "friendly" bacteria in the digestive tract which benefits digestive disorders such as IBS, colitis, gastritis, indigestion, gas, bloating, and constipation. Cranberry juice has also been known to help treat diaper rash by reducing pH levels in the diaper and thereby reducing irritation. Native Americans commonly ate their cranberries simmered in honey or maple syrup or sun-dried and mixed with nuts to last them through the winter months. Fresh cranberries can be added to salads, smoothies, fresh juices, and fruit and nut salads or cooked down into the classic cranberry sauce. Sun-dried cranberries are an excellent addition to trail mixes, hot or cold breakfast cereals, grain and vegetables dishes, and wholesome baked goods. Sun-dried cranberries can also be made into a medicinal tea by soaking in water overnight. Pure cranberry juice can be taken straight or mixed with spring water, coconut water, or apple or grape juice to receive its healing benefits. Cranberry extracts can also be found in capsule and tincture form online and in your local health food store for year-round use.

Best eaten: Cranberry can be used in cooking like we use lemon. These berries not only work well with sweets and tarts, but also go well with spicy, flavoured foods. Cranberry extracts have been shown to inhibit the growth of oral, colon, and prostate cancer cells, in addition to inducing cell death of breast cancer cells. Cranberries are loaded with unique antioxidants and other **phytonutrients** that may protect against heart disease, cancer and variety of other diseases. **Phytonutrients** in cranberries may also be involved in inhibiting the spread of cancer cells throughout the body. A urinary tract infection is usually caused by bacteria or germs which get into the urinary tract and multiply, causing redness, swelling and pain. But there's an easy way to promote good urinary tract health, and it starts by taking the cranberry glass-a-day challenge. Cranberries contain powerful antioxidants called **proanthocyanidins** or condensed tannins, and they prevent certain bacteria from sticking to the urinary tract, " says registered dietitian Lois Ferguson. "And when bacteria can't stick, bacteria cannotcause infection.

Nutritional facts per cup Vitamin C: 24 per cent Manganese: 20 per cent Vitamin K: 7 per cent Iron: 2 per cent Vitamin A, Calcium: **1 per cent**

Benefits of the Cranberry

Cranberries are a rich source of the flavonoid **quercetin** which can inhibit the development of both breast and colon cancers.

Preliminary studies show that drinking cranberry juice is good for the health of the heart.

Research indicates that cranberries are an excellent source of antioxidants which may protect against cancer, heart disease and other diseases.

Found to decrease production of cavity and plaque producing bacteria in your mouth.

Also found to reduce the bacteria associated with peptic stomach ulcers.

In clinical studies, cranberries have been shown to help maintain a healthy urinary tract.

Cranberries are especially beneficial to the eyes (they significantly improve symptoms of cataracts, macular degeneration, and diabetic retinopathy).

Evidence on how cranberry juice fights bacteria that cause urinary tract infections.

Cranberries are part of a heart-healthy diet. Cranberries are a fat free, cholesterol free and low sodium food. Whole cranberries are a good source of dietary fibre, and all cranberry products contain flavonoids and **polyphenolics**, natural compounds that offer a wide range of potential health benefits.

Urinary Tract Health: Many people know that cranberries help maintain urinary tract health. Research is now showing just how cranberry juice promotes urinary tract health and why cranberries are the only food with this benefit.

Drinking cranberry juice can block urinary infections by binding to bacteria so they can't adhere to cell walls. While women often drink unsweetened cranberry juice to treat an infection, there's no hard evidence that works.

A compound Howell discovered in cranberries, **proanthocyanidine,** prevents plaque formation on teeth; mouthwashes containing it are being developed to prevent periodontal disease.

In some people, regular cranberry juice consumption for months can kill the *H. pylori* bacteria, which can cause stomach cancer and ulcers.

Preliminary research also shows:

Drinking cranberry juice daily may increase levels of HDL, or good cholesterol and reduce levels of LDL, or bad cholesterol.

Cranberries may prevent tumors from growing rapidly or starting in the first place.

Extracts of chemicals in cranberries prevent breast cancer cells from multiplying in a test tube; whether that would work in women is unknown.

BLUEBERRIES

This little berry has so many health benefits that we don't know where to start. Its rich vitamin C content aids blood circulation and provides minerals and salts, which help the body to fight ageing. Also, as it is high in potassium – found in all fruits and veggies – fluid balance in cells is regulated, thus combating puffiness and fluid retention. Have one of the highest antioxidant capacities among all fruits, vegetables, spices and seasonings. It makes memory sharp, improves digestive system, regulates blood sugar and cholesterol.

The highest antioxidant capacity of all fresh fruit: Blue Berries, being very rich in anti oxidants like Anthocyanin, vitamin C, B complex, vitamin E, vitamin A, copper (a very effective immune builder and anti-bacterial), selenium, zinc, iron (promotes immunity by raising haemoglobin and oxygen concentration in blood) etc. boost up your immune system and prevent infections. Once your immunity is strong, you won't catch colds, fever, pox and all such nasty viral and bacterial communicable diseases.

Neutralizes free radicals which can affect disease and aging in the body: Blueberries are laden with antioxidants and rank number 1 in the world of anti oxidants. This is mainly due to presence of Anthocyanin, a pigment responsible for the blue color of the blue berries. The abundance of vitamin-C is also a big factor for this as well. Experts say that blueberries have high levels of proanthocyanidins, which are linked to a reduced risk of prostate cancer. Blueberries are also said to be effective in reducing the risk of heart disease, Type 2 Diabetes and age-related memory loss.

Great antioxidants

Dietary fibre

Flavonoids

Fighting inflammation

Lowers cholesterol

Improves insulin sensitivity

Immune boosting system

Improves glucose control

Lowers risk of heart disease

Aid in reducing Belly

Helps promote urinary tract health

Been proved to preserve vision

Constipation and Digestion

Acai berries (Euterpeoleracea) may destroy cancer cells, particularly those associated with leukemia, are grown on the palm trees in the Amazon rainforest of northern Brazil. The name of the game with acai berries is pure antioxidant and nutrient power. Each acai berry measures about the size of small sized grape, 2-3 cm in diameter. Immature fruits appear dark-green initially, which turn deep-purple later upon attaining maturity.

Health Benefits of Acai Berry

Acai berry has very good levels of anti-oxidants, minerals, and vitamins that have health benefiting and disease preventing properties.

Unlike other berries and fruits, acai has high caloric values and fats. 100 grams of berries provide about 80-250 calories depending up on the preparation and serving methods.

Acai berry contains many polyphenolic anthocyanin compounds like resveratrol, cyanidin-3-galactoside, ferulic acid, delphinidin, petunidin as well as astringent pro-anthocyanidin tannins like epicatechin, protocatechuic acid and ellagic acid. These compounds act as anti-aging, anti-inflammatory, anti-cancer functions by virtue of their anti-free radical fighting actions. In addition, tannins found to have anti-infective, anti-inflammatory, and anti-hemorrhagic properties.

Ellagic acid in acai has anti-proliferative properties by virtue of its ability to directly inhib it DNA binding of certain carcinogens (nitrosamine toxins) in the food.

Acai berry is also rich in medium chain fatty acids like oleic acid (omega-9) and linoleic acid (omega-6). These compounds help reduce LDL-cholesterol level and raise good HDL-cholesterol levels in the body and thus help cut down heart disease risk.

Acai pulp has good levels of dietary fibre. Adequate fibre in the diet helps clear cholesterol through the stools.

Berries contain a good amount of minerals like potassium, manganese, copper, iron, and magnesium.

Acai berries contain antioxidants, fibre and heart-healthy fats. They may have more antioxidant content than other commonly eaten berries, such as cranberries, blueberries and strawberries.

The good: This food is very low in Cholesterol and Sodium. It is also a good source of Dietary fibre, and a very good source of Vitamin A.

The bad: This food is high in Saturated Fat.

They fight leukemia: A well-known study, done by the University of Florida, found that extracts of the acai berry destroyed human cancer cells grown in a lab.

They reduce inflammation. One of the best things that acai berries can do for you, due to the large amounts of anthocyanins they contain, is reduce inflammation associated with chronic diseases. Heart disease, cancer, diabetes, arthritis, fatigue syndromes, digestive discomforts, aches, and pains are all helped by reducing inflammation.

They shield your heart against disease: The pulp of acai berries has deep healing agents that contain antioxidants and fibre that reduces cholesterol – and keeps your digestive system healthy to boot.

Goji Berries

Goji berries (*Lycium barbarum*) are richly immersed in antioxidant power. Goji berries, which are the commercial name for wolfberries, have been used for thousands of years in Chinese medicine. Unique among fruits because they contain all essential amino acids, goji berries also have the highest concentration of protein of any fruit. They are also loaded with vitamin C, contain more carotenoids than any other food, have twenty-one trace minerals, and are high in fibre. Boasting 15 times the amount of iron found in spinach, as well as calcium, zinc, selenium and many other important trace minerals, there is no doubt that the humble goji berry is a nutritional powerh

Battle cancer: The phytochemicals in goji berries may have powerful anticancer effects. A 1994 study published in the*Chinese Journal of Oncology* stated that goji berries have a positive effect on treatments when used in conjunction with other cancer therapies.

Support weight loss: Goji berries contain natural compounds that are lipotropic, meaning they help carry fat away from the liver and burn those extra calories.

Protect your heart: Goji berries have compounds to lower cholesterol, are natural defenders against free radical damage, and release levels of*homocysteine,* a protein associated with heart disease and inflammation.

Prevent age-related eye problems: Goji berries have a high level of antioxidants, like beta carotene and zeaxanthin, which are important for vision. Zeaxanthin protects the eyes, specifically the retina, and may reduce the risk of age-related macular degeneration.

Boost your libido: This amazing superfood not only raises your spirits, but it also raises your libido! Goji berries raise testosterone levels, and, therefore, your sex drive goes up.

Elderberries

Elderberries (*Sambucus nigra*) are dark berries that grow in areas with subtropical climate. They are also called Sambucus. Besides food, these berries are an excellent way to treat various health conditions. The berries are black or very dark blue, and have a sharp, sweet flavour that makes them highly preferred for desserts, syrups, jams, jellies, spreads, and as the base for various cocktails and beverages.

Health Benefits

Antioxidants: Moreover, studies have shown that these berries protect the cells from damage, which often leads to chronic medical conditions, such as cancer and heart disease. Elderberries contain more antioxidants, vitamin A and vitamin C than other types of berries, like cranberries and blueberries. Therefore, this fruit is able to cleanse the body from reactive molecules and free radicals that damage the body cells.

Antiviral properties: Elderberries are able to protect the cells from infections caused by proteins that cause viruses. So, elderberries also have antiviral qualities. According to research, elderberries improve the function of the immune system, which destroys the flu virus. Moreover, they speed up the healing process in case of influenza virus, by removing the infection by destroying pathogens and inducing sweating.

Diabetes control: Elderberries protect people suffering from diabetes from osteoporosis and oxidation. Elderberries are also able to reduce the levels HbA1c or glycosylated hemoglobin. Other benefits of elderberries are enhancing bone density, decreasing oxidation lipids and restoring the levels of glutathione, an antioxidant that the liver uses to cleanse the body from toxins.

High blood pressure: They are a natural diuretic. Therefore, they are an excellent supplement for lowering high blood pressure.

Digestive Health: Elderberries are packed with dietary fibre that can help eliminate constipation, reduce excess **gas**, and generally increase the health of your gastrointestinal system. Fibre can also help to increase the nutrient uptake efficiency in your gut so you get more out of your **food!**

Cardiovascular Protection: The high fibre levels help to eliminate excess cholesterol from the system and make room for "good" cholesterol that the body needs. This can help eliminate the chances of developing atherosclerosis and other cardiovascular issues. The high levels of potassium in elderberries also protects the heart by relaxing the tension of blood vessels and arteries. As a vasodilator, potassium can significantly reduce blood pressure and keep your heart healthy!

Respiratory Health: When it comes to clearing up a sore throat, a cough, cold, bronchitis, or any other issue that affects your respiratory system, elderberry juice might be your best choice. Like many cough syrups, elderberries contain active ingredients (**bioflavanoids** like **anthocyanins**, to be exact) that can soothe inflammation and irritation and also act as an expectorant and clear out phlegm that can trap foreign agents in your glands. Elderberry juice is even recommended for people with asthma.

Immune System: It does have certain antibacterial and anti-infectious qualities, and it is very commonly used to ward off influenza during bad seasons where it seems that everyone is catching it. Furthermore, elderberries can strengthen the immune system against itself, protecting against the effects of autoimmune disorders, even alleviating certain symptoms and associated pain of **AIDS.**

Diabetes Aid: Some of the active antioxidant ingredients in elderberries work directly on the pancreas to regulate insulin and glucose levels, either providing stability for people who suffer from diabetes or helping non-diabetics to avoid developing this terrible condition.

Bone Health and Inflammation: While the antioxidant and anti-inflammatory compounds found in elderberries can help alleviate joint pain and soreness from inflammation, the high levels of essential minerals will help promote bone strength and the development of new bone tissue. Osteoporosis is a condition that millions of people face in the future, but increasing bone density during your younger decades can delay the onset considerably.

NUTS

Eating one ounce of mixed nuts (raw unpeeled walnuts, almonds and hazelnuts) a day may benefit those who are at high risk for heart disease and type 2 diabetes, a new study has suggested. Scientists have for the first time found a link between eating nuts and higher levels of **serotonin** in the bodies of patients with metabolic syndrome (MetS), who are at high risk for heart disease. **Serotonin** is a substance that helps transmit nerve signals and decreases feelings of hunger, makes people feel happier and improves heart health. Brazil nuts are a very rich source of selenium, a mineral that works very well with Vitamin E to stop oxidative **stress** and cell damage caused by free radicals; thus, slowing down the ageing process. But should be eaten in moderation (two nuts a day) because of its high fat content. Dry fruits are rich source of nutrients such as vitamins and minerals. Consumption of dry fruits daily enhances the overall bioavailability of nutrients. In addition, mixing dry fruits with some fresh fruits would help you get vitamin C which in turn favours absorption of iron inside the stomach. Almonds and walnuts are a great source of **omega** and **fatty acids** and **anti-oxidants** that help in pain control. Add these nuts to your salad or just munch on a handful as an afternoon snack. Walnuts and almonds are the healthiest of all nuts. Nuts are fortified with the B group of vitamins, magnesium, zinc and omega oils that helps to keep **cortisol** (a stress hormone) levels low. Also, they are a good source of energy, balance out sugar cravings and aids metabolism. The mono-unsaturated fats present in nuts help in curbing overeating. But there's no need to go overboard with nuts as they have high calories. Nuts also provide magnesium, which helps maintain bone structure and chromium, which helps to ensure proper insulin function. These contain zinc, needed for growth and wound healing and manganese, which protects against free radicals. All nuts are a good source of vitamin E, an important antioxidant. They are high in fibre and photochemicals, both of which help protect against cancer and other chronic diseases. They make our skin glow and are an excellent source of protein. They contain plant sterols, mono-unsaturated and polyunsaturated fats, which lower the LDL (bad) cholesterol.

When eaten in moderation, nuts are not fattening. Besides **antioxidants,** nuts are rich sources of good fats which tend to lower the bad cholesterol, are good for skin and hair and are highly beneficial for health. These can be sliced and sprinkled over salads, cereal/oatmeal porridge and can be roasted and had as snacks.

Researchers have found that consumption of tree nuts (almonds, Brazil nuts, cashews, hazelnuts, macadamias, pecans, pinenuts, pistachios and walnuts) was associated with a better nutrition profile and diet quality; lower body weight and lower prevalence of metabolic syndrome; and a decrease in several cardiovascular risk factors compared to those seen among non-consumers. They further reported that one serving (28g or 1ounce) of tree nuts per week was significantly associated with 7 per cent less MetS. Interestingly, while overall nut consumption was associated with lower prevalence of MetS, tree nuts specifically appear to provide beneficial effects on MetS, independent of demographic, lifestyle and other dietary factors (Anonymous, 2013). Brazil nuts, uniquely rich in selenium, fibre, and phytochemicals, can help fight inflammation, improve the immune system and prevent tumour growth. Researchers at Imperial College London and Norwegian University have founded that eating 20 grams of nuts a day, may lower people's risk of coronary heart disease by nearly 30 per cent, cancer by15 per cent and premature death by 22 per cent, as nuts and peanuts are high in fibre, magnesium and polysaturated fats- which are beneficial for cutting cardiovascular disease (Anonymous, 2016).

One thing all nuts have in common is their ability to enhance your energy in a short period of time. But if you look closer walnuts are a good source of **Omega fatty acids**. Just a handful of nuts and you're on a roll, but you are also stronger in controlling negative moods like anxiety and anger. They are rich in.

Protein, fibre, vitamin E and flavonoids.

Mono-unsaturated fat

Lowers LDL Cholesterol.

ALMOND

Unshelled Nuts

With Shell

Almonds (*Prunus dulcis*) are among the most nutrient-rich tree nuts and just a handful a day is a great way to ensure you're getting more of the good things your body needs and you want. Almonds are delicious, have long been revered as symbol of wellness and health. The king of all nuts with the highest iron content is the humble almond. One ounce of almonds every day provides 6 per cent of iron. In fact, the good news is that almonds are the cheapest of all the other nuts and dry fruits available to us. The nuts are enriched with many health-benefiting nutrients that are essential for optimum health. Almonds are an excellent source of **vitamin E,** magnesium, manganese, and a good source of fibre, copper, phosphorous, riboflavin and small amount of calcium. Plus, a one-ounce serving has 13 grams of good unsaturated fats, just 1 gram of saturated fat, and is always **cholesterol** free. And when compared ounce for ounce, almonds are the tree nut highest in protein, fibre, calcium, vitamin E, riboflavin, and niacin. The tiny almond is a storehouse of nutritional goodness. Each ounce contains 6 grams of protein and 3 grams of fibre, not to mention healthy doses of iron, zinc, copper, calcium, magnesium, Vitamin E and array antioxidants. Snacking on almonds is a great way to help lower cholesterol and high blood pressure, and can smooth out spikes in blood sugar in diabetics. Calcium is an important nutrient found in almonds. A single serving of almonds can contain up to a quarter of one's daily calcium needs. Such high levels of calcium make almonds a tooth and bone strengthening food as well.

Our diet needs to provide adequate amounts of all of these 'little helpers'. Almonds are a power packed nut which provide you with all the good variants of nutrition like carbohydrates, fats, minerals, proteins and vitamins. Since time immemorial, almonds have been cherished, not only for their delicate flavour and delectable crunch, but also for their nutritional benefits. They are a simple, tasty choice that can make a big difference in the way you feel every single day, not a small feat. And that means you can feel good about what you're eating while enjoying their delicious crunchy flavour. Almonds promote a healthy heart in many ways. The cholesterol-lowering effects of almonds can also help almond snackers maintain healthy hearts. **Alpha-tocopherol vitamin E** found in Almonds are one of the leading food sources of the important antioxidant, which helps neutralize harmful free radicals in your body and help you feel your best day and night. So what are free radicals? They're unstable molecules that form as your body burns oxygen. Free radicals can damage your cells, tissues, and even your DNA. So make a healthy resolution this nutrition week and keep yourself healthy for the rest of the year.

Soaked almonds help in digestion. It releases the enzyme Lipase, which is beneficial for fat digestion. They are a good source of antioxidants, which in turn resists free radical damage and prevent ageing. They contain Vitamin B17, folic acid that are vital for fighting cancer and reducing birth defects. They are rich in mono-unsaturated fats that curb your appetite and make you feel full.

An ounce (or a handful) of almonds is rich in proteins, fibres, vitamins and trace minerals. They are also less in sugar. So, consuming almonds every day has a multitude of health benefits. They help protect the heart against diseases, as they are rich in magnesium, which is essential for a healthy heart. Additionally, they are good for reducing cholesterol levels and aid in weight loss too. They promote gastrointestinal health and help combat diabetes too, as the sugar content in almonds is less.

Eat almonds and make yourself stronger forever.

Here's a piece of advice **for men** who want a healthy sex life - start taking more almonds.

Sin Chew Daily reported that a study showed that **almonds** have a high quantity of a type of amino acid called arginine, which helps to relax blood vessels and improve blood circulation. However, the study added that if almond was not a person's cup of tea, other food items that are high in arginine were beans, salmon and wheat, reports the *Star Online.*

Since ages almonds have been considered as an anti-ageing food because of the high content of Vitamin E. It is great for healthy skin, hair and nails. The body needs Vitamin E to protect the cells from the onslaught of free radicals generated by air pollution, peroxides and ultra violet rays. Experts say that eating around a dozen almonds daily will help. Almond milk, just like coconut milk is a great alternative for vegans. Almond milk is a good source for proteins. It is also low in calories unlike cow's milk. This milk contains vitamin C, magnesium, mono-unsaturated fats, manganese, copper, riboflavins. It boosts energy and lowers heart diseases.

Almonds are a great source of essential fatty acids. These provide the raw material for a man's healthy production of hormones. Additionally, the smell of almonds arouses passion in females. You can add to the sexual ambience by lighting some almond-scented candles to build up her mood.

Almonds are packed with skin-boosting vitamin E and protein. Also, they are high in fibre, which stops you feeling hungry all the time. Not to mention almond taste great when eaten raw, roasted, with salt they are just perfect. Researches at Purdue University also found that, despite almonds are relatively high in calories, they actually do not contribute to bell fat.

Libido dietary tip: Eat almonds raw (with no added salt or sweetness). Alternately, crush some fresh almonds and sprinkle them on your salad to get the energy you need.

They are very low in Cholesterol and Sodium but a good source of Riboflavin, Magnesium and Manganese, and a very good source of Vitamin E (Alpha Tocopherol). All kind of nuts are **a**n excellent source of essential fatty acids, nuts are known to give your passion a hike. Rich in several kinds of minerals, they are pretty good for reproduction and sexual health. Almonds, in particular, are known to arouse enthusiasm in females simply through aroma. **Full of zinc**, nuts of all kinds are known to perk your sexual desire by helping you store energy which will help you last longer. "All kinds of nuts provide one with energy that's directly proportional to your stamina while having sex, " suggests Sanjana. So, feel free to consume your daily dose of nuts, including almonds, pistachios, walnuts, groundnuts *etc.* After all, stamina and energy are the driving force behind a satisfying sex life.

One ounce of almonds (about 20 to 24 shelled whole almonds) provides 35 per cent of your daily value for vitamin E. Vitamin E may help promote healthy aging. A study reported in the *Journal of the American Association* (June 26, 2002) suggests a diet rich in foods containing vitamin E may help protect some people against **Alzheimer's disease.**

Nuts also are important for what they DON'T offer:

Cholesterol. Nuts are cholesterol-free.

Sodium. Unless salt is added to nuts, they naturally contain, at most, just a trace of sodium.

Good for brain: Almond is a source of many nutrients which help in development of the brain. Almond induces high intellectual level and has been considered as an essential food item for growing children. Many mothers give almonds soaked in water to their children daily in the morning (2-3 pieces of soaked almonds are good enough, you can also remove the outer shell if it causes allergy to you).

Regulates Cholesterol

Regular consumption of almonds helps to increase the level of high density lipoproteins (HDL) and reduce the level of low density lipoproteins (LDL), thereby effectively controlling cholesterol levels. LDL cholesterol is called bad cholesterol.

Good for heart: Studies suggest that eating a handful of almonds on a regular basis can actually help lower bad cholesterol levels, while maintaining good cholesterol levels. Besides, snacking on almonds also increases the levels of alpha-1 HDL, the form of cholesterol that helps promote a healthy heart.

Mono-saturated fat, protein and potassium contained in almonds are good for the heart. Vitamin E acts as an anti-oxidant and reduces the risk of heart diseases. The presence of magnesium in almonds helps to avoid heart attacks. Almonds help reduce C-reactive protein which causes artery-damaging inflammation. Almond is also a source of folic acid. They therefore help to reduce the level of homocystein, which causes fatty plaque buildup in arteries. Experts say that eating even an ounce of almonds a day can reduce the risk of potential heart disease.

Skin care: The benefits of almond for skin care are well known and hence a massage with almond oil is often recommended for new born babies. Almond milk is also added in some soaps as almonds help in improving the complexion of the skin. Almond oil or moisturisers that are rich in almonds is very good for the skin. Almond oil is rich in olein glyceride and linoleic acid which help prevent the occurrence of acne, pimples, and blackheads. Almonds are also rich in Vitamin E which keeps the skin soft and supple. Almond oil is also used on babies to tone their skin and to aid in muscle growth.

Regulates blood pressure: Potassium present in almond helps to regulate blood pressure. Almonds are very low in sodium which also helps in containing blood pressure

Prevention of cancer: Almond improves the movement of food through colon, thereby preventing colon cancer.

Protection against diabetes: Almonds also help in reducing the rise in sugar and insulin levels after meals. They are low in cabs, but high in healthy fats, protein and fibre. Besides, they are loaded with magnesium, a mineral that most diabetics do not get enough of. Studies suggest that high levels of magnesium may be beneficial for prevention of type 2 diabetes and also metabolic syndrome. Both of which are major health concerns today.

Good in pregnancy: Almond contains folic acid. Folic acid helps to reduce the incidence of birth defect in newborn babies.

Weight loss: Snacking on a few almonds in moderation helps you feeling full and satisfied for longer periods, thanks to its fibre, protein and fat content which help curb the urge to over eat, Studies also suggest that almonds that almonds play an important role in reducing the body's absorption of calories, there by making them the perfect weight loss friendly snack.

Unsweetened almond milk helps one to reduce weight. The mono-saturated fat contained in almonds satisfies appetite and prevents over eating. Studies have revealed that almond rich low calorie diet is good for obese people to assist in shedding their weight.

Prevention of constipation: Almonds are rich in fibre. Like most other fibre rich food, almonds also help in preventing constipation. Make sure you drink good amount after eating almonds.

Boosts energy: The presence of manganese, copper and Riboflavin helps in energy production.

Best muscle building foods: Almonds provide a good source of protein and fat, but it's their vitamin E that is most beneficial to your muscles. The powerful antioxidant fights free radicals and helps you recover quicker from your workouts.

Improves bone health: Raw almonds are rich in phosphorus, magnesium and calcium, all of which are good for bones and teeth.

Learn more: Just a hand full of these nuts a day provides enough recommended levels of minerals, vitamins, and protein. Besides, almond oil extracted from the nuts has been used in as base or carrier oil in medicine, **aromatherapy** and in pharmaceuticals.

For hair: Almond oil is useful in treating almost all kinds of hair ailments. It not only arrests hair fall, but is also useful in the treatment of dandruff, and delays greying. In addition, using almond oil, will make your hair healthier and shiny too. Almond oil is easily available in the market these days, and it can be used either by itself or in combination with other oils, in order to maintain the health of your hair.

Health Benefits

Raw almonds are abundant source of anti-oxidants, that help protect the body against oxidative stress, which contributes to ageing and cancer. However, as these powerful antioxidants are largely concentrated in the skin of the almonds, eat them whole to get the best out of these power nuts.

Almond nuts are rich in dietary fibre, vitamins, and minerals and packed with numerous health promoting phyto-chemicals; the kind of well-balanced food ensure protection against diseases and cancers.

These nuts are rich source of energy and nutrients. They are especially, rich in mono-unsaturated fatty acids like **oleic** and **palmitoleic acids** that help to lower **LDL** or "bad cholesterol" and increase HDL or "good cholesterol."

The nuts are an excellent source of **vitamin E**; contain about 25 g per100 g. Vitamin E is a powerful lipid soluble antioxidant, required for maintaining the integrity of cell membrane of mucus membranes and skin by protecting it from harmful oxygen free radicals.

Almonds are free in gluten and therefore, are a popular ingredient in the preparation of gluten free food formulas. Such formula preparations are in fact healthy alternatives in people with wheat food allergy and celiac disease.

Nuts are packed with many important B-complex groups of vitamins such as riboflavin, niacin, thiamin, **pantothenic acid**, vitamin B-6, and folates.

Almonds are also rich source of minerals like manganese, potassium, calcium, iron, magnesium, zinc, and **selenium**.

The sweet almond oil is obtained from the nuts is an excellent emollient; helps to keep skin well protected from dryness. It has also been used in cooking, and as "carrier or base oil" in traditional medicines in aromatherapy, in pharmaceutical and cosmetic industry.

Almonds are low in saturated fat and contain many other protective nutrients - calcium and magnesium - for strong bones, vitamin E and compounds called **phytochemicals**, which may help protect against cardiovascular disease and even cancer.

Almonds are **phytochemica**l powerhouse.

Reduce heart attack risk and lower **cholesterol.**

20-25 almonds (approximately one ounce) contain as much calcium as 1/4 cup of milk, a valuable tool in preventing **osteoporosis**.

Almonds are the best whole food source of vitamin E, in the form of **alpha-tocopherol**, which may help prevent cancer.

If you're pregnant, or thinking about it, almonds are a great source of the folic acid you need!

Almonds contain more magnesium than oatmeal or even spinach. Build strong bones and teeth with the phosphorus in almonds. Most of the fat in almonds is mono-unsaturated, also known as the "good" fat.

Intake of 6 almonds (soaked overnight) is also helpful in keeping a check on diabetes.

Benefits of Soaked Almonds

Why soaked almonds are better - Firstly, the brown peel of almonds contains tannin which inhibits nutrient absorption. Once you soak almonds the peel comes off easily and allows the nut to release all nutrients easily (Sharma, Gargi, 2016).

How to soak? Soak a handful of almonds in half a cup of water. Cover them and allow them to soak for 8 hours. Drain the water, peel off the skin and store them in plastic container. These soaked almonds will last you for about a week.

Help digestion - Soaking almonds helps in releasing enzymes which in turn help with digestion. Soaking almonds releases enzyme lipase which is beneficial for digestion of fats.

Help weight-loss - The mono-unsaturated fats in almonds curb your appetite and keep you full. So feel free to snack on them because they'll help you avoid binge-eating and trigger weight-loss.

Keep heart healthy, reduce bad cholesterol (low density lipoprotein) and increase good cholesterol (high density lipoprotein).

Good source of antioxidants: Vitamin E present in soaked almonds works as an antioxidant which inhibits free radical damage that prevents ageing and inflammation.

Fight Cancer: Soaked almonds contain Vitamin B17 which is vital for fighting cancer. Flavonoid present in almonds suppresses tumor growth.

Help in **lowering** and maintaining glucose levels and regulating high blood pressure. (Eat nuts to control blood sugar and fat.)

Soaked almonds contain folic acid which reduces birth defects.

Caution

Almond nut allergy, although not so common as other tree nut allergies like cashew, pistachio etc., may cause hypersensitivity cross-reactions in some people to food substances prepared using the nuts. The type and severity of symptoms may vary and may include vomiting, diarrhea, pain abdomen, swelling of lips, and throat leading to breathing difficulty, and chest congestion. Therefore, caution should be exercised in those with nut allergic syndrome while using food preparations that contain nut products.

WALNUT

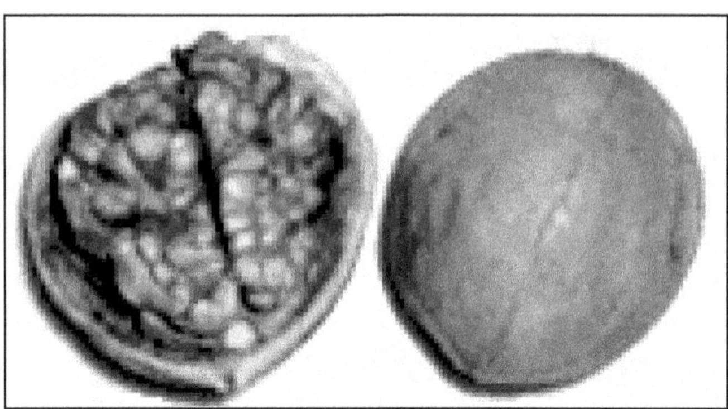

Walnuts (*Juglans regia*) are revered since ancient times as symbol of intellectuality since their kernels have convoluted surface inside the shell resembling as that of brain! It is believed that royal Kings and Queens used to eat raw walnuts every day to keep their body healthy and curvy. According to experts, eating walnuts on a daily basis also helps in making skin look younger. The nuts are enriched with many health-benefiting nutrients, especially Ω-3 fatty acids that are essential for optimum health. One ounce of walnuts (about 14 shelled walnut halves) is all that is needed to meet the dietary recommendation for **omega-3 fatty acids.** Each nut is roughly spherical in shape, about the size of medium sized **lemon** weighing about 10-15 g enclosing edible kernel. The walnut kernel consists of two uneven, corrugated lobes, off white in colour and covered by a paper thin, light brown skin. The lobes are partially attached to each other. Oil extracted from the nuts, apart from cooking, has also been used in as base or carrier oil in medicine, and aromatherapy. Walnuts are a rich source of Vitamin E and having a handful can help your skin look great. Walnuts are a very good source of manganese and a good source of copper, two minerals that are essential cofactors in a number of enzymes important in antioxidant defenses. For example, the key oxidative enzyme *superoxide dismutase*, which disarms free radicals produced within cell cytoplasm and the mitochondria (the energy production factories within our cells) requires both copper and manganese. Research has shown that having walnuts daily can reduce the risk of heart disease and can help in lowering bad cholesterol (LDL) and the C-reactive protein. When having a meal high in saturated fats, having a handful of raw walnuts can limit the ability of the harmful fat to damage arteries. Walnuts contain less than 4 grams of saturated fat per 50 grams. However, all nuts are high in calories, so, having them in moderation is essential. Walnuts are rich in antioxidants, **omega-3-fats**, **alpha linoleic acid** and abundant in omega-3-fatty acids, natural phytosterols and antioxidants, walnuts make for an ideal anti-breast cancer food.

Walnuts also contain an antioxidant compound called ellagic acid, which blocks the metabolic pathways that can lead to cancer. Ellagic acid not only helps protect healthy cells from free radical damage, but also helps detoxify potential cancer-causing substances and helps prevent cancer cells from replicating. In a study of over 1, 200 elderly people, those who ate the most strawberries (another food that contains ellagic acid) were three times less likely to develop cancer than those who ate few or no strawberries. It reduces vata, aggravates Kapha and pitta, promotes intelligence, promotes digestive functions, oleation, anti -toxin, expectorant. It also increases physical strength, potency and is a bulk promoting herb. Its paste is beneficial for complexion, anti –dermatoses, relieving oedema and is an effective pain reliever (Acharya Balkrishna, 2012). Scientists discovered that these edible seeds cut the risk of dying from cancer by 40 per cent and from cardiovascular disease by at least 55 per cent. In general, nut eaters in the research had a 39 per cent lesser risk of death and walnut eaters in particular a 45 per cent reduced threat (Anonymous, 2013). Research at the Yale-Griffin Prevention Research Centre, Connecticut, found that daily intake of 56 walnuts as a snack or with a meal improves endothelial function in overweight adults with visceral adiposity, and can ward off diabetes and heart disease in at risk individuals (Anonymous, 2013). Scientific studies show that incorporating walnuts in a healthy diet reduces the risk of all major diseases like cardiovascular diseases, high blood pressure, diabetes, cancer, asthama, arthritis and add a glow to the skin, luster to the hair, prevents aging and memory loss (Setalvad, 2014).

Just a handful of walnuts provides:

1. 2.5grams ofALA, the plant-based source of omega-3 fatty acid.
2. 4 grams of protein
3. 2 grams of fibre
4. 10 per cent of your daily value of magnesium and phosphorus

Health Benefits of Walnuts

Cures infections: Walnut oil is known to be a remedy to treat fungal infections. Athlete's foot and Candida are some of the infections which are known to get treated with the application of walnut oil. Apply the oil topically on the affected area. You can also mix this oil with other herbal anti-fungal like garlic to enhance its effectiveness.

Treats psoriasis: Psoriasis is known to be a painful and persistent skin condition that can be eased with the use of walnut oil. Add this oil to your bath or use it topically on your skin to treat this condition.

Excellent source of Omega-3: Walnut oil is known to be an excellent plant source of Omega-3. It is also associated with numerous health benefits which includes protection against heart disease, some types of cancer and other inflammatory disorders like eczema *etc.* (Meghna Mukherjee, 2013).

The nuts are rich source of energy and contain many health benefiting nutrients, minerals, antioxidants and vitamins that are essential for optimum health.

They are rich in mono-unsaturated fatty acids like **oleic acid** and an excellent source of all important **omega-3** essential fatty acids like *linoleic acid, alpha linolenic acid* (*ALA*) *and arachidonic acids.* Regular intake of walnuts in the diet help to lower total as well as LDL or "bad cholesterol" and increase HDL or "good cholesterol" levels in the blood. Diet rich in mono-unsaturated fatty acids and omega-3 fatty acids help to prevent coronary artery disease and strokes by favoring healthy blood lipid profile.

Eating 25 g each day provides about 90 per cent of RDI (recommended daily intake) of **omega-3 fatty acids**. Studies have revealed that n-3 fatty acids by their virtue of anti-inflammatory action helps to lower the risk of blood pressure, coronary artery disease, strokes and breast, colon and prostate cancers.

Walnuts are rich source of many phyto-chemical substances that may contribute to their overall anti-oxidant activity, including *melatonin, ellagic acid, vitamin E, carotenoids, and poly-phenolic compounds.* These compounds have potential health effects against cancer, aging, inflammation, and neurological diseases.

Walnuts have highest levels of popyphenolic antioxidants than any other common edible nuts. Eating as few as six to seven average sized nuts a day could help scavenge disease causing free radicals from the body.

Walnuts are also excellent source of **vitamin E**, especially rich in *gamma*-tocopherol; contain about 21 g per 100 g. These nuts are packed with many important B-complex groups of vitamins such as riboflavin, niacin, thiamin, pantothenic acid, vitamin B-6, and folates.

They also very are rich source of minerals like **manganese, copper**, potassium, calcium, **iron,** magnesium, zinc, and selenium.

Walnuts oil has flavorful nutty aroma and has an excellent astringent properties; helps to keep skin well protected from dryness. It has also been used in cooking, and as "carrier or base oil" in traditional medicines in massage therapy, aromatherapy, in pharmaceutical and cosmetic industry.

Eating a handful of nuts daily may help prevent weight gain and possibly promote weight loss. The fat, protein and fibre in nuts help you feel full longer, so you may eat less during the day.

Women in a Harvard School of Public Health, reported eating 5 or more 1 ounce servings of nuts/peanuts per week reduced their risk of Type 2 diabetes by almost 30 per cent compared to those who rarely or never ate nuts. Women in the study who ate five tablespoons of peanut butter each week reduced their risk for Type 2 diabetes almost 20 percent. (*Journal of the American Medical Association,* Nov. 27, 2002.)

Melatonin**,** a hormone produced by the pineal gland, which is involved in inducing and regulating sleep and is also a powerful antioxidant, has been discovered in walnuts in bio-available form, making them the perfect evening food for a natural good night's sleep.

Walnuts are full of **omega-3 fatty acid**, which has numerous benefits. They also contain alphalinoleic acid, which strengthens the bones.

Munch a handful of walnuts a day and you will have enough recommended levels of minerals, vitamins, and protein.

Walnuts which are rich in omega-6 fatty acids and arginine help in the production of nitric oxide, thus causing relaxation of arterioles and increasing the blood flow.

Sprouted walnuts have higher benefits.

Soaked walnuts are less acidic.

Walnut oil helps relieve joint pains and improves skin quality.

Walnut is the only nut rich in polysaturated fatty acids like oleic acid and an excellent source of all important omega-3 fatty acids like alpha-linolenic acid. These have maximum omega -3 fatty acids than any other nut. Therefore these are known to reduce bad (LDL) cholesterol and improve HDL (good) cholesterol making them a very powerful anti oxidant which fights free radical damage.

Walnts are one of the best plant sources of protein.

The anti-inflammatory benefits and Omega -3 fatty acids in watnuts help fight asthma, rheumatoid arthritis and inflammatory skin diseases such as eczema and psoriasis (Setalvad, 2014).

Caution

Walnut allergy is a type of hypersensitivity reaction to food substances prepared using these nuts. It is due to prior sensitization of the immune system by allergens in the nuts, which may lead to severe physical symptoms like pain abdomen, vomiting, swelling of lips and throat leading to breathing difficulty, and chest congestion. Therefore, it is advised to avoid any food preparations that contain these nut products in known case of walnut allergic individuals.

CASHEW NUT (*Anacardium occidentale*)

Cashews are a crunchy, yet delicately sweet nut that is treasured around the world for its flavour and health benefits. Cashews are high in protein and are a rich source of vitamin B-complex and essential minerals such as iron, selenium, zinc, and copper. They are also packed with anti-cancer compounds called proanthocyanidins that have the ability to starve tumors and stop cancer cells from dividing. Cashews are highly beneficial for lowering blood pressure, preventing heart attacks, preventing gallstones, reducing the frequency of migraines, providing energy boosts, strengthening bones, and promoting healthy skin and hair. Cashews have a lower fat content than most other nuts and contain a high amount of oleic acid which is a heart healthy mono-unsaturated fat that is great for the cardiovascular system. They are also excellent for the nervous system and they are known help keep muscles and nerves relaxed and free from tension and constriction. They can also help the body become more flexible and aid in elasticity of the tendons, muscles, and joints. Cashews have the ability to satiate hunger which decreases overeating and aids in weight loss and weight management. They are also good for overall eye health and they contain a bioflavonoid called zea-xanthin that helps to prevent age-related macular degeneration. For the maximum health benefits try to find raw cashews that are unsalted. Cashews can be eaten as a snack or added to salads, and vegetable or rice dishes. Cashews can also be ground into a nut butter and used to spread on celery sticks, apples, cucumbers, or mixed with dried coconut, honey and spices for a delicious and healthy desert. Good quality cashews can be readily found online or at your local health food store.

According to various scientific researches, two handfuls of cashews give the same effect as a prescribed dose of the antidepressant Prozac. Namely, cashews contain a large amount of the **amino acid L-tryptophan**, a substance that is converted into serotonin in the brain, which in turn is an important neurotransmitter that stimulates the feelings of joy and happiness. This very effect of serotonin is what Prozac and other antidepressants try to mimic.

Other important benefits that cashew nuts bring are: reduced blood pressure, cancer prevention, healthy bones and nerves, improved digestion, prevention of gallbladder and kidney stones, healthy heart, a good night's sleep, healthy gums and teeth they also have contents high in vitamins and antioxidants.

So, whenever you feel down, stressed or depressed, eat two handfuls of natural cashews and increase the serotonin level in your organism. Shortly after you will start to feel better.

A piece of advice – cashews hide a lot of fats. Although the amount of fats they contain is considerably lower than in other nuts, a handful of this delicious snack, which is about 20 grams, has 111 calories. So, if you're trying to lose weight, make sure that the amount of cashews you eat, to benefit your spiritual health, replaces one of your daily snacks.

Health Benefits

These nuts possess an abundance of calories. About 50 grams of these nuts contain 275 calories. In addition to that, they are rich in vitamins, minerals, soluble dietary fibre and numerous phytochemicals that prevent severe diseases, including cancers.

Cashews are also filled with mono-unsaturated fatty acids, such as palmitoleic and oleic acids, which are great for the heart's health. They belong in the group of essential fatty acids which can decrease the level of harmful cholesterol (LDL) and also increase the level of good cholesterol (HDL) in our blood.

Scientific studies have proven that the well known Mediterranean diet, which is based mostly on mono-unsaturated fatty acids, can help organism system protect itself from coronary artery disease and strokes by supporting the blood lipid profile.

These tasty nuts are also rich in essential minerals. Some of the most useful minerals that can be found in cashew nuts are potassium, magnesium, copper, zinc, selenium and iron.

By consuming just a handful of cashews daily you will provide the necessary amount of the above mentioned minerals and protect your organism from deficiency-related diseases. Selenium is a highly important micronutrient and it works like a co-factor for antioxidant enzymes, such as Glutathione peroxides, which is known as one of the strongest antioxidants in the human body. Copper also acts as a co-factor for many important enzymes such as cytochrome c-oxidase and superoxide dismutase. Zinc, on the other hand, plays the same role with enzymes responsible for regulation of the progress of digestion, DNA synthesis and gonadal function.

Cashews can be quite useful due to their numerous essential vitamins, such as pyridoxine (vitamin B6), pantothenic acid (vitamin B5), thiamin (vitamin B11) and riboflavin. Just 3.5 Oz (100 grams) of these nuts will provide about 30 per cent of the daily recommended levels of B6. B6 can significantly lower chances of developing homocystinuria and sideroblastic anemia.

On the other hand, niacin improves protection against dermatitis or pellagra. In addition, aforementioned vitamins are crucial for fat, protein and carbohydrates metabolism at a cellular level.

Furthermore, cashews also contain a very small quantity of zeaxanthin, which is a very important pigment of flavonoid antioxidant. Zeaxanthin is selectively absorbed inside the retinal macula in the eyes. It is believed that it provides antioxidant and protective UV ray filtering features and also that it protects us from ARMD also known as "age-related macular degeneration" in old people.

Besides All of the above-mentioned benefits of Cashew Nuts, they also are considered to be one of the best alternative treatment for depression.

How Treat Depression Naturally

Cashew nuts are more than great source of tryptophan, which is a vital amino acid our body needs to take it throughout the food. This amino acid helps children develop and grow, regulates mood, balances our behavior, improves sleep and can significantly lower the level of stress, anxiety and depression.

Just two handfuls of cashews contain somewhere between 1.000 and 2.000 mg of tryptophan. This compound has been proven to be as equally successful as prescription antidepressants. The best thing about treating depression with cashews is that you will surely avoid side effects that are usually caused by antidepressants.

Cashew Nut Allergy

Beside allergies caused by eating walnuts, almonds and other nuts, cashew nut cause allergy in both adults and children and is one of the commonest food allergies. Same as peanuts, cashews can lead to occurrence of some very severe allergic reactions, even if the individual has consumed just a small quantity of cashews.

AVOCADO

Mature and Ripe Avocadoes

Avocado (*Persea americana*): The Aztecs referred to the avocado tree ahuacatl as a "testicle tree." Avocados, like **bananas,** mature on the tree but ripen only after their harvest. Once ripen, they turn dark green or deep purple and yield to gentle pressure. Inside, cream colour flesh has butter-like consistency with bland taste yet pleasant aroma. The fruit features centrally placed single brown colour seed. On an average, each fruit weighs about 300-700 g, although heavier avocados are quite common in the markets. Avocados contain high levels of **folic acid**, which helps metabolize proteins, thus giving you more energy. They contain vitamin B6 (a nutrient that increases male hormone production) and potassium (which helps regulate a woman's thyroid gland), two elements that help increase libido in both men and women. Loaded with potassium, avocados boost libido for both sexes. And they're rich in folic acid that provides energy and stamina - both of which are important once your libido is restored. Full of saturated fats, they are rich in folic acid, which helps in metabolizing in the body thus giving more energy. They are indeed excellent for heart health and if they are good for the heart, they have to be good for sex life. This creamy fruit not only helps lower bad cholesterol and reduce risk of laque build up to hence blood flow and fuel your brain power, Avacoda`s are a great alternative to sea food for those who are not fans of fish. It is an excellent fruit for expecting mothers. It is rich in useful nutrients, particularly folates. Avocado improves various functions of child development taking place in the womb. Research has shown that even a single avocado fruit eaten during a week, balances hormones, sheds unwanted birth weight and prevents cervical cancer in expecting mothers. Avocados receive a lot of flack for being high on fat content. However, these are a great source of good fats. Not a popular fruit in India, avacado is a fruit you can count on for boosting your energy. Its pear shape is sensual. Its green flesh contains high levels of vitamin E that maintains youthful vigour energy levels. Avocados provide nearly 20 essential nutrients including fibre, potassium, vitamin E and B and folic acid. Avocados also act as a nutrient booster by enabling the body to absorb more fat-soluble nutrients such as alpha and beta- carotene and lutein, in foods that are eaten with the fruit.

Everyone loves avocados, because they are high in vitamin E and are a rich source of skin-protective antioxidants, essential for glowing skin and shiny hair. These wonder fruits also have a high folate content, which is important for skin cell regeneration and will give you a more youthful complexion.

Libido dietary tip: Cut an avocado in half, scoop out the insides with your fingers, and then get your woman to lick it off. They are also known as nature's best moisturizers and are loaded with anti-ageing nutrients such as **Vitamin C** and E, and is also high in **Omega 3 fatty acids.** Though a medium-size avocado has 30 grams of fat, all of it is healthy mono-saturated fats that lower cholesterol and protect the heart. Deepshikha says, "Eating avocados can help stave off wrinkles and keep your skin looking soft and supple, because of the Vitamin E that helps to get your skin glowing. The buttery-fleshed **avocado** may not seem like a hydrating food, but these healthy gems replenish potassium and contain healthy fats and fibre, which help your body hold on to water. As a bonus, the mono-saturated fats in avocados contain **oleic acid**, which has been found to improve fat levels in the body and help control **diabetes** and cholesterol. Layer your sandwiches with avocado slices, toss diced avocado into your salads or simply scoop avocado out of its peel as a snack. Avocados are full of good energy-rich fats, rich in folic acid which helps in metabolizing proteins in the body thus giving you more energy. They are indeed excellent for your heart health and if they are good for your heart they have to be automatically good for your sex life. Isn't it? Natural conclusions aside, several studies over time have proved that increased overall health leads to higher sexual stimulation in men and women. Avocados are rich in mono-unsaturated fat, which is easily burned for energy. An avocado has more than twice as much potassium as a banana. For a delicious, creamy salad dressing, mix together avocado and fresh carrot juice.

Health Benefits

Avocados are rich in healthy fats. They also have a high content of mono-unsaturated folic acid that help lower cholesterol. The good fat in avocado increases metabolism and increases the production of testosterone a harmone which aids in fat loss

Maintain a healthy heart - Avocado contains vitamin B6 and folic acid, which help regulate homocysteine levels. It contains 25 milligrams per ounce of a natural plant sterol called beta sitosterol. Regular consumption of beta-sitosterol and other plant sterols has been seen to helpmaintain healthy cholesterol levels.

Control blood pressure - Avocados are also a great source of potassium, which helps in controlling blood pressure levels.

Promote eye health - Avocado is an excellent source of carotenoid lutein, which known to help protect against age-related macular degeneration and cataracts.

Fight free radicals - Avocados contain glutathione, a powerful antioxidants that helps fight free radicals in the body.

Cure bad breath - Avocados are one of the best natural mouth wash and bad breath remedies. It is cleanses intestine which is the real cause of coated tongue and bad breath.

Skin Care - The avocado oil is added in many cosmetics because of its ability to nourish the skin and make your skin glow.

It also aids in treating psoriasis, a skin disease that causes skin redness and irritation.

Avocados, like olives, are high in mono-unsaturated fats and calories. However, they are very rich in dietary fibre, vitamins, and minerals and packed with numerous health benefiting plant nutrients.

Their creamy pulp is very rich source of mono-unsaturated fatty acids like oleic and palmitoleic acids as well as omega-6 poly-unsaturated fatty acid linoleic acid. They rich in mono-unsaturated fatty acids help lower LDL or bad cholesterol and increase HDL or good cholesterol, thereby, prevent coronary artery disease and strokes by favoring healthy blood lipid profile.

Osteroporosis prevention: Half an avocado provides approximately 25 per cent of the daily recommended intake for vitamin K, which is essential for bone health.

Lower risk of depression: Food containing high levels of folate may help to decrease the risk of depression as folate helps to prevent the built up of homocysteine, a substance that can impaire circulation and delivery of nutrients to the brain excess homocysteine can also interfere with the production of the serotonin, dopamine and norepinephrine with regulate mode, sleep and appetite.

Improves digestion: Being high in fibre, approximately 6-7 grams per half fruit. Eating food with natural fibre can prevent constipation, maintain a healthy digestive tract and lower the risk of colon cancer.

Natural detoxification: Adequate fibre promotes regular bowel movements, which are crucial for the daily excretion of toxins through the bile and stool. Dietary fibre may also play a role in regulating the immune system and inflammation.

They are very good source of soluble and insoluble dietary fibre. 100 g fruit provides 6.7 g or about 18 per cent of recommended daily intake. Dietary fibres help lower blood cholesterol levels and prevent constipation.

Loaded with healthy fat and compounds that protect liver from damage *i.e.* protecting the liver from dangerous toxin galactosamine, which has been found to cause liver damage, similar to human viral hepatitis.

In addition, the fruit, like **persimmons,** contain high concentration of tannin. Tannin, a poly-phenolic compound, which was once labelled as anti-nutritional agent is in-fact, has beneficial anti-inflammatory, anti-ulcer and anti-oxidant properties.

Its flesh contains many health promoting flavonoid poly-phenolic antioxidants such as *cryptoxanthin, lutein, zeaxanthin, beta and alpha carotenes* in small amounts.

They are also a good sourde many vitamins *viz.* Vitamin A, E, and K are especially concentrated in its creamy pulp.

Avocados also excellent sources of minerals like iron, copper, magnesium, and manganese.

Fresh avocado pear is a very rich source of potassium. 100 g of fruit provides 485 mg or about 10 per cent of daily-required levels.

Avocado contains vitamin B6 and folic acid, which help regulate homocysteine levels. High level of homocysteine is associated with an increased risk of **heart disease.** Besides that, avocado also contains vitamin E, Glutathione and mono-unsaturated fat which help in maintaining a healthy heart.

Avocados are rich in a compound called **beta-sitosterol** which has been shown to be effective in **lowering blood** cholesterol **levels**.

Avocados are also a great source of potassium, which helps in **controlling blood pressure levels.**

Phytonutrient compound found in avocados, such as polyphenols and flavonoids have been found to have anti-inflammatory properties, thereby reducing the risk of inflammatory and degenerative disorders.

Avocado is an excellent source of carotenoid lutein, which known to help protect against age-related macular degeneration and cataracts. Lutein and zeaxanthin, two phytochemicals that are especially concentrated in the tissues in the eyes, where they provide antioxidant protection to help minimize damage, including from ultraviolet light. As the mono- unsaturated fatty acids in avocados also supports the absorption of other benefical fat-soluble antioxidants such as beta-carotene. Therefore including avocados as part of a healthy diet may help to reduce the risk of developing age related macular degeneration.

The mono-unsaturated (good) fats in avocados can reverse insulin resistance, which helping to **regulate the blood sugar levels.** Avocados also contain more soluble fibre, which keep a steady blood sugar levels.

Avocados are rich in folate, a B vitamin commonly known as folic acid. One cup of avocado provides you about 23 per cent of the Daily Value for folate. The high amount of folate in avocado is essential in the prevention of birth defects, such as neural tube defect and spina bifida.

The high levels of **folate** in avocados also protect against **stroke.**

Many studies have shown that avocado can inhibit the growth of **prostate cancer.**

The oleic acid in avocado is also effective in preventing **breast cancer.**

Avocados contain **glutathione**, a powerful antioxidants to helps fight free radicals in the body.

Being rich in antioxidants, avocado is beneficial in preventing aging symptoms. The glutathione in avocado may boosts immune systems, slows the aging process, and encourages a healthy nervous system.

Avocados are one of the best natural mouth wash and **bad breath remedies.** It is cleanses intestine which is the real cause of coated tongue and this unpleasant condition.

Avocado intake is linked with increased nutrient absorption. A study suggests that, when participants ate salad included avocados, they absorbed five times the amount of carotenoids (a group of nutrients that includes beta carotene and lycopene) than those who do not include avocados.

The avocado oil is added in many cosmetics because of its ability to nourish the skin and make your skin glow. It also aids in treating psoriasis, a skin disease that causes skin redness and irritation.

The avocado has 200 calories for 100 grams. Typically, fruits has approximately 60-80 calories for 100 grams. Due to the high amounts of calories, avocado is a best diet for people who want to gain weight. Avocado is a healthy source of calories, unlike many other calorie-dense foods that may contain excess saturated fats and sugar.

Caution: Raw unripe avocados concentrated with **tannins**. High tannin content makes them bitter and unappetizing. Very high levels of tannins in the food prevent minerals like iron, calcium and phosphorus and vitamins from absorption in the gut. Although very rare, avocados may result in allergic symptoms in some latex-sensitive persons. The symptoms may include itching in the throat, hives, runny nose, breathlessness *etc.* Often the symptoms are mild and self-limiting.

JAMUN

Jamun (*Syzygium cumini*) The fruit is acidic and astringent in nature, with a sweet taste. Due to its acidic nature, it is usually eaten with a sprinkling of salt. Children are fond of this fruit as it colours the tongue purple due to **anthocyanin**, a plant pigment. Itis a healthy fruit with absolutely no trace of sucrose. It is therefore, the only fruit with minimum calories. Not only the fruit, but the seed and also the leaves and bark of the jamun tree are believed to have medicinal properties. Many scientific researchers have shown that dried alcoholic extracts of the seeds of the fruit given to diabetic patients on a regular basis showed a reduction in the level of their blood sugar and **glycosuria**. Also a concoction made from the mixture of dried seeds and the bark is considered beneficial in the cure of diarrhea and dysentery. The bark of the tree has high astringent properties and is therefore used for gargles and as a mouthwash. Blackberries are nutritious and are often considered to be the best of the berry family. They are high in antioxidants, anthocyanin, salicylic acid, ellagic acid, and rich in fibre and Vitamins C and K. These berries have been shown to help reduce certain cancer risks and are good for cardiovascular health. They are also said to have anti-viral and anti-bacterial properties. The acidic, sour, sweet, and soothing fruit is used to treat diabetes, diarrhea and ringworm. The fruit also has blood purifying properties. The leaves of the jamun tree possess anti-bacterial properties and is used for making medicines for the strengthening of the teeth and gums. The bark of the tree has **antithelmintic** properties, and is used to formulate many herbal medicines. The juice of the jamun fruit is extremely soothing and has a cooling effect. It helps in the proper functioning of the digestive system. The leaves of the tree also help in controlling the blood pressure and gingivitis.

According to Ayurveda, Jamun cures pitta and skin diseases. Cures burning sensation, vomiting, it is a stabilizer, The fruit promotes digestive functions and body heat. It is digestive in nature, stimulates liver and enhances Vata. When eaten in excess quantities, it hardens stool. The seeds kernel improves digestion. It reduces urine and blood sugar levels. It also reduces amount of urine production. It normalizes heart beat, cures blood disorders and also controls boils, acne and eruptions on skin. It cures diarrhea due to pitta and clearsthe voice. It relives fatigue, asthma, cough, mouth and throat related disorders and gets rid of worms. Its squash is beneficial in curing vomiting, bleeding diarrhea and piles (Acharya Balkrishna, 2008).

The juice of ripe fruit or a decoction of it is administered in spleen enlargement, chronic diarrhea and urine retention.

The seed extract in liquid or powdery form are given to patients with diabetes mellirus or glycosuiria. In many cases blood sugar level drops quickly.

Dried alcoholic extract of Jamun seeds reduces blood sugar and glycosuria. Seeds contain an alkaloid, **jambosine** and **glycoside jambolin** or **antimellin** which halts the conversion of starch into sugar.

Nutrients in Jamun

The ripe fruit contains Glucose and Fructose which are the major forms of sugar. It also contains Vitamins C and A, riboflavin, nicotinic acid, choline, folic acid, malaic acid, sodium, potassium, calcium, phosphorus, manganese, zinc and iron. Anthocyanins are present in appreciable quantities and are the reason for the antioxidant activity of the fruit.

The stem and bark contains tannin, **gallic acid**, resin, **phytosterols.**

The seed contains the glycoside, jamboline, gallic acid and essential oils.

The leaves contain essential oils.

The flower contains **terpenoids.**

Medicinal Properties of Blackberry or Jamun

The extracts of the bark, seeds and leaves are used for the treatment of diabetes.

The leaves have antibacterial properties and used for strengthening teeths and gums.

Oral administration of dried alcohalic extracts of the seeds to diabetic patients was found to reduce the level of blood sugar and glycosuria, in trials conducted at CDRI, Lucknow.

The bark of black berry tree is astringent, digestive, diuretic, anthelmintic and is considered useful for throat problems

A decoction of the bark and powdered seeds is believed to be very useful in the treatment of diarrhea, dysentery and dyspepsia

The antibiotic activity of black berry extract has been widely studied and found useful against a number of microbial agents.

The fruit is also considered to be stomachic, carminative, antiscorbutic and diuretic.

Vinegar made from black berry fruit is administered in cases of enlargement of spleen, chronic diarrhea and urine retention

For ringworm treatment, water diluted juice is used as lotion

A decoction of bark is used in cases of asthma and bronchitis and are gargled or used as mouthwash for the astringent effect on mouth ulcerations, spongy gums, and **stomatitis.**

It is digestive and activates the liver and spleen.

It is a good remedy for urinary diseases and diabetes.

Regulates heartbeat.

Purifies blood, cures anemia and stops skin eruptions.

Stops diarrhea and dysentery.

Relieves throat affections and other respiratory diseases.

Removes worms.

Vinegar made from the fruit gives relief in colitis, indigestion, stomach diseases. It relieves gas and improves digestive power.

It breaks renal stones.

The bark, leaves, fruit and seeds are commonly used for treatments of various disorders.

The bark of jamun has astringent, carminative, diuretic, digestive and constipating properties.

It is good for sore throat, bronchitis, asthma, thirst and dysentery

SAPOTA OR CHIKOO

Sapota (*Achras sapota*): The fruit is sweet with smooth or grainy texture and light musky flavour. It contains 3-10 black, smooth, shiny "biconvex"/bean shaped, inedible seeds located at its centre. This delicious tropical fruit is rich in vitamins, minerals and health benefiting anti-oxidant *tannins*. The fruit is soft, composed of easily digestible pulp made up with simple sugars like fructose and sucrose that when eaten replenishes energy and revitalizes the body instantly. **It is a round fruit that resembles sometimes like a potato or a kiwi in rounded form. It is sweet in taste and is high calorie fruit. It is famous for its peculiar taste and is very commonly used in making shakes *etc.***

Health Benefits

Sapote contains a good amount of antioxidant vitamins like **vitamin A and vitamin C**. Vitamin A is essential for vision. It is also required for maintaining healthy mucus membranes and skin, **lung and oral health even at the old age also helping to enjoy younger happy life.**

Fresh ripen sapodilla is a good source of minerals like potassium, copper, iron and vitamins like folate, niacin and **pantothenic acid**. These compounds are essential for optimal health as they involve in various metabolic processes in the body as cofactors for the enzymes.

Chiku or sapota contains rich dietary fibre making it a good laxative, Fibres in ripe, sweet and tasty chiku fruits help to prevent constipation.

Taking sufficient quantity of ripe chiku in raw form or as chiku shake will protect colon system preventing chances of colon cancer.

Chiku contains natural antioxidant properties. Hence eating this seasonal fruits will be the most effective natural health cure medicine to prevent viral, bacterial and parasitic effects in human internal organ system.

Vitamin C natural medicinal property in chiku is useful for maintaining healthy and shining skin texture.

Eating this delicious and healthy seasonal fruits will develop body resistance to fight against many infectious diseases throughout the year.

One of the most important use of this herb is its **hemostatic** properties. This herb is very beneficial in stopping the blood loss. Helps in reducing bleeding in piles and injury *etc.*

Sapota has been found beneficial in conditions pertaining to infection. It helps in reducing viral as well as bacterial infections in the body. It also reduces inflammation making it very potent in reducing swellings and pain.

Sapota is a delicious food and is rich in glucose. This glucose helps in proving instant energy to the body. Hence it is very much recommended to athletes as it provides loads of energy.

This wonderful fruit is extremely beneficial in making your skin shiny and elegant thus you require minimum skin care products to enhance skin texture and complexion.

SITAPHAL

Useful part: Fruits, seeds, Roots, Leaves

Sitaphal (*Annona squamosa*) There are medicinal application of the custard apple tree. The bark and leaves contain annonaine, an alkaloid. A bark decoction is used to stop diarrhea, while the root is used in the treatment of dysentery. A decoction of the leaves is used as a cold remedy and to clarify urine. Annona squamosa Linn., family Annonaceae, is said to show varied medicinal effects, including insecticide, anti-ovulatory and abortifacient. The fruits of Annona are **Haematinic**, cooling, sedative, stimulant, expectorant, maturant, tonic. They are useful in anemia, burning sensation. The seeds are abortifacient and insecticidal and are useful in destroying lice in the hair. Leaves are used to overcome hysteria and fainting spells. Fruit is used in making of ice creams and milk beverages.

The fruits are very high in calorific value and are a rich source of minerals and Vitamins. They are used by athletes for their high energy content. The seeds are powerful insecticides and powdered seeds are used for removing head lice. It is also used as an effective pesticide in agriculture and horticulture. Apart from this, various parts of the plant are medicinally useful in a wide variety of diseases from tumor to cough. The leaves are shown to have anti-diabetic properties. It is also known for its hepato-protective powers and scientists have experimentally proven the efficacy of the alcoholic extract of the leaves and stem in malignant tumors.

Its efficacy as anti-depressant and anti oxidant has also been experimentally proven by various scientists. They are also used as anti-depressants, in epilepsy and in spinal cord disorders. Fruits are sweet, **haematinic** (A medicine which increases the **haemoglobin** content of the red cells in the blood), cooling, act as a sedative, stimulant and function as expectorant and tonic. Seeds are abortificient and insecticidal. Roots are powerful purgatives and are also used in dysentery. Acetogenins found in the seeds of Sitaphal would actually help fight plant pests. The Sitaphal pesticide is inexpensive, environment-friendly and highly effective in containing a variety of pests on a number of crops. The seed extracts also showed synergistic activity in combination with neem seed extract in pesticidal action.

It is nourishing, increases blood (Hb count), sweet, cooling, heart tonic, provides strength, helps muscle building and alleviates burning sensation, bleeding disorders and disorders of vaata.

Medicinal Uses

In India the crushed leaves are sniffed to overcome hysteria and fainting spells; they are also applied on ulcers and wounds and a leaf decoction is taken in cases of dysentery. Throughout tropical America, a decoction of the leaves alone or with those of other plants is imbibed either as an emmenagogue, febrifuge, tonic, cold remedy, digestive, or to clarify the urine. The leaf decoction is also employed in baths to alleviate rheumatic pain. The green fruit, very astringent, is employed against diarrhea in El Salvador. In India, the crushed ripe fruit, mixed with salt, is applied on tumors. The bark and roots are both highly astringent. The bark decoction is given as a tonic and to halt diarrhea. The root, because of its strong purgative action, is administered as a drastic treatment for dysentery and other ailments.

Health Benefits

The numerous vitamins and nutrients present in custard apple provide protection from various medical conditions such as heart attacks, strokes and cancer.

If you have a fluctuating blood pressure then a regular intake of this fruit is effective in keeping it under control.

Being a rich source of Vitamin A, custard apples give the consumer a glowing and healthy skin and shiny and lustrous hair.

Looking for a natural way to enhance your eye sight? Then the custard apple acts as a storehouse of various vitamins making your vision sharper and brighter.

Intake of custard apple helps in curing indigestion problems. Since this is indirectly linked to constipation, these fruits inject copper into the body which ensures the production of stools.

The high amount of magnesium in custard apples maintains water balance in the body. This promotes the removal of various acids from the joints and lowers the risk of rheumatism and arthritis.

Custard apples are packed with iron and are the simplest and fastest cure for treating anemia.

For extremely underweight individuals, custard apples are an ideal way to put on a few kilos. Also, the sugar fibres present in them make the making metabolic system more efficient.

The skin of custard apple is helpful in fighting against tooth decay and gum pain.

These fruits are excellent in treating vomiting, gout, lice and vitamin B6 deficiencies.

The custard apple fruit is a sweet fruit that has many benefits. Some of these benefits are:

It is low in calorie content and therefore good for maintaining optimum health.

It contains Vitamin A which is good for the skin, hair, eyes.

It helps in digestion as it is a rich source of dietary fibre.

It is a coolant, stimulant and expectorant.

It has healing properties as well – the flesh can be applied on boils, abscesses and ulcers.

The magnesium in custard apples helps to protect the heart from cardiac diseases. It can relax the muscles as well.

Custard apples can also be used to treat dysentery and diarrhea.

It is good for anemic people.

The potassium present in a custard apple helps fight muscle weakness.

It is also a rich source of magnesium. It therefore helps to equalize the water balance in the body and this helps in removing acids from the joints. It is thus beneficial for rheumatic and arthritic patients.

Olive

Olive (***Olea europaea***) contains a compound called **Oleuropein** which can cleanse liver of the toxins. It is the **Oleuropein aglycone**, the most abundant polyphenol present that is responsible for its characteristic bitter, pungent taste. Oleuropein has anti-oxidant, anti-inflammatory, anti- ageing, anti- cancer, anti-viral and anti- microbial properties. It is a skin protector, as it acts as a free radical scavenger. This polyphenol can also help in the management of **Alzheimer's** disease. It has healing properties due to its hypotensive; anti-rheumatic, diuretic and anti-pyretic effects. It is helpful in prevention of diabetic complications associated with oxidative stress. It can help in reduction of coronary heart stress disease and certain cancers. Olive oil is one of the best sources of mono-unsaturated fats and has the advantage of being less susceptible to oxidation. In addition, oleic acid, a fatty acid abundant in olive oil, appears to also protect from oxidation of LDL. In addition to bolstering the immune system and helping to protect against viruses, olive oil has also been found to be effective in fighting against diseases such as:

Cancer prevention: Black olives are a great source of vitamin E, which has the brilliant ability to neutralize free radicals in body fat. Especially when working with the stable mono-unsaturated fats found in olives, vitamin E can make cellular processes safer. When such processes such as mitochondrial energy production are not well protected, the free radicals produced can cause oxidation, damaging a cell's mitochondria, and preventing the cell from producing enough energy to supply its needs. If the DNA of a cell is damaged, it may well mutate and become cancerous. Studies have shown that a diet supplemented with olive oil leads to a lower risk of colon cancer, almost as low a risk as a diet rich in fish oil (Anonymous, 2013).

The phyto-nutrient in olive oil, oleocanthal, mimics the effect of ibuprofen in reducing inflammation, which can decrease the risk of breast cancer and its recurrence. **Squalene** and **lignans** are among the other olive oil components being studied for their possible effects on cancer.

Cardiovascular benefits: When free radicals oxidize cholesterol, blood vessels are damaged and fat builds up in arteries, possibly leading to a heart attack. The antioxidant nutrients in black olives impede this oxidation of cholesterol, thereby helping to prevent heart disease. Olives do contain fat, but it's the healthy **mono-unsaturated** kind, which has been found to shrink the risk of **atherosclerosis** and increase good cholesterol. Olive oil helps lower levels of blood cholesterol leading to heart disease.

Oxidative Stress: Olive oil is rich in **antioxidants,** especially vitamin E, long thought to minimize cancer risk. Among plant oils, olive oil is the highest in mono-unsaturated fat, which doesn't oxidize in the body, and it's low in **polyunsaturated** fat, the kind that does oxidize.

Blood Pressure: Recent studies indicate that regular consumption of olive oil can help decrease both systolic and diastolic blood pressure.

Diabetes: It has been demonstrated that a diet that is rich in olive oil, low in saturated fats, moderately rich in carbohydrates and soluble fibre from fruit, vegetables, pulses and grains is the most effective approach for diabetics. It helps lower "bad" low-density lipoproteins while improving blood sugar control and enhances insulin sensitivity.

Obesity: Although high in calories, olive oil has shown to help reduce levels of obesity.

Rheumatoid Arthritis: Although the reasons are still not fully clear, recent studies have proved that people with diets containing high levels of olive oil are less likely to develop rheumatiod arthritis.

Osteoporosis: A high consumption of olive oil appears to improve bone mineralization and calcification. It helps calcium absorption and so plays an important role in aiding sufferers and in preventing the onset of Osteoporosis.

Bone and connective tissue: The anti-inflammatory abilities of the mono-unsaturated fats, vitamin E and polyphenols in black olives may also help dull the asperity of asthma, osteoarthritis, and rheumatoid arthritis. Most of the suffering in having one of these three bone maladies is brought about by high levels of free radicals. Olive oil also contains a chemical called oleocanthal, which acts as a painkiller. Research has found that oleocanthal inhibits inflammation by the same means that drugs like Ibuprofendo.

According to a new study from France, older individuals who consume olive oil daily may be able to protect themselves from a stroke. The study which is part of the Three-City Study, an ongoing multicenter study of vascular risk factors for dementia, was published in the online issue of Neurology. Olive Oil Might Help Prevent Strokes:

Olive Oil Diet Reduces Risk of Type 2 Diabetes

Traditionally a low fat diet has been prescribed to prevent various diseases such as heart disease and diabetes. While studies have shown that high fat diets may increase the risk of certain diseases such as cancer and diabetes, it appears that it is the type of fat that counts rather than the amount of fat. We now know that a diet rich in mono-unsaturated fats such as the ones found in olive oil, nuts and seeds actually protects from many of these chronic diseases.

Olive Oil Keeps the Heart Young:. A diet rich in olive oil may be able to slow down the aging of the heart. It is a known fact that as we grow older the heart also goes through a normal aging process. The arteries may not function as well as they did and this can lead to a number of health problems. However, researchers discovered that a diet rich in olive oil or other mono-unsaturated fats

Olive Oil Fights Osteoporosis: Osteoporosis is a disease characterized by a decrease in bone mass, which in turn causes the architecture of bone tissue to become fragile. This can then increase the possibly of fractures, making even the slightest of knocks potentially fatal for sufferers. The disease is recognized as being particularly prevalent among post-menopausal women for whom a decrease in the production of estrogen then weakens bone structures and most commonly affects the ribs, wrists, and hips. For this study, scientists were particularly interested in how a supplementation of olive oil could be used to help women in this category.

Digestive tract health: Frequent consumption of both vitamin E and the mono-unsaturated fats in black olives is associated with lower rates of colon cancer. These nutrients help prevent colon cancer by neutralizing free radicals. Olive oil's protective function also has a beneficial effect on ulcers and gastritis. Olive oil activates the secretion of bile and pancreatic hormones much more naturally than prescribed drugs, thereby lowering the incidence of gallstone formation. A cup of black olives also contains 17 per cent of the daily allowance of fibre, which promotes digestive tract health by helping to move food through the system at a healthier pace. This keeps any one part of the digestive tract from having to work too hard and supports the ideal balance of chemicals and populations of microorganisms required for a healthy digestive system.

Eye health: One cup of black olives contains ten per cent of the daily recommended allowance of vitamin A which, when converted into the retinal form, is crucial for healthy eyes. It enables the eye to better distinguish between light and dark, thereby improving night vision. Furthermore, Vitamin A is believed effective against cataracts, macular degeneration, glaucoma and other age-related ocular diseases.

Good source of iron: Black olives are very high in iron. The ability of red blood cells to carry oxygen throughout the body is due to the presence of iron in the blood. If we suffer from a lack of iron, our tissues don't get enough oxygen, and we may feel cold or weak. Iron also plays a vital role in the production of energy. It is a necessary part of a number of enzymes, including iron catalase, iron peroxidase, and the cytochrome enzymes. It also helps produce carnitine, a nonessential amino acid important for the utilization of fat. To top it all off, the proper function of the immune system is dependent on sufficient iron.

Skin and hair health: Black olives are rich in fatty acids and antioxidants that nourish, hydrate and protect. Chief among those is vitamin E. Whether applied topically or ingested, vitamin E has been shown to protect skin from ultraviolet radiation, thus guarding against skin cancer and premature aging. You can gain a healthy, glowing complexion by washing your face in warm water, applying a few drops of olive oil to vulnerable spots, and letting it work its magic for 15 minutes before rinsing it off. In fact, you can moisturize with olive oil before any bath, and even condition your hair with it by mixing it with an egg yolk and leaving it before rinsing and washing.

Best Muscle Building Foods (Olive oil)

The mono-unsaturated fat in olive oil helps prevent muscle breakdown and protects joints. Olive oil is rich in mono-unsaturated fats and omega-3 fats. These good fats are heart healthy. Olive oil is anti-inflammatory and helps with muscle pain and inflammation, aiding recovery, making it a perfect muscle building food (Moghul, 2013).

AMLA

Indian Gooseberry (*Phyllanthus emblica*) is rich in vitamin C. Winter brings along its share of colds and cough. To cure this, prepare a mixture of two tablespoons of amla with an equal amount of honey. Have this three to four times a day, people suffering from Vitamin C deficiency can greatly benefit by eating amlas, as it is one of its most potent sources. It has 20 times more vitamin C compared to orange. It also contains gallic acid, tannic acid, carbohydrates, albumen, cellulose and calcium. On the other hand, Vitamin C, taken in the form of supplements, is not ea blems, amla contains high fibre content and will ease your problem. Because they are also rich sily absorbed by the body, unlike when taken in a natural form. Battling with constipation pro in antioxidants, which protect against the formation of free radicals in the body, amlas, when eaten frequently, help to prevent cancer. To seek relief from painful mouth ulcers, simply gargle with a mixture of water and amla juice. Are joint pains getting you down? Have amlas, as they contain anti-inflammatory properties which will help in reducing the swelling in the joints caused by arthritis.

The fruit has stool binding properties and is a blood and urine purifier, It promotes the taste and is beneficial in curing diarrhea, burning sensation, jaundice, bleeding diathesis, gout, piles, constipation, indigestion, breathing disorders, cough *etc.* It improves eyesight, sexualpotency and promotes longevity.

Amla is one of the best anti-inflammatory herb. It prevents aging and promotes longevity.

It is very useful in skin diseases as it has sheet virya in potency. It inhibits pitta and thus helps in getting relief from all the skin disorders caused by pitta dominance. Regular intake of Amla promotes glow on skin and delays wrinkles or loosening of skin.

It improves texture of the hair.

It also prevents premature graying of hairs and dandruff.

It is very useful in improving eyesight. It also counters diseases like reddening, itching and watering of eyes.

Amla is very useful in controlling blood sugar level.

It also helps maintain the functioning of the liver.

Amla cleanses the mouth and strengthens the teeth.

It balances stomach acids because it improves digestion but does not heat the body.

It nourishes the brain and mental functioning.

Amla has powerful antioxidants, which protects from harmful free radicals that can increase the risk of a wide range of serious diseases including cancer.

It is atridosa healer. Its acid suppresses Vata, mild coolness suppresses Pitta and arid astringence suppresses Kapha.

It acts like an anti-dermatoses and anti-pyretic.

It is healthy for heart and is a blood purifier.

It stimulates sexual desire and stabilizes the womb.

It cures cough and cold (kapha, pitta, nashak), cures burning sensation due to dermatoses. It is beneficial for eyes and digestive system and also cures difficult and incurable fever.

Indian Gooseberry along with harad and pippali cures all types of fevers.

Indian Gooseberry is bit sweet and sour and pungent in taste. Its juice is good for eyes (Acharaya Balkrishna, 2006).

DRAGON FRUIT

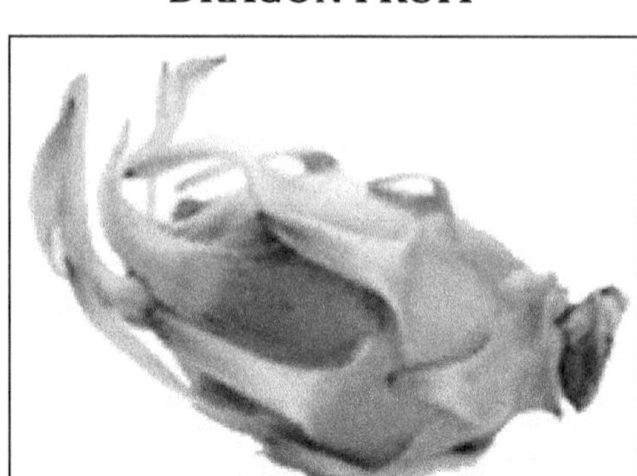

Dragon fruit (*Hylocereus undatus),* also referred as *Kamalan,* originates in South America, it is now grown in Thailand, Sri Lanka, Bangladesh and the Philippines, where it's loved for its many health benefits. Said to be rich in vitamin C, fibre and calcium, the pitaya - or dragon fruit as it's also known - gives you a bigger bang for your buck than similar super fruits such as acai berries. Already available in Waitrose and Tesco, the bright pink fruit is full of edible black seeds and boasts flesh that tastes like a cross between a kiwifruit and a pear. Its production commenced in 1990's in Sangli in Maharashtra and in some North East Indian states.

Along with a mega-dose of vitamin C, pitaya is also thought to help lower cholesterol and is packed with skin-boosting antioxidants. The tiny black seeds are a great source of healthy **Omega-3 fat** and there's plenty of magnesium and calcium in the flesh.

Pitaya is even believed to help regulate blood sugar levels.

Dragon fruits are high in **antioxidants**, which help to fight **carcinogenic** free-radicals from forming in the body. In addition, they are a good source of Vitamin C, and are rich in minerals, especially calcium and phosphorus. They are also low in calories and high in fibre, while the seeds are have high polyunsaturated fatty acids. Aside from its nutritional content, the fruit is also said to help excrete heavy metal toxins from the body and lower cholesterol and blood pressure. Dragonfruits are also known to be a natural laxative.

This nutritional information is based on the edible flesh of 1 dragon fruit. The average weight of a dragon fruit is about 3.6 oz. (100 g), although only 2 oz. of this actually edible. The serving information also includes the nutrition from the seeds. This assumes that you'll chew the seeds so that you can digest them.

Dragonfruits contain phytoalbumins, which have antioxidant properties that help prevent the formation of cancer cells. In addition, dragonfruits are also known to increase the excretion of heavy metal toxins from the body.

Lycopene is also present in dragonfruits, and this is the pigment responsible for their red color. Lycopene is said to prevent prostate cancer. They are among the many wonder fruits that are said to provide multiple health benefits. In addition, dragonfruits help protect the environment because they absorb carbon dioxide at nighttime, and then release oxygen to purify the air.

Health Benefits

As dragon fruit is a good source of antioxidants, it prevents free radicals and protects from causing cancer and other health detriments from entering your body.

This fruit helps in neutralizing the toxic substances such as heavy metals. Besides, high blood pressure and cholesterol levels can be reduced by consuming dragon fruit.

Regular consumption of this aids in fighting against cough and asthma.

This has high amounts of vitamin C and helps in healing bruises and cuts quickly, and enhancing the immune system.

Vitamin B2 present in dragon fruit acts like a multivitamin and helps to improve and recover the loss of appetite.

Vitamin B1 in this fruit helps in increasing the energy production and also in metabolizing carbohydrates.

Due to the presence of vitamin B3, it helps in lowering bad cholesterol levels. It improves the appearance of your skin by smoothing and moisturising it.

It improves eyesight and prevents **hypertension**.

As it is a good source of phosphorus and calcium, it reinforces bones, helps in tissue formation and forms healthy teeth.

Regular consumption of dragon fruit decreases weight, thereby creates a well-balanced body.

Dragon fruit is also helpful in reducing blood sugar levels in people suffering from type 2 diabetes.

Tips

Dragon fruits taste best when served chilled.

To eat dragon fruit, slice it into half and scoop out the flesh with a spoon. The skin is not eaten.

The middle of the fruit is the sweetest part. Hence, save it and eat it at the last.

Dragon fruit is eaten either raw or mixed in juices and cocktails.

It is also used as preserves and spreads like jam, puree and cordial, salads, sherbets and sorbets, fruit pizza and other beverages.

CHERIMOYA

Cherimoya (*Annona cherimola*) is among the healthiest fruit with high nutrition. The custard apple is prickly and has a yellow green, some like brownish skin. There is a white fleshy meat hiding in the inside, which is edible. Ripe fruits turn pale green to light brown colour, and emanate fragrant sweet aroma that can be appreciated from a distance. Inside it features, cream colour pulp with black colour, smooth seeds embedded all over. All of you need to do is cut the fruit in half and start eating the juicy and ripe flesh with a spoon.

It is rich in Vitamin C which increases immunity to fight common disease and infection. Due to it's rich antioxidants it has anti-cancer property. It is full of vitamin C, B1, B2 and minerals like iron, calcium and phosphorus. The fruit is helpful in treating inflammation of the bladder and kidney.

The fruit is harvested while still hard given the fact that they mature and get ripe on a room temperature or when exposed to sun. When ripe, the fruit loosens up under the pressure of fingers just like Kiwi and Avocado.

Health Benefits

Immunity: Cherimoya contains a high amount of Vitamin C. Vitamin C proves the best nutrient to increase immune system function. It provides immunity from the common disease like cold and flu. It is also effective to prevent any infections.

Inflammation: Cherimoya contains a high amount of Vitamin C, a powerful antioxidants. It is effective to protect from inflammatory free radicals.

Cardiovascular Health: It contains a high amount of dietary fibre which proves best nutrient that promote cardiovascular health. Also cherimoyahave proved to reduce bad cholesterol (LDL) and increase good cholesterol (HDL) level in the blood. Thus it improves the blood flow towards the heart and thus it reduces the risk of the heart attack, stroke and hypertension.

Cancer: It is effective against cancer due to it's rich antioxidant content. These antioxidants provide cherimoya anti-cancer property. Cancer cells are formed due to free radical produced due to oxidative stress. But antioxidants are effective to neutralize this free radical effect. Also cherimoya is rich in fibre; it cleanses toxic from colon and liver. So it is effective to prevent colon and liver cancer. It also protects from breast cancer.

Brain Health: It is a high source of Vitamin B6, which protect from Parkinson's disease. Due to rich Vitamin B6 it increases brain function. It is effective to relieve stress and thus reduce tension.

Rich in Potassium: It contains high amount of potassium. Good amount potassium Protect from Osteoporosis: It contains calcium which improves bone health. Thus it protects from osteoporosis.

Caution

If you are eating only fruit then there is no risk or side-effects. But it is risky to eat seeds of cherimoya fruit. Cherimoya seed is considered as toxic to health so avoid eating cherimoya seed. It is not proved yet but avoid eating cherimoya seed.

DRUMSTICK

Many of the nutritional and medicinal properties of drumstick (*Moringa oleifera*) have long been known in India. Moringa" is derived from the Tamil word "Murungai". It is thought to have been originated from Northwest India.

The quickly growing, perrennial drumstick tree is a common sight in the backyards of homes in South India. Drumstick is now widely cultivated as an important crop in India, Ethiopia, the Philippines and the Sudan, West, East and South Africa, tropical Asia, Latin America, the Caribbean, Florida in the US, and the Pacific Islands.

All parts of this tree are useful and have long been used for nutritional, medicinal, and industrial purposes. The drumstick pods or fruits are used as a vegetable in curries and soups and very popular in Indian food. Crushed drumstick leaves are used as a domestic cleaning agent; powdered seeds are used for clarifying honey and sugarcane juice, and for purifying water. Moringa seeds produce oil, also known as Ben oil, which is a sweet non-sticky oil that doesn't become rancid. This oil is used in salads, for lubricating machines, and in perfumes and hair-care products.

The seeds are also eaten green, roasted, powdered and steeped for tea or used in curries. This tree has in recent times been advocated by organizations such as Trees for Life as an outstanding indigenous source of highly digestible protein, calcium, iron, Vitamin C, and carotenoids suitable for use in regions of the world where malnourishment is a major concern.

Incredibly, every part of the plant (the bark, the pods, the roots, the fruits and the flowers) are rich in nutrients and edible. The entire tree can be used and it is very beneficial.

It is loaded with minerals, purifies blood, cures respiratory problems and enhances sexual health.

It is a rich source of highly digestive proteins, calcium, iron, Vitamin C, and carotenoids.

Drumstick soup helps to control any kind of chest congestion, coughs and sore throats.

Juice extracted from drum stick leaves, when mixed with milk and offered to children will greatly helps to strengthen their bones as it is said to be a great source of Calcium.

Drumstick leaves juice gives a good glow to your skin and face also drumstick is said to be a great blood purifier.

Drumstick leaves are used to treat many problems such as the wheezing of asthma, bronchitis, and tuberculosis.

Soup made with drumstick leaves and flowers are highly beneficial in preventing infections such as throat, chest and skin.

Drumstick-leaf juice is also very beneficial for pregnant women as it can help them to overcome sluggishness of the uterus, ease delivery, and reduce post-delivery complications.

Inhaling steam of water in which drumsticks have been boiled helps to control asthma and other lung problems.

Because of the high calcium, iron, and vitamins, drumstick leaves can be used as a wonderful tonic for infants, growing kids and teens to promote strong and healthy bones.

Pregnant women should often eat drumsticks as it helps to ease any kind of pre and post-delivery complication.

Why are Moringa Trees so Useful?

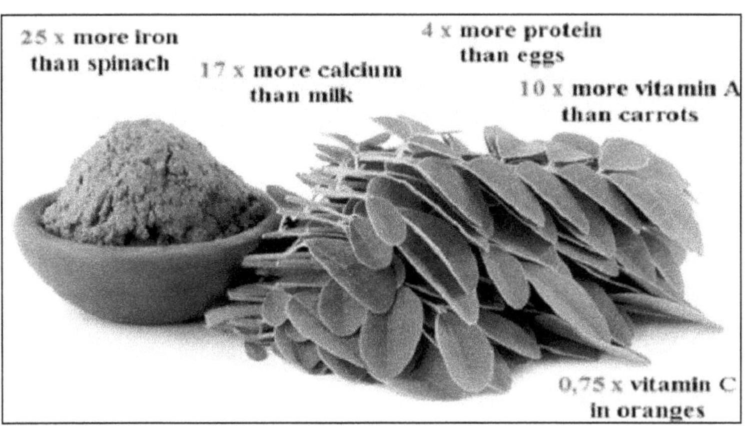

The importance of this plant can be understood if it is compared with some other healthy products and foods.

The edible oil called Ben oil produced from moringa seeds has the same nutritional value as olive oil. The moringa seed oil is loaded with antioxidant. This oil is also special since its shelf life is indefinite.

We all know that spinach is one of the best sources of iron. However, the leaves of moringa tree contain three times more iron than the green leafy vegetable spinach.

Most of us drink coffee in the morning in order to wake up quickly. However, moringa tea is better. The tea of moringa leaves will boost your energy naturally. Moreover, this marvelous tea will keep your energy levels high for a long time.

Carrots are loaded with vitamin A and therefore they are great for the eyes. However, moringa contains four times more of this vitamin per gram than carrots. Moringa is an excellent source of beta-carotene as well.

Many people have a good habit of taking multivitamins. The moringa supplements are made from moringa pods and leaves. These supplements contain no synthetic ingredients so it is believed that they are better than most supplements and vitamins.

Yogurt is a recommended source of protein. However, moringa tree contains two times more protein per gram than yogurt.

Bananas are loaded with potassium. Moringa plant contains three times more potassium per gram than bananas.

Oranges and similar citrus fruits are an excellent source of vitamin C. This vitamin is 7 times more present in moringa.

Moringa even contains 4 times more calcium than milk per 1 gram. So, moringa tree is excellent for the bones.

Researchers Prove that Moringa has Amazing Anti-cancer Effects

Moringa is excellent against cancer since it contains many different anti-cancer compounds like isoquercetin, rhamnetin and kaempferol. According to researchers' lab tests, moringa's anti-cancer properties are effective in the treatment of melanoma, ovarian, lung and liver cancer.

Even though a long time needs to pass until moringa is accepted as a cure for cancer, these tests and studies are an excellent starting point for the future of medicine. On the other side, it is an interesting fact that moringa is already used in herbalism and other alternative therapies.

Additional Benefits

Moringa trees are able to purify water.

These trees contain nine important amino acids, 46 antioxidants, 27 vitamin and a lot of minerals.

Moringa improves lactation in humans.

Moringa plant can be used in the treatments of chronic diseases, wounds, bites and other health problems.

Moringa trees make your hair vibrant and healthy and cleanse the skin effectively.

The moringa tree consumption has no side effects. It can be consumed both by adults and children.

TAMARIND (*Tamarindus indica*)

Tamarind is a tropical fruit. It is a hard wood tree, known scientifically as a *Tamarindus indica*. It is native to Africa but is also grown in Indian sub-contitent and other tropical countries. The tree produces bean like pods filled with seeds surrounded by a fibrous pulp. The pulp of the young fruit is green and sour. As it ripens, the juicy pulp becomes paste like and more sweet- sour. Tamarind is some times referred to as the 'date of India'. It is used in many dishes around the world and has many medicinal properties. In beverage form, it was normally used to treat diarrhea, constipation, fever and peptic ulcers. The bark and the leaves were used to promote wound healing.

Tamarind is effective in helping to reduce blood pressure and cholesterol. It contains a good amount of fibre, which helps to reduce cholesterol by removing LDL from arteries. Potassium found in Tamarind helps reduce the blood pressure because it is a well known Vasodilator.

It contains high levels of vitamin C, which helps in the neutralization of free acids. It can aid in weight loss and weight maintenance. Tamarind contains Hydroxy citric acid, which has been found in enzyme, found in the body that is known to store fat. Studies have shown that eating Tamarind can suppress the appetite. It also increases Serotonin- Neurotrans Mitters. It has been shown to be a neutral laxative and recent studies have proven this to be true. Full of fibre when eaten as a fruit, the tamarind can increase the effectiveness of your digestive tract. Fibre can also bulk up stool which makes it move through the intestinal tract smoother and easily. Research has found that the consumption of tamarind helps to stimulate the activities in the bile which dissolve food faster, because tamarind has a lot of fibre, the gastric juice are stimulated which speeds up digestion.

The polyphenols in tamarind have anti –oxidant and anti –inflammatory properties. These can protect against diseases such as heart diseases, cancer and diabetes. The seed extract may also help lower blood sugar, while the pulp extract may help lose body weight and reverse fatty liver disease.

Tamarind is high in magnesium, which has many health benefits and plays a role in more than 600 body functions. Research studies has found that tamarind can also help with chronic diarrhea as well. Tamarind has very high quantities of thiamine which is responsible for improving nerve function, and it also aids with muscle development as well. Tamarind is a B-complex that helps reflexes (Nandi, Partha, 2020).

Helps with lowering blood pressure and cholesterol

Helps with weight loss and maintenance.

Natural laxative

Contains thiamine

May be helpful for diabetics.

It has anti-microbical, anti –fungal, anti viral and anti –bacterial.

1. Increases the Natural Defenses of the body
2. Provides nourishment to the eyes and the brain.
3. Promotes

Chapter 3

Vegetables

Vegetables contain various medicinal and **therapeutic** agents. There are a large array of sedative and soporifics or sleep inducing in the vegetable kingdom. Vegetables like onion, radish and celery exercise a tonic effect and are excellent for nerves. Certain vegetables are highly beneficial in the treatment of various diseases. Carrots are good for blood. White crisp juicy stalks of celery serve a much better medicine in case of **rheumatism** or nervous dyspepsia than any nerviness that relieves nerve disorder. A dish of spinach or dandelion will be beneficial in the treatment of kidney troubles. Lettuce can be used as a food remedy for insomnia. Onion can be used with advantage in the treatment of cough, cold, influenza, constipation, scurvy and hydrophobia. The leaves of fenugreek are highly valuable in the treatment of indigestion, flatulence and sluggish liver. Garlic can be beneficially used in heart diseases, **hypertension**, **hypoglycemia**, diabetes and even in fatal form of meningitis. It has been effectively used in long blood cholesterol and preventing blood clotting. Vegetable are alkaline in nature and are ideal foods to build immunity. They are high in antioxidants, promote good digestive health and help to cleanse and restore the body's pH balance.

Fibres in the vegetables act as the mechanical intestinal expanders, draw more water and protein in them, help easy expulsion of the waste in the form of stool. They prevent habitual constipation and keep the entire intestinal track free from harmful germs. Fibres in the form of cellulose help the elimination of cholesterol. Dietary fibres is very useful for the human body because people who have a diet rich in dietary fibre have low incidence of diseases like coronary heart disease and cancer of intestinal tract, piles, obesity, diabetes, constipation, hiatus hernia, diverticulitis, irritable bowel-syndrome, dental caries and gallstones. Beet root, cabbage, carrots, cucumber, green peas and beans are especially valuable in this.

They are useful in case of **arteriosclerosis**, high blood pressure and constipation. But when there is inflammation in the intestine, vegetables having less cellulose content such as tomatoes, lettuce, potatoes and vegetable juices should be taken. Pectin food in the vegetables such as brinjal, radish, pumpkin and beet root absorb water, kill certain bacteria and toxins and eliminate them from the body. Garlic, onion, radish and mint contain pectin as well as **anti-microbic** qualities.

Vegetables supply trace elements which are necessary for human organism. Iodine for instance is essential for thyroid hormone, which regulates much physical and mental activities, cobalt for increasing the number of blood corpuscles and zinc for proper growth. Parsley, sage, rosemary and thyme have many times more anti-oxidants than typical salad veggies. Also various herbs have been associated with different things like lowering blood pressure and helping to control cholesterol levels. Both fresh and dried herbs taste great and add an anti-oxidant punch. Green leafy vegetables are useful because they are high in anti-oxidants and magnesium. According to scientists, eating one and a half extra servings of green leafy vegetables cut the risk of diabetes by 14 percent. **Emodin** present in Chinese Rhubrab (*Rheum palmatum*) and Knot weed (*Polygonum cuspidatum*) inhibits an enzyme called 11-Beta-HSDI which plays a role in resistance to insulin, the hormone that helps clear excess sugar from the blood.

Vegetables can provide widely accessible sources of essential vitamins (particularly A, C, niacin, riboflavin and thiamine), and minerals (calcium and iron), as well as supplementary protein and calories (given with each vegetable). Some vegetables, such as roots, tubers and leafy greens, are capable of producing proteins and calories at rates (per hectare, per day) comparable to those of most efficient stable cereal crops. The greatest advantage with vegetables is that they promote intake of essential nutrients from other foods by making them more palatable. They provide dietary fibre to improve digestion and health, and are essential for properly balanced diet. In many regions around the world, such as India or Southeast Asia, certain cultures survived only on vegetarian meals, either for economic or religious reasons.

India has a long tradition of being predominantly a vegetarian country, as our great sages and saints from millennium have preached vociferously its usefulness. They classified our diet into three categories *viz.* Sattvik, Rajasik, and Tamasvik. Sattvik food which include leafy vegetables, they stressed that it encourages intellectual pursuit, clarity of mind, purer thought, forbearance and keep the nervous system calm and quite. They provide not only the energy rich food like minerals and vitamins. They not only adorn the table, but also enrich health from the most nutritive menu and tone up the energy and vigor of man. They are the cheapest sources of natural nutritive foods. They not only benefit us in protecting against some degenerative diseases but also play a key role in neutralizing the acids produced during digestion of proteinous and fatty foods, also provide roughages which promote and help in preventing constipation *etc.* The only things that brings down **cholesterol** and **homocysteine** levels are vegetables, green leaves and citrus by providing B vitamins, folic acid and vitamin B6 and B12 which help convert **homocysteine** to other amino acids that are not harmful. According to the recent studies, people who adopt a diet free of meat, eggs and dairy products can drop their **homocysteine** levels by 20 per cent in one week. This is because folic acid, a vitamin -B found in whole grains, green vegetables, beans and other plant- based foods, helps convert homocysteine to another, more useful amino acid (Manek Gandhi, 2005).

Spinach, kale and collards, appear to help keep the brain young and may slow the mental decline Vegetables are rich sources of vitamin A, thiamine, niacin, folic acid and β- carotene. Vitamin A (β- carotene) is essential for the normal functioning of the visual process and the structure of the eye, and a prolong deficiency of the vitamin can lead to blindness. They do not contain active vitamin A compound retinol but only carotenoids, which are converted into active retinol in the body. Carrots, sweet potato, amaranth, basella, spinach are important sourcesof vitamin-A. Some vegetables are rich in calcium, phosphorus and iron. The cruciferous vegetables such as broccoli, cabbages (green savoy, red), Chinese cabbage, collard greens, mustard greens, kale, khol rabi and cauliflower are high in vitamins, **monoterpenes,** thiols and indoles. Cauliflower has high levels iron, boron, thiol, indoles and sulphur compounds. The greatest benefits found in eating cruciferous vegetables may be due to the concentration of potassium, folate, fibre and dietary flavonoids found in these foods. The nutritional composition of vegetables is influenced by genetic as well as environmental factors, such as temperature, light, moisture and the nutritional status of the soil in which the vegetable is grown. It is also influenced by the cultural practices, stage of maturity, post harvest handling, storage conditions, as well as cooking and canning also influence the nutrient content of vegetables (Roy and Chakraborti, 1993). A study published in journal of Neurology by a researcher, Martha Clare Morris, at the Rush Institute for Healthy Aging at Chicago's Rush University Medical Center, has found that eating green leafy vegetables including sometimes associated with growing old. The researchers were of the view, that this may be because they contain healthy amounts of vitamin E, an antioxidant that is believed to help fight chemicals produced by the body that can damage cells. Vegetables generally contain more vitamin E than fruits, which were not linked with slowed mental decline in the study. Vegetables also are often eaten with healthy fats such as salad oils, which help the body absorb vitamin E and other antioxidants (Anonymous, 2006.). Similarly another researcher reported at a meeting of the American Society for Reproductive Medicine, that men hoping to boost their fertility may want to eat more fruits and veggies. The more produce a man consumed, the less sluggish his sperm, the new research suggests (Anonymous, 2006). Vegetable like carrot help produce female hormones needed for ovulation, whereas, men should eat garlic for sperm development, quality and mobility.

Broccoli is a good alkaline vegetable and can be consumed in plenty by just boiling it and adding some salt and lemon to it for flavour. Brinjal though not a green vegetable is still an alkaline vegetable. It can be consumed in a variety of ways.

Brinjals with a few potatoes can make a lovely alkaline vegetable dish. The alkaline vegetable list is unending. Cabbage, Lettuce, mushrooms, onions, peas, pepper bells, parsley, radish, cucumber and almost all vegetables are alkaline vegetables. So go out there and skip the meat section and walk into health and vigour. Broccoli and spinach- both the vegetables have high vitamin-C content. It has been found that vitamin-C prevents spots caused by clumping of pigments and red lines caused by damaged capillaries. Carrots have constituents like carbohydrates, protein, fat, calcium, iron, fibre, vitamins B_1, B_2, B_6, C, K, biotin, potassium and thiamin which together provide the body important enzymes, vitamins and minerals. Carrots contain beta-carotene that slows down the process of ageing (Indian Syndicate, Tribune, 16th April, 2010). Plant chemicals in vegetables like tomatoes and pumpkins *etc.* seem to be able to inhibit **angiogenesis** so that a single cancer cell or cluster of cancer cells is never able to grow enough to cause any mischief. Some plants contain tumour suppressor proteins- which help to curb the growth of cancer cells. Studies have shown that men who eat cooked tomato products two to three times a week reduce their prostrate cancer risk by about 50 percent. Eating of vegetables such as brussels sprouts, cabbage, cauliflower and broccoli significantly cut the risk of developing colorectal cancer, a new research has found. Mean while, the report also found that both total fruit and vegetable intake- especially apples and dark yellow vegetables- lower the risk of distal colon cancer (Anonymous, 2011).

Carrots and sweet potatoes may prove to be a vital weapon against breast cancer in the early stages of the disease, says Scientists. A study has found that a nutrient in carrots and sweet potatoes, called **retinoic acid**, can reverse the early changes in cells that lead to breast cancer in women. The chemical, which affects cell growth and proliferation, can also rejuvenate the skin, and a weak version of it is used in anti –wrinkle face creams. Watermelon, carrots and potatoes are considered to be good for the body since they help in rejuvenating our bodies.

A vegetarian diets are healthful, nutritionally adequate and may provide health benefits in the prevention and treatment of certain diseases. Well- planned vegetarian diets are appropriate for individuals during all stages of the life cycle, including pregnancy, lactation, infancy, childhood and adolescence and for the athletes. Vegetarian also appear to have lower–low density lipo protein cholesterol levels, lower blood pressure and lower rates of hypertension and type 2 diabetes than non- vegetarians. Further more, vegetarians tend to have a lower body mass index and lower overall cancer rates. Features of a vegetarian diet that may reduce risk of chronic disease include lower intake of saturated fat, cholesterol and higher intake of fruits, vegetables, whole grains, nuts, soy products, fibre and phytochemicals. A vegetarian diet containing more potassium, complex carbohydrates, fibre, calcium, magnesium, vitamin C, all of which may have favourable influence on blood pressure. It can be a great benefit to start lowering your blood pressure naturally. A new study has suggested that sticking to a vegetarian diet can help kidney disease patients avoid accumulating toxic levels of phosphorus in their bodies. Kidney disease patients must limit their phosphorus intake, as high levels of the mineral can lead to heart disease and death.

Vegetable diet will lower the risk of developing cardio - vascular disease, **osteoporosis,** kidney disease, higher blood pressure, obesity and cancer of colon of lungs. Foods like fruits and vegetables, grains and beans have very little fat and no cholesterol but gives us fibre and other nutrients *viz.* magnesium, potassium, vitamin C and vitamin E, folate, carotenoids that our own body need. Many vegetables and beans like spinach, pinto beans, chick peas have a lot of iron and vitamin B12. Vegetarian diet is associated with a lower risk of death from **Ischemic heart** disease. Many green vegetables contain potassium which greatly helps calm one's nerves when stressed. Also a diet rich in green vegetables helps the body feel lighter and feel great. Vegetables like basil, black pepper, chilies, cumin, ginger and turmeric stimulate hormone (sex) like Dopamine which enhances mood and confidence.

It won't be wrong if we tag spinach as a natural **Viagra.** This green leafy vegetable is loaded with sexual benefits. It is rich in Vitamin E, which is a major catalyst in the production of sex hormones in the body. It is also rich in manganese, which facilitates the production of the female hormone estrogen. A deficiency of magnesium also affects a woman's fertility levels, informs nutritionist Aishwarya Rajan. Green leafy veggies are also loaded with zinc, which is known for its libido and sperm production qualities men.

Now, celery may not be the food of choice that you dream of when thinking sex, but it is a great source food for sexual stimulation. Celery contains **androsterone**, an odorless hormone released through male perspiration and turns women on. Celery is best eaten raw. Wash and cut some, and munch away in bed just before you get down and dirty. Spinach, zucchini, broccoli and brusselsprouts contain a caroteinoid called lutein and a pigment called zeaxanthinan. Both defend your cells from free radicals and help keep your eyes lustrous. Avoid overcooking for best benefits. Steam or saute, instead. Orange and other vegetablesthat are rich in antioxidants and fibre work against under-eye puffiness.

Reds

What to eat: Tomatoes, beets, strawberries, watermelon and red peppers.

Why: The same fruits and veggies responsible for staining your best dress shirts are also believed to fight prostate cancer, cardiovascular disease, diabetes and – listen up, guys – male infertility. Crimson-coloured crops contain varying amounts of lycopene and anthocyanin, two naturally occurring chemicals in plants that are as rich in antioxidants as they are in difficult-to-pronounce syllables. Antioxidants, of course, are powerful molecules that cruise around your body, bonding to and safely defusing other less stable molecules (called free radicals, man!), which, if left unchecked, could cause you some serious cellular damage.

Oranges and Yellows

What to eat: Pumpkins, carrots, yellow squash, lemons and sweet corn

Why: Sure, a tall glass of Tang can deliver your daily dose of vitamin C, which aids in the healing of wounds and the synthesis of collagen. But actual oranges and similarly shaded foods also provide you with the pigments alpha- and beta-carotene. "Beta-carotene is a precursor to vitamin A, " "The body converts this compound into vitamin A, which in turn promotes healthy vision, strong bones and smooth skin." Got psoriasis? Eat more oranges.

Greens

What to eat: Spinach, broccoli, kale, Brussels sprouts and cabbage.

Why: Milk may do the body good, but spinach may do the body even better. Greens are actually packed with higher and more absorbable concentrations of calcium than dairy products. They also contain the **phytochemicals** lutein and **zeaxanthin**, which are important for vision. "And that's to speak nothing of the chlorophyll in greens, which is a great detoxifier. Kelp in particular is high in magnesium, an important nutrient in over 325 chemical activities in your body!" As a rule of thumb, the darker the green, the more chock-full of nutrients it is.

Whites

What to eat: Cauliflower, potatoes, mushrooms, onions and garlic.

Why: When you feel the need to chill out, reach for an onion. The aromatic bulb, like many whitish veggies, is rich in the compound **allicin**. This powerful antioxidant is known to combat high blood pressure and high LDL levels. Pale fruits and veggies are also packed with nutrients that are believed to stimulate your body's B and T cells, which in turn boost your overall immune system. And let's not forget about bananas and potatoes, which are high in heart-healthy potassium.

Blues and Purples

What to eat: Beets, purple cabbage and eggplant.

Why: Once upon a time, blueberries were largely ignored by nutritionists because of the fruit's low level of vitamin C. Now, the same group of experts is tripping over itself to recommend that you eat 1 to 2 cups of the fruit every day. Why? "They're high in **anthocyanins**, which can reduce the risk of high blood pressure and improve heart health." Blue fruits and veggies are also high in fibre and packed with **antioxidants**, and have been shown to reduce the risk of some male cancers. What can't they do? They can't make you fat, since they're really low in calories too.

How much Vegetables Should be in our Daily Diet?

Eat at least 5-7 servings of fresh vegetables every day. Seasonal vegetables should be encouraged. Include variety in vegetable type and colour in your diet.

Yellow and orange colour vegetables are rich in Vitamin A, α and β carotenes, zea-xanthins and crypto-xanthins, where as dark green vegetables are very good source of minerals and phenolic flavonoid anti-oxidants.

Selection of Vegetables

Whenever possible, go for organic farm vegetables to get maximum health benefits. They are not very expensive if you can find them from the nearby local farm owners. Organic verities tend to have smaller in size but have rich flavour and feature good amounts of vitamins, minerals and stuffed with numerous health benefiting anti-oxidants.

1. In the markets, however, always buy small quantities so that they should last within a day or two. There is no point in eating unfit greens!
2. Buy that feature freshness, bright in colour and flavour and feel heavy in your hands.
3. Look carefully for blemishes, spots, fungal mold and signs of insecticide spray. Buy whole vegetables instead of section of them (for example, pumpkin).

Health Benefits of Vegetables

Vegetables, like fruits, are low in fat but contain good amounts of vitamins and minerals. All the Green-Yellow-orange vegetables are rich sources of calcium, magnesium, potassium, iron, beta-carotene, vitamin B-complex, vitamin-C, vitamin A, and vitamin K.

As in fruits, vegetables are also home for many **antioxidants** that; **firstly,** help body protect from oxidant stress, diseases and cancers, and **secondly,** help body develop capacity to fight against these by boosting immunity.

In addition, vegetables contain soluble as well as insoluble dietary fibre known as non-starch polysaccharides (NSP) like cellulose, gums, mucilage, pectin *etc.,* that absorb excess water in the colon and retain good amount of moisture in the fecal matter, thereby helps its easy passage out of the body. Thus helpful in conditions like hemorrhoids, chronic constipation, rectal fissures *etc.*

Vegetable nutrition has widely drawn the attention of fitness conscious as well as food scientists alike for their proven health benefits. Many vegetables are very low in calories, such as *celery* which is even as less than 10 cal per 100 g and here is the long list of vegetables whose calorie is less than 20 per 100 g: Bottle gourd, Bitter melon, Cabbage, **Chinese cabbage,** Bok choy, Eggplant, Endive, Spinach, Summer squash, Swiss chard, *etc.* Scientific studies have shown that low calorie but nutrient rich foods help body stay fit and disease free.

And, also our body spends considerable amount of energy during digestion of foods which is known as **BMR** or **Basal metabolism rate**. So just imagine when you add more vegetable nutrition in the diet, in fact you set to lose more weight than you would gain Right!. This is the concept behind the **"negative calorie foods"**.

Best Muscle Building Fruits and Vegetables

Naturally available muscle building foods that will ensure that your time spent in the gym isn't a waste of sweat. They are Almonds, Banana, Pineapple, Papaya, olive oil, Quinoa, Spinach, Broccoli, watercress, Turmeric, Cucumber and Sweet potato *etc.*

How to Use vegetables?

First thing- you need to do immediately after shopping, wash them, especially green leafy vegetables. Rinse in salt water for few minutes and gently swish in cool water until you are satisfied with cleanliness. This way you ensure they are free from dust, sand and any residual chemical sprays. Use them early while fresh because firstly, certain vegetables have very short shelf life and secondly, the health benefiting properties of a vegetable declines with time. However, if you need to store them, then place in plastic wrappings or in zip pouches in order to preserve nutrition for short periods until you use them.

Chemical Constituents of Vegetables

Vegetable	*Part*	*Chemical Compound*
Asparagus	Sprouts	Lutein, Zeaxathin, Carotenes, Crypto-xanthins
Amaranth	Root	A-spinastevol, Saponins.
Aloe vera		Barbiloine (glucoside).
Artichoke	Bitter Principle	Cyanarin and Sesquiterpene, lactones (Cholesterol reduction action)
	Anti-oxidants	Silymarin, Caffeic acid and Ferulic acid
Arugula	Phyto-chemicals	Indoles, Thiocyanates, Sulforaphane and Isothiocyanates
		di-indolyl- methane (DIM) Anti- bacterial and Anti viral
Ajwain (seeds)	Aromatic fragrance	**Thymol,** pinene, cymene, limonene and terpinene
Beets (red)	Pigment	Betanin and Betacyanin
	-	β-xanthin
Beet (yellow)	Root	Glycine Betaine
Bell pepper	Capiscum	Capsaicin- Aggravates Gastro-esophageal reflux condition (GER).
Black pepper		Piperin, Piperidin, piperetine, Chavicine.
Bitter Gourd	-	Polypeptide-P (plant insulin known to lower blood sugar levels), Mamoridisine.
	Hypoglycemic agent	Charantin (increases glucose uptake and glycogen synthesis in liver cells *etc.*

Vegetable	*Part*	*Chemical Compound*
Broccoli	Phyto- nutrients	Thiocyanates, indoles, sulforaphane, isothiocyanates
	Flavonoids	β-carotene, cryptoxanthin, lutein and zeaxanthin
	Flowers head	Omega-3-fatty acids
Brussels Sprouts	i. Flavonoid	Thiocyanates, zeaxanthin, sulforaphane and isothiocyanates
	i. Anti-bacterial, viral	di-indolyl-methane (DIM)
	iii. Glucoside	**Sinigrin** (helps protect colon cancers by destroying pre-cancerous cells)
	vi. Carotenoid	**Zeaxanthin**
		Goitrogens
Butternut squash	Flavonoids	**α and β-carotens**, cryptoxanthin-β and lutein
	Seed	**Tryptophan** (converts to health benefiting GABA neuro-chemical in the brain).
Basil	Flavonoids	Orientin and Vicenin
	Oils	Eugenol*, citronellol, linalool, citral, limonene, terpineol, Zeaxanthin**
Brinjal	Deep blue/purple	Anthocyanins
Borage	Fatty acid	Gamma- linolenic acid (GLA) ***
Burdock	Rot	Inulin, glucoside-lappin, mucilage
Black pepper	Pungent	Peperine#, sabinene, pinene, terpenene, limonene, mercene
Cabbage	Thiocyanates	Indol-3-carbinol, lutein, sulforaphane, isothiocyanate, zeaxanthin, goitrogen
Cauliflower	Phyto-chemical	Indole-3-carbinol ##, di- indolyl-methane, goitrogen
Carrot	Root	Falcarinol (destroys pre-cancerous cells in the tumers
Celery	Leaves and seeds	Limonene (75-80 per cent), sesquiterpenes like β- seliene (10 per cent) and humulene
Chives	Thio-sulfides	Diallyl disulfide, diallyl trisulfide and allyal propyl disulfide+
Coriander	Leaves and seeds	Borneol, linalool, cineole, cymene, terpineol, dipentene, phellandrene, pipene and terpinolene, Coriandol
	Leaves/stem tips	Quercetin, kaempferol, rhamnctin and epigenin
		Petroselinic acid, linoleic acid, oleic acid, palmitic acid, linalool, α-pinene, geraniol, camphene and terpine
	Seeds	Linaloon and Decanoic acid.

Vegetable	*Part*	*Chemical Compound*
Caraway	Seeds	Carvone, limonene, carveol, pinen, cumuninic-aldehyde, fururol and thujone
Cardamom	Seeds	Pinene, sabinene, myrcene, phellandrene, limonene, 1, 8-cineole, p-cymene, linalool, ciniol
Cloves	Essential oils	Euginol, acetyl-eugenol, β-caryophyllene, vanillin, crategolic acid, eugenin, kaempferol, eugenitin, oleanolic acid, stigmasterol, campesterol, Emitil, sesquiterpenes
Cumin		Cuminaldehyde (4-isopropylbenzaldehyde), pyrazines, 2-methoxy-3-sec-butyl-pyrazine, 2-ethoxy-3-isopropyl pyrazine and 2-methoxy-3-methyl pyrazine. Thymoquinone-checks proliferaton of cells responsible for prostate cancer.
Cinamon		Eugenol@, cinnamate, linalool, β-caryophyllene, methyl chavicolCinnamaldehyde (flavour and smell)
Collards green	Leaves	Di-indolyl-methane (DIM) and Sulforaphane
Cucumber		B-carotene and α- carotene, Zeaxanthin and lutein
Capers	Flower buds	Rutin++ (rutoside), Quercetin
Dandelion	Roots-Bitter crystalline compound.	Taraxain, Taraxacerin (acrid resin), Inulin and Levulin (therapeutic properties)
Dil	Monoterpenes	
Endive		B-carotene, folic acid, bitter glycosides and inulin
Fennel	Oils	Anethole!, estragole and fenochone (fenchyl acetate)
Fenugreek	Seeds	4-hydroxy-isoleucine !! (Facilitates insulin secretion)
Green beans	Phenolic compounds	Zeaxanthin, lutein, β-carotene, folates, Oxalic acid
Garlic	Thio-sulfinites	Diallyl-disulfide, diallyl-trisulfide, Germanium, allyl propyl disulfide+++
Ginger	Roots	Gingerol^, Zingerone^^, shogaol, farnesene, Gingerine (bitterness).
Horse radish	Pungency	Allyl isothiocyanate, 3-butenyl isothiocyanate, 2-propenyl glucosinlate (sinigrin), 2-pentyl isothiocyanate and phenyl ethyl isothiocyanate
Kale	Leaves	Sulforaphane, indole-3-carbinol, di- indolyl-methane (DIM).
Knol Khol		Isothiocyanates, sulforaphane, indole-3-carbinol, Goitrogens

Vegetable	*Part*	*Chemical Compound*
Leeks		Diallyl disulfide, diallyl trisulfide and allyal propyl disulfide
Lettuce	Leaves	B-carotene, zeaxanthin, folate
MustardMint	i. Seed	Brassicasterol, campesterol, sitosterol, avenasterol, stigmasterol also sinigrin, myrocin, erucic, eicosenoic, oleic, palmitic acid, Sinigrin, Sinapin
	ii. Green	Indoles, sulforaphane, di-indolyl-methane
	Oil	Menthol, menthone, menthol acetate (analgesic).
Okra	Pod	B-carotenes, xanthin, lutein
Oregano	-	Carvacrol
Onion	Bulb (Allyl sulfide gas)	Allium, allyl-disulfide conver to Allicin, Quercetin, pyridoxine, chromium
		Phyto-chemical quercetin which restricts growth of malignant cells.
Parsley	Oil	Eugenol
	Antioxidants	Apiin, apigenin, crisoeriol and luteolin zea-xanthin~
Purslane	-	Omega-3-fatty acid
	Alkaloids pigment	Betalain - β-cyanins (reddish), β- xanthins (yellow)
Pumpkin	Fruit	α and β carotenes, cryptoxanthin, lutein, zeaxanthin
	Seed	Tryptophan (health promoting amino acid) selenium, zinc
Peas	Pods (phytosterols)	B- sitosterol, carotenes, lutein, zeaxanthin
Potato	Toxic alkaloid	Solanine and Chaconine
Parsnip		Falcocarinol, falcarindiol, panaxydiol and methy-falcarindol
Radish		Goitrogens"
Rhubarb	Leaf blade	Oxalates
Rosemary	Flower tops	Rosmarinic acid
	Volatile oils	Cineol, camphene, borneol, bornyl acetate, α-pipene.+*
Saffron	Flavour	Picrocrocin and Safranal
	Colour	Crocin (dicarboxylic acid) gives saffron its golden yellow hue.

Vegetable	*Part*	*Chemical Compound*
Shallot Sage Spinach$		Diallyl-disulfide, diallyl-trisulfide, allyl propyl disulfide Thujone****, Salvigenin Omega-3-fatty acid, lutein, zeaxanthin, β-carotene, oxalic acid.
i. Savory ii. Sweet corn iii. Sweet potato iv. Swiss chard		i. Thymol and carvacrol<, linalool, camphene, caryophyllene, terpineol, myrcene ii. Ferulic acid iii. oxalic acid, β-carotene iv. Omega -3- fatty acid.
Thyme	Oil Antioxidants	Thymol, Zeaxanthin, lutein, pigenin, naringenin, luteolin, thymonin
Turmeric	Pigment	**Curcumin-** Anti tumor, i arthritic, i amyloid, i septic and anti inflammatory properties
Tomato	Fruit	Lycopene- A flavonoid anti oxidant
Turnip	Roots/tops	Oxalic acid
Tarragon		Estragole (methyl chavicol), cineol, ocimene, phellandrene
Water cress	Stems (Pappery flavor	**Gluconasturtiin-** On hydrolysis provide 2-phenethyl isothiocyanate (PEITC)
Watermelon		Is believed to be cancer preventive inhibition of phase 1 enzymes. Citrulline.
Yam		Vit. B6, Thiamin, riboflavin, folic acid, pantothemic acid, niacin

*: Anti–inflammatory function. **: Yellow flavonoid carotenoid compound is selectively absorbed into the retinal maculaluteua where it found to filter harmful UV rays from reaching retina. ***: Linolenic acid is Omega-6- fatty acid that plays vital role in restoring of joint health, immunity, healthy skin and mucus membranes. #: Amine alkaloid- which gives strong spicy pungent character to pepper. ##: Acts as a anti-estrogent agent. +: Convert to allicin by enzymatic reaction when its leaves are disrupt (crushing, cutting). ++: Strengthens capillaries and inhibits platelet clump formation in the blood vessels. Both these actions of rutin help in smooth circulation of blood in very small vessels. @: A phenylpropanoids class of chemical compound, which gives pleasant, sweet aromatic fragrance. !: Anethole-antifungal and anti bacterial properties. !!: Present in the fenugreek seed, has faciliator action on inulin secretion. +++: That can form allicin by enzymatic reaction, which is activated by disruption of bulb. - Allicin- (volatile compound) has anti bacterial, anti viral and anti fungal activities. ^: Helps improve the intestinal motility and has anti-nflammatory, painkiller (analgesic), nerve soothing, anti-pyretic as well as anti bacterial properties. ^^: Zingerone- Which gives pungent character to the ginger root, has been found to effective against *E. coli* induced diarrhea, especially in children. ": May cause swelling of the thyroid gland. Should be avoided in individuals with thyroid dysfunction. ?: These toxic substances are destroyed when cooked at high temperature (over 170°C). ~: Helps prevent age related macular-degeneration (ARMD) in the retina of the eye in the old age population, through its anti-oxidant and ultra-violet light filtering functions. +*: Are known to have rubefacient, anti

inflammation, anti-allergic, anti- fungal and anti septic properties. ****: It enhances concentration, attention, span and quickens the senses. It helps to deal with grief and depression.,: Inhibits the growth of several bacteria strains like *E. coli* and *Bacillus cereus*. $: Reheating of spinach left over may cause conversion of nitrates into nitrites and nitrosamines by certain bacteria that thrive on preprepared nitrate rich foods such as spinach and many other green vegetables. These poisonous compounds may be harmful to health, especially to children. Oxalic acid: A naturally occurring substance found in some vegetables which may crystallize as oxalate stones in the urinary tract in some people.

ASPARAGUS

Asparagus Spears

Asparagus officinalis is widely consumed worldwide and has long been used as an herbal medicine due to its anticancer effects. An excellent food for pregnant women, asparagus is full of fibre, folate and Vitamin B6 qualities which help in maximizing heart's health. These long veggie sticks are also pretty low in sodium and high in potassium content, which makes them ideal for the healthy working of our large intestine. The wonder vegetable has very high levels of vitamin E, which plays a major role in skin cell restoration. It is rich in folic acid and contains magnesium that are famed for their mood-lightening properties. Baby asparagus stalks are tender, crunchy and tasty all the same time making them the perfect for any time snack. Consumption of asparagus helps make your skin more firm and clears blemishes on it. Its shoots contain a white crystalline substance, **asparagines** (an amino acid) which is used in the medicine as diuretic in cardiac dropsy and chronic gout. Among vegetables, asparagus is the leading supplier of **folic acid** for human nutrition, which is necessary for blood cell formation, growth and prevention of liver disease. Folic acid plays a significant role in the prevention of neutral tube defects, such as spina bifida, that cause paralysis and death in infants. Asparagus is low in calories, contains no fat or cholesterol, is very low in sodium and is a good source of potassium, fibre, thiamin, B6 and one of the richest source of rutin, a chemical that strengthens capillary walls (Michigan Asparagus Council, 2005), making asparagus a wise choice for health- conscious consumers. Asparagus rhizomes and root is used ethnomedically to treat urinary tract infections as well as kidney and bladder stones. It has also aphrodisiac properties (this belief is at least partially due to the phallic shape of the shoot). Ingestion of asparagus may bring on an attack of gout in certain individuals, due to the high level of purines. According to modern physicists it inhibits stimulation of uterus and normalizes its functioning. The plant is also a diuretic and can aid the elimination of water through urination, a special boon in cases of water retention (oedema). Asparagus can be used to treat many kidney conditions, but it should not be used in cases of inflammations. It is also helpful in cleansing cholesterol from arteries and hence is useful in treating vascular problems such as hypertension and arteriosclerosis. Asparagus is high in the B vitamin folate that helps increase your production of histamine. Histamine is important for increasing the sex drive in men and women.

According to **Bhawa Prakash** it is heavy, cool and bitter and nutritive. It enhances intelligence, promotes digestion, pacifies vata and pitta disorders and cures reduced semen production. It also cures reduction of breast milk. According to **Acharya Sushart** asparagus is beneficial in dryness and promotes physical strength. It purifies semen, enhances intelligence and promotes digestive fire. While according to **Dhanvantari nighantu**, Shatawar can cure even a very weak person. It promotes physical strength in the patients and enables the patient to fight the disease. It enhances the immune system and suppresses vata pitta, produces semen, is cool and sweet. It is a semen enhancing herb, Shatavari, sariva, arand, punnarva and hansraj all these suppress pitta and vayu and cures respiratory disorders.

According to **Acharya Charak**, it is beneficial for physical strength and arresting aging (Acharya Balkrishna, 2008).

Health Benefits

Asparagus's biggest talent is its ability to encourage the body to flush out toxins, due to its natural diuretic abilities. Asparagus is both cleansing and anti-inflammatory to the body. It has the antioxidant glutathione, which can lower your risk factor for heart disease and cancer. It's useful for all inflammatory conditions, such as arthritis and irritable bowel syndrome.

Asparagus is a very low calorie vegetable. 100 g fresh spears give only 20 calories. More calories will be burnt to digest than gained, the fact, which fits in to the category of low calorie or negative calorie vegetables.

In addition, the shoots have good levels of dietary fibre. Dietary fibre helps control constipation conditions, decrease bad, "LDL" **cholesterol** levels by binding to it in the intestines, and regulates blood sugar levels. In addition, high fibre diet helps prevent colon-rectal cancer risks by preventing toxic compounds in the food from absorption.

Its shoots have long been used in many traditional medicines to treat conditions like dropsy and irritable bowel syndrome.

Fresh asparagus spears are good source of anti-oxidants such as *lutein, zeaxanthin, carotenes,* and *crypto-xanthins*. Together, these flavonoid compounds help remove harmful oxidant free radicals from the body protect it from possible cancer, neuro-degenerative diseases, and viral infections.

Fresh asparagus are rich in folates. 100 g of spears provide about 54 mcg or 14 per cent of RDA of folic acid. **Folates** are one of the important co-factors for DNA synthesis inside the cell. Scientific studies have shown that adequate consumption of folates in the diet during pre-conception period and during early pregnancy help prevent neural tube defects in the newborn baby.

The shoots are also rich in B-complex group of vitamins such as thiamin, riboflavin, niacin, vitamin B-6 (pyridoxine), and **pantothenic acid** those are essential for optimum cellular enzymatic and metabolic functions.

Fresh asparagus also contains fair amounts of anti-oxidant vitamins such as vitamin-C, vitamin-A and vitamin-E. Regular consumption of foods rich in these vitamins helps body develop resistance against infectious agents and scavenge harmful, pro-inflammatory free radicals from the body.

Its shoots are also good source of vitamin K; provides about 35 per cent of DRI. Vitamin K has potential role bone health by promoting **osteotrophic** (bone formation) activity. Adequate vitamin-K levels in the diet helps limiting neuronal damage in the brain; thus, has established role in the treatment of patients suffering from **Alzheimer's** disease.

Asparagus is good in minerals especially copper and iron. In addition, it has small amounts of some other essential minerals and electrolytes such as calcium, potassium, manganese, and phosphorus. Potassium is an important component of cell and body fluids that helps controlling heart rate and blood pressure by countering effects of sodium. Manganese is used by the body as a co-factor for the antioxidant enzyme, *superoxide dismutase*. Copper is required in the production of red blood cells. Iron is required for cellular respiration and red blood cell formation.

Asparagus is high in **glutathione**, an important anti-carcinogen.

It also contains rutin, which protects small blood vessels from rupturing and may protect against radiation.

Asparagus is a good source of vitamins A, C and E, B-complex vitamins, potassium and zinc.

It also has antifungal, anti-inflammatory and diuretic properties.

The biological functions of asparagus can help alleviate alcohol hangover and protect liver cells (Anonymous, 2012).

Caution

In general asparagus is well tolerated in general population and allergic reactions are quite rare to occur. Ingestion of young shoots may give offensive smell to urine. This is due to the metabolism of *asparagusic acid*, which breaks down into various sulfur-containing degradation products such as methanethiol, sulfides.etc. The condition, however, is harmless.

TUBER CROPS

Tuber crops species like Dioscorea, Alocasia, Sweet potato, Elephant foot yam and Coleus are known to contain medicinal properties. **Beta-carotene** and **atocopherol** (vitamin E) are known to have anti- cancer and anti-oxidant properties. Sweet potato varieties with high content of these principles may be developed to enhance their pharmaceutical potential. Many yam and aroid tubers contain high levels of mucilage. Tuber rich in mucilage shall be exploited for extraction of mucilage for industrial as well as medicinal application. Researchers have found that white potatoes are the largest and most affordable source of potassium of any vegetable or fruit.

POTATO (*Solanum tuberosum*)

Research has shown that it is the peel that **antioxidants** are contained. These antioxidants can help revitalize body cells as well as fight infection. This potato fibre has also parts that lower cholesterol levels. Also the layer of potato lying just below the skin is very rich in protein and mineral salts. There is a "Saying that "cook the peel to eat and use the rest to feed the chicken." Potato fibre has also parts that lower cholesterol level. Potato is perhaps the most alkaline vegetable. It is rich in soda and potash but low in lime. Therefore, it is helpful in maintaining the alkaline reserve of body and is natural antidote for an overdose of acidic foods. They are a highly starchy food and when eaten the with skin or peel a good source of vegetable protein, potassium, vitamin C, iron, phosphorus, niacin and enzymes - although old potatoes are low in vitamin C. Medicinally, potatoes can help to relieve **arthritis** and reduce water retention due to their high potassium content. They also neutralize body acids and potato juice can be used to treat stomach and duodenal ulcers. The fresh juice is sometimes used to reduce **hypertension** and promote intestinal flora.

Nutritionally, potatoes are best known for their carbohydrate content (approximately 26 grams in a medium potato). The predominant form of this carbohydrate is **starch.** A small but significant portion of this starch is resistant to digestion by enzymes in the stomach and small intestine and so reaches the large intestine essentially intact. This resistant starch is considered to have similar physiological effects and health benefits as fibre; it provide bulk, offers protection against colon cancer, improves glucose tolerance and insulin sensitivity, lowers plasma **cholesterol** and **trigylyceride** concentrations, increases satiety, and possibly even reduces fat storage (Cummings *et al.,* 1996). The amount of resistant starch in potatoes depends much on preparation methods. Cooking and then cooling potatoes significantly increases resistant starch. For example, cooked potato starch contains about 7 per cent resistant starch, which increases to about 13 per cent upon cooling (Englyst *et al.,* 1992). Individuals who consumed potatoes (baked, boiled and roasted) had higher intakes of potassium and vitamin C and consumed more total vegetables in a day compared to those who did not consume potatoes.

Potatoes contain vitamins and minerals which are vital to human nutrition. Potatoes consist of about 80 per cent water, 20 per cent solid matter and have a high nutritional value. Starch makes up about 85 per cent of this solid mass and the rest is protein. As humans can subsist healthily on a diet of potato and milk, the later supplies vitamin A and vitamin D. A normal sized potato (150g) with skin intact provides 27 mg of vitamin C (45 per cent of the Daily Value DV), 620mg of potassium (18 per cent of DV), 0.2mg vitamin B6 (10 per cent DV) and trace amounts of thiamine, riboflavin, folate, niacin, magnesium, phosphorus, iron and zinc. Potassium is essential to the body because of its role in attaining optimal muscle performance and improving the nerve's response to stimulation. Iron is essential in helping the body convert food to energy as well as resist infection, is also present. Potato contains no fat or cholesterol and mineral sodium. What they do have is natural fibre in the skins, all those vitamins and minerals and great flavour. And 170grams of potato contains 2 grams of highly digestible protein, almost as much as half a glass of milk, making it a great foundation for a whole meal.

Moreover, the fibre content of a potato with skin (2grams) equals that of many whole grain breads, pastas and cereals. Potatoes also contain phytochemicals such as carotenoids and polyphenols. The notion that "all of the potato`s nutrients" are found in the skin is a misnomer. While skin does contain approximately half of total dietary fibre, more than 50 per cent of the nutrients are found within the potato itself. The cooking method used can impact significantly the nutrient availability of the potato. Almost all the protein content of a potato is contained in a thin layer just under its skin. This is evident when the skin of a boiled potato is carefully peeled, it appears as a yellowish film. For maximum utilization of this small, but available dietary source of protein, potatoes should be consumed whole or peeled or peeled after cooking.

One medium sized potato with skin contains 620mg of potassium, followed by one medium stalk of broccoli at 540mg. Foods that are a good source of potassium and low in sodium- such as potatoes- may reduce the risk of high blood pressure and stroke. Potatoes are also an excellent source of vitamin C and bring important B vitamins and fibre. A 5.3 ounce potato has a mere 100 calories and contains no fat, cholesterol or sodium.

Potato	1 medium	720mg
Broccoli	1medium stalk	540mg
Tomato (Plum roma)	1 medium	360mg
Tomato (other)	1 medium	360mg
Celery	2medium stalks	350mg
Cantaloupe	1\4medium	280mg

Sweet pepper	1 medium	270mg
Onion	1 medium	240mg
Sweet corn	1 medium ear	240mg
Water melon	2 coups (diced)	230mg
Cucumber	1\3 medium	170 mg
Lettuce (Iceberg)	1\6 medium head	120mg

Carbo Facts

One of the most popular nutritional myths today concerns the dietary evils of carbohydrates.

Despite the media hype, scientific evidence indicate that when consumed in normal amounts.

Carbohydrates DO NOT cause weight gain and obesity.

Carbohydrates DO NOT automatically cause blood sugar "spikes" and insulin" surges".

Carbohydrates DO NOT cause diabetes, heat disease or cancer.

Despite the popular notion, the majority of nutrients are not found in the skin, but in the potato itself. Nonetheless, leaving the skin on the potato retains all the nutrients, the fibre in the skin and makes potato easier to prepare.

Potatoes and blood sugar level: Unlike the general perception of high glycemic index food, potatoes have a medium glycemic index and they do not raise the blood sugar levels to crazy. Their glycemic index is a decent 56. However, the forms in which they are consumed make them fattening such as deep fried, shallow fried and so on.

Potatoes and recommended amount: The moderate amount recommended is 1 medium-sized potato a day.

Potatoes and Calories

1 large potato (260gms) will have 278 calories and 63 grams i.e. 21 per cent of carbohydrate. Calories from potatoes result in weight gain only if takenregularly in largeamounts. Also, as stated earlier, the form in which they are consumed matters. The best preparations are baked and steamed.

Potatoes and Top Health Benefits

Potential health benefits of potatoes are: Weight gain for the lean: They could result in weight gain as most calories come from a high percentage of carbohydrates in potatoes.

Facilitates digestion: Potatoes contain simple carbohydrates that are easy to digest and thus facilitate digestion. They are a source of instant energy, but they should be consumed only in moderate quantities to avoid other health complications.

Helps skin care: Potatoes contain vitamin-C and B-complex and minerals like potassium, magnesium and zinc which are good for the skin.

Prevents scurvy: The vitamin-C present in potatoes can help prevent scurvy which is caused due to lack of vitamin-C.

Relieves inflammation: Potato is a simple carbohydrate especially in steamed and mashed form and is easily digested and absorbed through our digestive system. It relieves inflammation of the intestines and the digestive system.

Keeps brain function active: Carbohydrates are a major and easiest form of glucose and the good news is that potatoes are high in carbohydrates. By providing instant calories, potatoes maintain instant replenished levels of glucose in the blood which keeps the brain active instantly.

Heart disease: Usually, potatoes should not be given during treatment and immediately after treatment of heart disease. This is because, usually, heart diseases are related with obesity and this would be a time to eat a lean diet devoid of simple carbs. Although, as per your cardiologist's recommendation, potatoes can be taken in a healthy balanced manner.

Potatoes surpass bananas in their Potassium content: Take a large potato and bake it with the skin on. What you have is a large spud ready to be spiced with the ingredients of your choice and containing an astounding supply of potassium.

At a mind blowing 1, 600 milligrams of potassium, potatoes pack in almost half of the potassium amount recommend for an adult for an entire day. Compared to the more common choice of bananas, that's nearly *four times* the potassium that is to be found in a medium sized banana.

Potatoes are packed with dietary fibre: One large potato, eaten with the skin on, can fetch you a quarter of your daily recommended fibre intake. That is seven grams of dietary fibre. However, the key is to keep the skin on as the fibre content drops exponentially if you peel off the potato.

A fibre rich diet enables you to stay fuller for longer periods, reducing the urge for inter-meal snacks. As a result, contrary to the popular perception of equating potatoes with weight gain, eating whole potatoes can actually assist you in losing some flab. Fibre also reduces the risk of heart attacks, cholesterol problems and blood sugar imbalances.

Potatoes can Compete with Oranges in Delivering Vitamin C: A whole potato, eaten with the skin on, can give you half of your daily recommended Vitamin C supply - that equals 29 milligrams of the vitamin or more than a third of what is found in a full ripe orange, considered your best bet as a Vitamin C source but not quite available all the year round like the much more modest potato.

Vitamin C is a highly recommended mineral that plays an active role as an antioxidant in the body. Antioxidants are essential for a number of reasons - they prevent ageing and do wonders for your skin, hair and overall health, help heal wounds faster, act as a barrier against the onslaught of various infections and according to some studies, even help battle the scourge of cancer.

Potatoes are a source of another essential mineral – Manganese: Despite not being referred to as commonly as others in our usual health conversations, Manganese is a vital nutrient for the human body. One big potato spud with the skin on contains 33 per cent of your daily recommended amount of this mineral. Manganese plays a pivotal role in metabolism, helps the body in digesting proteins, carbohydrates and cholesterol and also aids in bone formation. Manganese also assists in regulating the functioning of the nervous system, helps in the formation of the thyroxine hormone for the thyroid gland and also aids in the secretion of various sex hormones. Manganese has several antioxidant properties too that can prevent cancer as well as a host of heart diseases.

Potatoes are a rich source of Vitamin B6: A single potato with the skin on can fetch you as much as 46 per cent of your daily recommended dose of Vitamin B6. Vitamin B6 has several crucial functions to perform in your body - right from your cardiovascular system to your digestive health to your immune response efficiency to the health of your muscular and nervous systems. Vitamin B6 lowers the blood pressure and keeps in check the blood cholesterol levels. It also helps produce vital brain hormones and prevents blood platelets from sticking together.

Health Benefits

Potatoes are one of the richest sources of starch, vitamins, minerals and dietary fibre. Contains very low fat and no cholesterol.

Both soluble and insoluble fibre in them increases the bulk of the stool, thus, it helps prevent constipation, decrease absorption of dietary cholesterol and there by lower plasma LDL cholesterol. Additionally, the rich fibre content also helps protect from colon cancer.

The fibre content helps slow absorption of starch in the gut and thereby keeping blood sugar levels within normal range. For the same reason, potato is still favored source of carbohydrates in diabetics.

The tubers are one of the richest sources of B-complex group of vitamins such as Vitamin B6, Niacin, Pantothenic acid and folates.

Fresh potato skin as well as flesh are good source of antioxidant vitamin; vitamin-C. Regular consumption of foods rich in vitamin-C helps body develop resistance against infectious agents and scavenge harmful, pro-inflammatory free radicals.

Red and *russet potatoes* contain good amount vitamin A, and flavonoids like carotenes and zeaxanthins.

Recent studies at Agricultural research service (by plant genetics scientist Roy Navarre) suggests that flavonoid antioxidant, *quercetin* present in potatoes has anti-cancer and cardio-protective properties.

Eat potatoes as they resist the accumulation of calcium in the kidney, thus preventing the formation of kidney stones.

Caution

Potatoes may contain toxic alkaloids, *solanine* and *chaconine.* These alkaloids present in the greatest concentrations just underneath the skin and increase proportionately with age and exposure to sun light. Cooking at high temperatures (over 170 °C) partly destroys these toxic substances. When consumed in sufficient amounts, these compounds may cause headache, weakness, muscle cramps and, in severe cases loss of consciousness and coma; however, poisoning from potatoes occurs very rarely. Exposure to light also causes green discoloration; thus giving a visual clue as areas of the tuber that may have more toxins; however, this does not provide a definitive clue, as greening and solanine accumulation can occur independently to each other. Some varieties contain greater solanine concentrations than others.

SWEET POTATO

Sweet Potatoes (*Ipomoea batatas*) leaves (top greens) are also edible; in fact, the greens contain more nutrients and dietary fibre than some green leafy vegetables like **spinach** (for example, 100 g sweet potato leaves provide 1028 IU of vitamin A). Grown under earth, sweet potatoes are full of ant-defending qualities as they are rich in manganese, Vitamin A and Vitamin C. Good for our digestive system, they are full of iron and fibre which helps in giving us good amounts of energy. Outstandingly rich in vitamin A (many times more so than potatoes), sweet potatoes also provide vitamin C, calcium, iron, potassium and sodium, but they must be baked or boiled in their skin to retain these nutrients. Due to their exceptional vitamin A content, sweet potatoes can be used to improve night vision. They are also reported to increase milk secretion in nursing mothers, remove toxins from the body and treat underweight and diarrhoea. Spinach - High in vitamins A and C and folate, it's also a good source of magnesium. The plant compounds in spinach may boost your immune system while the carotenoids found in spinach – beta carotene, lutein and zeaxanthin are protective against age-related vision diseases, such as macular degeneration and night blindness, as well as heart disease and certain cancers.

Sweet Potatoes - With their deep orange-yellow color, sweet potatoes are high in the antioxidant beta carotene which is converted to vitamin A in your body. It helps slowing down the aging process and reduce the risk of some cancers. In addition to being an excellent source of vitamins A and C, sweet potatoes are also a good source of fibre, vitamin B-6 and potassium. And like all vegetables, they're fat-free and relatively low in calories with one-half of a large sweet potato having just 81 calories.

It is a rich source of **lutein**. Its use as a leafy vegetable is novel in most countries and potentially may serve a market niche given its enhanced nutritional status. Lutein is a carotenoid capable of delaying blindness-related muscular degeneration. The content of lutein in sweet potato ranged from 0.38 – 0.58mg. g^{-1} fresh weight. Lutein is not synthesized donovo and must be ingested from outside sources. Dark green vegetables have been identified as major edible source of lutein, by decreasing order, are kale, spinach, lettuce and broccoli (Johnson, 2002). Sweet potatoes are an excellent source of carotenoid antioxidants. They contain calcium, are high in vitamins A and C and contain thiamine.

Sweet potatoes have a slightly lower carbohydrate content than regular potatoes. One medium potato contains 26 grams of **carbohydrates**, while one medium sweet potato contains 23 grams of carbohydrates.

Sweet potatoes contain almost twice as much fibre as other types of potatoes. Contributing close to 7 grams of fibre per serving. The high fibre content gives them a 'slow burning quality'. Which means their caloric energy is used more slowly and efficiently than a low-fibre carbohydrate. They are an excellent source of vitamin A and a good source of potassium and vitamin C, B6, riboflavin, copper, pantothetic acid and folic acid. A medium sweet potato has 28 per cent more potassium than a banana. The high amount of water that is found in sweet potatoes helps you lose weight. Like fibre, water takes up a lot of room in your stomach. Therefore, eating foods that contain high amounts of water makes you feel full and will prevent overeating and snacking betweenmeals. In fact, they're so good for you that almost everyone should incorporate them into their diet. There are three main reasons that these orange tubers (that's what all potatoes are called) are so healthy. And these reasons can benefit everyone's overall health. They contain antioxidants. Antioxidants prevent or delay cell damage, and can strengthen your body's own ability to fight off infection. They contain anti-infammatory nutrients. Anything that is anti-inflammatory prevents or reduces inflammation in your body. Chronic or "out of control" inflammation leads to poor health. People with arthritis and many autoimmune diseases benefit greatly from the anti-inflammatory nutrients in sweet potatoes. They contain blood sugar regulating nutrients. Some people's bodies cycle between blood sugar levels that are either too high or too low. This lack of stabilization is hard on the body.

Health Benefits

Best muscle building foods: Sweet potatoes are an excellent source of vitamin A and a good source of potassium and vitamin C, B6, riboflavin, copper, pantothetic acid and folic acid. A medium sweet potato has 28 per cent more potassium than a banana. They replenish energy stores and fuel the muscle-building process (Moghul, 2013).

Sweet potatoes are far superior than the run-of-the-mill white potato. The orange variety contains beta carotene, which makes them filled with robust antioxidant, antiviral, and anticancer abilities. They're also full of fibre and the vitamin E they contain is healthy for the skin.

They are high in vitamin B6. Vitamin B6 helps reduce the chemical homocysteine in our bodies. Homocysteine has been linked with degenerative diseases, including the **prevention of heart attacks**.

They are a good source of vitamin C: While most people know that vitamin C is important to help ward off cold and flu viruses, few people are aware that this crucial vitamin plays an important role in bone and tooth formation, digestion, and blood cell formation. It helps accelerate wound healing, produces collagen which helps maintain skin's youthful elasticity, and is essential to helping us cope with stress. It even appears to help protect our body against toxins that may be linked to cancer

They contain Vitamin D: Vitamin D, which is critical for immune system and overall health at this time of year. Both a vitamin and a hormone, vitamin D is primarily made in our bodies as a result of getting adequate sunlight. You may have heard about seasonal affective disorder (or SAD, as it is also called), which is linked to inadequate sunlight and therefore a vitamin D deficiency. Vitamin D plays an important role in our energy levels, moods, and helps to build healthy bones, heart, nerves, skin, and teeth, and it supports the thyroid gland.

Sweet potatoes contain iron: Most people are aware that we need the mineral iron to have adequate energy, but iron plays other important roles in our body, including red and white blood cell production, resistance to stress, proper immune functioning, and the metabolizing of protein, among other things.

Sweet potatoes are a good source of magnesium, which is the relaxation and anti-stress mineral. Magnesium is necessary for healthy artery, blood, bone, heart, muscle, and nerve function, yet experts estimate that approximately 80 per cent of the population in North America may be deficient in this important mineral.

Nutritious sweet potatoes are low in calories (provide just 90 cal/100 g, on comparison with starch rich cereals) and contains no saturated fats and cholesterol; but are rich source of dietary fibre, anti-oxidants, vitamins and minerals.

They are storehouse of starch, a complex carbohydrate, which raises the blood sugar levels *slowly* on comparison to simple sugars; therefore, recommended as a healthy food supplement even in diabetes.

They are excellent source of *flavonoids* like beta-carotene and vitamin A. The valueis one of the highest among root vegetables category. These compounds are powerful natural antioxidants. Vitamin A is also required by the body to maintain integrity of healthy mucus membranes and skin. It is also vital nutrient for vision. Consumption of natural vegetables and fruits rich in flavonoids helps to protect from lung and oral cavity cancers.

The tubers are packed with many essential vitamins such as pantothenic acid (vitamin B5), pyridoxine (vitamin B-6) and thiamin (vitamin B-1), niacin, and riboflavin. These vitamins are essential in the sense that body requires them from external sources to replenish. These vitamins function as co-factors for various enzymes during metabolism.

They also contain good amounts of minerals like iron, calcium, magnesium, manganese, and potassium that are very essential for body metabolism.

Be careful: eating too many may cause abdominal swelling and indigestion.

Sweet potatoes are also high in sugar and therefore should be used sparingly.

Sweet potatoes are not related to the potato nor the yam–they are actually a member of the morning.

Sweet potato leaves (top greens) are also edible. In fact, the greens contain more nutrients and dietary fibre than some of the green-leafy vegetables.

Caution

Sweet potatoes contain *oxalic acid,* a naturally occurring substance found in some vegetables that may crystallize as oxalate stonesin the urinary tract in some people. It is, therefore, individuals with known history of oxalate urinary tract stones may have to avoid eating them. Adequate intake of water is therefore advised to maintain normal urine output in these individuals to minimize stone risk. Since sweet potatoes can cause indigestion and abdominal swelling, they should be eaten in moderate servings.

CRUCIFEROUS VEGETABLES

Cruciferous vegetables such as cabbage, cauliflower and broccoli are rich in **Indole-3-Carbinal** and **isothiocyanates**, which inhibit carcinogenesis in the lungs and esophagus and reduces breast cancer by metabolizing the female hormone **estrogen.** Green leafy vegetables, which are rich in chlorophyll and chlorophyllides are antimutagenic and antitumorigenic in humans. Globble down green vegetable such as spinach, broccoli, brussel sprouts, parsley to keep your mind sharp, thanks to their high amounts of folic acid. Feeding of vegetables induces enzymes of xenobiotic mechanism and thereby accelerates the metabolic disposal of xenobiotics. **Glucosinolates** (thioglucosides) are sulphur containing compounds in brassica vegetables. The glucosinolates induce **physiological** changes in humans including carcinogensis inhibition. The greatest benefits found from eating cruciferous and other green leafy vegetables is the concentration of potassium, folate, fibre, and dietary flavonoids found in these foods. The chemical called **Allyl isothio cyanate** (AITC) is created when some vegetables of brassica family are chopped, chewed, cooked, processed and digested, AITC- is a break- down product of **Sinigrin**- chemical compound in cruciferous vegetables. AITC - sabotages the uncontrolled cell division of colon cancer cells. Sinigrin- a chemical compound found in brassicas, is converted by processing or eating in to Ally-isothiocynate (AITC). Brassica crops are a source of natural antioxidants. In ayurvedic medicine, cabbage leaves are prescribed for cough, fever, skin diseases, peptic ulcers, urinary discharges and hermorrhoids. The seeds are diuretic, laxative, stomachic and antihelminthic. The chemoprotective components include ascorbic acid, tocopherols, carotenoids, isothiocyanates, indoles and flavonoids. Fresh cabbage is reported to contain a heat- labile anti –peptic cancer components, vitamin U (98- 100). Cabbage juice has ulcer-healing capabilities, but should be used in conjunction with a doctor's prescribed therapy. However, patients with gas trouble should avoid it.

A new study, conducted on mice, has found that compounds extracted from green vegetables like broccoli and cabbage could be a potent drug against melanoma, a type of skin cancer. The study found that these compounds, when combined with selenium, target tumours more safely and effectively than conventional therapy. "There are currently no drugs to target the proteins that trigger melanoma", said Gavin Robertson, associate professor of pharmacology, pathology and dermatology, Penn State College of Medicine (ANI, 2009). These vegetables are good for memory.

Cruciferous vegetable (belonging to the plant family Cruciferae) *viz.,* cauliflower, brussels sprouts, cabbage and kale; broccoli are rich in sulforophane and indoles, which are shown to regulate cell growth in multiple ways and help fight a range of cancers, including breast, bladder, lymphoma, prostate and lung cancer (Ipshita Miitra, 2013).

Cruciferous vegetables are the gold standard in immune-boosting vegetables. Cruciferous vegetables have a special chemical composition: They have sulfur containing compounds that are responsible for their pungent flavours. When they're broken down by biting, blending, or chopping, a chemical reaction occurs that converts these sulfur-containing compounds into isothiocyanates (ITCs). ITCs prevent and knock out cancer and have infinite proven immune-boosting capabilities.

They contain **anti-viral** and anti-bacterial agents that keep you disease free. Adding the following cruciferous vegetables to your daily plate is like taking an anti-cancer pill: arugula, beet greens, bok choy, broccoli, Brussels sprouts, cabbage, cauliflower, collards, horseradish, kale, **kohlrabi**, mustard greens, radishes, red cabbage, turnip greens, or watercress. A recent research found that higher intake levels of cruciferous vegetables were associated with a 10 per cent reduction in prostrate cancer risk.

Apart from being full of precious vitamins and minerals, cruciferous vegetables have been shown to be twice as effective in the battle against cancer. Studies have shown that as cruciferous vegetable intake goes up 20 per cent, cancer rates in a population drops by 40 per cent Cruciferous vegetables are as under:

Cabbage	Brussels sprouts	Mustard greens	Raddish
Cauliflower	Kale	Water cress	Turnip
Broccoli	Collards greens	Arugula	Bok choy

Anti-cancer Benefits

These vegetable contain glucosinolates and Sulphur containing compounds. In different areas of the cell, they also contain an enzyme myrosinase. When we bend, chop or chew these vegetables, we break up the plant cells, bringing these two compounds into contact. Their contact initiates a chemical reaction that produces Isothiocyanates (ITCs), which are powerful anti-cancer compounds. ITCs protect cells from DNA damage, inactivate carcinogens, kill cancer cells and prevent tumors from growing. They also have anti-bacterial, anti-viral and anti inflammatory effects. ITCs rich cruciferous vegetables protect against cancer.

One serving per day of cruciferous vegetables reduced the risk of breast cancer by over 50 per cent.

One or more servings of cabbage per week reduced risk of pancreatic cancer by 38 per cent.

Just 3 serving of cruciferous vegetables per week decreased prostate cancer by 41 per cent.

Why it is Better to eat them Raw.

It is good to cook most of the vegetables since heat breaks down cell walls and releases anti-oxidants. This is not the case with cruciferous vegetables. Heating these reduces the conversion of glucosinolates to isothiocyanates, the powerful anti-cancer compound. Heat kills the cancer fighting substances of crucifers. During cooking process, their water soluble vitamins are also wasted. It is absolutely necessary for you to add heat, it is best to apply just a light steam without letting the vegetables touch water. A light steam will keep more of their precious nutrients intact.

CABBAGE (*Brassica oleracea capitata*)

A family of cruciferous vegetables that include cabbage, broccoli, brussel sprouts, cauliflower and kale. They are rich in vitamins and minerals, particularly sulphur, the 'beauty mineral' which produces collagen and boosts the immune system against germs and viruses. They also contain compounds with anti- cancer properties, such as dithiolthiones and indoles which protect against breast andcolon cancer. This family of vegetables also has a beneficial effect on the liver.

Crucifer's well known cancer fighting properties are thought to result from their high level active phytochemicals called **Glucosinolates,** which our bodies metabolize into powerful anti-carcinogens called **isothiocyanates.** Isothiocyanates offer the bladder in particular, significant protection, most likely because the majority of compounds produced by isothiocyanate metabolism travel through the bladder enroute to excretion in the urine, suggested the researchers. Those eating the most cruciferous vegetables were found to have a 29 per cent lower risk of blader cancer compared to participants eating the least of this family of vegetables. Cabbage raw or cooked, inhibits the conversion of sugar and other carbohydrates in to fat. Hence, it is of great value in weight reduction.

This leafy vegetable is a delight for people wanting to lose weight. It is packed with vitamin A which protects your eyes and skin, vitamin C, vitamin E, vitamin B which helps in boosting energy metabolism. Cabbage extracts have been proven to kill certain viruses and bacteria. Cabbage boosts the immune system's ability to produce more antibodies. Cabbage provides high levels of iron and sulfur, minerals that work in part as cleansing agents for the digestive system. Cabbage shredded and boiled 150 grams contains 33 calories.

Cabbage is a wise vegetable. It reminds people that youth is health. It is very good cleaner of stomach if taken raw and unsalted. It prevents and cures ulcers. Some cancers are also prevented by the green leafy vegetables. A combination of vitamins that the cabbage contains helps you always feeling young. It keeps your blood vessels stronger and its tartonic acid content inhibits the conversion of sugar and other carbohydrates in to fat. A meal of cabbages gives even the child enough nutrition and roughage, as 100 g of cabbage yields 27 kilo-calories of energy which is a lot more than most other foods of equal measure can give. It is said to be great for the skin and eyes as it is full of vitamin A. The phytonutrients in this vegetable help the body fight against the free radicals. It also helps lungs, stomach and colon prostate cancer. Those interested in body building may want to eat cabbage as it is full of iodine that is needed for muscle development.

Cabbage contains a lot of vitamin C and keeps the teeth and gums healthy. It is now highly regarded as one of the most nutritious vegetables available today, and is thought to have strong anti-aging and anti-cancer properties. Studies show that it should be brought to the table at least 2 or 3 times a week. It is an excellent source of vitamin C and Beta-carotene. These anti-oxidants not only slow down the process of aging, but also cut down the risk of cataracts. Cabbage also reduces risk of heart disease and stroke, alleviates **rheumatisms** and skin problems. All this, while being quite low on calories. To top it all, this is truly a miracle vegetable for cancer patients. Cabbage is high in both fibres and chemicals, whose combined action is known to prevent colon cancer. This is because fibres help the intestines stay healthy by promoting transit movements. Similarly, the chemicals present inhibit tumour growth and protects cells against free radicals. Some of these chemicals are also believed to reduce the incidence of breast, uterus and ovarian cancers as well. Moreover, juiced cabbage promotes healing of ulcers (Gita Angara, 2007).

Cabbage is recommended in natural medicine practice for improving digestion, treating constipation, preventing the common cold and alleviating depression. It also contains a factor called Vitamin U, which is a remedy for ulcers and raw cabbage juice has been reported to assist in the cure of both peptic and duodenal ulcers. The recommended method is to drink half a cup of freshly made cabbage juice two or three times a day between meals, on empty stomach. Many of the healing properties of cabbage as blood purifier are due to its high sulphur content. Grated cabbage can be made into a poultice to be applied externally in the treatment of wounds, varicose veins and leg ulcers. Cabbage (purple) is a great store of **vitamin C** and **vitamin K**. The rich deep colour of this **vegetable** is due to a high concentration of anthocyanin polyphenols (strong dietary antioxidants, possessing anti-inflammatory properties), making it have even more phytonutrients than a green cabbage. A few studies also show that anthocyanins may help reduce the risk of heart disease, **diabetes** and certain types of cancer. It is rich in anti-oxidants which makes skin supple and clear.

Cabbage forms a core ingredient for the most loved and relished Indian food. But this layered vegetable is a great source for phytochemicals that cut down free radicals, which damage the body and the skin. Essentially, any fruit and vegetable that breaks down free radicals is a secret to natural glowing skin. Raw cabbage is the preferred way to consume cabbages for those going through dialysis.

Cabbage is a good source of fibre and very low in calories. It contains a substantial amount of amino acid glutamine which our bodies need to produce HGH (human growth hormones). Stanford University's School of of Medicine conducted extensive research and found that the glutamine in fresh cabbage juice relieves peptic ulcer symptoms.

All cabbage varieties have sulfuropane in their leaves which scientists have discovered may help as a detoxification agent and promote liver and colon health. They all contain phytonurtients which can help our bodies fight off the attacks of free radicals. Red cabbage has more phytonutrients, as well as up to 8 times the Vitamin C as green cabbage. Varieties also contain a key antioxidant called anthocyanin, an ingredient needed for brain function.

But there are a few things to keep in mind before rushing to grab a bite of the healing cabbage. Here are a few do's and don'ts when it comes to cooking, storing and eating cabbage:

Do's	***Don'ts***
Add a few seeds of caraway/cumin or fennel (or both) before or while cooking cabbage, as it helps to digest the vegetable as also to avoid flatulence.	Avoid cutting the vegetable in advance, or you could lose vitamins, especially vitamin C. As the leaves in the inside are usually well tight and clean, you should just remove the exterior leaves and rinse the whole cabbage under running water. You can then cook it.
Shredded in small pieces, raw cabbage can be added to any salad, or it can be a salad itself, with a few pieces of apples, raisins and cheese.	Do not cook cabbage in aluminium pot, this discolours the vegetable and also alters its flavour.
When shredded it can be cooked with dal or cereals.	Don't miss out on your iodine intake. Cabbage reduces **iodine absorption**, so if you do eat more than 2 or 3 times a week, ensure your iodine intake through other means.
A whole cabbage can be stored in a refrigerator or cold place for at least two weeks, as it keeps well.	

Health Benefits

Cabbage is a good source of fibre and very low in calories. It contains a substantial amount of amino acid glutamine which our bodies need to produce HGH (human growth hormones). Stanford University's School of Medicine conducted extensive research and found that the glutamine in fresh cabbage juice relieves peptic ulcer symptoms.

All cabbage varieties have sulfuropane in their leaves which scientists have discovered may help as a detoxification agent and promote liver and colon health. They all contain phytonurtients which can help our bodies fight off the attacks of free radicals. Red cabbage has more phytonutrients, as well as up to 8 times the Vitamin C as green cabbage. Varieties also contain a key antioxidant called anthocyanin, an ingredient needed for brain function.

Fresh, dark green-leafy cabbage is incredibly nutritious and low in fat and calories (Provides just 25 cal/100 g).

It is storehouse to many phyto-chemicals like *thiocyanates, indole-3-carbinol, lutein, zeaxanthin, sulforaphane* and *isothiocyanates*. These compounds are powerful anti-oxidants and known to help protect against breast, colon, and prostate cancers and help reduce LDL or "bad cholesterol" levels in the blood.

Fresh cabbage is an excellent source of natural antioxidant, vitamin C. Provides about 61 per cent of RDA, more than that of in the **oranges.** Regular consumption of foods rich in vitamin C helps body develop resistance against infectious agents and scavenge harmful, pro-inflammatory free radicals.

It is also rich in many essential vitamins such as pantothenic acid (vitamin B5), pyridoxine (vitamin B-6) and thiamin (vitamin B-1). These vitamins are essential in the sense that our body requires them from external sources to replenish.

It also contains good amount of minerals like potassium, manganese, iron, and magnesium. Potassium is an important component of cell and body fluids that helps controlling heart rate and blood pressure. Manganese is used by the body as a co-factor for the antioxidant enzyme, *superoxide dismutase.* Iron is required for the red blood cell formation.

Cabbage is very good source of vitamin K, provides about 63 per cent of RDA levels. Vitamin-K has potential role in bone metabolism by promoting osteotrophic activity in them. So enough vitamin K in the diet gives you healthy bones. In addition, vitamin-K also has established role in curing **Alzheimer's** disease patients by limiting neuronal damage in their brain.

Cabbage is rich in the following nutrients:

Vitamin A: responsible for the protection of your skin and eyes.

Vitamin C: an all important anti-oxidant and helps the mitochondria to burn fat.

Vitamin E: a fat soluble anti-oxidant which plays a role in skin integrity.

Vitamin B: helps maintain integrity of nerve endings and boosts energy metabolism.

Cabbage extracts have been proven to kill certain viruses and bacteria.

Cabbage boosts the immune system's ability to produce more antibodies.

Cabbage provides high levels of iron and sulphur, minerals that work in part as cleansing agents for the digestive system.

Cabbage is an excellent source of vitamins K and C: 1 cup (150 grams) of shredded, boiled cabbage contains 91 per cent of the RDA for vitamin K and 50 per cent for vitamin C.

Cabbage is also a very good source of dietary fibre, manganese, vitaminB6 and folate, and a good source of thiamin, riboflavin, calcium, potassium, Vitamin A, tryptophan, protein and magnesium.

It also aids in draining out excess fluids from your body, which in turn naturally detoxifies your system.

Caution

Cabbage may contain "**goitrogens**", a plant based compounds, especially found in cruciferous vegetables like cauliflower, broccoli *etc.*, may cause swelling of thyroid gland and should be avoided in individuals with thyroid dysfunction. However, they may be used liberally in healthy persons.

CAULIFLOWER (*Brassica oleracea* Botrytis)

The allicin in cauliflower is known to promote a healthy heart and reduce the risk of strokes. It contains selenium and Vitamin C, both of which work together to strengthen the immune system. It helps in maintaining healthy cholesterol level. The high amount of fibre in cauliflower improves colon health and helps prevent cancer. Cauliflower even acts as a blood and liver detoxifier. People suffering from ailments like arthritis, asthma, constipation, high blood pressure and kidney bladder disorders can benefit from the consumption of cauliflower. Packed with rich nutrients, cauliflower or cabbage flower is one of the commonly used flower-vegetable. The flower heads contains numerous health benefiting phtyo nutrients such as indole-3-carbinol, sulforaphane *etc.* that help prevent prostate, ovarian and cervical cancers. Botanically, it is a member of the cruciferous or brassicaceae family; has got similar nutritional and phyto-chemistry profile with **broccoli and cabbage**. Several cultivars exists other than common snow-white variety including green, orange, purple, and romanesco heads. Cauliflower lacks the green chlorophyll found in other vegetables; cabbage, broccoli and kale.

The simplest way to prepare cauliflower is to boil it and add pepper and salt. But this flowery vegetable is an enemy for toxic compounds in your body. The cauliflower is rich is indoles, glucosinolates and thiocyanates that bumps off all the toxic waste in your body. Cauliflower contains Sulforaphane, a compound that has been shown to have anti-cancer effects. Sulforaphane are released when cauliflowers are broken down, *i.e.* on chewing it before swallowing. This compound sucks and destroys certain cancer cells without having your healthy cells. Several Researches on this flowery vegetable have revealed that it has compounds to resist cancer; it has the ability to eliminate cancer enzymes.

Cauliflower helps to increase the body's ability to detoxify itself, it boosts the body's antioxidant system, and helps to balance the levels of inflammation to anti-inflammation in the body. In helping with these systems cauliflower helps to reduce the risk of several types of cancer. Cauliflower is a good source of Vitamin C and Vitamin K which are antioxidants along with glucosinolates, which are also antioxidants. Antioxidants are beneficial in helping the body fight free radicals and in reducing inflammation.

Health Benefits

Prevent cancer: Researches have shown that cauliflower helps keep off various kinds of cancers *viz.*, lung, colon, breast and ovarian, with the most recent addition being bladder cancer. Cauliflower contains several compounds that have been shown to inhibit the growth or spread of cancer cells, including sulforaphane, indoles and isothiocyanates. Combining cauliflower with **turmeric** brings curcumin into the battle, another potential cancer fighter. Combining cauliflower and turmeric on your plate is also just plain tasty.

It Contains Anti-Inflammatory Compounds: Nutrients found in cauliflower, including indole-3-carbinol, have been noted for their anti-inflammatory properties. The omega-3 fatty acids and vitamin K found in cauliflower can also help manage the chronic inflammation that contributes to diseases like arthritis.

It Promotes Heart Health: Those same anti-inflammatory properties can also aid in preventing against the inflammation that can exacerbate the decreased diameter of blood vessels. These conditions, in combination with other factors, can lead to heart attacks, strokes and high blood pressure.

It's Packed With Vitamins and Minerals: Cauliflower, like its cruciferous cousins, is practically a natural multivitamin. Cauliflower contains B-vitamins, vitamin K, Omega-3 fatty acids, proteins, phosphorus, potassium, vitamin C and manganese. Take a look at your multi-vitamin bottle next time – or just have some more cauliflower.

It Supports Brain Health: Choline has been shown to might also combat age-related memory decline and help to resist vulnerability to toxins during childhood.

Good for detoxification of liver: It contains **glucosinolate**s and **thiocyanates** (including **sulforaphane** and **isothiocyanates**), which help to increase the liver's ability to neutrolise potentially toxic substances that could lead to cancer if left unattended. The presence of enzymes like **glutathione transferase**, **glucuronosyl transferase** and quinine reductase also help in the detoxification. **The glucosinolates in cauliflower** can support the liver and trigger enzymes in your body to help with detoxification. The antioxidants and nutrients occurring in cauliflower also contribute to Phase 1 and Phase 2 detoxification. Cauliflower might not contain the same amount of glucosinolates as other members of the crucifer family, but it still has a significant amount.

It Contains Antioxidants and Phytonutrients: Beta-carotene, kaempferol, quercetin, rutin, cinnamic acid and other antioxidant compounds can be found in cauliflower. These antioxidant compounds can help your body's cells defend against reactive oxygen species (ROS). Without these chemicals (or other antioxidants), your body is less effective at fighting the free radicals and toxins that can cause damage.

Provides Digestive Benefits: Cauliflower is a source of dietary fibre, which is an important component of digestive health. It has also been suggested that sulfurophane and glucosinolate can help regulate the growth of bacteria in your stomach and prevent damage to the stomach and intestine linings.

Shield against rheumatoid arthritis: Cauliflower, which is rich in vitamin C, is said to providea shield against inflammatory polyarthritis, a form of rheumatoid arthritis affecting two or more joints (Anonymous, 2013).

Very low in calories. 100 g of fresh cauliflower has only 26 calories. However, it is very low in fat and contains no cholesterol.

Its florets contain about 2 g of dietary fibre per 100 g; providing about 5 per cent of recommended value.

Cauliflower contains several anti- cancer phyto-chemicals like **sulforaphane** and plant sterols such as **indole-3-carbinol** which appears to function as an anti-estrogen agent. Together these compounds have proven benefits against prostate, breast, cervical, colon, ovarian cancers by virtue of their cancer cell growth inhibition, cytotoxic effects on cancer cells.

Also, **Di-indolyl-methane** (DIM), a lipid soluble compound present abundantly in brassica group of vegetables has found effective as immune modulator, anti-bacterial and anti-viral compound by potentiating Interferon-Gamma receptors and its production. DIM has currently been found application in the treatment of recurring respiratory papillomatosis caused by the Human Papilloma Virus (HPV) and is in Phase III clinical trials for cervical dysplasia.

Fresh cauliflower is excellent source of **vitamin C**; 100 g provides about 48.2 mg or 80 per cent of daily recommended value. Vitamin-C is a proven antioxidant helps fight against harmful free radicals, boosts immunity and prevent from infections and cancers.

It contains good amounts of many essential B-complex group of vitamins such as folates, pantothenic acid (vitamin B5), pyridoxine (vitamin B6) and thiamin (vitamin B1), niacin (B3) as well as vitamin K. These vitamins are essential in the sense that body requires them from external sources to replenish and required for fat, protein and carbohydrates metabolism.

It is also good source of minerals such as manganese, copper, iron, calcium and potassium. Manganese is used in the body as a co-factor for the antioxidant enzyme *superoxide dismutase. Potassium* is an important intracellular electrolyte helps counter the hypertension effects of sodium.

The allicin in cauliflower is known to promote a healthy heart and reduce the risk of strokes.

It contains selenium and Vitamin C, both of which work together to strengthen the immune system.

It helps in maintaining healthy cholesterol level. The high amount of fibre in cauliflower improves colon health and helps prevent cancer.

Cauliflower even acts as a blood and liver detoxifier.

People suffering from ailments like arthritis, asthma, constipation, high blood pressure and kidney bladder disorders can benefit from the consumption of cauliflower.

Cauliflower contains 'indole3-carbinol', a substance that can prevent breast and other female cancers.

Being rich in nutrient folate, cauliflower helps in improving cell growth and cell replication mechanism.

The milk, sweet and delicious nutty flavour of cauliflower is at its best when it is in season from December through March

Caution

Like other members of the brassica/cruciferous family, prolong/excessive use of cauliflower may cause swelling of thyroid gland and thyroid hormone deficiency. This is due to the presence of certain plant compounds known as ***goitrogens*** in these group of vegetables. It is therefore advised to avoid, especially in individuals with thyroid dysfunction. However, these vegetables may be used liberally in healthy person.

BROCCOLI (*Brassica oleracea* Italica)

Cauliflower Head

When one talks about vegetables which can fight diseases effectively, broccoli ranks amongst the highest. Broccoli heads are rich source of phyto-nutrients that help protect from prostate cancer and stroke risks. It is actually a flower vegetable and known for its notable and unique nutrients that are found to have disease prevention and health promoting properties. They are a storehouse of antioxidants which help in fighting rectal, lungs and stomach cancers. They are also rich in folate, Vitamin C and beta-carotene which help in boosting your immunity to fight against colds and flues. Broccoli heads are rich source of phyto-nutrients that help protect from prostate cancer and stroke risks. It is actually a flower vegetable and known for its notable and unique nutrients that are found to have disease prevention and health promoting properties. It is considered a wonder food for its ability to help prevent cancer. It has the powerful anti cancer component ever detected. Scientists have isolated an ingredient in broccoli that kindles the activity of critical enzyme in the cell known to guard against tumour. The newly discovered chemical in broccoli is **Sulfora phane**- is by far the most powerful inciter. Besides anti-carcinogenic properties, broccoli is rich source of vitamins, minerals, proteins, *etc.* It has about 130 times more Vitamin A content than cauliflower and 22 times more than cabbage. It is a rich source of **Sulphoraphane,** a compound associated with reducing the risk of cancer. Broccoli which is known for its anti cancer properties but it could also boost the immune system in older people and slow down the effects of ageing according to the new research, as **sulforaphane** present in cruciferous vegetables was found to activate a number of antioxidant genes and enzymes in immune cells. These prevent free radicals from damaging cells (Methew, 2008). When one talks about vegetables which can fight diseases effectively, **broccoli** ranks amongst the highest. They are a storehouse of **antioxidants** which help in fighting rectal, lungs and stomach cancers. They are also rich in folate that helps in reducing stress and irritability, Vitamin-C and **beta-carotene** which help in boosting your immunity to fight against colds and flues. It is considered as one of nature's superfoods.

Broccoli is exceptionally nutritious, containing an abundance of sulphur, iron and cholorophyll, which purifies the blood; it is also rich in vitamins A, B-Complex and especially C. In fact, it contains more vitamin C than oranges - one cup provides as much as 70 mg of vitamin C.

It provides stress relief, helps enhance mood with regard to depression, panic, anxiety and other emotional disorders. A team in Britain has found that **sulforaphane**, a compound present in broccoli, boosts the production of enzymes which protect the blood vessels and encourages a reduction in high levels of molecules which cause significant cell damage and may help counter processes linked to the development of vascular disease in diabetes as high glucose levels (**hyperglycaemia**), are associated with diabetes. Their study found that a 73 per cent reduction of molecules in the body called **Reactive Oxygen Species** (ROS). **Hyperglycaemia** can cause levels of ROS to increase nearly three fold and such high levels can damage human cells. **Sulforaphane** present in broccoli activates a protein in the body called **nrf2**, which protects cells and tissues from damage by activating protective antioxidant and detoxifying enzymes (Anonymous, 2008). **Sulforaphane** may offer special protection to those with colon cancer-susceptible genes, suggests a study conducted at Rutgers University and Published online in Journal Carcinogenesis. Sulforaphane (SFN) is the hydrolysis product of **glucoraphanin**, particularly high in young sprouts of broccoli and cauliflower. SFN can also be obtained by eating cruciferous vegetables such as Brussels sprouts, broccoli, cauliflower, cabbage kale, collards, arugula, broccoli raab, turnip, radish, mustard, knolkhol and water cress. SFN- is therefore able to prevent, delay or reverse **preneoplastic** lesions, as well as to act on cancer cells as a **therapeutic** agent.

Broccoli contains high levels of calcium, pantothenic acid, vitamin A, thiols and indoles and sulphur compounds that have antimicrobial activity. Direct sulforaphane (glucosinolate), a naturally occurring compound found in cruciferous vegetables and broccoli sprouts may even be more versatile and effective than vitamin C and E in protecting cells from oxidative damage. It plays a key role in increasing the body's natural antioxidant defense systems and may function as a powerful indirect antioxidant with a longer lasting effect than other antioxidants such as vitamin C and vitamin E. Other sources are broccoli, cabbage and cauliflower, kale, knolkhol, collards, radish, rutabaga and turnips. Cruciferous vegetables contain high concentration of carotenoids, which are believed to be chemopreventive and associated with decreased risk for various human cancers in epidemiological studies. Researchers at the University of California, Los Angeles, found that **diindolymethane** (DIM) a compound resulting from digestion of cruciferous vegetable and genistein, an isoflavone in soy, reduce the production of two proteins needed for breast and ovarian cancers to spread (Reuters, 2007).

Researchers put broccoli sprouts to the test and found that they contain high levels of **glucoraphanin** also known as **sulforaphane glucosinolates** (SGS). In addition to being a cancer preventative, SGS boosts production of phase 2 enzymes, which are part of the body's anti-oxidant system. This extra boost lowers blood pressure and the body's inflammatory responses, leading to better cardiovascular health over all. As just 5 grams of broccoli sprouts contain concentrations of SGS equal to that found in 150grams of mature broccoli. That makes you just one salad a day away from better health. Known for its great cancer - combating ability, broccoli was found to contain **Sulphoraphane,** a sulphur -based compound, which helps to kill cancer causing substances in food. When sulphoraphane is released in the gut, it steps up production of powerful enzymes that destroy carcinogenic substances such as those found in heavily barbecued meat. Another naturally occurring compound found in broccoli **-13C-** was found in US study to increase DNA repair proteins, thus preventing genetic information damage and lowering the risk of developing cancer. The vegetable is recommended for those suffering from heart disorders and strokes. Being packed with iron it is ideal for anemia patients as well.

According to a study in cancer prevention research, a journal of the American Association of Cancer Research, that regular eating of 70g, three-day old broccoli sprouts can suppress a bacteria (*H. pylori*) linked to stomach cancer, gastritis and ulcers. Helicobactor pylori infections are one of the most common stomach infections affecting people globally and are a major cause of stomach cancer.

Broccoli has recently entered the public awareness as a preventive dietary agent. This study supports the emerging evidence that broccoli sprouts may be able to prevent cancer, in humans not just in laboratory animals, said Jed Fahey, a faculty research associate in the department of pharmacology at John Hopkins School of Medicine (Anonymous, 2009).

In this study, researchers enrolled 48 Helicobacter infected Japanese men and women and randomly assigned them to eat 70 grams of fresh broccoli sprouts for eight weeks or even equivalent amount of alfalfa sprouts. Researchers assessed the severity of *H.pylori* infection at enrolment and again at 4 and 8 weeks using standard breath, serum and stool tests. *H. pylori* were significantly lower at 8 weeks on all measures among those patients who had eaten broccoli sprouts, while they remained the same for patients who had eaten alfalfa sprouts (Anonymous, 2009).

Researchers at the University of East Anglia have discovered that a compound sulforaphane, found mainly in the vegetable, slows down the destruction of cartilage in joints linked with osteoarthritis. The study found that the vegetable might have health benefits for people with osteoarthritis and even protect them from developing the disease (Anonymous, 2013). The leafy vegetable is a nutritional wonder. Adding it to your pasta or salad will not enhance the taste but also provide your body with essential nutrients. Here's why you should eat broccoli.

Health Benefits

Good for eyes: According to several studies, the carotenoid lutein in broccoli can help prevent age-related macular degeneration and cataracts. Moreover, vitamin A present in the vegetable helps form retinal, the light-absorbing molecule that is essential for both low-light and colour vision.

Regulates blood pressure: Broccoli should be a must-have for those suffering from high BP as the magnesium and calcium in the veggie helps regulate blood pressure.

Helps nervous system: Containing a high amount of potassium, broccoli helps maintain a healthy nervous system. It also helps the growth of your muscles.

For healthy bones: The high levels of calcium and vitamin K in broccoli can help you maintain strong bones and also prevent osteoporosis.

Boosts immune system: Including broccoli in your diet can help boost your body's defence mechanism as it contains a large dose of beta-carotene and minerals, like zinc and selenium.

Detoxifies skin: Thanks to the glucoraphanin that it contains, broccoli is good for repairing skin, especially from sun damage.

Broccoli, as well as cabbage and sprouts, contains a strong cancer-fighting chemical called sulphoraphane, which is said to reduce men's risk of developing bladder cancer, prostate cancer and colorectal cancer.

Broccoli is very low in calories, provides just 34 cal per 100 g. However, it is rich in dietary fibre, minerals, vitamins, and anti-oxidants that have proven health benefits.

Fresh Broccoli is a storehouse of many phyto-nutrients such as *thiocyanates, indoles, sulforaphae, isothiocyanates* and flavonoids like beta-carotene cryptoxanthin, lutein, and zeaxanthin. Studies have shown that these compounds by modifying positive signalling at molecular receptor levels help protect from prostate, colon, urinary bladder, pancreatic, and breast cancers.

Fresh vegetable is exceptionally rich source of vitamin-C. Provides 89.2 mg or about 150 per cent of RDA per 100 g. Vitamin-C is a powerful natural anti-oxidant and immune modulator, helps fight against flu causing viruses.

Furher, it contains very good amounts of another anti-oxidant vitamin, vitamin-A. 100 g fresh head provides 623 IU or 21 per cent of recommended daily levels. Together with other pro-vitamins like beta-carotene, alpha-carotene, and zeaxanthin, vitamin A helps maintain integrity of skin and mucus membranes. Vitamin A is essential for vision and helps prevent from macular degeneration of retina in the elderly population.

It is also a good source of minerals like calcium, manganese, iron, magnesium, selenium, zinc and phosphorus.

Fresh broccoli heads are an excellent source of folates; contains about 63 mcg/100 g Studies have shown that consumption of fresh vegetables and fruits rich in folates during pre-conception and pregnancy helps prevent neural tube defects in the offspring.

Broccoli leaves (green tops) are an excellent source of **carotenoids** and vitamin A; contain these compounds several times more than in the roots.

This flower vegetable is also rich source of other vitamin-K and B-complex group of vitamins like Niacin (vit B-3), pantothenic acid (vit. B-5), pyridoxine (vit. B-6) and riboflavin. The flower heads also have some amount omega-3 fatty acids.

Broccoli contains twice the vitamin C of an orange.

It has almost as much calcium as whole milk–and the calcium is better absorbed.

It contains selenium, a mineral that has been found to have anti-cancer and anti-viral properties.

Broccoli is a modest source of vitamin A and **alpha-tocopherol** vitamin E.

It also has **antioxidant** properties.

Caution

Like other members of the cruciferous family, broccoli contain "goitrogens" which may cause swelling of thyroid gland and therefore, should be avoided in individuals with thyroid dysfunction. However, it may be used liberally in healthy person.

BRUSSELS SPROUTS (*Brassica oleracea* Gemmifera)

Brussels Sprouts Heads

Brussel sprouts are small leafy green buds resemble like miniature cabbages in appearance. The buds are exceptionally rich in protein, dietary fibre, vitamins, minerals and antioxidants which work wonders to get rid of many health troubles.

Plant phytonutrients found in brussel sprouts enhance the activity of body's natural defenses to protect against diseases, including cancers, Scientists have found that sulforaphane, one of the powerful glucosinolate phytonutrients found in brussel sprouts and other cruciferous vegetables, boosts the body's detoxification enzymes, potentially by altering gene expression, thus helping to clear potential carcinogenic substances more quickly. Brussels sprouts are excellent for pregnant women, these little veggies are full of B-vitamin and folic acid which keeps women away from uteral tube defects. They are also quite rich in omega 3 fatty acids, potassium, fibre and Vitamin C and K.

Brussels sprouts are similar to cabbage in flavour and nutrient content and provide an excellent source of vitamins C, B1 and β-carotene. Similarly they are also rich in potassium, calcium, and sulphur but unlike cabbage and broccoli, which can be enjoyed both raw and cooked, Brussels sprouts are palatable only as a cooked vegetable. Excellent for pregnant women, these little veggies are full of B-vitamin and folic acid which keeps women away from uteral tube defects. They are also quite rich in omega 3 fatty acids, potassium, fibre and Vitamin C and K.

Brussels sprouts stronger protective effects are thought to be due to the fact that this cruciferous vegetable contains 2-3 times the amount of **glucosinolates** than are found in red cabbage. **Glucosinolates** increase phase 11 glucuronidation activity, one of the primary pathways through which toxins made even more dangerous by phase 1 are rendered water soluble and ready for elimination from the body.

Cruciferous vegetable like Brussels sprouts, cabbage, cauliflower, broccoli kale help prevent cancer, when these vegetables are cut, chewed or digested, a sulfur- containing compound called singrin is brought into contact with enzyme **myrosinase**, resulting in the release of glucose and breakdown products, including highly reactive compounds called **isothiocyanates**. Isothiocyanates are not only potent inducers of the liver's phase 11 enzymes, which detoxify carcinogens but research recently conducted at the Institute for Food Research in the U. K. shows one of these compounds allyl- isothiocyanate, also inhibits mitosis (cell division) and stimulates **apoptosis** (programmed cell death) in human tumour cells.

Brussels sprouts contain goitrogens, naturally-occurring substances in certain foods, can interfere with the functioning of the thyroid gland. Therefore, individually with already existing and untreated thyroid problems should avoid brussel sprouts for this reason.

Health Benefits

Brussel sprouts are an incredibly nutritious vegetable that are rich in antioxidants, phytonutrients, vitamins C, A, and E, and alkalizing minerals such as calcium, copper, iron, and manganese. They are also a good source of omega-3 fatty acids and vitamin K which are essential for proper brain and nerve function and are vital for aiding cognitive issues such as brain fog, memory loss, lack of concentration, ADD, ADHD, dementia, and Alzheimer's disease. Brussel sprouts are also known to help stabilize the DNA within white blood cells which creates a stronger, more resilient immune system. Brussel sprouts also contain powerful anti-cancer compounds which are shown to particularly help fight and prevent colon, breast, prostate, lung, bladder, liver, endometrial, and ovarian cancer. They can help stimulate the kidneys to release more water, which can reduce bloat and edema and make it easier for dislodged fatty wastes to be flushed out of the body. Brussel sprouts are also good for stimulating sluggish glands and for promoting a cleaning effect on adipose cells and tissues making it an ideal weight loss food. They contain high amounts of **glucosinolates** which can significantly reduce inflammation and is excellent for prevention and relief from rheumatoid arthritis, fibromyalgia, lupus, strokes, heart attacks, arteriosclerosis, and bursitis. Brussel sprouts are excellent for helping to lower cholesterol, prevent constipation, stop overeating, and decrease *H.pylori* in the digestive tract. To obtain the most health benefits from brussel sprouts, consider eating them raw (they can also be juiced) or steamed until soft. Try making a delicious salad with chopped romaine, arugula, tomato, avocado, onion, steamed brussels sprouts, fresh lemon juice and minced garlic. It is a nutritious and satisfying dish that will nourish your body, mind, and soul.

One of the nutritious vegetable that should be considered in weight reduction programs. 100 g Brussels sprouts just provide 45 calories but contain 3.8 g of dietary fibre and no cholesterol.

In fact, Brussels sprouts are storehouse of flavonoid anti-oxidants like *thiocyanates, indoles, lutein, zeaxanthin, sulforaphane* and *isothiocyanates*. Together these phytochemicals offers protection from prostate, colon, prostate and endometrial cancers.

Di-indolyl-methane (DIM), a metabolite of *indole-3-carbinol* has been found to be an effective immune modulator, anti-bacterial and anti-viral agent through its action of potentiating "Interferon-γ" receptors.

In addition brussel sprouts contain glucoside, sinigrin. Early laboratory studies suggest that *sinigrin* helps protect from colon cancers by destroying pre-cancerous cells.

Brussels sprouts are an excellent source of vitamin C; 100 g sprouts provide about 142 per cent of RDA. Along with other antioxidant vitamins, vitamin A and E; it protects body by trapping harmful free radicals.

Zeaxanthin, an important dietary carotenoid in sprouts, is selectively absorbed into the retinal macula lutea in the eyes where it is thought to provide anti-oxidant and protective light-filtering functions from UV rays. Thus, it helps prevent retinal damage, "age related macular degeneration disease" (ARMD), in the elderly.

Sprouts are good source of another anti-oxidant vitamin A, provides about 754 IU per 100g. Vitamin A is required for maintaining healthy mucus membranes and skin and is also essential for acuity of vision. Foods rich in this vitamin offer protection against lung and oral cavity cancers.

It is one of the excellent vegetable sources for vitamin-K; Vitamin K has potential role bone health by promoting **osteotrophic** (bone formation and strengthening) activity. Adequate vitamin-K levels in the diet helps limiting neuronal damage in the brain; helps prevent or at least delay onset of **Alzheimer's** disease.

The sprouts are notably good in many B-complex group of vitamins such as niacin, vitamin B-6 (pyridoxine), thiamin, pantothenic acid, *etc.* that are essential for substrate metabolism in the body.

They are also rich source of minerals like copper, calcium, potassium, iron, manganese and phosphorus. Potassium in an important component of cell and body fluids that helps controlling heart rate and blood pressure by countering effects of sodium. Manganese is used by the body as a co-factor for the antioxidant enzyme *superoxide dismutase*. Iron is required for cellular oxidation and red blood cell formation.

Brussels sprouts are incredibly nutritious vegetable that offers protection from vitamin A deficiency, bone loss, iron deficiency anemia, and believed to protect from **cardiovascular** diseases and, colon and prostate cancers.

Caution: Brussels sprouts may contain **goitrogens** which may cause swelling of thyroid gland and should be avoided in individuals with **thyroid dysfunction**. However, they may be used liberally in healthy person.

CHINESE CABBAGE (*Brassica rapa*)

Chinese Cabbage with Dark Green Leaves

In structure, Chinese cabbage resembles **collards**and could be described as a non-heading cabbage (Acephala group). It is basically a small plant which grows upright from the ground with smooth white romaine lettuce like stalks which spread in the end to fine, glossy green oval or round leaves. Fully grown up plant may reach about 12-18 inches in length. leafy Chinese cabbage is one of the popular mainland crop in China, Philippines, Vietnam, India and other regions; nonetheless this humble brassica family vegetable has gained popularity even in the western world for its sweet, succulent nutritious stalks.

Brassica campestris group can be further categorized according to the colour of the petiole in its leaves. White petiole varieties include joi choi, pak-choy white, prize choi, lei choi, taisai, canton pak choi *etc.* Green petiole types are Chinese pak choi green, mei qing choi *etc.*

Health Benefits

It is one of the rare vegetable that is very low in calories. However, it is very rich source of many vital phyto-nutrients, vitamins, minerals and **anti-oxidants**.

100 g of bok choy provides just 13 cal. It is one of the recommended vegetable in the zero calorie or negative calorie category of foods which when eaten would add no extra weight to the body but in-turn facilitate calorie burns and reduction of weight.

Like other brasisca family vegetables, bok choy contains certain anti oxidant plant chemicals like thiocyanates, indole-3-carbinol, lutein, zeaxanthin, sulforaphane and isothiocyanates. Along with dietary fibre, vitamins these compounds help to protect against breast, colon and prostate cancers and help reduce LDL or "bad cholesterol" levels in the blood.

Fresh pak choi is an excellent source of water soluble **antioxidant, vitamin-C** (ascorbic acid). 100 g provides 45 mg or 75 per cent of daily requirements of vitamin C. Regular consumption of foods rich in vitamin C helps body develop resistance against infectious agents and scavenge harmful, pro-inflammatory free radicals.

It has more vitamin A, carotenes, and other flavonoid polyphenolic anti-oxidants than cabbage, cauliflower *etc.* Just 100 g of fresh bok choy provides 4468 IU or 149 per cent of daily required levels vitamin A.

Pak choi is very good source of **vitamin K,** provides about 38 per cent of RDA levels. Vitamin-K has potential role in bone metabolism by promoting **osteotrophic** activity in bone cells. Therefore, enough vitamin K in the diet makes your bone stronger, healthier and delay osteoporosis. Further, vitamin-K also has established role in curing **Alzheimer's** disease patients by limiting neuronal damage in their brain.

Fresh bok choy has many vital B-complex vitamins such as pyridoxine (vitamin B6), riboflavin, pantothenic acid (vitamin B5), pyridoxine and thiamin (vitamin B-1). These vitamins are essential in the sense that our body requires them from external sources to replenish.

It also contains good amount of minerals like calcium, phosphorous, potassium, manganese, iron and magnesium. Potassium is an important component of cell and body fluids that helps controlling heart rate and blood pressure. Manganese is used by the body as a co-factor for the antioxidant enzyme *superoxide dismutase.* Iron is required for the red blood cell formation.

Chinese cabbage has anti-inflammatory properties.

It is an excellent source of folic acid.

Chinese cabbage is low in calories and low in sodium.

It is also high in vitamin A and a good source of potassium.

KALE (*Brassica oleracea acephale*)

Kale Leaves

Similar to cabbage, Kale reduces cholesterol and is great for digestion. Although relatively unheard of by most people, kale is now fast emerging as a kind of super-food. Whether it is consumed as a soup, cooked with food or in a smoothie, this slightly bitter vegetable is chock full of antioxidants and help to protect you from free radical damage, which is said to cause premature ageing and disease. If you like cabbage, then chances are you will like the flavour, texture of kale. It is in fact a type of cabbage though on examination, you will find that its leaves resemble crisp lettuce leaves. Be sure to cook it, however, as consumed raw, the vegetable is hard to digest.

Kale is rich in numerous health benefiting polyphenolic flavonoid compounds such as lutein, zeaxanthin, and beta-carotene, and vitamins than found in any other green leafy vegetables.

It is high in calcium.

It is widely cultivated across Europe, Japan, United States and in some regions of India for its "frilly"leaves. It features distinctive very long, curly, blue-green leaves with embossed surface resembling dinosaur skin, giving its name as dinosaur kale. It is considered to be a highly nutritious vegetable, with powerful **anti-oxidant** properties and is anti-inflammatory. Kale is very high in beta -carotene, vitamin-K, vitamin-C, lutein, zeaxanthin and reasonably rich in calcium. Because of its high vitamin-K content, patients taking anti-coagulants such as Warfarin are encouraged to avoid this food since it increases vitamin -K concentration in the blood, which is what the drugs are often attempting to lower. This effectively raises the effective dose of the drug. Kale, as with **broccoli** and other brassicas, contain sulforaphane, a chemical believed to have potent **anti-cancer** properties, particularly when chopped. Kale, as with broccoli, it contains sulforaphane (particularly when chopped or minced), a chemical with potent anti-cancer properties. Kale is also a source of indole-3-carbinol, a chemical which boosts DNA repair in cells and appears to block the growth of cancer cells.

In addition to its unique organosulfur compounds, kale is well known for its carotenoids, especially **lutein** and **zeaxanthin.** These carotenoids act like sunglass filters and prevent damage to the eyes from excessive exposure to ultra violet light. Studies have shown the protective effect of these nutrients against the risk of cataracts, where increased eye cloudiness leads to blurred vision. In one study, people who had a diet history of eating lutein-rich foods like kale-had a 50 per cent lower risk for new cataracts.

Kale also emerged from our food ranking system as an excellent source of traditional nutrients, including vitamin A, vitamin C, vitamin B6, and manganese. It is also a very good source of dietary fibre, calcium, copper, vitamin B6 and potassium. This combination of vitamins, minerals and phyto-nutrients makes kale a healthy superstar. Both vitamin A and beta-carotenes are important vision nutrients.

Kale being an excellent source of vitamin C as just one cup of this cooked vegetable supplies 88.8 per cent of the daily value of vitamin C. Vitamin C is the primary water soluble antioxidant in the body, disarming free radicals and preventing damage in the aqueous environment both inside and outside cells. Inside cells a potential result of free radical damage to DNA is cancer. Especially in areas of the body where cellular turnover is especially rapid, such as digestive system, preventing DNA mutations translates into preventing cancer. This is why a good intake of vitamin C is associated with a reduced risk of colon cancer. Vitamin C, which is also vital for the proper function of a healthy immune system, is good for preventing colds and may be helpful in preventing recurrent ear infections.

Kale's health benefits continue with its fibre; a cup of kale provides 10.4 per cent of the daily value for fibre, which has been shown to reduce high cholesterol level, thus helping to prevent **atherosclerosis.** Fibre can also help out by keeping blood sugar level under control, so kale is an excellent vegetable for people with diabetes. Kale's fibre binds to cancer causing chemicals, keeping them away from the cells lining the colon, preventing yet another line of protection from colon cancer.

Kale is also very good source of calcium. Calcium is one of the nutrients needed to make healthy bones and dairy products are a heavily promoted source of this nutrient, but unlike dairy products, kale is not a highly allergenic food, nor does it contains any saturated fat-plus. A cup of kale supplies 93.6 mg of calcium (9.4 per cent of the daily value for this mineral) for only 36.4calories, in contrast, a cup of 2 per cent cow's milk provides 296.7mg of calcium, but costs is high; 121.2 calories and 14.6 per cent of days suggested limit on saturated fat.

Health Benefits

Kale is very versatile and nutritious green leafy vegetable. It is widely recognized as an incredibly nutritious vegetable since ancient Greek and Roman times for its low fat, no cholesterol but health benefiting anti-oxidant properties.

Kale, like other members of the brassica family, contains health-promoting phytochemicals, sulforaphane and indole-3-carbinol that are appears to protect against prostate and colon cancers.

Di-indolyl-methane (DIM), a metabolite of *indole-3-carbinol* has been found to be an effective immune modulator, anti-bacterial and anti-viral agent through its action of potentiating "Interferon-Gamma" receptors.

Borecole is very rich source of β-carotene, lutein and zeaxanthin. These flavonoids have strong anti-oxidant and anti-cancer activities. Beta-carotene is converted to vitamin A in the body.

Zeaxanthin, an important dietary carotenoid, is selectively absorbed into the retinal macula lutea in the eyes where it is thought to provide antioxidant and protective light-filtering functions. Thus, it helps prevent retinal detachment and offers protection against "age related macular degeneration disease" (ARMD) in the elderly.

It is very rich in vitamin A, 100 g leaves provide 512 per cent of RDA. Vitamin A is required for maintaining healthy mucus membranes and skin and is essential for vision. Foods rich in this vitamin offer protection against lung and oral cavity cancers.

It is one of the excellent vegetable sources for vitamin-K; 100 g provides about 700 per cent of recommended intake. Vitamin K has potential role bone health by promoting osteotrophic (bone formation and strengthening) activity. Adequate vitamin-K levels in the diet helps limiting neuronal damage in the brain; thus, has established role in the treatment of patients suffering from Alzheimer's disease.

100 g of fresh leaves contain 120 mg or 200 per cent of daily-recommended levels of vitamin C. Scottish curly leaf variety yet has more of this vitamin, 130 mg/100g. Vitamin C is a powerful antioxidant, which helps body develop resistance against infectious agents and scavenge harmful oxygen free radicals.

This leafy vegetable is notably good in many B-complex group of vitamins such as niacin, vit. B-6 (pyridoxine), thiamin, pantothenic acid, *etc.* that are essential for substrate metabolism in the body.

It is also rich source of minerals like copper, calcium, sodium, potassium, iron, manganese, and phosphorus. Potassium is an important component of cell and body fluids that helps controlling heart rate and blood pressure by countering effects of sodium. Manganese is used by the body as a co-factor for the antioxidant enzyme, *superoxide dismutase*. Iron is required for cellular oxidation and red blood cell formation.

Kale eases lung congestion and is beneficial to the stomach, liver and immune system.

It contains lutein and zeaxanthin, which protect the eyes from macular degeneration. It also contains indole-3-carbinol, which may protect against colon cancer.

Kale is an excellent source of calcium, iron, vitamins A and C, and chlorophyll.

Kale provides rich nutrition ingredients that offer protection from vitamin A deficiency, osteoporosis, iron deficiency anemia, and believed to protect from cardiovascular diseases and, colon and prostate cancers.

Caution

Because of its high vitamin K content, patients taking anti-coagulants such as warfarin are encouraged to avoid kale since it increases the vitamin K concentration in the blood, which is what the drugs are attempting to lower. This effectively raises the effective dose of the drug and causes toxicity.

Its leaves contain 0.2 g/100 g of oxalic acid, a value far less than some other comparable greens such as spinach (0.97 g/100) and purslane (1.31 g/100 g). It may be used, however with caution, even in individuals with known oxalate urinary tract stones are advised to avoid eating other greens and certain vegetables belonging to amaranthaceae and brassica family because of their high oxalic acid content. Adequate intake of water is therefore advised to maintain normal urine output.

COLLARD (*Brassica oleracea* L. *Acephala*)

Collard Leaves

Collard greens are highly nutritious staple green "cabbage-like leaves" vegetable. Widely considered to be healthy food, collards are good source of vitamin C and soluble fibre and contains multiple nutrients with potential anti-cancer properties, such as **diindolylmethane** and **sulforaphane.** Researchers at the University of California at Berkeley have recently discovered that **3, 3-diidolylmethane** in brassica vegetables such as collard greens is a potent modulator of the innate immune response system with potent anti-viral, anti-bacterial and anticancer activity.

Health Benefits of Collard (Greens)

Wonderfully nutritious collard leaves are very low in calories (provide only 30 cal per 100 g) and contain no cholesterol. However, these green leaves contain very good amount of soluble and insoluble dietary fibre that helps control LDL cholesterol levels and also; offers protection against hemorrhoids, constipation as well as colon cancer diseases.

Widely considered to be healthful foods, collards are rich in invaluable sources of phyto-nutrients with potent anti-cancer properties, such as di-indolyl-methane (DIM) and sulforaphane that have proven benefits against *prostate, breast, cervical, colon, ovarian* cancers by virtue of their cancer cell growth inhibition and cytotoxic effects on cancer cells.

Di-indolyl-methane has also found to be effective immune modulator, anti-bacterial and anti-viral properties by potentiating Interferon-gamma receptors and production.

The leaves are also an excellent source of folates, provides about 166 mcg *or* 41.5 per cent of RDA. Folates are important in DNA synthesis and when given during peri-conception period can prevent neural tube defects in the baby.

Fresh collard leaves are also rich in vitamin-C. Provides about 59 per cent of RDA per 100 g. Vitamin-C is a powerful natural anti-oxidant that offers protection against free radical injury and flu-like viral infections.

Collard greens are also an excellent source of vitamin-A (222 per cent of RDA per 100 g) and flavonoid poly-phenolic anti-oxidants such as *lutein, carotenes, zea-xanthin, crypto-xanthin etc.* These compounds are scientifically found to have antioxidant properties. Vitamin A also required maintaining healthy mucus membranes and skin and is also essential for vision. Consumption of natural fruits rich in flavonoids helps to protect from lung and oral cavity cancers.

This leafy vegetable contain amazingly high levels of vitamin-K, provides staggering 426 per cent of recommended daily levels per 100 leaves. Vitamin K has potential role in the increase of bone mass by promoting osteotrophic activity in the bone. It also has beneficial effect in Alzheimer's disease patients by limiting neuronal damage in their brain.

Collards are rich in many vital B-complex groups of minerals such as niacin (vitamin B-3), pantothenic acid (vitamin B-5), pyridoxine (vitamin B-6) and riboflavin.

The leaves and stems are good in minerals like iron, calcium, copper, manganese, selenium and zinc.

Caution

Like other members of the brassica family, collards may contain goitrogens which may cause swelling of thyroid gland and therefore, should be avoided in individuals with thyroid dysfunction. However, it may be used liberally in healthy person.

Should be used sparingly with people suffering from oxalate kidney stones.

100 g of raw collard greens provide more than 500 mcg of vitamin K well above daily recommended value; it is therefore, should be used cautiously in people taking anticoagulants like warferin.

KNOLKHOL (*Brassica oleracea* Gongylodes)

Knol Khol – White Vienna

It is grown for its large, edible stem which has a slight bitter –sweet flavour similar to turnip. Knolkhol improve blood circulation, strengthen digestion and help to stabilize **blood sugar** levels, making it beneficial for diabetics and **hypoglycaemics**. It can also help to relieve painful urination, stop internal bleeding and purify the body from toxins and alcohol. The juice is considered a remedy for nosebleeds. The edible stem attains maturity and ready for harvest in 55-60 days after sowing. Approximate weight is 150 g. The tuber is tolerant to cracking and has good standing ability for up to 30 days after maturity. Knol-knols have similar taste and texture as that of a **broccoli** stem or cabbage, but milder and sweeter. The younger stems have crispy, pleasant taste and rich flavour.

Knol khol is a small, bulbous vegetable encased in two harder shalls of leaves. It offers amazing health benefits. Apart from containing high levels of glucosinalates that prevent cancer, it is also full of nutrients and minerals like copper, potassium, iron, manganese and vitamin C, B compounds and potassium.

Health Benefits

It aids in weight loss, improves metabolism, helps the immune, cardiovascular and digestive system, prevents anemia and osteoporosis and prevents mascular degeneration.

When eaten raw, it has a super crisp texture and a mild peppry taste.

Mildly sweet, succulent Kohlrabi is notably rich in vitamins and Dietary fibre; but has only 27 calories per 100 g, negligible amount of fat, and zero cholesterol.

Fresh kohlrabi stem is rich source of vitamin-C; provides 62 g per 100g weight that is about 102 per cent of RDA. Vitamin C (ascorbic acid) is a water-soluble vitamin and powerful anti-oxidant. It helps body maintain healthy connective tissue, teeth, and gum. Its anti-oxidant property helps body protect from diseases and cancers by scavenging harmful free radicals from the body.

Kohlrabi, like other members of the brassica family, contains health-promoting phytochemicals such as isothiocyanates, sulforaphane and indole-3-carbinol that are appears to protect against prostate and colon cancers.

This stem vegetable is especially contain good amounts of many B-complex group of vitamins such as niacin, vitamin B-6 (pyridoxine), thiamin, **pantothenic acid** *etc.* that acts as co-factors to enzymes during various metabolism inside the body.

Knol-knol notably has good levels of minerals; copper, calcium, potassium, manganese, iron, and phosphorus are especially available in the stem. Potassium is an important component of cell and body fluids that helps controlling heart rate and blood pressure by countering effects of sodium. Manganese is used by the body as a co-factor for the antioxidant enzyme, *superoxide dismutase.*

In addition, its creamy color flesh contains small amounts of vitamin A and carotenes.

Knolkhobleaves or tops, like turnip greens, are also very nutritious greens abundant in carotenes, vitamin A, vitamin K, minerals, and B-complex group of vitamins.

Kohlrabi, which belongs to the cabbage family, is an excellent source of vitamin C and potassium.

It is also high in fibre.

Kohlrabi helps to stabilize blood sugar and is therefore useful hypoglycemia and diabetes.

It can also be effective against edema, candida and viral conditions.

Caution

Kohlrabi may contain **goitrogens**, a plant based compounds found in cruciferous vegetable like cauliflower, broccoli *etc.*, may cause swelling of thyroid gland and should be avoided in individuals with **thyroid dysfunction**. However, knol-knols may be eaten liberally in healthy person.

MUSTARD GREENS (*Brassica juncea*)

Spicy, crunchy mustard greens or leafy mustard is indeed one of the most nutritious green leafy vegetables. The greens actually have more vitamin A, carotenes, vitamin K, and flavonoid anti-oxidants than many commonly consumed fruits and vegetables. Actually, its young tender green leaves that is harvested when the plant reaches about 60cm in height and used as green leafy vegetable. Completely grown plant reaches about 120-150cm in height and bears golden yellow coloured flowers. Fresh mustards feature dark green colored broad leaves with flat surface and may have either toothed, frilled or lacy edges depending up on the cultivar type. Its light green stem branches out with many laterals.

Mustard Greens are one of the most nutritious green leafy vegetables and are packed with health promoting and disease preventing properties. They are exceptionally high in vitamins A, K, and C as well as minerals such as zinc, selenium, calcium, magnesium, and iron. Mustard greens are very beneficial for arthritis, osteoporosis, anemia, asthma, and cardiovascular disease. They are also essential for aiding cognitive disorders such as brain fog, memory loss, and Alzheimer's disease. Mustard Greens contains compounds called **dithiolthiones** which have strong anti-cancer properties and are particularly beneficial in the protection against breast, colon, prostate, lung, bladder, and ovarian cancers. Mustard greens also have powerful anti-inflammatory properties and are an ideal food for helping auto-immune disorders such as Chronic Fatigue Syndrome, Lyme disease, Fibromyalgia, IBS, Endometriosis, Graves disease, and Multiple Sclerosis. Due to their incredibly high vitamin A content, mustard greens are amazing for eye and vision problems including age related macular degeneration, blurry vision, weak nighttime vision, and dry, itchy eyes. Mustard greens can also help in preventing hair loss and strengthening the roots of your hair. They are also great for keeping your bones and muscles strong. Mustard greens are often used as a detoxifying food and as an overall tonic for the body, specifically the liver. Mustard greens have a spicy, rich flavor and are great juiced with celery and/or cucumbers or added to salads, wraps, nori rolls, soups, stews, or stir-fry.

Mustard seeds are a medicinal spice that have been used therapeutically for thousands of years. Mustard seeds are rich in antioxidants, phyto-nutrients, and vitamins and minerals such as vitamin A, C, K, E, B-complex, calcium, iron, selenium, and zinc. They are also high in essential fatty acids such as omega-3, omega-6, and oleic acid. Mustard seeds spicy and pungent flavor are known to stimulate the appetite and can increase salivation by as much as eight times. They help to promote digestion and neutralize toxins which prevents indigestion and bloating. Mustard seeds are highly beneficial for lowering blood pressure and cholesterol, speeding up metabolism, reducing the frequency of migraines, and preventing atherosclerosis and cancer. Mustard seeds have anti-fungal and anti-septic properties which makes them good for cleansing the digestive tract and detoxifying the body. They are one of the best treatments for skin diseases as its rids the blood of excess impurities and slows the activities of the sebaceous glands which throw off cellular debris. Mustard seed powder works exceptionally well in the bath as it increases the blood flow to the skin and removes toxins from the body. Mustard oil creates a "warming" sensation on the skin which helps to relieve muscle pain, arthritic pain, and rheumatism. Mustard oil is also known to stimulate hair growth when applied to the scalp and can be used a poultice or hot compresses to help reduce the impact of a sprain or strain to the body. Mustard oil can also be massaged into the chest and sinuses to help loosen up congestion in the lungs and sinuses. In ancient times, mustard seeds were believed to be so powerful and potent that they were strategically placed around the home to help ward off evil spirits. Mustard seeds can be used whole, ground, or powdered in soups, sandwiches, potato, vegetable, and rice dishes. If you love prepared mustard, consider making it fresh at home by grinding/crushing mustard seeds and mixing with some apple cider vinegar, sea salt, and spices of choice. Raw honey can also be added for additional flavor and health benefits. Mustard seeds, mustard powder, and mustard oil can be found online or at your local health food store. Mustard seeds used as spice are being used extensively in cooking as well as in oil production all over South-Asian region.

Health Benefits

Mustard greens like spinach are the storehouse of many phytonutrients that have health promotional and disease prevention properties.

Mustards are very low in calories (26 kcal per 100 g raw leaves) and fats, but rich in dietary fibre; recommended in cholesterol controlling and weight reduction programs. However, its dark green leaves contain very good amount of dietary fibre that helps control **cholesterol** level and also help protect against hemorrhoids, constipation as well as colon cancer diseases.

Fresh mustard greens are an excellent source of several vital **anti-oxidants** and minerals like vitamin C, vitamin A, vitamin E, carotenes as well essential minerals such as calcium, iron, magnesium, potassium, zinc, selenium, and manganese.

The greens are supposed to be one of the highest among leafy vegetables which provide **vitamin K**. 100 g of fresh leaves contain about 497 mcg or about 500 per cent of daily requirement of vitamin K 1 (phylloquinone). Fresh leaves are also very good source of **folic acid**. 100 g provide about 187 mcg (about 47 per cent of RDA) of folic acid. This water-soluble vitamin has an important role in DNA synthesis and when given before and early pregnancy help prevent neural tube defects in the baby.

Mustard greens are rich source of anti-oxidants **flavonoids**, indoles, **sulforaphane**, carotenes, lutein and **zeaxanthin.** Indoles, mainly **di-indolyl-methane (DIM)** and **sulforaphane** have proven benefits against prostate, breast, colon and ovarian cancers by virtue of their cancer cell growth inhibition, cytotoxic effects on cancer cells.

Fresh mustard leaves are excellent source of **vitamin-C**. Provides 70 mcg or about 117 per cent of RDA per 100 g. Vitamin-C (ascorbic acid) is a powerful natural anti-oxidant that offers protection against free radical injury and flu-like viral infections.

The leaves are also an excellent source of **vitamin-A** (provide 10500 IU or 350 per cent of RDA per 100 g). Vitamin A is essential nutrient for maintaining healthy mucus membranes and skin and is also essential for vision. Consumption of natural fruits rich in flavonoids helps to protect from lung and oral cavity cancers.

Regular consumption of mustard greens in the diet is known to prevent arthritis, osteoporosis, iron deficiency **anemia** and believed to protect from cardiovascular diseases, **asthma** and colon and prostate cancers.

Mustard greens are an excellent anti-cancer vegetable.

They may also be beneficial for colds, arthritis or depression.

While mustard greens sold in the United States are relatively mild in flavour, some mustard green varieties, especially those in Asia, can be as hot as a jalapeno pepper depending on their mustard oil content.

Caution

Like **spinach,** reheating of mustard green leftovers may cause conversion of nitrates to nitrites and **nitrosamines** by certain bacteria that thrive on prepared nitrate-rich foods. These poisonous compounds may be harmful to health.

Phytates and dietary fibre present in the mustard greens may interfere with the bioavailability of iron, calcium and magnesium.

Because of its high vitamin K content, patients taking anti-coagulants such as warfarin are encouraged to avoid this food since it increases the vitamin K concentration in the blood, which is what the drugs are often attempting to lower. This effectively raises the effective dose of the drug.

Mustards contain oxalic acid, a naturally occurring substance found in some vegetables, which may crystallise as oxalate stones in the urinary tract in some people. It is therefore, people with known oxalate urinary tract stones are advised to avoid eating vegetables belong to brassica family.

Mustard greens may also contain goitrogens, which may interfere with thyroid hormone production and can cause thyroxin hormone deficiency in individuals with thyroid dysfunction.

BROCCOLI RABE (*Broccoli rabe*, Rapini)

It is a bitter dark leafy green vegetable that resembles broccoli but with thin stalks. While it appears to have baby-like broccoli heads surrounded by leaves on long thin stems (hence the name), the florets are anything but a toned-down version of broccoli. In fact, they are bitter and pungent and more closely resemble **mustard greens** with a zesty flavour and intense emerald colour. Rabe is a member of the Brassiceae tribe (or cruciferous vegetables) along with mustards and **cabbages,** and shares a subspecies with the **turnip.** Broccoli rabe (pronounced "robb") is a non-heading variety of broccoli with long, thin leafy stalks topped with small florets.

Nutrition

It gives 60 calories (40 from fat), 4.5g total fat, 0.5g saturated fat, 0mg cholesterol, 210mg sodium, 3g total carbohydrate (0g dietary fibre, 0g sugar), 2g protein per serving.

Because of its impressive nutritional profile that includes beta- carotene, vitamin C, calcium, fibres and **phytochemicals**, specially indoles and aromatic i**sothiocynates**, broccoli and its kin may be responsible for boosting certain enzymes that help to detoxify the body. These enzymes help to prevent cancer, diabetes, heart diseases, **osteoporosis** and high blood pressure. Broccoli along with onions, carrots and cabbage may also help to lower blood cholesterol. At U. S. Department of Agriculture's Regional Research Center in Philadelphia two researchers, Dr. Peter Hoagland and Dr. Philip Pfeffer, discovered that vegetables contain certain pectin fibre called calcium pectate that binds to bile acids, holding more cholesterol in the liver and releasing less into the blood stream. They found broccoli equally as effective as some **cholesterol** lowering drugs.

Broccoli's wealth of the trace minerals, chromium, may be effective in preventing adult-onset diabetes in some people. Diabete's expert, have found that chromium boosts the ability of insulin to perform better in people with slight glucose intolerance. This food is low in Saturated Fat. It is also a good source of Pantothenic Acid, and a very good source of Dietary Fibre, Protein, Vitamin A, Vitamin C, Vitamin E (Alpha Tocopherol), Vitamin K, Thiamin, Riboflavin, Niacin, Vitamin B6, Folate, Calcium, Iron, Magnesium, Phosphorus, Potassium, Zinc and Manganese.

Broccoli Rabe also contains a **phytochemical** called lutein. Lutein is an antioxidant that protects the retina of the eye from oxidative damage caused by harmful free radical molecules. Studies have found that people who get the most lutein from their diet have a 20 to 50 per cent lower risk of cataracts compared with people who consume the least.

Lutein-rich vegetables such as Broccoli rabe, spinach, kale, romaine lettuce may even slow the onset of age-related macular degeneration, a disease that attacks the center of the retina and diminishes the ability to see fine details.

Broccoli Rabe is also good for your bones thanks to its sizable vitamin K content. A recent study from Tufts University found that women who consumed the most vitamin K had significantly better bone mineral density compared to those whose diets contained the least.

Health Benefits

Broccoli Rabe is highly nutritious and contains many of the same cancer-fighting elements as its cruciferous relatives. It is high in **phytochemicals**such as **sulforaphane** and indoles, which are believed to help the body naturally defend itself against certain cancers. Like all Brassicas, it's a rich source of **glucosinolates,** which our body converts to cancer-fighting sulforophanes and indoles. Studies show that these compounds are particularly effective against stomach, lung, and colon cancers, and promising research hints at protective effects against breast and prostate cancers as well.

A 3 1/2-ounce serving of broccoli rabe provides more than half your daily requirement of antioxidant-rich vitamins A and C, both of which fight off dangerous free radicals that can cause damage to your body's cells. The bitter green is also a good source of folate (a B vitamin that protects against birth defects and heart disease), not to mention potassium, fibre, and calcium.

Rapini has cancer-fighting properties: Rapini contains **powerful cancer fighting phytochemicals** called indole-3-carbonol (I3C). I3C helps prevent cellular damage caused by free radicals, maintains a healthy hormonal balance for both men and women and even reduce yeast infections in the body.

Rapini helps keep bones strong:-Rapini's bone-strengthening properties are the outcome of its vitamin K content. One half cup serving contains 169 micrograms of vitamin K1, a daily dose enough to keep your bones from thinning.

Rapini lowers the risk of heart disease: Rapini contains strong anti-inflammatory nutrients, such as folate and vitamin C. Both nutrients **reduce**

homocysteine, a type of amino acid that can damage the arteries causing coronary heart disease.

Detox and heal with rapini: Rapini is one of the many cruciferous vegetables that contain sulphur. Sulphur contains a specific compound called methyl solfonyl methane (MSM) that assists detoxification of the liver. **Dr. Usha** reported the effectiveness of MSM to reduce inflammation in arthritis patients.

Improve insulin sensitivity with rapini: Rapini contains two grams of fibre for every cup. Specifically, the fibre in rapini is soluble fibre, **which slows transit and digestion time** in the GI tract. Soluble fibre dissolves in water to form a gel-like material that lowers blood cholesterol and glucose levels. Combining rapini with a high carbohydrate meal (such as a pasta dish) will reduce the insulin response, which in turn will prevent both hyper and hypoglycemia.

BRANJAL (*Solanum melongena*)

Brinjal Fruits

Eggplant or Brinjal is a very low calorie vegetable and has healthy nutrition profile; good news for weight watchers! The veggie is popularly known as **aubergine** in the western world. Deep purple ones are considered to be most nutritious. The brinjal owes its colour to the presence of **iodine** and is recommended for people who suffer from **goiter** and **thyroid disorders**. Brinjal is high in **bioflavanoids**; plantains have high folic acid, vitamin C and potassium. Brinjal is reported to stimulate the **intrapeptic metabolism** of blood cholesterol. Leaf and fruit, fresh or dry produce a marked drop in blood **cholesterol** level. The de-cholesterolizing action is attributed to the presence of poly-unsaturated fatty acids (**Linoleic and Lenolenic**) which are present in flesh and seeds of the fruit in higher amounts (65.1 per cent). The presence of magnesium and potassium salts also help in **de-cholesterolizing** action. Aqueous extracts of fruit inhibit choline esterase activity of human plasma.

It may not be a very glamorous food, but eggplant is packed with fibe and contains the whole gamut of B vitamins, which give you all the energy you need. Its deep purple colour is evidence that it has powerful **antioxidants** to protect brain cells and control lipid levels.

It is high in dietary fibre, low in fat contains an array of nutrients such as vitamins A, B1, B3, B6, folic acid and the minerals iron, calcium, magnesium, copper, manganese, phosphorus and especially potassium. It is also rich in **anthocyanin** flavonoids, particularly one called **nasunin,** a potent antioxidant and free- radical scavanger known to protect the cell membranes, whose integrity is crucial for proper metabolic activity and general health. **This** unique antioxidant in them like **nasunin** is a complex compound which defends your brain cells against damage. Many studies have proved that they are good for lessening your risk of stroke and dementia as they are packed with fibre and potassium. The best quality about eggplants is that they are pretty low in calories, thus they have a positive effect on one's heart health. Eggplant is a very versatile vegetable. It can be baked or roasted for an entree or mixed into a dip or side dish.

In addition, the mucilaginous fibre of brinjal can help to lower **cholesterol** levels, while its potassium content is useful in the prevention of water retention. It is reputed to reduce bleeding and has traditionally been used for the treatment of haemorrhoids and bleeding in general. It is an astringent food and is useful in the treatment of diarrhea. It is also very suitable for inclusion in weight loss programes as one cup of cooked cubes contain only 27 calories, 3mg of sodium and a trace of fat, but due to its high fibre content, it provides a feeling of fullness.

Dry fruit is reported to contain **goitrogenic** principles. In native medicines, role of brinjal in treatment of liver diseases, cough due to allergy, **rheumatism,** colilithiasis, leucorrhea and internal worms has been mentioned (Gopalkrishnan, 2007).

Nandkarni (1927) has given the following uses of the brinjal. Pierced all over with a needle and fried in oil, the fruit is employed as cure for toothache. It has also been recommended as an excellent remedy for those suffering from liver complaints. The green leaves of brinjal plant are a source of the appetizer, aphordisiac, cardiotonic and beneficial in 'Vata and Kaph' *etc.* In Unani system of medicine, roots of brinjal are used to alleviate pain. Brinjal fruit is used as cardiotonic,

Laxative muturant and reliever of pain. The vegetable is high in water content and is very good resource of potassium. It is supposed to possess many medicinal properties, although it is not proven. It is still being used as a remedy for cancer, hypertension and diabetes.

Nutritional Profile and Medicinal Value of Brinjal

Reduced Risk of Cancer

When used properly, eggplant can provide many health benefits, including being important in the treatment and prevention of cancer. Eggplant has been found to be especially useful in the treatment ofcolon cancer due to the high amount of fibre found within eggplant.

Fibre is important in the treatment of colon cancer because it is a relatively porous nutrient, and because of this, as it moves through the digestive tract, it has the tendency to absorb toxins and chemicals that can lead to the development of colon cancer. For best results, individuals who are interested in reducing their risk for the development of color cancer should be sure to include the skin of the eggplant during consumption. Research has found that the skin of eggplant may contain more fibre that the actual eggplant itself (Moghul, 2013). The purple pigments in brinjals contain phenolic flavonoid phyto-chemicals (called anthocyanin) that have potential health effects against cancer and other problems like aging, inflammation and neurological disorders.

Eggplants Help Weight Loss

As stated above, eggplant contains high amounts of fibre, making it a great food in the fight against cancer. But, the fibre found in eggplant has other uses-namely, its ability to be a useful tool for people who are trying to lose weight.

Fibre is a relatively "bulky" food, meaning that it takes up a lot of room in the stomach. Therefore, by eating eggplant in a salad or appetizer before a meal, dieters are likely to have a greater feeling of satiety, and generally eat fewer calories (thereby achieving a substantial weight loss with time). In addition, fibre is slow digesting, and takes a long time to move from the stomach to the digestive tract. Because of this, the eggplant keeps dieters feeling full for a longer period of time- and therefore, they won't be as tempted to snack between meals, which will again aid in weight loss.

Eggplants for better skin tone: Finally, eggplant contains a high amount of water-and similarly, aside from fat and bone, our bodies are composed of almost primarily water. Water has important roles in a variety of pathways through the body, but has been found to be especially important in the maintenance of healthy skin and hair.

Individuals who are dehydrated are more likely to exhibit hair that is thin, dry and has split ends, along with skin that appears to be flaky, dry, with a greater number of lines and wrinkles.

Consuming adequate amount of water through either water itself or through food items such as eggplant can not only improve the quality of your hair and skin, but also the general performance of your body. For best results, eat the eggplant raw-research has found that cooking it removes some of the water that is so beneficial (Moghul, 2013).

Cholesterol: Brinjals are effective in curing high blood cholesterol, hyper-cholesterol-lemia. It is rich in antioxidant monoterpenes and helps block the formation of free radicals to maintain cholesterol.

Flatulence: Roasted brinjal soup taken with **asafeotida** and garlic is very good for getting rid off gas and congestion within the body.

Hypertension: The presence of potassium, an important intra-cellular electrolyte, helps to counteract the hypertension effects of sodium.

Metabolism: Brinjals contain good quantity of vitamins B-complex (B1, B3, B5 and B6). These vitamins are essential for fat, protein and carbohydrate metabolism.

Dietary fibre: This vegetable is extremely low in fat and calories and is rich in soluble fibre content. 100 gram bringals contain just 24 calories and 9 per cent of RDA of fibre.

Digestion: Meshed brinjal and tomato soup taken with salt and pepper seasoning helps promote digestion and enhance appetite.

Insomnia: If you have problem getting sound sleep, eat baked brinjal with honey daily before bed time. It will cure your insomnia after sometime.

Enlarged spleen; It is a valuable remedy for treating enlarged spleen caused by malaria. Consume soft baked brinjal with raw sugar on empty stomach for curing it.

Stones: Eating brinjals helps in destroying stones in their early stages.

Wind humorand phlegm can be cured by taking roasted, peeled brinjals with a dash of salt.

Health Benefits

Eggplant is very low in calories and fats but rich in soluble fibre content. 100 g provides just 24 calories but contributes about 9 per cent of RDA of fibre.

Research studies at the Institute of Biology of *São Paulo State University*, Brazil showed that eggplant is effective in the treatment of high blood cholesterol.

It contains good amounts of many essential B-complex groups of vitamins such as **pantothenic acid** (vitamin B5), pyridoxine (vitamin B6) and thiamin (vitamin B1), niacin (B_3). These vitamins are essential in the sense that body requires them from external sources to replenish and required for fat, protein and carbohydrates metabolism.

It is also good source of minerals like manganese, copper, iron and potassium. Manganese is used as a co-factor for the antioxidant enzyme, *superoxide dismutase.* Potassium is an important intracellular electrolyte helps counter the hypertension effects of sodium.

The peel or skin (deep blue/purple varieties) of aubergine has significant amounts of phenolic flavonoid phyto-chemicals called **anthocyanins.** Scientific studies have shown that these anti-oxidants have potential health effects against cancer, aging, inflammation, and neurological diseases.

Caution

Brinjal contains very small amount of nicotine than any other edible plant with a concentration of 0.01mg/100g. However, the amount of nicotine from eggplant or any other food is negligible and therefore shall not warrant against its usage. So, enjoy!

TOMATO (*Lycopersicon esculentum*)

Technically called a fruit, tomatoes are mostly eaten as a well-liked vegetable. It is one of the natures' attractive deobstruents. It removes disease particles from the body and even cleans the kidneys of toxins. Tomato is rich in calcium, phosphorus, vitamin C and carbohydrates. It actually triggers off alkaline reactions in the body though it tastes acidic. Its acidic taste is due to the presence of a little citric acid, and a little malic acid. Because of its alkaline nature it acts as an anti-acid. Tomato is helpful to those who are diabetic. It is said to be effective in controlling sugar in the urine in addition to being a good food because of its low carbohydrate concentration. The most important feature is the presence of glucoalkaloides, **Tomatine** [C_{50} H_{83}[81] O_{21} N and traces of **Solanine** [C_{45} H_{71} O_{15} N]. The former is of a significant therapeutic value. Tomatine has an antibiotic property against fungal affliction of skin, particularly in cancer of mouth, stomatitis and chelilosis *etc.* **Tomatine** acts as a blood purifier and precipitating agent for cholesterol. Being rich source of vitamin C, it is a good anti scorbutic. Tomatoes help in the formation of gamma aminobutyric acid (GABA), a compound that can help bring down the blood pressure. Tomatoes have lycopene and red orbs which are well known for their cancer-defending properties. They are full of **Vitamin** A and Vitamin K which help in keeping your blood pressure levels under control.

Tomatoes have a good amount of vitamin A, and so are good for eyes and the optic system. The vitamin A found in tomatoes is ten times the amount found in a similar quantity of oranges. Though tomato juice contains a great deal of citric acid, it is not acid forming but gets alkaline within the body. It acts as an eliminator, and helps to remove uric acid from the body. They are also good for those who want to trim off that excess weight. Tomato juice helps digestive system to perk up and add pep to the liver functions. Sluggish liver, excessive formation of gas *etc.* are taken care of by the tomato. **Asthmatic** patients should ensure an all year supply of tomatoes for they reduce the congestion and keep a check on the increased secretion of mucous. Naturopaths believe that they also help in reducing spasms. Harvard

Medical School found that consumption of tomato lower the risk of suffering from cancer and also provide general resistance. In tomato, the red colour is due to **lycopene** which is again a precursor for vitamin A, is a powerful **antioxidant** that can protect against harmful free radicals, and anthocyanin pigment in purple aubergines is also water insoluble. Eating over 10 tomatoes/week has shown to reduce cancer by 35 per cent and is also effective in reducing cholesterol. The red colour of tomato protects the human body from the ravages not only of cancer but cardiovascular diseases. Tomatoes have been ranked high as a protector against acute appendicitis. Research findings have revealed that eating tomatoes or tomato sauce four days a week reduces the risk of prostrate cancer and tomatoes have been ranked high as a protector against appendicitis (Anonymous (1993). Ronald G. Butterry, an Agricultural Research Service Scientist in Albany, California found that tomatoes kept in the refrigerator lose their freshness fast and further stated that such tomatoes had less of aroma- imparting chemical **(Z)-3-hexanol** than did tomatoes kept at room temperature (Khan, 1989).

Tomatoes contain modest levels of many nutrients including provitamin A, vitamin C and potassium as well as other components such as flavonoids and lycopene that may contribute to cardiovascular health (Willcox, *et al.,* 2003). Recent study demonstrated that a clarified tomato juice inhibits platelet aggregation in vitro and exvivo following consumption of tomato by the people with type 2 diabetes. These results support epidemiological data showing that consumption of tomato products are inversely associated with risk of cardiovascular disease. A number of good characteristics are also attributed to **lycopene** including the potential ability to reduce the risk of developing cancer. Lycopene is an effective antioxidant which is also potentially linked to preventing cardio-vascular diseases. Lycopene content is normally correlated to colour, cartenoids and dry matter. These aspects are important quality factors of any cultivar released in to the market. However, the antioxidant levels in tomatoes can be influenced by environmental factors such as light and temperature, genetics and cultural practices. Tomato can be an effective alternative to drugs in lowering cholesterol and blood pressure and in preventing heart disease.

Scientists recently found that people in northern Italy who ate seven or more servings of raw tomato every week had 60 per cent less chance of developing colon, rectal or stomach cancer than those who ate two servings or less. Almost a day's requirement of calcium is present in a medium sized tomato and the protein content is almost the same as mother's milk. The vitamin C content of a tomato helps to form and maintain **collagen** which holds together all cells in the body. Collagen is concentrated in the ligaments, walls of the blood vessels, base of bones, developing teeth and gives all these structures strength and elasticity. Vegetables such as tomato, onion and cabbage have compounds that help to keep **cholesterol** down, arteries clear, bolster resistance to cancer in numerous ways. Tomatoes are high in vitamin A and C and are naturally low in calories. Being an excellent source of lycopene, which is the pigment that makes tomatoes red and has been linked to the prevention of many types of cancer. Lycopene is an antioxidant which fights free radicals that can interfere with normal cell growth and activity. These free radicals are what can potentially lead to cancer, heart disease and premature ageing. The best source of lycopene are found in processed tomato products, such as ketchups and tomato products. A medium sized tomato has about as much fibre as a slice of whole - wheat bread and only about 35 calories. It is recommended to have at least 16 milli grams of lycopene per day as they reduce the number of free radicals in the body and also help to retain the moisture in people who are used to sitting in air- conditioned environs.

Tomatoes are also the richest source of the exceptionally potent antioxidantlycopene, a substance that prevents cancer, particularly cancer of the prostate. Tomatoes also have high levels of beta carotene, an antioxidant that supports the immune system. They have high dietary fibre and taste delicious raw or cooked.

Tomatoes contain **lycopene**, one of the most powerful natural anti-oxidants, which, especially when tomatoes are cooked, has been found to help prevent prostate cancer in men, also reduces the chance of stomach and **colorectal** cancer. Lycopene is considered somewhat of a natural miracle antioxidant that may help to stop the growth of cancer cells. Tomato extract branded as **lycomato** is now also being promoted for treatment of high blood pressures. Not only tomatoes taste great but they also help the skin healthy. Lycopene has also been shown to improve the skins ability to protect against harmful U V rays, as it works as sunscreen from within. Natural genetic variation in tomatoes and their wild relatives has given a genetic treasure trove of genes that produce lycopene, carotene, **anthocyanin** and other anti-oxidants. Tomato varieties are available with double the normal vitamin C, 40 times normal vitamin A, high levels of anthocyanin and two to four times the normal amount of **lycopene** (numerous available cultivars with high crimson gene). Researchers at University of Toronto in Canada, who carried out the study, claimed that drinking two glasses of tomato juice containing 15mg of lycopene could be enough to strengthen brittle bones. Research conducted by Karin Ried (University of Adelaide- South Australia) revealed that if more than 25 milligrams of Lycopene is taken daily, it can reduce LDL (bad) cholesterol by up to 10 per cent. Half a litre of tomato juice taken daily or 50 grams of tomato paste, provide protection against heart disease. According to them, the key ingredient –lycopene, the anti- oxidant, already credited with cutting risk of prostrate cancer in men and protecting against heart diseases and led to a significant drop in levels of N- telopedtide in women drinking either type of juice or taking the capsule. Due to potassium and vitamin B, tomatoes help to lower blood pressure and to lower high cholesterol levels, which in turn, could help prevent strokes, heart attacks and other potentially life threatening heart problems. Because of **vitamin A**, tomatoes are also an excellent food to help improve your vision, and help prevent night blindness. Vitamin A is good for eyes, skin, bones and keeps your hair strong and shining.

There are many reasons to toss a tomato in your diet, it adds to the content of collagen in the skin, thus preventing it from sagging. Also, tomatoes have higher antioxidants, this helps reduce fine lines on the skin. Also, eating tomato sauce and tomato juice can help clear up acne.

Tomatoes are a staple in every kitchen but hardly will you hear any one extolling its cosmetic benefits. Whether you want to cure large pores or reduce acne and rashes or sooth a nasty sun burn, or simply to revive the glow on dull skin, tomatoes are beneficial in many home made beauty treatments. Not only do tomatoes taste great but they keep the skin healthy. It is therefore, necessary to consume tomato as they have lycopene, which is an **antioxidant** and hence works as a sun screen from within. These antioxidants make tomato an anti- ageing product as they help in fighting cellular damage and reddening of skin. It is recommended to have at least 16 milligrams of **lycopene** per day as they reduce the number of free radicals in the body and also help to retain the moisture in the people who are used to sitting in air conditioned environs.

Table 2.1: Plant Material Sources of Lycopene (mg/kg)

Vegetables (Fresh wt.)	*Scientific Name*	*Lycopene Content*
Asparagus (Cooked)	(*Asparagus officinalis*)	0.30
Carrots Orange	(*Daucus carota*)	0.02
Tomato Ripe	(*Lycopersicon esculentum*)	25.73
Red Cabbage	(*Brassica oleracea*)	0.20
Red Carrots	(*Daucus carota)*	61.00
Red peppers	(*C. annum*)	03.08
Water mellon	(*Citrulla lanatus*)	45.32
Swede	(*Brassica rapa*)	0.50

Health benefits of this rich easily available vegetable.

Anti-oxidant: Tomatoes contain a lot of vitamins A and C, mostly because of beta-carotene, and these vitamins act as an anti-oxidant, working to neutralize dangerous free radicals in the blood stream. Free radicals in the body can be flushed out with high levels of Lycopene, and the tomato is so amply loaded with this vital anti-oxidant that it actually derives its rich redness from the nutrient. These dangerous free radicals can cause cell damage. Also, remember that cooking destroys much of vitamin C, so stick to raw tomatoes for these benefits. So, do remember to carry a tomato-cucumber sandwich in your tiffin box.

Vision: Because of all that vitamin A, tomatoes are also an excellent food to help improve your vision. This also means tomatoes can help your eyes be better about night blindness. Tomatoes are full of vitamins A, C and E. They also contain anti-inflammatories, potassium and mineral salts, which are beneficial to the body. Vitamin A helps improve one's vision and makes your hair healthy.

Cancer: Various studies have shown that because of all that lycopene in tomatoes, the red fruit helps to lessen the chances of prostate cancer in men, and also reduces the chance of stomach cancer, colorectal cancer and help ward off breast cancer by absorbing damaging elements in the body. They also help control cholesterol. Lycopene is considered somewhat of a natural miracle anti-oxidant that may help to stop the growth of cancer cells. And, interestingly enough, cooked tomatoes produce more lycopene than do raw tomatoes, so enjoy that tomato soup!

Heart troubles: Due to potassium and vitamin B, tomatoes help to lower blood pressure and to lower high cholesterol levels. This, in turn, could help prevent strokes, heart attack and other potentially life-threatening heart problems.

Skin care: Because of high amounts of lycopene, tomatoes are excellent for skin care. Make sure the inside of the tomato skins are against your skin, and let this be there for at least 10 minutes. Then wash off. You face will be cleaner and give a shiny look. Tomatoes are good for the skin since they control sun damage and keep wrinkles at bay. They also act as an anti-aging agent and lend a glow to one's skin if consumed regularly. You can either have one raw tomato every day, or just rub one on your face on a regular basis.

Hair: Remember all that vitamin A in tomatoes? Well, it's good for keeping your hair strong and shiny, and it is also good for your eyes, teeth, skin and bones.

Bones: Tomatoes have a fair amount of vitamin K and calcium, both of which help to strengthen bones.

Tomatoes are one of the low calorie vegetables containing just 18 calories per 100 g. They are also very low in any fat contents and have zero cholesterol levels.

The antioxidants present in tomatoes are scientifically found to be protective against cancers including colon, prostate, breast, endometrial, lung, and pancreatic tumors.

Lycopene, a flavonoid antioxidant, is the unique p**hytochemical** present in the tomatoes. Red varieties are especially concentrated in this antioxidant. Together with carotenoids, it has the ability to protect cells and other structures in the body from harmful oxygen free radicals. Studies have shown that *lycopene* prevents skin damage from ultra-violet (UV) rays and offers protection from skin cancer.

Tomatoes are also known as superfoods because of their numerous benefits. Tomatoes contain lycopene, which, studies say help reduce the risk of colorectal cancer, prostate cancer, heart disease and lower cholesterol – all common ailments in men.

Zeaxanthin is another flavonoid compound present abundantly in this vegetable. Zeaxanthin helps protect eyes from "age related macular disease" (ARMD) in the elderly persons by filtering harmful ultra-violet rays.

The vegetable contains very good levels of vitamin A, and flvonoid anti-oxidants such as α and β-**carotenes**, xanthins and lutein. Altogether, these pigment compounds are found to have **antioxidant** properties and are take part in vision, maintain healthy mucus membranes and skin, and bone health. Consumption of natural vegetables and fruits rich in flavonoids is known to help protect from lung and oral cavity cancers.

In addition, they are also good source of antioxidant **vitamin-C,** consumption of foods rich in vitamin C helps body develop resistance against infectious agents and scavenge harmful free radicals.

Fresh tomato is very rich in potassium. 100 g contain 237 mg of potassium and just 5 mg of sodium. Potassium is an important component of cell and body fluids that helps controlling heart rate and blood pressure caused by sodium.

They contain moderate amounts of many vital B-complex vitamins such as folates, thiamin, niacin, riboflavin as well some essential minerals like iron, calcium, manganese and other trace elements.

Tomatoes are rich in lycopene, flavonoids and other phytochemicals with anti-carcinogenic properties

Tomatoes are an excellent source of vitamin C (the vitamin C is most concentrated in the jelly-like substance that surrounds the seeds)

They also contain vitamin A and B-complex vitamins, potassium and phosphorus

A tomato grown in a hothouse has half the vitamin C content as a vine-ripened tomato

Drink tomato juice with salt and pepper every morning on an empty stomach for controlling diabetes.

They also act as blood purifiers and thereby keep pimples at bay and your system clean. Tomatoes also clean up urinary tract infections.

Tomatoes have high water content and are used in salads that keep you refreshed and hydrated.

Cooked tomatoes are perhaps more beneficial as they contain more lycopene. They also maximize anti-aging effects. Lycopene also improves bone mass, which is beneficial in fighting osteoporosis. Tomatoes help strengthen the bones as they contain calcium.

Benefits of Tomato Juice

Packed with a host of minerals and vitamins, tomato juice contains Vitamin A, K, B1, B2, B3, B5 and B6. It also comprises minerals such as iron, magnesium as well as phosphorous. Not only are these vitamins and minerals excellent for your health, but they're also great for your skin and hair.

The lycopene present in the juice is said to lower the risk of prostate cancer. The ample amount of fibre and water content present in tomatoes is said to be good for the colon.

The juice is also said to be soothing for those who suffer from digestive problems such as constipation and diarrhoea.

Some experts say that tomato juice is an effective way to lower cholesterol levels. It comprises generous amounts of fibre which breaks down LDL cholesterol, and contains niacin as well – effective in fighting high cholesterol levels.

The antioxidants present in tomato juice help fight diseases that cause inflammation like diabetes, asthma and heart disease.

Caution: Excessive consumption of tomatoes might result in excess build up of Lycopene in the system, which in turn will affect the ability to fight back several diseases. As a result, the overall immunity goes down. Thus leaving our body vulnerable to many infections that may prove harmful.

OKRA (*Abelmoschus esculentus*)

Okra Plant with Fruits

Lady finger, also known as okra or gumbo in English and bhindi in Hindi, is a green vegetable that is packed with nutrients. An annual, the vegetable is produced for its green pods which contain large amounts of vitamins A and C, calcium, phosphorus, potassium and magnesium. Okra is the number one vegetable source of manganese. The plant is also rich in mucilaginous substances, which lubricate the intestines and are recommended for soothing duodenal ulcers and reducing high **cholesterol** levels. Internally, the pods feature small, round, mucilaginous white colored seeds arranged in vertical rows. The pods are harvested while immature and eaten as vegetable. High in vitamins, minerals and fibre, it has several health benefits and is great for your hair and skin. Here are some of the reasons to make ladyfinger a regular addition to your diet.

Health Benefits

Prevents diabetes: Ladies' fingers contain plenty of fibre and mucilage, which control the absorption of sugar in your small intestine and thereby regulate your blood sugar levels. In case you're wondering what mucilage is, it is the gluey substance that gives ladies' fingers their sliminess.

Helps your bowels: Ladies' fingers have laxative properties that not only improve your bowel movements, but also soothe irritable bowels and heal peptic ulcers. They also trap any toxins and excess bile that are in your digestive tract and eject them from your system before they can be absorbed into your bloodstream.

Improves digestion: The combination of fibre and mucilage in ladies' fingers also make them very good for digestion. They ease the movement of food through your gut and help to reabsorb water. This excess water prevents constipation, gas and bloating in your digestive tract.

Improves your skin: Having regular bowel movements vastly improves the condition of your skin, making you less likely to suffer from acne, psoriasis and other skin conditions. The vitamins in ladies' fingers also help to keep your skin young and prevent any pigmentation.

Conditions your hair: Boil some lady finger pods in water until you see the transparent mucilage. Apply this on your hair to make it bouncy, or on your scalp as a conditioner to smooth your hair. Massage it into your scalp after you have shampooed, then rinse it off with water. Apart from moisturising your scalp, it also prevents dryness and itchiness and helps fight dandruff.

Enables weight loss: Ladies' fingers contain plenty of nutrition while still being low in calories. In fact, they are among the lowest calorie vegetables and they contain no saturated fats, making them the perfect addition to your diet if you would like to lose weight. In fact, they even protect you from trans fats. Eating ladies' fingers regularly is recommended to help ward off obesity.

Reduces cholesterol: Not only do ladies' fingers not contain any cholesterol, they also help to lower your cholesterol by trapping it in your digestive system and preventing it from being absorbed. This helps to reduce atherosclerosis and lowers the risk of heart disease.

Reduces gut flora: The fibre in ladies' fingers helps to reduce gut flora by encouraging the growth of beneficial bacteria, which are known as probiotics. These bacteria are similar to the ones that are produced when you eat yogurt, and they help with the synthesis of Vitamin B.

Prevents cancer: Ladies' fingers contain plenty of antioxidants that boost your immunity, protect your body from harmful free radicals and prevent the mutation of cells that lead to cancer. Eating ladies' fingers is thought to reduce the risk of developing lung cancer, colon cancer and oral cavity cancers.

Improves vision: Ladies'fingers contain plenty of Vitamin A and flavonoid antioxidants like beta-carotene, xanthin and lutein, all of which improve your vision and prevent problems like cataract or glaucoma.

Strengthens bones: These pods are very high in Vitamin K and folate, which is a type of a Vitamin B. Both these nutrients improve bone health by strengthening your bones, increasing their density and preventing conditions like osteoporosis.

Reduces asthma: Eating ladies' fingers regularly can improve your asthma and reduce the severity of attacks, because in addition to antioxidants the pods also contain plenty of Vitamin C.

Prevents cold and cough: The antioxidants and Vitamin C in ladies' fingers also helps to boost your general immunity, protecting you from infections and preventing you from catching a cold or cough.

Improves pregnancy: The folate content in ladies' fingers makes them a good food to eat during pregnancy, since this nutrient has been linked to a decreased incidence of neural tube defects in babies

Prevents anaemia: Ladies' fingers are high in iron, which improves the production of red blood cells in your body and prevents anaemia.

Improves mental health: Eating ladies' fingers also helps to prevent fatigue, weakness and depression.

Very low in calories, provides just 30 cal per 100 g and contains no saturated fats or cholesterol; but is a rich source of dietary fibre, minerals, vitamins; recommended in cholesterol controlling and weight reduction programs.

The rich fibre and mucilaginous content in Okra pods helps smooth peristalsis of digested food particles and relieve constipation condition.

The pods contain healthy amounts of vitamin A, and flavonoid anti-oxidants such as beta carotenes, xanthin and lutein. It is one of the *green* vegetable with highest levels of these anti-oxidants.

Fresh pods are good source of folates, consumption of foods rich in folates, especially during pre-conception period helps decrease the incidence of neural tube defects in the offspring.

Okra is rich in antioxidants and vitamin C, so it is a decent immune booster food. It also has other vital minerals such as calcium, iron, magnesium, manganese, battle against unsafe free radicals and support the immune system.

It helps the natural production of vitamin B complex and facilitates the propagation of probiotics.

Okra is very beneficial in cases of diabetes since it regulates glucose levels. The kind of fibre found in Okra (Eugenol), aids in settling glucose by controlling the rate at which sugar is ingested from the intestinal tract.

Okra has its role in the process of building the structure of blood vessels.

Okra is rich in vitamin C, has anti-inflammatory and antioxidant properties, and it can shorten the duration of asthma manifestations and prevent deadly attacks.

The high cell reinforcements in Okra help in securing the immune system against unsafe free radicals and avoid transformation of cells. Also Okra cleans out the intestinal tract with its insoluble fibre, reducing the risk of colon-rectal tumor.

Okra provides major effects in treating respiratory problems and has been shown to reduce the frequency of asthma attacks in patients.

It aids red blood cell production.

The fibre in Okra helps the feeling of satiety, and the copious supplements sustain you.

The same mucilage in okra which helps ease digestion can also heal and relieve the pain of stomach ulcers.

Okra is an excellent source of vitamin K, and this vitamin is crucial for reinforcing bones and counteracting osteoporosis.

It contains soluble fibre pectin and thus helps lower the serum (bad) cholesterol and avoids atherosclerosis.

Okra can be used to repair body tissues, heal psoriasis and eliminate pimples and in the treatment of different skin disorders.

Okra can be of great help in the treatment of genital issues such as syphilis, leucorrhoea, dysuria, extreme menstrual bleeding, and gonorrhea.

Okra has mucilaginous and rich fibre content so it binds poisons, guarantees simple solid discharges, enhances stool mass, encourages fitting assimilation of water, and lubricates the intestines with its common purgative properties.

Okra eases fatigue, general misery and weakness.

The pods are also an excellent source of anti-oxidant vitamin, vitamin-C; provides about 36 per cent of daily recommended levels. Consumption of foods rich in vitamin-C helps body develop immunity against infectious agents, reduce episodes of cold and cough and protects body from harmful free radicals.

The veggies are rich in B-complex group of vitamins like niacin, vitamin B-6 (pyridoxine), thiamin and pantothenic acid. The pods also contain good amounts of vitamin K. Vitamin K is a co-factor for blood clotting enzymes and is required for strengthening of bones.

The pods are also good source of many important minerals such as iron, calcium, manganese and magnesium.

Okra is rich in antioxidants and vitamin C, so it is a decent immune booster food. It also has other vital minerals such as calcium, iron, magnesium, manganese, battle against unsafe free radicals and support the immune system.

It helps the natural production of vitamin B complex and facilitates the propagation of probiotics.

Okra is very beneficial in cases of diabetes since it regulates glucose levels. The kind of fibre found in Okra (Eugenol), aids in settling glucose by controlling the rate at which sugar is ingested from the intestinal tract.

Okra has its role in the process of building the structure of blood vessels.

Okra is rich in vitamin C, has anti-inflammatory and antioxidant properties, and it can shorten the duration of asthma manifestations and prevent deadly attacks.

The high cell reinforcements in Okra help in securing the immune system against unsafe free radicals and avoid transformation of cells. Also Okra cleans out the intestinal tract with its insoluble fibre, reducing the risk of colon-rectal tumor.

Okra provides major effects in treating respiratory problems and has been shown to reduce the frequency of asthma attacks in patients.

It aids red blood cell production.

The fibre in Okra helps the feeling of satiety, and the copious supplements sustain you.

The same mucilage in okra which helps ease digestion can also heal and relieve the pain of stomach ulcers.

Okra is an excellent source of vitamin K, and this vitamin is crucial for reinforcing bones and counteracting osteoporosis.

It contains soluble fibre pectin and thus helps lower the serum (bad) cholesterol and avoids atherosclerosis.

Okra can be used to repair body tissues, heal psoriasis and eliminate pimples and in the treatment of different skin disorders.

Okra can be of great help in the treatment of genital issues such as syphilis, leucorrhoea, dysuria, extreme menstrual bleeding, and gonorrhea.

Okra has mucilaginous and rich fibre content so it binds poisons, guarantees simple solid discharges, enhances stool mass, encourages fitting assimilation of water, and lubricates the intestines with its common purgative properties.

Okra eases fatigue, general misery and weakness.

The jel like substance known as mucilage which sticks to cholesterol in the body. The mucilage then exits the body during excretion, carrying the cholesterol along with it.

CAPSICUM (*Capsicum annum*) Peppers

Capsicum Annum Plant with Fruits

Bell peppers (Capsicum). Whichever colour of bell peppers you may go for - red, orange or yellow- they are full of heart-healthy properties because of **folic acid** and **lycopene** present in them. Again, they have been substantially proved to lower your risk again cancer too. Unlike their fellow members, sweet peppers have characteristic bell shape with crunchy, thick fleshy skin. On comparison to other capsicum members, bell (sweet) peppers have very mild or zero hotness. The other important difference is that they are used worldwide as vegetables instead of spices. The Scientists at Central Institute for Medicinal and Aromatic Plants (CIMAP) have isolated 'Capsaicin' the pigment responsible for pungency in the fruits of family Capsicum annum, which they reported has the capacity to cure common ailments like pain, arthritis, sprain and nerve reception. **Capsaicin** induces the process of apoptosis that destroys potential cancer cells and reduces the size of leukemia tumour cells considerably. It can be concluded that apart from setting our tongues on fire, chilli peppers can scare cancer pathogens off too.

Bell peppers contain an impressive amount of vitamin C with up to as much as six times as oranges. They are also packed with vitamin A and beta carotene which can help boost the immune system, improve vision, and help protect the eyes against cataracts. They are also an excellent source of potassium, fibre, thiamine, beta carotene, folic acid, zeaxanthin, and lycopene and have been shown to help prevent blood clot formation and reduce the risk of heart attacks and strokes. They are excellent for helping to lower cholesterol levels and they contain anti-cancer compounds that can help lower the risk of prostate, breast, lung, and colon cancer. Peppers are highly beneficial for the brain and can help to strengthen memory and concentration skills as well as reduce brain fog and confusion. Bell peppers are an ideal weight loss food as their fibre helps to curb the appetite while helping to keep you energized throughout the day. Green peppers are technically an "unripe" pepper. Even though green peppers are edible, the red, orange, and yellow bell peppers contain significantly higher levels of vitamins, mineral, and antioxidants. They are so sweet, crunchy, and juicy that they are a perfect snack to munch on and are a fantastic addition to salads, wraps, nori rolls, hummus, and dips.

Its colourful fruit – red, green and yellow peppers or capsicum – provide an excellent source of beta carotene and vitamin C; they also contain high amounts of calcium, phosphorus, sodium and potassium. They also contain two other important carotenoids, **lutein** and **zeaxanthin,** both of which can help protect against age-related mascular degeneration (AMD), which can lead to a loss of central vision in the elderly. Peppers can improve appetite and digestion, promote circulation and reduce swellings.

Capsaicin (Methyl vanillyl nonenamid) the phytochemical present in chili, impedes carcinogens such as nitrates and cigarette smoke from attaching to cellular material, thus preventing formation of cancer cells. It may also kill the bacteria that can cause ulcers. Capsaicin is shown to be a potent stressor agent and produces a rise in **corticosterone** level When applied to human skin capsaicin produces **erythema** and burning without blistering. Intragastrically capsaicin promotes ulceration. Capsaicin at a dilution of 1: 1000 showed bacteriostatic activity against several organisms. **Oleorsin** red pepper- obtained from the longer, moderately pungent chilies chiefly used for flavoring.

Capsaicin wool, prepared by dissolving the **Oleoresin** in ether and pouring it on absorbent cotton wool, is useful in rheumatoid affections. **Capsicum** tincture has been employed externally in the treatment of chilblains. Chili, in small doses, is a powerful stimulant, carminative and helps in increasing the metabolism. Capsaicin is a thermogenic food, so it causes the body to burn calories for 20 minutes after you eat the chilies. It stimulates the secretion of saliva and gastric juice, and increases peristaltic movements and motility of stomach. It is an excellent remedy in atonic and **flatulence** dyspepsia and dipsomania. Most of the **capsaicin** in a pungent (hot) pepper is concentrated in blisters on the epidermis of the interior ribs (septa) that divide the chambers of the fruit to which the seeds are attached. The amount of capsaicin in hot peppers varies significantly between varieties and is measured in Scoville heat units (SHO). Capsaicin in tropical form is promoted mainly for pain caused by conditions such as **arthritis** and general muscle soreness. There is some evidence that capsaicin may be useful in managing post-surgical pain from **mastectomy**, thoracotomy (chest surgery), amputation and other surgery related to main stream cancer treatment. Researchers have found that capsaicin may provide temporary relief for pain from mouth sores caused by chemotherapy and radiation. Some proponents claim that capsaicin has anti-oxidant properties which help to fight the carcinogen nitrosamine (a cancer causing agent). An anti-oxidant is a compound that blocks the action of activated oxygen molecules, known as free radicals that damage cells. Still others claim that it may prevent DNA damage and cancer of the lungs from cigarette smoke. **Capsaicin** is also active ingredient in several Tropical medications for temporary pain relief. A new study has found that chili peppers which usually make people 'tear up' may help people 'clear up' certain types of sinus inflammation. The study showed that participants who used a nasal spray with capsicum, reported a faster onset of action or relief, on an average within minute of using the spray.

Over the years the Capsicum annum or Capsicum frutescens has been used by alternative medicine practitioners as a remedy for a variety of conditions, such as upset stomach, menstrual cramps, head aches, shingles (herpes zoster), diarrhea, loss of appetite, stomach ulcers, poor digestion, sore throat, itching, alcoholism, motion sickness, tooth ache, malaria, yellow fever. While as some petitioners also claim it can prevent colds, heart diseases and stroke, increase **sexual potency**, foster weight loss and strengthen the heart, although available scientific evidence does not support these clams. Oleoresin of capsaicin (OC) is used in spray form as a way to incapacitate people in threatening situations. Usually this pepper spray causes burning of the eyes, nose, mouth and skin.

Health Benefits

Red bell pepper is bursting with vitamin C, making it a powerful immune builder. Red bell peppers high level of beta carotene turns into vitamin A, making it a strong defense against disease. Although green and yellow peppers are certainly healthy, they're more superfoodish. Although they both have similar amounts of vitamin C, red bell peppers have quite a bit more of the superstar beta carotene.

Capsicum contains impressive list of plant nutrients that are known to have disease preventing and health promoting properties. Unlike other chili peppers, it is very low in calories and fats. 100 g provide just 31 calories.

Sweet (bell) pepper contains small levels of health benefiting an alkaloid compound **capsaicin**. Early laboratory studies on experimental mammals suggest that capsaicin has anti-bacterial, **anti-carcinogenic**, **analgesic** and **anti-diabetic** properties. When used judiciously it also found to reduce triglycerides and LDL cholesterol levels in obese individuals.

Fresh bell peppers, red or green, are rich source of vitamin-C. This vitamin is especially concentrated in red peppers in highest levels. 100 g fresh red pepper provide about 127.7 mcg or about 213 per cent of RDA. Vitamin-C is a potent water soluble antioxidant. It is required for the collagen synthesis in the body. Collagen is the main structural protein in the body required for maintaining the integrity of blood vessels, skin, organs, and bones. Regular consumption of foods rich in vitamin C helps body protect from scurvy; develop resistance against infectious agents (boosts immunity) and scavenge harmful, pro-inflammatory free radicals from the body.

It also contain good levels of vitamin-A. 100 g of sweet pepper has 3131 IU or 101 per cent of vitamin A. In addition, it contains anti-oxidant flavonoids such as α and β **carotenes**, lutein, zeaxanthin, and **cryptoxanthin**. Together, these antioxidant substances in capsicum helps to protect body from injurious effects of free radicals generated during stress and diseases conditions.

Bell pepper has adequate levels of essential minerals. Some of main minerals in it are iron, copper, zinc, potassium, manganese, magnesium, and selenium. Manganese is used by the body as a co-factor for the antioxidant enzyme superoxide dismutase. Selenium is anti-oxidant micro-mineral that acts as co-factor for enzyme superoxide dismutase.

Capsicum is also good in B-complex group of vitamins such as niacin, pyridoxine (vitamin B-6), riboflavin, and thiamin (vitamin B-1). These vitamins are essential in the sense that body requires them from external sources to replenish. B-complex vitamins facilitate cellular metabolism through various enzymatic functions.

Capsicum is recommended for those suffering from constipation, because of its laxative effect.

Caution

The pungent level in bell peppers is almost zero "Scoville heat units" (SHU). Sometimes the seeds and central core may contain some **capsaicin**, which when eaten causes severe irritation and hot sensation to mouth, tongue and throat.

Note some of these points while handling capsicum annum members in general:

Capsaicin in chilies, especially **cayenne peppers**, initially elicit inflammation when it come in contact with the mucus membranes of oral cavity, throat and stomach, and soon produces severe burning sensation that is perceived as 'hot' through free nerve endings in the mucosa. Eating cold yogurt helps reduce the burning pain by diluting capsaicin concentration and preventing its contact with stomach walls.

Avoid touching eyes with pepper contaminated fingers. Rinse eyes thoroughly in cold water to reduce irritation.

They may aggravate existing **gastro-esophageal reflux** (GER) condition.

Capsaicin, the active ingrident in chili, does cause tissue inflammation so that mucosa of the stomach or intestine might be damaged by a sufficiently large dose (John Prescott, 2011).

CHILLI (*Capsicum annum*)

Green Chillies in Plant

Chili plant is a perennial small shrub with woody stem growing up to a meter height and bears white colored flowers. The pods are very variable in size, shape, colour and pungency. Depending on the cultivar type, they range from the mild, fleshy, Mexican bell peppers to the tiny, fiery, finger-like chili peppers, commonly grown in Indian subcontinent. The hotness of chili is measured in "Scoville heat units" (SHU). On the Scoville scale, a sweet bell pepper scores 0, a jalapeño pepper around 2, 500-4, 000 and a Mexican habañeros 200, 000 to 500, 000. Chilies can be picked up while they are green or when they reach complete maturity and dried in the plant. Usually, the fruits are picked up by hand when they are matured and turned red. They are then left to dry which causes them to shrivel. Chilies have strong spicy taste that comes to them from the active alkaloid compounds *capsaicin, capsanthin* and *capsorubin*.

Its taste is extremely used as a local stimulant for the tonsils, internally it is irritant and large doses produce gastroenteritis. Its fruit is digestive, stimulant, prevents heart diseases and curative for rheumatic troubles. According to boffins, capsaicin, the stuff that gives chili, peppers their kick, may cause weight loss and fight build up by trigging certain beneficial protein changes in the body. As laboratory studies have hinted that capsaicin may help fight obesity, by decreasing calorie intake, shrinking fat tissue and lowering fat levels in the blood (India times. com) #. Foods containing chilies are said to be as foods that burn fat. Chili peppers activate sympathetic nervous system, which helps control most of the body's internal organs, to expend more energy, so that the body burns more calories when the same food is eaten with chili peppers. Eating chilies is associated increases in metabolic rate and thermogensis. Chilies contain capsaicin, a chemical which also increases blood flow and triggers the release of mood-enhancing endorphins that naturally pump up your libido.

According to Zhiming Zhu of third Military Medical University (China) chili might set your mouth on fire but it also leads blood vessels to relax. Long term dietary consumption of capsaicin could reduce blood pressure in genetically hypertensive rats. Chili also contain a compound called **dihydrocapsiate** that is a fat buster. The capsaicin in chilies also offer great pain relief and thanks to the burning sensation associated with them are a great help in relieving nasal congestion. It is recommended that you consume chilies three times a day to enjoy their complete health benefits. Needless to say this should not be a problem given that green and red luscious chilies are present in almost everything we eat.

Chillies are rich in Vitamin C, Vitamin B6, Vitamin K1, Vitamin A, Potassium, manganese and Copper, excellent for negating the effects of harmful free radicles in the body. They are also packed with antioxidants such as Capsanthin, Violaxanthin, Lutein, Capsaicin and Ferulic acid. Dry chillies benefits might out weight those of immature green chillies, as the antioxidant content is higher in mature red Chillies (Shah, Bonny 2020).

Capsaicin is the main bioactive compound which is responsible for giving Chillies their spicy flavor. Capsaicin boosts the metabolism, increases body heat and thereby the amount of calories burned daily. Further, Capsaicin can also help reduce carvings by keeing you feel full for longer. While it is commonly believed that spicy foods lead to stomach ulcers, one study has shown that when taken in moderation, Capsaicin can decrease the risk of gastric ulcers, thus aiding digestive health. Capsaicin binds with pain receptors, which are nerve endings that sense but does not cause any real burning.

According to AYURVED- it suppress Kapha, Vata, promotes bile and is lekhan (scraping). It purifies blood, cures gastric problems and sleeplessness, stimulates the heart and cleanses the urinary system. It is anti-pyretic and especially effective in controlling dangerous fever. Due to its hot, sharp and bitter nature, it induces saliva secretion thus helps digestion, but in large amounts may cause burning sensation. Food scientists say that capsaicin actually causes your body to heat up. However, researchers have found growing evidence that the body- heat generating power of peppers might even lend a hand in our quest to lose those extra inches accumulating around the waistline.

#-Appears in AC'S monthly Journal of Proteome Research.

Health Benefits

Weight loss: For one, that extra spice can also cause you to lose the flab. As per research, chillies have capsaicin, a compound that gives a thermogenic effect and thus makes the body to burn more calories after you have had the meal.

Prevents cancer: As per research, it has been found that capsaicin in chillies has the ability to kill some cancer cells. In addition, chillies may also help to battle common colds and prevent stroke and obesity.

Better heart health: Chillies are also known to reduce cardiovascular risk. They lower incidences of heart attack and stroke as hot chillies lessen damaging effects of LDL (bad cholesterol). Capsaicin is also said to help fight inflammation, which is a major factor in heart problems.

Lower blood pressure: A study has shown that a compound in chillies has the ability to lower blood pressure. It also was shown to induce blood vessels to relax.

Reduces anger levels: Spicy foods are said to boost the production of serotonin (feel -good hormones). They thus help ease depression.

Chilli pepper contains chemical compounds that are known to have disease preventing and health promoting properties.

Chillies contain health benefiting an alkaloid compound *e.g.* capsaicin which gives strong spicy pungent character. Capsaicin has anti-bacterial, anti-carcinogenic, analgesic and anti-diabetic properties. It also found to reduce LDL cholesterol levels in obese individuals.

Fresh chili peppers, red or green, are rich source of vitamin-C.

They are also good in other antioxidants like vitamin A, and flvonoids like beta carotene, alpha carotene, lutein, zeaxanthin, and cryptoxanthins. These antioxidant substances in capsicum helps to protect body from injurious effects of free radicals generated during stress, diseases conditions.

Chillies contain good amount of minerals like potassium, manganese, iron, and magnesium. Potassium in an important component of cell and body fluids that helps controlling heart rate and blood pressure. Manganese is used by the body as a co-factor for the antioxidant enzyme superoxide dismutase.

Chilies are also good in B-complex group of vitamins such as niacin, pyridoxine (vitamin B-6), riboflavin and thiamin (vitamin B-1). These vitamins are essential in the sense that body requires them from external sources to replenish.

Medicinal Uses

Chilli peppers contain chemical compound **capsaicin.** Capasicin and its co-compounds used in the preparation of ointments, rubs and tinctures for their astringent, counter-irritant and analgesic properties.

These formulations have been in use in the treatment of arthritic pain, post herpetic neuropathic pain, sore muscles etc.

Capsaicin has anti-bacterial, anti-carcinogenic, analgesic and anti-diabetic properties. It also found to reduce LDL cholesterol levels in obese persons.

Eating raw chillies helps in dissolving blood clots and aids in digestion.

It acts as a natural pain reliever of wounds.

The vitamin C in these chillies helps in the cure of wounds.

It helps in maintenance of bones and teeth

Intake of green chillies increase your metabolism by burning your calories.

Caution

Chili peppers contain active component in them, capsaicin, which gives strong spicy pungent character which when eaten causes severe irritation and hot sensation to mouth, tongue and throat.

Capsaicin in chilies initially elicit inflammation when it come in contact with the delicate mucus membranes of oral cavity, throat and stomach, and soon produces severe burning sensation that is perceived as 'hot' through free nerve endings in the mucosa. Eating cold yogurt helps reduce the burning pain by diluting capsaicin concentration and also preventing its contact with stomach walls.

Avoid touching eyes with chili contaminated fingers. Rinse eyes thoroughly in cold water to reduce irritation.

Chilies may aggravate existing gastro-esophageal reflux (GER) condition.

Certain chemical compounds like aflatoxin (fungal mold), found in spoiled chilies have been known to cause stomach, liver and colon cancers.

ONION (*Allium cepa*)

Onion-Spanish red

Including onions in your daily food intake, gives your body's immunity levels a royal boost. They are full of peptide known as GPCS which helps in keeping your body's calcium level under control. Packed with Vitamin C and folate, onions promote good **cardiovascular** and gastrointestinal health.

The healing properties of onions are now scientifically established and are attributed to their many important components, such as **quercetin**, **mustard oils** and sulphur- containing compounds. Onions can reduce blood stickiness, preventing blood clots and heart attacks. Crude extracts of onions have been shown to lower high blood pressure and cholesterol levels. Onions are also used to expel phlegm and alleviate coughs, colds and inflammations of the nose and throat. Onion soup can help cure hangovers and is good for coughs, colds and bronchitis.

It contains an acid volatile oil (0.05 per cent) with a pungent smell. The oil is rich in sulphur and contains a variety of aliphatic disulphide, dipropyl disulphide, methyl propyl disulphide and their trisulphide. Onions contain an enzyme called "**alliinase**" which converts anlliins to disulphide oxides. These oxides are allicin type compounds. The bactericidal action of onion and garlic juice is due to allicin type compounds. The allicins and disulphides interact with -SH (Thiol) compounds like cysteine and prevent their incorporation into proteins, which inhibits the growth of bacteria. The **hypoglycemic** [blood sugar lowering) effects of onion and garlic extracts has long been recognized. Recently, another anti hypoglycemic agent present in onion has been identified *viz.* **diphenylamine**, a non sulphur compound to have hypoglycemic action. Diphenyl amine found in onion is also effective against diabetes. Compound, dially disulphide limits the amount of cartilage damaging enzymes when introduced to human. The doctors at Kings College London have further found that those whose consume a healthy diet with a high intake of fruits and vegetables particularly alliums such as garlic there was less evidence of early **osteoarthritis** in the hip joints.

The dry onion skin, are 'rich in compounds that are beneficial for human health, according to the researchers, the brown skin and external layers are rich in fibre and flavonoids, while the discarded bulbs contain sulphurous compounds and fructans.

According to the study, the brown skin could be used as a functional ingredient high in dietary fibre (principally the non-soluble type) and phenolic compounds, such as quercetin and other flavonoids (plant metabolites with medicinal properties). The two outer fleshy layers of the onion also contain fibre and flavonoids.

Eating fibre reduces the risk of suffering from cardiovascular disease, gastrointestinal complaints, colon cancer, type-2 diabetes and obesity", added the researcher. Phenolic compounds, meanwhile, help to prevent coronary disease and have anti-carcinogenic properties. The high levels of these compounds in the dry skin and the outer layers of the bulbs also give them high antioxidant capacity.

Charak, Sushruta and other leading lights of Ayurveda have described onion to be carminative, digestive, liver stimulant, anti- inflammatory, expectorant, diureti, emmanagogue (that enhances the menstural flow) and aphordisiac in effect. Scientific studies have also found that onion is an anti-oxidant, anti- rheumatic, anti-tumour, lipid lowering and bronchodilatory agent. Many of its therapeutic effects match to those of garlic. The use of onion has been found to be effective in conditions like anorexia, indigestion, flatulence, bronchitis, asthma, otalgia, osteoporosis and skin blemishes. Ancient reference indicate that regular use of onion enhances glow on one's face and keeps premature changes of age at bay. Researchers have found that under controlled studies onion inhibit platelet aggregation and increased the bleeding time. It was observed that an addition of onion to a fatty diet can protect against changes in serum cholesterol. It tends to keep fibrinolytic activity at normal levels in ischemic heart patients, thereby preventing further risk of arterial thrombosis. Reasonable consumption of the allium group of herbs like onion and garlic has been found to be protective against arthritic conditions and cancer of the stomach and the colon (Vatsyayan, 2010). Onion, garlic, shallot, leek chives, bunching onion *etc.* regulate the blood sugar in the body to moderate levels thereby, promote the death of the cancer cells in colon, breast, lung and prostate cancers.

According to traditional medicinal literature they are a source of many vitamins and are useful in fever, dropsy, catarrh, and chronic bronchitis, mixed with common salt, onions are a domestic remedy in colic and scurvy. Rosted or otherwise they are applied as a poultice to indolent boils, bruises, wounds *etc.* to relieve hot sensation and applied to the navel for dysentery and fever. Raw onion has a antiseptic value though out the alimentary canal that is more effective than when roasted or cooked. Eaten raw it is also a diuretic and emmenagogue. Warm juice is dropped into the ear to relieve each ache, applied hot to the soles of feet in convulsive disorders. It is sniffed in epistaxis. It is applied to eyes in dimness of vision and locally to ally the irritation of insect bites and scorpion stings and of skin diseases. It is given to as an antidote in tobacco poisoning. Mixed with mustard oil in equal proportions it is good application for rheumatic pains and other inflammatory swellings. Onion eaten with unrefined sugar stimulate the growth of children. Oil contained in the bulbs is a stimulant, diuretic and expectorant, Mixed with vinegar, onions are useful in case of sour throat. Cooked with vinegar they are given in jaundice, spleenic enlargement, and dyspepria. In malaria fevers they are eaten twice a day with remarkable relief. Rosted onions mixed with cumin, vinagar candy, butter oil are a demulcent of great benefit in piles ((Nandkarni, 1954). Fresh onion juice is moderately bactericidal. The essential oil (0.05 per cent of bulbs) contain a heart stimulant, increases pulse volume, and frequency of systolic pressure and coronary flow and stimulates the intestinal smooth musculature and the uterus.

As per Bhava Prakash, onion is tasty, eradicates gastric problems, promotes physical strength, improves pacifies and is heavy. Red onion is cool, alleviates pitta, improves digestive functions and works like sedative. Onion seeds are aphrodisiac, cure tooth infections and urinary disorders. As per Ayurveda, onion pacifies vata, enhances pitta and kapha, relieves pain, cures oedema and swelling, subsides inflammation and cures skin disorders. It increases the rajas and tama guna and so is amedhya (reduces intellect). It enhances digestive functions, improves digestion, anuloman and purifies urinary system and increases potency. It is spermatogenic, coagulates blood, emenogogue, aphrodisiac and improves physical strength and cures itching (Acharya Balkrishna, 2008). Scientists in Hong Kong have discovered that red onions help remove bad cholesterol- which can cause heart attacks and strokes-from the body. At the same time, red onions retain the body's good cholesterol, which help and protect against heart disease.

Health Benefits

Onions are high in vitamin C, folic acid, biotin, chromium, calcium and the richest food source of quercitin which is a potent antioxidant that has been shown to lower cholesterol, blood pressure, and triglycerides as well as help to prevent blood clots, asthma, sinus infections, bronchitis, atherosclerosis, and diabetes. They also contain powerful anti-cancer properties which have been found to help slow and reverse tumor growth within the body. Onions are particularly beneficial for stomach, colon, prostate, breast, lung, bladder, and ovarian cancer. They contain valuable sulphur compounds that are known to significantly strengthen the immune system, brain, and nervous system. These sulphur compounds also act as a heavy metal detoxifier and can help to safely remove mercury, cadmium, arsenic, and lead from the body. Onions can help to treat colds, coughs, bacterial infections, angina, and bronchial spasms. They can also help to stabilize blood sugar levels and provide relief to the liver when processing glucose and insulin. Onions have powerful antibacterial properties and have been shown to be able to destroy many disease causing pathogens such as E.coli and salmonella. They also act as a natural diuretic and help reduce bloating, water retention, and edema. Onions can help purify the digestive tract and help to stop putrefactive and fermentation processes in the gastr-ointentinal tract. They are also known to help re-grow hair as well as add volume and shine. Onions that are eaten raw provide the most nutritional benefits and are a delicious addition to salads, sandwiches, and wraps. Onions come in a wide range of varieties including red, yellow, white, and sweet. Using onions in your daily meals will help to boost your immune system and keep you healthy and strong.

Onions are rich in **quercetin**, a powerful antioxidant that may reduce the risk of cancer. Like garlic, onions also contain the amazing compound **allicin**. Red and purple onions contain **anthocyanins**, the same **antioxidants** that make berries so robust in healing powers. In addition to being extraordinary at preventing and healing cancer, the quercetin contained in onion makes them a safe therapy for allergies; it also helps prevent heart disease and reduce high blood pressure.

Onions are very low in calories and fats; but rich in soluble dietary fibre.

Onion phyto-chemical compounds ***allium*** and ***Allyl disulphide*** convert to **allicin** by enzymatic reaction when the bulb disturbed (crushing, cutting *etc.* these compounds have anti-mutagenic (protects from cancers) and anti-diabetic properties (helps lower blood sugar levels in diabetics).

Allicin reduces cholesterol production by inhibiting *HMG-CoA reductase* enzyme in the liver cells. Further, it also found to have anti-bacterial, anti-viral, and anti-fungal activities.

Allicin also decreases blood vessel stiffness by release of nitric oxide (NO) ; thereby bring reduction in the total blood pressure. It also blocks platelet clot formation and has fibrinolytic action in the blood vessels which, helps decrease overall risk of coronary artery disease (CAD), peripheral vascular diseases (PVD), and stroke.

They are rich source of **chromium,** the trace mineral that helps tissue cells respond appropriately to insulin levels in the blood; thus helps facilitate insulin action and control sugar levels in diabetes.

Onions are also good source of **antioxidant** flavonoid quercetin, which is found to have anti-carcinogenic, anti-inflammatory, and anti-diabetic functions.

They are also good in anti-oxidant vitamin, vitamin-C and mineral manganese which is required as co-factor for anti-oxidant enzyme *superoxide dismutase.*

Onions are also good in B-complex group of vitamins like pantothenic acid, **pyridoxine**, folates and thiamin.

Onions contain sulphur compounds as thiosulfinates which exhibit antimicrobial properties.

Onion is an aphrodisiac food, it helps in strengthening the reproductory organs.

Onions are an excellent antioxidant, and they contain anti-allergy, antiviral and **antihistamine** properties.

Sulfur compounds in onions help to detoxify the body.

Onions aid in cellular repair.

Onions are a rich source of **quercetin**, a potent **antioxidant.**

To obtain the maximum nutritional benefits, onions should be eaten raw or lightly steamed

SPRING ONIONS

Nutritionally, green onions have a combination of the benefits of onions and greens. They are an excellent source of vitamin K and vitamin C, and a very good source of vitamin A too. Spring onions can be added to dal and make it a much tastier dish. Similarly, you can add it to vegetables like cauliflower and potatoes and it will make an excellent dish. Mushrooms go very well with spring onions and interestingly spring onions can be added to soy nuggets while making a Chinese dish. Spring onions are used in salads as the flavor tends to be milder than other onions. It is used widely in oriental **food** both as an ingredient and as garnish. Spring onion is a nutritious plant, and therefore it provides a host of **health** benefits to us. It is a rich source of vitamins and minerals which aids in curing the various ailments. It is seen to have helped in reducing the harmful impacts of various diseases.

Some of the health benefits of spring onions are as follows:

Spring onion lowers the **blood sugar** level.

It is a support against gastrointestinal problems.

It is often used as a medicine for common cold.

It is used as an appetizer as it helps **digestion.**

It speeds up the level of blood circulation in the body.

Health benefits of Onions

Onions are very low in calories (just 40 cal per 100 g) and fats; but rich in soluble dietary fibre.

Onion phyto-chemical compounds *allium* and *Allyl disulphide* convert to allicin by enzymatic reaction when the bulb disturbed (crushing, cutting *etc.*). Studies have shown that these compounds have anti-mutagenic (protects from cancers) and anti-diabetic properties (helps lower blood sugar levels in diabetics).

Laboratory studies show that *allicin* reduces cholesterol production by inhibiting *HMG-CoA reductase* enzyme in the liver cells. Further, it also found to have anti-bacterial, anti-viral, and anti-fungal activities.

Allicin also decreases blood vessel stiffness by release of nitric oxide (NO) ; thereby bring reduction in the total blood pressure. It also blocks platelet clot formation and has fibrinolytic action in the blood vessels which, helps decrease overall risk of coronary artery disease (CAD), peripheral vascular diseases (PVD), and stroke.

They are rich source of chromium, the trace mineral that helps tissue cells respond appropriately to insulin levels in the blood; thus helps facilitate insulin action and control sugar levels in diabetes.

They are also good source of antioxidant flavonoid *quercetin*, which is found to have anti-carcinogenic, anti-inflammatory, and anti-diabetic functions.

They are also good in anti-oxidant vitamin, vitamin-C and mineral manganese which is required as co-factor for anti-oxidant enzyme *superoxide dismutase*. In addition, *isothiocyanate* anti-oxidants in them help provide relief from cold and flu by exerting anti-inflammatory actions.

Onions are also good in B-complex group of vitamins like pantothenic acid, pyridoxine, folates and thiamin. Pyridoxine or vitamin B-6 helps keep up GABA levels in the brain, which works against neurotic conditions.

Caution

Raw onions can cause irritation to skin, mucus membranes and eyes. This is due to release of *allyl sulphide* gas while chopping or slicing them. The gas when mixed with moisture (water), convert to sulfuric acid. Allyl sulphide is concentrated more at the ends, especially at the root end. Its effect can be minimized by immersing the trimmed bulb in cold water for few minutes before you chop or slice it.

GARLIC (*Allium satium*)

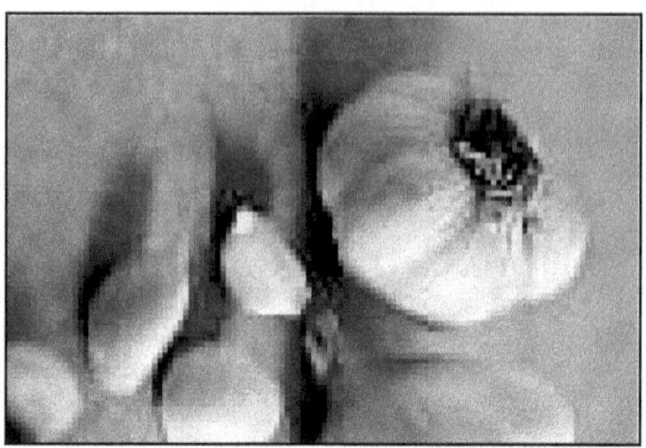

Vegetables belonging to Allium family like garlic bulb, Chinese leek, Chinese chive, scallion and shallot bulb have been shown to possess strong antioxidant activity. Whereas garlic is likely the most important herb widely known for its medicinal properties. The purported health benefits include chemoprotective, antibiotic and cholesterol lowering properties. It is known both for its culinary flavour and medicinal value, can protect against radiation damage to tissues. The findings indicate that in future garlic may protect against natural back ground and manmade source of radiation *i.e.* patients undergoing radiotheraphy. Scientists speculate that eating garlic regularly could help traffic policemen and industrial workers, who are particularly exposed to lead poisoning. Garlic is a common spice food, reduces blood sugar, unhealthy fats and blood cholesterol in man. It has been reported to reduce hypertension as it helps break the cholesterol in the blood thus preventing the hardening of arteries. It have anti- bacterial and anti tumour properties. Those who are figure conscious, garlic reduces the production of body fat. Most of the effects of garlic have been linked to its **Sulphydryul** compounds - the non volatile non protein sulphur amino acid that are responsible for its pungent flavour. These include S-Allyl cysteine Sulphoxide (SACS). This chemical either increase the secretion of insulin by the pancreas or enhance the transport of glucose to the peripheral tissues. Continuous SACS intake, the concentration of serum, lipids, blood glucose, activities of various enzymes and liver glucose decreases considerably. The results established strong anti- diabetic and the already proved hypoglycemic effects of SACS. SACS-reduces the body lipids by decreasing their synthesis and increases their excretion through the intestinal tract. SACS, has potent- anti hyper chloles- trotemic effect which considerably lowers the blood cholesterol level in body and extra cholesterol synthesis in the liver. Garlic helps lower cholesterol, acts as a natural antibiotic and blood thinner. It has the simple effect of Asprine. It is effective fat burning food. Garlic is an excellent source of manganese, a very good source of vitamin B6 and vitamin C and a good source of selenium. Garlic oil is rich in **Sulphydryl** compounds that are known to protect against tissue damaging effects of irradiation, chiefly by destroying a harmful group of chemicals called free radicals. It works as vasodialator and reduces hypertension. Anti septic properties are attributed to garlic which can ward off germs and diseases thus improving one's general health. It stimulates stomach and gall bladder secretions and kills intestinal worms. Chinese doctor used intravenous garlic treatment for cryptococcal meningitis- in which the fungus can invade the nervous system. It is known that some bacteria convert food into nitrosamines which is a carcinogenic. Potent organosulphur compounds in garlic inhibit bacterial growth in the stomach. Garlic compounds were found to slow the growth of breast cancer cells and prostrate cells. Garlic also inhibit growth of tumors of colon, rectum, esophagus and skin in rodents. It also reduces the stickiness of platelets that clamps together to from clots, which can lead to heart attacks. Garlic may also cut down on thickening and hardening of arteries known as atherosclerosis by preventing them from adhering to the lining of blood vessels. An animal study, involving rats, conducted by scientists from the Indian Institute of Chemical Biology in Kolkata, has found evidence to prove that garlic helps fight arsenic poisoning, caused by drinking contaminated ground water. Garlic proved effective because it is rich in sulphur, which scavenged arsenic from tissues and blood (Kounteya Sinha, 2008).

Modern scientific research confirms these ancient uses of garlic, including its ability to lower cholesterol and blood pressure. It is the sulphur-containing compounds of garlic that lend the herb its spicy aroma and are responsible for many of its healing properties. These compounds lower cholesterol by stimulating the release of bile by the gall bladder and by decreasing the production of cholesterol in the liver. In addition, garlic compounds gently lower blood pressure by slowing the production of hormones related to blood pressure. This wonder drug aids digestion, relieves earache, acts as an expectorant and as an antiseptic. Garlic also increases the flow of urine thereby killing thread worms and roundworms in your body. Besides, its ability to expel mucous by liquefying it, can help you control your cold, cough and other respiratory tract infections. This amazing herb has also demonstrated the ability to protect against a variety of environmental toxins. Garlic's sulphur compounds are potent antioxidants which protect cell membranes and DNA (deoxyribonucleic acid) from damage."Lipid peroxide plays a detrimental role in all cancers, including skin carcinogenesis, " a paper submitted at an International Symposium on New Frontiers in Hermatology and Oncology, said the report, titled "Garlic Suppresses oxidative Stress during DMBA- induced skin carcinogensis. As the tests done on Swiss Albino mice showed that garlic suspension inhibited the oxidative modification of lipids, thus protecting cells from injury by oxidized molecule. It is said that the chemo-preventive action was observed in mice in which garlic treatment was performed before and after the induction of skin carcinogenesis (Anonymous, 2009). While you do not want to be stinking of garlic during a passionate lip lock, but allicin, an ingredient present increases blood flow to the sexual organs. This 'wonder-drug' contains more than 100 biologically useful chemicals. It also has various anti-bacterial, anti-viral, anti-fungal and anti-oxidant properties.

"Garlic injection delays formation of papillomas (kind of rash) in animals and simultaneously decrease the size and number of papillomas which was also reflected in skin histology of mice treated, " the report, by the researchers at Department of Cancer Chemo prevention, Chittrangan National Cancer Institute (Anonymous, 2009). Garlic oil possesses more than 20 substances that have significant potential for propecting heart from diabetes- induced cardiomyopathy.

The good: This food is very low in Saturated Fat, Cholesterol and Sodium. It is also a good source of Calcium, Phosphorus and Selenium, and a very good source of Vitamin C, Vitamin B_6 and Manganese.

Garlic is one of the world's most potent medicines, and its potent smell is what makes it so powerful. The active ingredient allicin turns into **organosulfurs**, which are the compounds that keep the cells safe from all the destructive cellular processes that can cause major chronic diseases. Garlic is a natural antiseptic; it prevents cancer, fights infection, and prevents colds. Research also states that garlic may prevent or decrease chronic diseases associated with age, such as **atherosclerosis**, stroke, cancer, immune disorders, brain aging, cataracts, and **arthritis**. A 2014 review also affirmed that garlic may help in lowering blood sugar levels. Therefore, garlic has shown to provide blood sugar lowering effects. Apart from that it also improves the health of the cardiovascular system by reducing levels of bad cholesterol, triglycerides and blood lipids. It also decreases blood pressure, prevents cancer cell growth, has a strong anti-bacterial and anti-fungal effect and has a anti-tumor effect. So apart from being a healthy solution to Type-2 diabetes, garlic is also healthy for over all health.

Therapeutic and Medicinal Values

Garlic is a wonder 'drug' contains more than 100 biologically useful chemicals. It is a hot stimulant, carminative, anti-rheumatic and has antoxidant properties. Garlic oil is a powerful antiseptic. It is used as vermifuge for expelling round worms and has long been recommended for cure of a number of ailments, *viz.* wounds, foul ulcers, pneumonia, bronchitis, atopic dyspepsia and gastro-intestinal disorders. Insecticidal, anti-bacterial, anti-fungal, anti-tumor, hypoglycemic, hypolipidemic, anti-atherosclerotic, fibrinolytic and anti-platelet, -aggregation effects of garlic have been reported by modern scientists. It is best used in conditions like chest congestion, raised lipids, sciatica, degenerative disease of nervous system and various types of joint pains such as rheumatoid arthritis and **osteoarthritis**. It is also a major immuno- strengthening and 'rasayan' drug used in geriatrics. A controlled and standardised dose of garlic has been found to be effective in hypertensive and atherosclerotic patients with asignificant decrease in the blood cholesterol level. But uncontrolled use of onion and garlic may also lead to anemia as all these vegetables contain compounds which remove cysteine from the system and inhibit protein synthesis. Therefore, only 50-80g of onion and 7-10 g per day of garlic is advised. However, these vegetables may enhance the effect of insulin and reduce hyperlipidemia, obesity and resistance to insulin treatment. Also, these vegetables may prevent the incidence of altheroselerosies and related diseases of the heart. Researchers have reported that men who ate the most vegetables containing allium - the pungent, sulfur- based compound blamed for the antisocial effects of garlic and onions had a 50 per cent lower risk of having prostate cancer than those who ate the least. Garlic is most important than onion in these aspects. Garlic is very good for cleansing the body of toxic wastes and should be added to most of the meals for it`s alkalising effect. It helps in lowering cholesterol, bronchial congestion, aids in digestion, controls blood pressure, arthritis, cancerous growths and whooping cough. Garlic kills viruses responsible for colds and the flu, according to tests by James North, a microbiologist at Brigham Young University, Eat garlic when you feel a sore throat coming on, he says, and you may not even get sick. (Eat garlic when you`re stuffed up, too: It acts as a decongestant). Other studies suggest that garlic revive up immune functioning by stimulating infection- fighting **T- cells**. Eating one or two cloves of raw garlic a day is recommended to people with chronic or recurrent infection, says Andrew Weil of the University of Arizona College of Medicine, author of Natural Health, Natural Medicine. Garlic oil because of its selective bactericidal action, is a part of the current revival of folk type medicines. Garlic oil acts selectively in food systems such as salami to prevent the growth of unwanted bacteria while permitting desirable microbes to flourish and produce the desired flavour and colour. Ayurveda describes garlic as a digestive, carminative, cardiac stimulant, expectorant, analgesic, anti-oxidant and an aphrodisiac agent. It contains numerous secondary metabolites and sulphur containing compounds which are responsible for its peculiar colour and numerous healing properties. Eating green chilies with garlic is an old (tried and tested) way of enjoying sex for a longer period. Peel off its top layers, crush cloves and then fry in butter and your partner is ready to be a nutter.

Health Benefit

This 'wonder-drug' contains more than 100 biologically useful chemicals. It also has various anti-bacterial, anti-viral, anti-fungal and anti-oxidant properties. It contains sulphur compounds that help to activate liver enzymes and two strong and powerful components viz allicin and selenium. Here's a list of some of it's proven properties.

Here's a list of some of it's proven properties.

Increases blood circulation: Garlic helps reduce blood coagulation, thereby, reducing the likelihood of blood clots and the risk of having a heart stroke.

Maintains blood pressure: According to several studies, garlic lowers the amount of cholesterol in your body and also helps control your blood pressure.

Stimulates the immune system: Winter months mean the increased likelihood of contacting flu and common cold. Having garlic in your diet may reduce your chances of catching infection as the spice has antiseptic, anti-fungal and nutritive properties that boosts the immune system.

Supports recovery from cancer: Since it contains **germanium**, an anti-cancer agent, it is known to kill intestinal parasites related to **cancer.**

Prevents allergies: Killing fungi like warts, garlic helps strengthen your body's defenses against allergies. It even helps regulate your blood sugar levels (Krutika Behrawala, 2014).

While garlic is known for the role it plays in protecting the heart, it is said that men who consume garlic regularly have lower cholesterol levels.

Studies have shown that taking 600 mgs of garlic daily, lowers cholesterol levels considerably (Anonymous, 2013).

Garlic's ability to fight infections and bacteria makes it an effective cure for warts and skin problems.

It's a great **aphrodisiac**. It helps improve blood circulation and also helps prolong an erection.

Regulates blood sugar levels by increasing the release of insulin in diabetics.

Garlic's antibacterial and anesthetizing properties can help cure toothaches.

Strong flavoured, garlic cloves contain many noteworthy minerals, vitamins, anti-oxidants, and phyto-nutrients that have proven health benefits.

Its bulbs contain organic thio-sulfinites such as *diallyl disulfide, diallyl trisulfide* and *allyl propyl disulfide* that can form **allicin** by enzymatic reaction, which is activated by disruption of bulb (like crushing, cutting *etc.*). Allicin reduces cholesterol production by inhibiting *HMG-CoA reductase* enzyme in the liver cells.

Allicin also decreases blood vessel stiffness by release of nitric oxide (NO) ; thereby bring reduction in the total blood pressure. It also blocks platelet clot formation and has fibrinolytic action in the blood vessels, which helps decrease the overall risk of coronary artery disease (CAD), peripheral vascular diseases (PVD) and stroke.

Consumption of garlic is associated with possible decrease in the incidence of stomach cancer.

Allicin and other essential volatile compounds in the garlic also found to have anti-bacterial, anti-viral, and anti-fungal activities.

Garlic is an excellent source of minerals and vitamins that are essential for optimum health. The bulbs are one of the richest sources of potassium, iron, calcium, magnesium, manganese, zinc, and selenium. **Selenium** is a heart-healthy mineral, and is an important cofactor for anti-oxidant enzymes in the body. Manganese is used by the body as a co-factor for the antioxidant enzyme, *superoxide dismutase*. Iron is required for red blood cell formation.

It contains many flavonoid anti-oxidants like carotene beta, zea-xanthin, and vitamins like vitamin-C. Vitamin C helps body develop resistance against infectious agents and scavenge harmful, pro-inflammatory free radicals.

Studies have shown that taking 600 mgs of garlic daily, lowers cholesterol levels considerably.

Garlic juice when applied on face for an hour at least once a week, kills the acne causing bacteria. But do not keep it for too long as it can aggravate sensitive skin.

Garlic helps remove toxins, stimulates blood circulation and normalises intestinal flora.

The allicin in garlic has been shown to lower blood pressure. Garlic has been proven to lower LDL cholesterol and raise HDL cholesterol.

Garlic is not only known to help in digestive disorders but it is also known to naturally detox your body. Eating garlic stimulates the liver into creating detoxifying enzymes that help clean out toxins which remain in your digestive system.

If sore throat is accompanied by fever have raw garlic throughout the day. It's a natural antibiotic, effective against bacteria and viruses.

Garlic contains 17 amino acids. Amino acids are essential to nearly every bodily function, and make up 75 per cent of the human body.

Garlic not just is good for heart as it brings down cholesterol levels. It also fights cough and cold and is a pro-biotic and keeps the gut healthy. It is regarded as a heart tonic by maintaining the fluidity of blood and strengthens the heart. **Allcin**, a sulphur-containing compound, is one of the key components of garlic, which is known for its cholesterol lowering, anti-clotting and blood-pressure- lowering properties Eating half to one clove of garlic a day may lower blood pressure by 9 percent, provided taken regularly "**Agoene**" a breakdown product of allicin, may reduce the risk of heart attacks by preventing formation of blood clots (Khosla, Ishi, 2015).

Undesirable effects: The sulfide compounds in the garlic metabolized to allyl methyl sulfide, which is excreted through sweat and breathe producing unpleasant odour and breath (halitosis).

Caution

Its cloves contain allicin that acts as blood thinner. It is therefore, advised to avoid in patients on anticoagulants like ***warfarin*** as the resultant combination might cause excessive bleeding.

Garlic-in-oil, as in the preparations of pickles favours growth of *clostridium botulism,* which may result in a condition known as **botulism** (paralysis of nervous system). It is therefore, advised that garlic preparations should be preserved inside the refrigerator and should be used as quickly as possible.

Side-Effects

Raw garlic is very strong, so eating too much could produce problems, for example irritation of or even damage to the digestive tract.

There are a few people who are **allergic to garlic**. Symptoms of garlic allergy include skin rash, temperature and headaches. Also, garlic could potentially disrupt anti-coagulants, so it's best avoided before surgery.

SHALLOT (*Allium cepa* var. *aggregatum*)

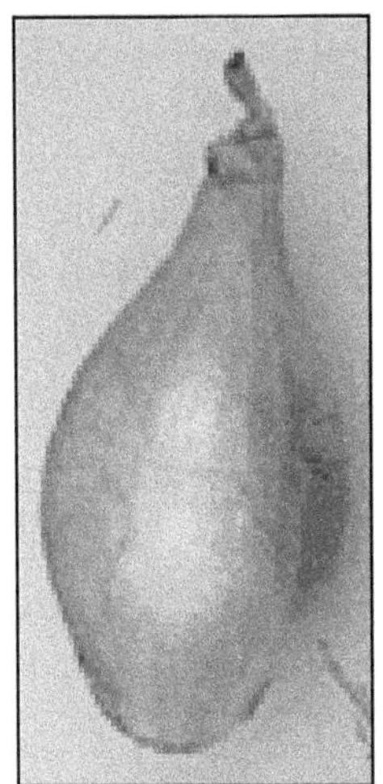

Shallots or *eschalots* are long slender bulbs in the Allium family of root vegetables. In general, they differ from onions in being smaller, and grow in clusters of bulbs from each plant root system. Like in other allium species, shallots top greens and flower heads are also eaten in many parts of world. The bulbs have less sharp in pungency than that of **onions and garlic** which makes them one of the favorite vegetable all around the world

Health Benefits

Shallots have better nutrition profile than onions. Have more anti-oxidants, minerals, and vitamins on weight per weight basis than onions.

They are rich source of flavonoid anti-oxidants such as quercetin, kemferfol *etc*. Further, they contain sulphur anti-oxidant compounds such as *diallyl disulfide, diallyl trisulfide* and *allyl propyl disulfide*. These compounds convert to allicin through enzymatic action following disruption of their cell surfaces by actions like crushing, cutting.etc.

Research studies show that *allicin* reduces cholesterol production by inhibiting *HMG-CoA reductase* enzyme in the liver cells. Further, it also found to have anti-bacterial, anti-viral, and anti-fungal activities.

Allicin also decreases blood vessel stiffness by releasing vasodialator chemical nitric oxide (NO) and thereby help bring reduction in the total blood pressure. Further research studies suggest that allicicn blocks platelet clot formation in the blood vessels that helps decrease overall risk of coronary artery disease (CAD), peripheral vascular diseases (PVD), and stroke.

The phyto-chemical compounds *allium* and *Allyl disulphide* in onion have anti-mutagenic (protects from cancers) and anti-diabetic properties (helps lower blood sugar levels in diabetics).

Shallots have several fold more concentration of vitamins and minerals than in onions, especially vitamin A, pyridoxine, folates, thiamin, vitamin C.etc. Pyridoxine (B-6) helps raises GABA chemical levels in the brain that help sooth nervous irritability. In addition, 100 g fresh shallots have 1190 IU (35 per cent RDA) of vitamin A. Vitamin A is a powerful antioxidant that helps protect from lung and oral cavity cancers.

They are also good in minerals and electrolytes than onions particularly **iron**, calcium, copper, potassium, and phosphorus

Caution

Although less in severity than onions, raw shallots can cause irritation to skin, mucus membranes, and eyes. This is due to release of *allyl sulphide* gas while chopping or slicing them which when mix with wet surface becomes sulfuric acid. Allyl sulphide is concentrated more at the ends, especially at the root end. Its effect can be minimized by immersing the trimmed bulbs in cold water for a few minutes before you chop or slice them.

LEEK (*Allium ampeloparsum* var. *porrum*)

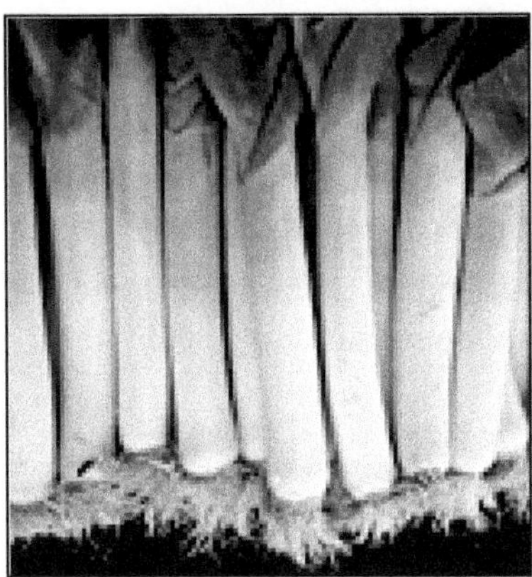

Leek Stalks

Delicate flavoured leeks are cylindrical, leafy stalks obtained from the onion-like plant in the *Allium* family. They are biennial, tall, slender plants with long cylindrical stem composed of overlapping leaves. They are commonly used as vegetables in many parts of Europe, America and Asia. It has mild astringent qualities which makes it helpful in the treatment of diarrhea and internal bleeding.

Health Benefits

Leeks contain many noteworthy flavonoid anti-oxidants, minerals, and vitamins that have proven health benefits.

Leeks are low in calories. 100 g fresh stalks contain 61 calories. Further, their elongated stalks provide good amounts of soluble and insoluble fibre.

Though leeks contain proportionately less thio-sulfinites than that in garlic, they still possess significant amounts of these anti-oxidants such as *diallyl disulfide, diallyl trisulfide* and *allyl propyl disulfide*. These compounds convert to allicin by enzymatic reaction when the stalk disturbed (crushing, cutting *etc.*).

Laboratory studies show that *allicin* reduces cholesterol production by inhibiting *HMG-CoA reductase* enzyme in the liver cells. Further, it also found to have anti-bacterial, anti-viral and anti-fungal activities.

Allicin also decreases blood vessel stiffness by release of nitric oxide (NO) ; thereby bring reduction in the total blood pressure. It also blocks platelet clot formation and has fibrinolytic action in the blood vessels which, helps decrease overall risk of coronary artery disease (CAD), peripheral vascular diseases (PVD), and stroke.

Leeks are great source of minerals and vitamins that are essential for optimum health. Their leafy stems indeed contain several vital vitamins such as pyridoxine, folic acid, niacin, riboflavin, and thiamin in healthy proportions. 100 g fresh stalks provide 64 μg of folates. Folic acid is essential for DNA synthesis and cell division. Their adequate levels in the diet during pregnancy can help prevent neural tube defects in the newborn babies.

In addition, leeks are one of the good source of vitamin A (1667 IU or 55 per cent of RDA per 100 g) and other flavonoid phenolic anti-oxidants such as carotenes, xanthin, and lutein. They also have some other essential vitamins such as vitamin C, K, and vitamin E. Vitamin C helps body develop resistance against infectious agents and scavenge harmful, pro-inflammatory free radicals.

Further, its stalks have small amounts of minerals such as potassium, iron, calcium, magnesium, manganese, zinc, and selenium.

CHIVES (*Allium schoenoprasum*)

Chives in Field

The sweet, mild-onion flavored chives are fresh top greens of lily family vegetables. The leaves are round and hollow, similar to onions, but smaller in diameter, somewhat appear like grass from the distance. They should not be confused to green **onions,** which are top greens of young, immature onion plants or **scallions,** which are top greens of *Allium fistulosum* (welsh onion). Unlike in onions and garlic, their small underground bulbs have unpleasant taste and not used in cooking. Flower stems, which rise directly from the base, grow slightly taller than leaves and bears small clusters of mauve or purple flower heads. Chives are one of the most sought after ingredients used for flavoring and garnishing recipes in many parts of the world. They have subtle onion flavor, besides imparting bright green color to the recipes. Generally, fresh leaves tied in bunches are sold along with other leafy-greens.

Health Benefits

Chives are very low in calories; 100 g of fresh leaves provide just 30 calories, but they contain many flavonoid anti-oxidants, fibre, minerals, and vitamins.

Like in scallions, they contain more plant derived dietary fibre than fellow allium members like **onions, shallots, leeks.**etc. 100 g fresh leaves provide 2.5 g or 7 per cent of daily-recommended levels of dietary fibre.

Chives too possess thio-sulfinites anti-oxidants. Thio-sufinites such as *diallyl disulfide, diallyl trisulfide* and *allyl propyl disulfide* convert to **allicin** by enzymatic reaction when its leaves disrupted (crushing, cutting *etc.*). It also have anti-bacterial, anti-viral, and anti-fungal activities.

Allicin also decreases blood vessel stiffness by release of nitric oxide (NO) ; thereby bring reduction in the total blood pressure. It also blocks platelet clot formation and has fibrinolytic action in the blood vessels which, helps decrease overall risk of coronary artery disease (CAD), peripheral vascular diseases (PVD), and stroke.

Chives have exceptionally more vitamin A than any other allium family member vegetables. In addition, their green leaves have other flavonoid phenolic anti-oxidants such as carotenes, zeaxanthin, and lutein. Together, they help body protect from lung and oral cavity cancers.

They also have some other essential vitamins such as vitamin C and K. Chives are one of the richest sources of **vitamin K**, slightly more than that of scallions. Vitamin K has potential role in bone health by promoting osteotrophic (bone formation and strengthening) activity. Adequate vitamin-K levels in the diet helps limiting neuronal damage in the brain; thus, has established role in the treatment of **Alzheimer's disease.**

Fresh chives are rich source of folates. Folic acid is essential for DNA synthesis and cell division. Adequate folate levels in the diet during pregnancy may help prevent neural tube defects in the newborn babies.

Their leaves are packed with other B-complex vitamins as well as some essential minerals such as copper, iron, manganese, zinc, and calcium. The leafy greens contain several vital vitamins such as pyridoxine, pantothenic acid, niacin, riboflavin, and thiamin in healthy proportions.

BUNCHING ONION (*Allium fistulosum* L)

Bunching Onion Stalks

Bunching onion does not form a real bulb. Oriental bunching onion has a green leaf portion and a long blanched white stalk portion. The blanched portion can be from a few centimeter to 50 centimeters, depending the varieties. The long-stalk onions are blanched by earthing up during growth. The long-stalk onions are very tender and well flavoured, excellent for stir-fry, sukiyaki, tempura and many Japanese dishes. Plants can be harvested for vegetable use at any growing stage. It is thought that it improves eye sight and internal organ function, enhances metabolism, and prolongs life. It is further said that it is effective in aiding digestion, perspiration, recovery from the common cold, as treatment for head aches, wounds and sores. The odour of Japanese bunching onions is attributed to volatile allyl sulfides.

Health Benefits

They are very low in calories; 100 g of fresh leaves provide just 31 calories and flavonoid anti-oxidants, plant fibre, minerals, and vitamins that have proven health benefits.

Being greens they naturally contain more plant derived dietary fibre than fellow allium members like **onions, shallots, leeks** *etc.* 100 g fresh spring onions provide 2.6 g or 7 per cent of daily-recommended levels of dietary fibre.

Like in **leeks,** they possess proportionately less thio-sulfinites anti-oxidants than that in the garlics. Thio-sufinites such as *diallyl disulfide, diallyl trisulfide* and *allyl propyl disulfide* convert to **allicin** by enzymatic reaction when its leaves disrupted (crushing, cutting *etc.*

Allicin also decreases blood vessel stiffness by release of nitric oxide (NO); thereby bring reduction in the total blood pressure. It also blocks platelet clot formation and has fibrinolytic action in the blood vessels, which helps decrease overall risk of coronary artery disease (CAD), peripheral vascular diseases (PVD), and stroke.

Spring onions contain healthy composition of vitamin-A and other flavonoid phenolic anti-oxidants such as carotenes, zeaxanthin, and lutein. Together, they help body protect from lung and oral cavity cancers.

They also have some other essential vitamins such as vitamin C and K. In fact, scallions are one of the richest sources of **vitamin K**. 100 g of fresh greens provide **207 µg** or about 172 per cent of daily-recommended intake. Vitamin K has potential role in bone health by promoting osteotrophic (bone formation and strengthening) activity. Adequate vitamin-K levels in the diet helps limiting neuronal damage in the brain; thus, has established role in the treatment of **Alzheimer's disease.**

Spring onions are plentiful in B-complex vitamins as well as some essential minerals such as copper, iron, manganese, and calcium. The leafy greens contain several vital vitamins such as pyridoxine, **folic acid**, niacin, riboflavin, and thiamin in healthy proportions. 100 g fresh leaves provide 64 µg of folates.

LETTUCE (*Lactuca sativa*)

Lettuce Plant in Field

It is a rich source of cellulose and its intake increases the bulk of the intestinal contents, so it is highly effective in warding off constipation. As the leaves are rich in iron, chomping on salad leaves can increase the **hemoglobin** content of the blood. They act as a tonic for patients suffering from **anemia**. Eating lettuce together with other green vegetables like spinach, asparagus and cauliflower significantly increase the folic acid content of the body. The fresh and crisp leaves have a high juice content full of magnesium which is vital for the brain and the nerves. It is valuable source of vitamin A and folic acid. Folic acid and vitamin B12 are involved in the formation of healthy red blood cells that are so essential for carrying life giving oxygen to different parts of the body. In addition, folic acid is vital for pregnant women as it protects the unborn child from damage to spinal cord, and also lowers incidence problems as cleft lip and palate and vitamin B12 helps to maintain nervous tissues. Lettuce is a fat free and low calorie food. **Lactucarium** (or "Lettuce opium") is a mild opiate like substance that is contained in all types of lettuce. Both Romans and Egyptians took advantage of this property eating lettuce at the end of a meal to induce sleep. Depending on variety, the nutrients in lettuce vary. Romaine lettuce is said to be the most healthy one boasting of vitamin A, B and C, folic acid, manganese and chromium. It makes for excellent salad asit is low in calories. Lettuce is also a good source of iron which helps with blood formation.

Health Benefits

Lettuce leaves are the store house of many phyto-nutrients that have health promotional and disease prevention properties.

Vitamins in lettuce are plentiful. Fresh leaves are an excellent source of several Vitamin A and beta carotenes. Just 100 g of fresh, raw-lettuce provides 247 per cent of daily vitamin A, and 4443 mcg of beta-carotene (Carotenes convert to vitamin A in the body; 2 mcg of carotene is considered equivalent to 1 IU of vitamin A). These compounds have antioxidant properties. Vitamin A is required for maintaining healthy mucus membranes and skin, and is also essential for vision. Consumption of natural fruits and vegetables rich in flavonoids helps to protect body from lung and oral cavity cancers.

Zeaxanthin (1730 mcg per100) an important dietary carotenoid in lettuce is selectively absorbed into the retinal macula lutea where it is thought to provide antioxidant and protective light-filtering functions, thus it offers some protection against age related macular disese (ARMD) in the elderly.

It is a rich source of vitamin K, Vitamin K has potential role in the increase of bone mass by promoting osteotrophic activity in the bone. It also has established role in Alzheimer's disease patients by limiting neuronal damage in the brain.

Fresh leaves contain good amounts folates and vitamin C. Folates require for DNA synthesis and therefore, vital in prevention of neural tube defects in-utero fetus during pregnancy. Vitamin C is a powerful natural antioxidant; regular consumption of foods rich in vitamin C helps body develop resistance against infectious agents and scavenge harmful, pro-inflammatory free radicals.

It also contain good amounts of minerals like iron, calcium, magnesium, and potassium which are very essential for body metabolism. Potassium in an important component of cell and body fluids that helps controlling heart rate and blood pressure. Manganese is used by the body as a co-factor for the antioxidant enzyme *superoxide dismutase.* Copper is required in the production of red blood cells. Iron is essential for red blood cell formation.

It is rich in B-complex group of vitamins like thiamin, vitamin B-6 (pyridoxine), riboflavins.

Regular inclusion of lettuce in salads is known to prevent osteoporosis, iron deficiency anemia and believed to protect from cardiovascular diseases, ARMD, Alzheimer's disease and cancers.

Caution

Pesticides are commonly used in lettuce crops. The most common pesticides found in the leaves are organo-phosphates, Permethrin and DDT. Wash them thoroughly in cold water before consumption. However, the organic forms are believed to be free from these toxins and safe for consumption.

METHI (*Trigonella foenum-graecum*)

Methi Seeds

It is one of the oldest-known herbal remedies. The seeds are used as a spice and can be used to expel mucous from nasal passages. Its seeds are very rich in iron, calcium sulphur, chlorine and vitamin C. The most important nutrient contained in methi is beta carotene, the 21st century's wonder food. The high Sulphur and chlorine content gives it a cleansing property. Fresh tender leaves and shoots of fenugreek contain large quantities of fatty acids, proteins, fats, vitamins (A and C), carbohydrates and minerals such as calcium, phosphorus and iron and so are considered very nutritive. In fact it is equated to quinine in its efficacy in reducing fevers. Its alkaline nature makes it a coolant to the system. Therefore, it is that in instances of dyspepsia, indigestion, acidity, flatulence and even colic, fenugreek leaves come in very handy. The high iron content helps in the formation of blood. It is thus recommended for adolescent girls and young women particularly after childbirth, and its beta carotene content gives the micro nutrients so essential, for balanced growth. Its seeds have proven to be very useful to diabetic patients. A tea made from fenugreek seeds was traditionally known to increase milk secretion in nursing mothers. As fenugreek contains an alkaloid, the **trigonelline** and its seeds contain **saponins**, and these saponins are **diosegenin, gitogenin, trigogenin**, which are thought to stimulate production of male sex hormones in including testosterone which in turn causes the rise in male libido. Fenugreek has a variety of medicinal properties. Since ages it has been used to cure fatigue, small pox, gastric trouble, gout, flatulence, loss of appetite and irregular menses in women. It is also used in colic flatulence, dysentery, diarrhea, chronic cough, enlargement of spleen and liver, rickets and diabetes. National Institute of Nutrition Hyderabad has reported that fenugreek reduces blood sugar levels in diabetic patients. The alkaloids of seeds are said to stimulate the nervous system. The **diosgenin** present in its seeds is a starting material in the synthesis of sex hormones and oral contraceptives. Its seeds are also used by Indian women for promoting lactation. It is often ground and mixed with cotton seeds and is fed to the cattle to increase the milk production. In Java, it is used in hair tonic preparations and as a cosmetic. The mucilagenous watery extract of the seed when mixed with the dye imparts glossiness and durability. It is thus recommended for adolescent girls and young women, particularly after child birth and its beta carotene content gives the micro nutrient so essential, for the balanced growth. Its leaves contain several flatus and anti nutritional factors. Dry methi leaves contain so much of protein that nutritionists equate it to pulses in food value. Dried methi leaves also supplement lysine deficiencies. Lysine is an amino acid which is a fundamental constituent of all proteins.

Fenugreek seeds are said to resemble cod liver oil and can be taken instead of cod-liver oil for complaints of anemia, arthritis and diabetes. The alkaloids of the seeds are said to stimulate the nervous system and useful for diabetic patients. Besides, the seed also contains a lactating factor (Duggal, 1995). Studies have shown that adding fenugreek to our diet reduces fasting blood sugar and improves after meal glucose tolerance significantly. Fenugreek extract promotes insulin secretion and inhibits the rise of blood glucose, thus helping to reduce body fat production. Active ingredients in fenugreek seeds exhibits a specific effect on the Islets of Langerhans in the pancreas. These cells are directly responsible for insulin production. Most significantly, the effect of active ingredients is glucose dependent. The higher the level of blood glucose, the greater the insulin promoting response elicited by this active ingredient. Thus the active ingredient in fenugreek seeds exhibits a significant regulating effect, which corresponds with the insulin needs of the body at any given time, making this compound "adaptogenic." The saponins and flavonoids present in fenugreek extract has following benefits:

Powerful pancreatic stimulator.

A natural insulin secretion promoter.

Scientifically proven blood cholesterol and blood glucose lowering properties

Significant reductions in total cholesterol, LDL cholesterol and triglyceride levels, but not HDL cholesterol levels.

Proven non-toxic.

Its seeds was traditionally known to increase milk secretion in nursing mothers.

Used to lower blood sugar levels.

Fenugreek seeds contain compounds called saponins which are thought to stimulate production of male sex hormones including testosterone.

They are known to increase milk production and stimulate breast tissue growth.

Methi seeds help lower blood sugar and reduce the need for insulin in the body.

Methi subdues the assimilation of cholesterol and lowers the amount that is manufactured by the liver.

Health Benefits of Fenugreek Seeds

Its seeds are beneficial for diabetic patients. Galactomannan in fenugreek seeds is helpful to slow down the rate of sugar absorption in to blood. Fenugreek also contains amino acid responsible for inducing production of insulin.

Fenugreek helps induce labour, treat hormonal disorders and reduce menstrual pain to a certain extent.

For skin inflammations, like abscesses, boils, burns, cuts, eczema and **gout**, fenugreek is a big help.

It works as an effective tropical treatment for skin problems.

It is recognized as a treatment for diabetes- A glass of water in which a table spoon of fenugreek seeds has been soaked overnight is drunk each morning.

Fenugreek seeds are rich source of minerals, vitamins, and phytonutrients.

The seeds are very good source of soluble dietary fibre. Soaking the seeds in water makes their outer coat soft and mucilaginous.

Non-starch **polysaccharides** (NSP) which constitute major fibre content in the fenugreeks include *saponins, hemicelluloses, mucilage, tannin,* and *pectin*. These compounds help lower blood LDL-cholesterol levels by inhibiting bile salts re-absorption in the colon. They also bind to toxins in the food and helps to protect the colon mucus membrane from cancers.

NSPs (non-starch polysaccharides) increase the bulk of the food and augments bowel movements. Altogether, NSPs assist in smooth digestion and help relieve constipation ailments.

Amino-acid **4-hydroxy isoleucine** present in the fenugreek seeds has facilitator action on insulin secretion. In addition, fibre in the seeds help lower rate of glucose absorption in the intestines thus controls blood sugar levels. The seeds are therefore recommended in diabetic diet.

The seeds contain many phytochemical compounds such as ***choline, trigonelline diosgenin, yamogenin, gitogenin, tigogenin*** **and** ***neotigogens***. Together, these compounds account for the medicinal properties of fenugreeks.

Excellent source of minerals like copper, potassium, calcium, iron, selenium, zinc, manganese, and magnesium. Potassium is an important component of cell and body fluids that helps control heart rate and blood pressure by countering action on sodium. Iron is essential for red blood cell production and as a co-factor for cytochrome-oxidases enzymes.

Rich in many vital vitamins including thiamin, pyridoxine (vit. B-6), folic acid, riboflavin, niacin, vitamin-A and vitamin-C that are essential nutrients for optimum health.

Fenugreek leaves help in reducing the pain of the patient and help in restful sleep. It reduces the level of fever stabilizing the blood pressure and heartbeat of the patient (Anonymous, 2015).

Powdered fenugreek seeds are known to lower the levels of serum lipids such as total cholesterol and triglyceriods. Phytochemicals (saponins) in fenugreek have been claimed to aid in glucose, cholesterol metabolism and cancer protection. Anti-diabetic and cholesterol- lowering properties are also attributed to their dietary fibre constituent.

Its soluble fibre helps in reducing cholesterol especially LDL, controls blood sugarlevels and helps lose weight by suppressing appetite.

Due to the presence of mucilage, a compound found in it, fenugreek has a soothing effect on throat.

Fenugreek is traditionally known to increase breast size. It empowers the mammary glands and encourages the development of breast tissues. Italso contains phyto-estrogen that builds the level of prolactin in the body that helps increase the size of breasts.

It also aids in other uterine troubles like menstruals cramps, hot flushes and period distress.

It contains compounds that aid in hair health. It is also a great remedy for dandruff.

Consumption of fenugreek seeds enhances bowl movements and is a viable cure against digestive problem and heart burns. It helps in flushing out harmful toxins from the body and aids digestion.

It regulates blood sugar and controls diabetes. As the amino acid compounds in fenugreek seeds promote insulin discharge in the pancreas, which brings down the glucose levels in the body.

Fenugreek seeds destroy free radicals in our body, which cause wrinkles and dark spots and keep the skin free from pimples.

Fenugreek seeds have been known to be supportive in stimulating labor and uterine compressions. It additionally decreases labour pain also.

Its seeds have been used in many traditional medicines as laxative, digestive, and as a remedy for cough and bronchitis.

If used regularly, fenugreeks may help control cholesterol, triglyceride as well as high blood sugar (glycemic) levels in diabetics.

Booster of Male Potency: Fenugreek seeds are also great when it comes to helping men and their potency. Namely, they contain a substance called **diosgenin,** which is known to act as a natural Viagra. Even studies have proven that fenugreek seeds are a great aphrodisiac.

Fenugreek seeds added to cereals and wheat flour (bread) or made in gruel, given to the nursing mothers to increase milk synthesis.

One teaspoon of methi seeds soaked overnight in 100 ml of water is very effective in controlling diabetes.

Fenugreek seeds contain powerful anti oxidant that help guard healthy cells against damage from free radicals, unstable molecules that attack the body's healthy cells.

The antioxidants are natural compounds that seek and destroy free radicals in the body

Additional health promoting qualities attribute to the fenugreek seed includes antiseptic and expectorant.

How does Fenugreek Seed Work

Fenugreek seed is used to treat infections of the throat, lungs and airways. It has proven to be very effective in the fight against allergies and fever, it stimulates the functioning of the kidneys and lungs, and it helps warm up the body. Also, it helps reduce the cholesterol levels and it improves digestion. Fenugreek seeds can really help you enlarge your breasts. Because this seed contains a high concentration of phytoestrogens, it affects the growth of the breasts.

Put 2 tablespoons of fenugreek seed in a cup of water and leave it overnight. Use this water the next day to massage your breasts. Repeat the process on a daily basis.

Put a cup of cold water and 2 tablespoons of fenugreek seed in a small cooking pot. Cook the mixture until it comes to a boiling point, and then take it off the heat. Allow the tea to cool down, for about 20 minutes. Strain the tea, removing all seeds, and then add a little bit of lemon and raw honey. Drink! Just, be careful and do not drink more than two cups a day.

Pharmacological Profile

Analgesic effect

Anti adhesive properties

Anti-carcinogenesis effects

Anti-tumor activity

Anti-oxidant activity

Anti-platelet activity

Exercise recovrery effects

Hepatoprotective effect

Lipid- lowering effects

Galactagogue effects

SPINACH (*Spinacia oleracea*)

Spinach in Field

Spinach is an excellent source of iron, calcium, chlorophyll, beta carotene (provitamin A), vitamin C, riboflavin, sodium and potassium. It is a diuretic, natural laxative, intestine- toner and cleaner and can also be used to stop minor hemorrhaging such as nosebleeds. As it is rich in iron and chlorophyll, spinach builds the blood, while its sulphur content helps to clean the liver and relieve herpes irritations. In addition, its vitamin A content can help to prevent night blindness. Although spinach provides an exceptional source of calcium, it is also high in oxalic acid, which can partly interfere with the absorption of the calcium. It contains more nutrients than lettuce. It is a good source of calcium and so intrinsically good for growing children as well as older people whose bones need strengthening. It is a valuable gum toner and can actually aid bleeding gums and prevent cavities. However, it contains small amounts of oxalic acid so those who suffer from gout or liver dysfunctions should control their intake. Both carrots and spinach are rich in beta-carotene or vitamin-A, which appears to help prevent the built up of cholesterol on blood vessel walls. Cholesterol deposits eventually can block blood flow through an artery causing heart attack or stroke. Spinach contains iron, carotenoids and chlorophyll and is an especially good source of lutin. It is also loaded with vitamins like C and A that help with good eye sight and glowing skin. The high level of potassium makes it very beneficial for high blood pressure patients. Further it strengthens muscles and helps those suffering from anemia. Since it is rich in iron, spinach is nature's answer to anemia. It is good for eyes because of its vitamin A content. A regular cupful, with its nutrients intact, is a food proof remedy. Because of this, it is a must for pregnant mothers because it helps their own hemoglobin to remain steady. Even after the birth of the baby, continued intake improves the quality of the mother's milk. Two nutrients found in spinach and other leafy green vegetables offer some protection against the most common cause of blindness amongst the elderly. The two nutrients, Lutein and zeaxanthin, are both carotenoids- compounds that give many fruits and vegetables a yellow colour. They help ward off the condition apparently by allowing the eyes to filter harmful short-wavelength light and by curtailing other damaging effects to the macula, or the center of the eye's retina. Dark leafy greens viz spinach are a great source of iron- a nutrient that may help protect against the sleep robber known as restless legs syndrome. Well, it won't be wrong if we tag spinach as a natural Viagra. This green leafy vegetable is loaded with sexual benefits. It is rich in Vitamin E, which is a major catalyst in the production of sex hormones in the body. It is also rich in manganese, which facilitates the production of the female hormone estrogen. A deficiency of magnesium also affects a woman's fertility levels. Green leafy veggies are also loaded with zinc, which is known for its libido and sperm production qualities in men.

It is believed that spinach contains around 13 flavonoid compounds which keep us away from cancer, heart diseases and osteoporosis. 1/2 a cup of this lutein-rich food, daily, guards us against heart attacks.

Spinach is rich in numerous minerals and vitamins including iron, folic acid, potassium vitamins A, B, C, E and manganese. It's an excellent choice when it comes to prevention from a large number of diseases thanks to the high levels of zeaxanthin and lutein, which is why it should be regularly consumed.

But, you should never reheat spinach, since it contains nitrates, which turn into nitrosamines under high heat. Nitrosamines are dangerous and carcinogenic.

If you didn't know, remember this and never reheat spinach. Always consume it fresh and even if you decide to eat it the following day eat it cold. Spinach is a very nutritious and healthy vegetable, but you need to be careful not to reheat it again.

Best muscle building foods: Researchers at Rutgers University (2008) showed that the phytoecdysteroids contained in spinach may increase muscle growth up to 20 percent. This makes it a great muscle building food!

Health Benefits

Rich in vitamins B2 and B6, folate, Cu, Ca, Mag, K, Zn and Fibre.

It has anti-cancer and anti-inflammatory agent.

May decrease risk of cataracts.

May decrease risk of heart disease.

It has antioxidant-beta-carotene.

It protects cells from free radicals.

Prevents eye problems and age related macular degenation.

It is bone-supportive nutrient.

Spinach is rich in beta carotene, which the body transforms into vitamin A, triggering your immune response to keep you well. Spinach prevents cancer and heart disease and is rich in the disease-fighting mineral zinc. The vitamin- C helps you resist colds and infection and keeps your skin healthy; the B vitamins keep you calm and more energetic.

Spinach is store house for many phyto-nutrients that have health promotional and disease prevention properties.

Very low in calories and fats (100 g of raw leaves provide just 23 cal). It contains good amount of soluble dietary fibre; no wonder green spinach is one of the vegetable source recommended in cholesterol controlling and weight reduction programs!

Fresh 100 g of spinach contains about 25 per cent of daily intake of iron; one of the richest among green leafy vegetables. Iron is an important trace element required by the body for red blood cell production and as a co-factor for oxidation-reduction enzymes *cytochrome-oxidases* during the cellular metabolism.

Fresh leaves are rich source of several vital anti-oxidant vitamins like vitamin A, vitamin C; and flavonoid poly phenolic antioxidants such as lutein, zea-xanthin and beta-carotene. Together these compounds help act as protective scavengers against oxygen-derived free radicals and reactive oxygen species (ROS) that play a healing role in aging and various disease processes.

Zea-xanthin, an important dietary carotenoid, is selectively absorbed into the retinal macula lutea in the eyes where it is thought to provide antioxidant and protective light-filtering functions; thus helps protect from "age related macular disease" (ARMD), especially in the elderly.

Vitamin A is also required for maintaining healthy mucus membranes and skin and is essential for vision. Consumption of natural vegetables and fruits rich in vitamin A and flavonoids helps body protect from lung and oral cavity cancers.

100 g of Spinach provides 402 per cent of daily vitamin-K requirements. Vitamin K plays vital role in strengthening bone mass by promoting osteotrophic (bone building) activity in the bone. It also has established role in patients with Alzheimer's disease by limiting neuronal damage in the brain.

This greeny leafy vegetable also contain good amounts of many B-complex vitamins like vitamin- B6 (pyridoxine), thiamin (vitamin B-1), riboflavin, folates and niacin. Folates help prevent neural tube defects in the offspring.

100 g of farm fresh spinach has 47 per cent of daily recommended levels of vitamin C. Vitamin C is a powerful antioxidant which helps body develop resistance against infectious agents and scavenge harmful oxygen free radicals.

The leaves also contain good amount of minerals like potassium, manganese, magnesium, copper and zinc. Potassium in an important component of cell and body fluids that helps controlling heart rate and blood pressure. Manganese and copper are used by the body as a co-factor for the antioxidant enzyme *superoxide dismutase.* Copper is required in the production of red blood cells. Zinc is a co-factor in many enzymes that regulate growth and development, sperm generation, digestion and nucleic acid synthesis.

It is also rich source of omega-3 fatty acids.

Regular consumption of spinach in the diet helps prevent osteoporosis (weakness of bones), iron deficiency anemia and is believed to protect from cardiovascular diseases and colon and prostate cancers.

Spinach is high in brain friendly folate and B-vitamins. Folate reduces inflammation that harms brain functions.

L-tyrosine in spinach improves mental focus. Just as 1 cup of steamed spinach contains more than 65 per cent of your daily value (DV) for folate and more than 20 per cent of your DV for vitamin –B6.

Caution

People with a tendency to kidney stones should eat spinach sparingly since its high oxalic acid can tend to form calcium -oxalate kidney stones in susceptible individuals.

Reheating of spinach left-over may cause conversion of nitrates in to nitrites and nitrosamines by certain bacteria that thrive on pre-prepared nitrate-rich foods, such as spinach and many other green vegetables. These poisonous compounds may be harmful to health, especially in children.

Phytates and dietary fibre present in the leaves may interfere with the bio-availability of iron, calcium and magnesium.

Because of its high vitamin K content, patients taking anti-coagulants such as "warfarin" are encouraged to avoid spinach in their food since it interferes with drug metabolism.

Spinach contains oxalic acid, a naturally occurring substance found in some vegetables which may crystallize as oxalate stones in the urinary tract in some people. It is therefore, people with known oxalate urinary tract stones are advised to avoid eating certain vegetables belonging to amaranthaceae and brassica family. Adequate intake of water is therefore advised to maintain normal urine output.

It may also contain goitrogens which may interfere with thyroid hormone production and can cause thyroxin hormone deficiency in individuals with thyroid dysfunction.

TURMERIC (*Curcuma longa* L.)

Dried Turmeric-Root

The use of turmeric dates back nearly 4000 years to the Vedic culture. Most Indian recipes will always have a common ingredient i.e. a pinch of turmeric. In India, turmeric is considered as a healer's spice. Turmeric is a staple ingredient in most Indian cuisines. It is the king of spices when it comes to dealing with cancer diseases, besides it adding a zesty colour to our food on the platter. It contains the powerful polyphenol Curcumin that has been clinically proven to retard the growth of cancer cells causing prostrate cancer, melanoma, breast tumour, brain tumour, pancreatic cancer and leukemia amongst a host of others. Curcumin promotes Apoptosis- (programmed cell death\ cell suicide) that safely eliminates cancer breeding cells without posing a threat to the development of other healthy cells.

It has many therapeutic effects. Curcumin, the active component of turmeric is an object of research owing to its properties that suggest they may help to turn off certain genes that cause scarring and enlargement of heart. Regular intake may help reduce low-density lipoprotein or bad cholesterol and high blood pressure, increase blood circulation and prevent blood clotting, helping to prevent heart attacks.

Turmeric is often referred to as the spice of life inancient Indian lore. Being bitterand sharp it cures cough. It refines blood and normalizes blood cell levels. It is tasty and is an astringent. It accumulates urine and makes it clear. It is very effective in curing urinary diseases. Its paste relieves swelling and oedema. It is **anti-dermatoses**, cleanses and cures wounds and ulcers. Its fumes are useful for curing hiccoughs, bronchial asthma, and is an antidote to poison. It cures cold and cough, hepatitis, diarrhea, urinary diseases, jaundice, skin disorders and eye ailments (Acharya Balkrishna, 2008). Turmeric contains 5 - 8 per cent of a volatile oil and the main active components are **curcumin** – a fragrant compound oil and

turpenide. Curcumin derived from turmeric root can help reduce cancer risk among post- menopausal women exposed to hormone replacement therapy. Studies suggest that a combined oestrogen and progestin hormone replacement therapy increases post menopausal women's risk of developing progestin acceleration breast tumours. **Curcumin** prevents the appearance of gross morphological abnormalities in the mammary glands. Progestin accelerates the development of certain tumours by increasing production of molecule called **VEGF** that helps to supply blood to the tumour. And blocking the production of **VEGF** could potentially reduce the proliferation of breast cancer cells (Menopause- A journal of North America Menopause Society).

The new and ongoing research according to nutritionist have found that besides flavouring the palate, the most commonly used spices possibly reduces the risk of cancer and cardio-vascular diseases. Research now has revealed that its main component, "**curcumin** is a phyto-chemical and inhibits chronic diseases. Phytochemicals, otherwise known as bioactive chemicals have important physiological and bio-chemical impact on the body's processes. A new Saint Louis University study has revealed that a chemical that gives curry its zing holds promise in preventing or treating liver damage from an advanced form of a condition known as fatty liver disease called non-alcoholic steatohepatitis (NASH).

Besides turmeric, curry leaves, pepper, cumin (zeera), coriander, fenugreek (methi) have now been found to be able to prevent tissue damage and decrease glucose and cholesterol in the blood. Studies have shown that curcumin, the chemical in turmeric that gives it the yellow colour has effective anti-malarial qualities. The turmeric therapy in combination with other therapies is expected to redefine the whole anti-malaria fight. Scientists at the University of California, Los Angeles, have found that turmeric holds the potential to fight against Alzheimer's disease, one of the most common form of illness among the elderly across the world. Their studies indicated that turmeric could be more effective than not only many of the existing medications, but also the new therapies under test. Turmeric not only inhibited the accumulation of beta amyloid, a protein, in the brain of the Alzheimer's patients, but also broke up the existing plaques. It could be used for both treatment and prevention (Sunderajan, 2005). Turmeric curry once or twice a week could keep Alzheimer's disease and dementia at bay, thanks to a magic ingredient curcumin found in the spice according to an Indian American Phychiatrist.

Turmeric curry once or twice a week could keep Alzheimer's disease and **dementia** at bay, thanks to its magic ingredient curcumin, according to Murali Doraiswamy Psychiatry Professor at the Mental Fitness Laboratory of the Duke University Medical Centre (DUMC) North Carolina. He told delegates at Royal College of Psychiatrists annual meeting Liverpool that curcumin prevented the spread of **Amyloid plaques**. These plaques are thought to contribute to the degradation of the wiring in brain cells and lead to the subsequent symptoms of Alzheimer's disease. There is thus solid evidence that curcumin binds to plaques (IANS, 6th June2009). Researchers at Cork Cancer Research Centre in Ireland treated esophageal cancer cells with **curcumin-** a chemical in the spice, which gives curries a distinctive yellow colour and found that it started to kill cancer cells within 24 hours. The cells also began to digest themselves reports the BBC. Dr Sharon Mc Kenna said "Scientists have known for a long time that natural compounds have the potential to treat faulty cells that have become cancerous and we suspected that curcumin might have therapeutic value". They are of the opinion that it opens up the possibility that natural chemical found in turmeric could be developed into new treatment for oesophageal cancer (Anonymous, 2009). According to a study, curcumin that we take is quickly destroyed by the digestive juice in the gastro intestinal tract and very little actually gets into the blood. Therefore, to overcome this problem, a team of scientists is developing nano-sized capsule that would boost the body's uptake of curcumin and help fight several diseases. Trials are under way to test its safety and effectiveness in fighting colon cancer, **psoriasis** and Alzheimer's disease. Scientists are working on developing nano-sized capsules containing the curry ingredient in an effort to improve its absorption and effectiveness in the body (Anonymous, 2009). Researchers at UCLA'S Jonsson Comprehensive Cancer Centre have found that curcumin suppresses a cell- signaling pathway in human saliva that drives the growth of head and neck cancer. The study shows that curcumin can work in the mouths of patients with head and neck malignancies and reduce activity that promote cancer growth. "And it not only affected the cancer by inhibiting a critical cell signaling path way, it also affected the saliva itself by reducing pro- inflammatory cytokins within the saliva." (Anonymous, 2011). If you have an uneven skin tone and would do anything to get that perfect peaches and cream complexion, then turmeric's what you need. Turmeric evens out the skin tone by reducing the traces of pigmentation. Mix a little turmeric powder alongwith cucumber juice or lemon juice and apply on skin. Wash it off after 15 minutes. But remember to do this everyday to notice a significant change in your skin colour.

Turmeric makes for an awesome body scrub. This is why most Indian brides used to smear a paste of turmeric and besan on their bodies before their wedding to get skin that is super smooth, baby soft and glowing. Application of turmeric and a little milk cream alsomakes the skin supple and improves its elasticity. Indians have been using turmeric to beautify themselves since ages now.

If you have an uneven **skin** tone and would do anything to get that perfect peaches and cream **complexion,** then turmeric's what you need. Turmeric evens out the skin tone by reducing the traces of pigmentation. Mix a little turmeric powder alongwith cucumber juice or lemon juice and apply on skin. Wash it off after 15 minutes. But remember to do this everyday to notice a significant change in your skin colour. The study published in the European Review for Medical and Pharmacological Science Journal has shown that turmeric is more effective than popular pain killers at easing the agony of sports injuries, as turmeric was able to ease injured rugby player's discomfort just as much paracetamol or Ibuprofen, but with out their side effects. The major component of the spice stimulated the gallbladder to produce bile, instantly making digestative sytem more efficient. It is also known to reduce symptoms of bloating and gass. Lipopoly saccharide - a substance in turmeric with anti- bacterial, anti- viral and anti- fungal agent helps stimulate the human immune system.

Best muscle building foods: The explanation is that the curcumin found in turmeric promotes muscle growth and repair. Add this muscle building food to all your Indian recipes.

Health Benefits

The benefits for heart health arise because of **curcumin**, an active ingredient which has several properties including anti-oxidant, anti-clotting, anti-inflammatory and anti-proliferative. Several studies have shown the effect of curcumin in decreasing blood cholesterol levels. Anti-oxidant properties of curcumin may also help prevent cardiovascular complications among diabetics

Turmeric is also used in Indian traditional medicine systems. It contains the flavonoid curcumin, known for its anti-inflammatory properties. It detoxifies the liver, fights allergies, stimulates digestion and boosts immunity. Traditionally, women used to have a bath using a paste of turmeric and oil as it is excellent for skin.

It has outstanding ability to reduce inflammation. Turmeric protects the liver from damage while encouraging liver cells to regenerate. It also increases bile and acid production and keeps the body toxins free.

The root has been in use since antiquity for its anti-inflammatory (painkiller), **carminative, anti-flatulent** and **anti-microbial** properties in India for ages.

Curcumin, a poly-phenolic compound, is the principal pigment that imparts deep orange colour to the turmeric. In vitro and animal studies have suggested the curcumin may have **anti-tumor, antioxidant, anti-arthritic, anti-amyloid, anti-ischemic**, and **anti-inflammatory** properties.

This popular herb contains no cholesterol; but is rich in anti-oxidants and dietary fibre, which helps to control blood LDL or "bad cholesterol" levels.

It is very rich source of many essential vitamins such as pyridoxine (vitamin-6), choline, niacin, and riboflavin *etc.*, which are essential for optimum health. 100 g herb provides 1.80 mg or 138 per cent of daily-recommended levels of pyridoxine. Pyridoxine is used in the treatment of homocystinuria, sideroblstic anemia and radiation sickness. Niacin helps prevent "pellagra" or dermatitis.

Fresh root contains very good levels of vitamin-C. It is a water-soluble vitamin and a powerful natural anti-oxidant; helps body develop immunity against infectious agents, helps fight type-1 diabetes and remove harmful free oxygen radicals from the body.

Turmeric contains very good amounts of minerals like calcium, iron, potassium, manganese, copper, zinc, and magnesium. Potassium is an important component of cell and body fluids that helps controlling heart rate and blood pressure. Manganese is used by the body as a co-factor for the antioxidant enzyme, *superoxide dismutase.*

Iron is an important co-factor for *cytochrome oxidase* enzymes at cellular level metabolisms and required for red blood cells productions.

It's powder is used in all vegetables and curries. It contains curcumin that helps nip pain

It enhances the effectiveness of **chemotheraphy** and reduces tumour growth. A report in the journal Stem Cell Research and Therapy found a compound in the curry spice turmeric may hold the key to repairing the brains of people with neurodegenerative diseases such as Alzheimer's. A team in Germany say aromatic turmerone promoted the proliferation of brain stem cells and their development into neurons during laboratory tests on rats (Ananymous, 2014).

Medicinal Uses

Research studies have suggested that *Curcumin,* a poly-phenolic compound, found in this herb may inhibit the multiplication of tumor cells including multiple myeloma, pancreatic and breast cancer, **leukemia** and colon cancer (Ipshita Mitra, 2012).

It contains many health benefiting essential oils such as *termerone, curlone, curumene, cineole, and p-cymene.* These compounds have applications in cosmetic industry.

Curcumin, along with other antioxidants, has been found to have anti-amyloid and anti-inflammatory properties. Thus; is effective in preventing or at least delaying the onset of **Alzheimer's disease.**

This root herb contains no **cholesterol;** but is rich in anti-oxidants and dietary fibre, which helps to control blood cholesterol levels, preventing coronary artery disease and stroke risk.

Early laboratory studies have been suggestive that turmeric is liver protective, anti-depressant, anti-retroviral effects.

It has been in use since a very long ago as an important ingredient in traditional Chinese and ayurvedic medicines for its anti-microbial, anti-inflammatory, carminative, anti-flatulent and anti-microbial.

GINGER (*Zingiber officinale*)

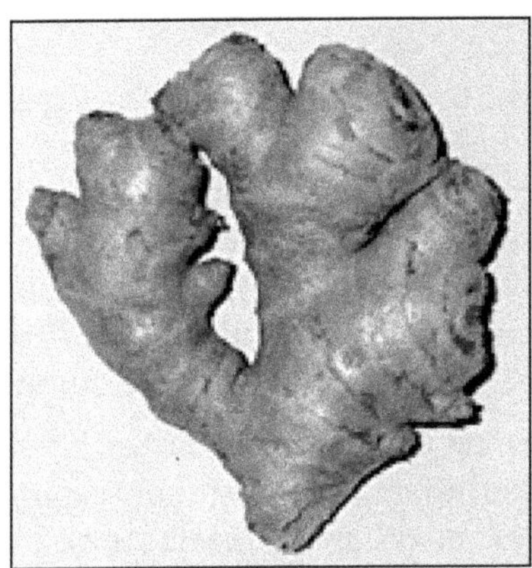

Ginger-Fresh Root

This humble spice boasts of medicinal qualities that help lowering cholesterol, boosts metabolism and kills cancer cells. It has long been used in food for its wonderful medicinal properties. If you are having cold, having a cup of 'kadak adraki chai' can give you instant relief. Similarly, on a cold winter morning, you can boil water with ginger, basil leaves, mint leaves and then add a spoon of honey. Even in traditional Indian dishes, ginger is added for giving the dish a unique flavor. Ginger's health benefits has made it an essential ingredient in Indian kitchens. It is the best source ofgingerol. It is known to increase the production of substances that protect the stomach lining, preventing the formation of ulcers; stimulates gastric activity, causing the stomach to empty more quickly, stimulates gall bladder, promoting healthy digestion and helps alleviate nausea caused by pregnancy and motion sickness. Ginger has anti-inflammatory properties and will also protect the stomach from irritation. Migraines and motion sickness can also be relieved by ginger juice. Ginger is stimulant that relieves flatulence, discharges mucus, strengthens digestion, stimulates glandular secretions and relieves vomiting.

The ginger is one of the most powerful foods when it comes to anti-inflammatory, anti-parasitic, anti-viral and anti-bacterial properties. It is full of vitamin C, many minerals and magnesium that have a positive effect over your health. In the past was used for better digestion, as a weapon against cardiovascular diseases, asthma, for relieving pain and improving the immune system.

According to Ayurveda, ginger (saunth) eradicates cold, cures swelling, and **oedema** and relieves pain. It strengthens nerves and cures vata. It quenches thirst, is purgative, enhances digestion and cures all digestive disorders; it makes wind go down and cures piles. It improves functioning of the circulatory system and purifies blood. Ginger is bitter and unctuous and so cures cough and **bronchial asthma**. It also stimulates sexual desire and is **anti-pyretic**.

British Journal of Nutrition has reported that an international team has carried out the study and found that curcumin, the major yellow constituent of turmeric, helps in reducing inflammation in many people suffering from bowel disease (Anonymous, 2009).

Our ancient Ayurvedic texts have described raw ginger as acrid, carminative, digestive and thermogenic agent whereas its dried form has been additionally cited to be possessing stomachic, expectorant, anthelmintic, rubefacient (counter-irritant if applied on the outer skin), anti-rheumatic, cardiac stimulant, aphrodisiac and anti-oxddant properties. Ginger has also been described as one of the best ama-pachak, meaning a substance helping to nullify toxins produced in the body due to improper digestion. Ginger increases sexdrive and stimulates sexual performance.

Dueto its anti-vata effect the use of ginger has been indicated in rheumatoid arthritis, chronic backache, inflammations, minor respiratory infections and cough and could. As a digestive aid no other herb seems to be a match to ginger, as it helps the digestion from the beginning till the end. Wide ranging, acute and chronic problems of the digestive tract such as anorexia, nausea, vomiting, flatulence, diarrhea, dysentery and colitis are known to be relieved by the judicious use of ginger (Vatsyayan, 2011).

Its root features knotty finger like projections that grow downward from the ground surface. Fresh raw root has silver gray outer surface. Cut sections feature creamy white, yellow, or red colored crunchy flesh depending upon the variety. The root often contains fibrils running through its center, especially in over-matured. Its pungent, spicy and aromatic smell is due to essential oils and phenolic compounds such as **gingerols** and **shogaols.** Ginger is the best pain killer having analgesic properties like the popular ibuprofen. It contains a quartet, gingerols, paradols, shogaols, and zingerone which are active ingredients to reduce pain. Drink ginger tea in monsoons and winter to get relief from that recurring pain. The bioactive compounds in ginger include **gingerol** and several other phenolic compounds. Gingerol is believed to relax blood vessels, stimulate blood flow and relieve pain. Ginger is also powerful anti-inflammatory agent, making it useful in heart diseases, diabetes and cancer. It has cholesterol lowering and anti-clotting properties

Health Benefits of Ginger Root

It can fight against infections and boost the immune system, because it contains many antioxidants.

It has warming properties and that is the reason why it is an efficient remedy against common cold, cold sores.

It also improves circulation and the delivery of oxygen, minerals and vitamins.

Relieves pain, flu, and headache and treats sore muscles.

Reduces the risk of heart attack.

Ginger has long been used as a natural treatment for colds and the flu. Many people also find ginger to be helpful in the case of stomach problems or food poisoning.

Ginger is good for curing digestive problems.

Ginger has anti-inflammatory properties and is a powerful natural painkiller.

Ginger has long been used as a natural heartburn remedy. It is most often taken in the form of tea for this purpose.

Research has shown that ginger may provide migraine relief due to its ability to stop **prostaglandins** from causing pain and inflammation in blood vessels.

Ginger is known to be beneficial in reducing the symptoms of morning sickness.

It is beneficial for treatment of **rheumatoid arthrities**.

To seek relief from menstrual cramps and heart burns – to sip on ginger tea.

It provides a great remedyfor cold and cough.

To curb motion sickness chew ginger candy.

Ginger aids in digestion. Either buy powdered ginger in capsule forms or add the herb to your recipes.

Ginger has been in use since ancient times for its **anti-inflammatory, carminative, anti-flatulent,** and **anti-microbial properties.**

The root contains many health benefiting essential oils such as *gingerol, zingerone, shogaol, farnesene and small amounts of* β*-phelladrene, cineol,* and *citral.*

Gingerols help improve the intestinal motility and has anti-inflammatory, painkiller (analgesic), nerve soothing, anti-pyretic as well as anti-bacterial properties. Studies have shown that it may reduce nausea induced by motion sickness or pregnancy and may help relieve migraine headaches.

Zingerone, which gives pungent character to the ginger root, has been found to be effective against E.coli induced diarrhea, especially in children.

This herb root is low in calories and contains no cholesterol, but is very rich source of many essential nutrients and vitamins such as pyridoxine (vitamin B6), pantothenic acid (vit. B-5) that are essential for optimum health.

It also contains good amount of minerals like potassium, manganese, copper, and magnesium. Potassium is an important component of cell and body fluids that helps controlling heart rate and blood pressure.

Ginger tea can be magical for a bad throat. It has now been found to be highly effective in treating asthma symptoms. Study shows that purified components of the spicy root also may have properties that help asthma patients breathe more easily (Sinah Kaunteya, 2013).

Relieves nausea

Motion sickness

Fights ovarian cancer

Lowers cholesterol

Prevents blood clots and arthritis

Medicinal Uses

Ginger root slices boiled in hot water with added lemon or orange juice and honey is a popular herbal drink in ayurvedic medicine to relieve common cold, cough, and sore throat.

It is also used as vehicle in many ayurvedic decoctions to mask bitterness and alter taste.

Gingerols increase the motility of the gastrointestinal tract and have analgesic, sedative, anti-inflammatory, and antibacterial properties. Studies have shown that it may reduce nausea caused by motion sickness or pregnancy and may relieve migraine.

BEANS (*Phaseolus vulgaris*)

Bean Plants with Pods

Broadly, tender beans are classified depending on their growth habits as "bush beans" which stand erect without support and "Pole Beans" that need climb supports. The other varieties of unripe fruits of beans family include Shell beans, Pinto or mottled beans, White beans, Red or kidney beans, Black beans, Pink beans and Yellow wax beans. All most all the varieties are available year around across the world.

Beans are not just vegetables- beans are a nutrient rich power house. Soybean is the best source of **genistein** (an iso-flavonoid), phytosterols and saponins. It suppresses the growth of cancer cells in the large intestine and enhances immunity against many cancers. Other compounds in soybean also reduce blood cholesterol level. Beans and other legumes provide a nice sampling of B-vitamins, including **B6,** BI2 and folic acid, all of which help the body regulate sleep cycles and produce relaxing **Serotonin**. In fact, studies have shown that boosting B vitamins may help people with insomnia. Beans have more fibre and less sugar. Hence they play an important role in bringing down the risk of diabetes while simultaneously controlling the body's cholesterol level.

Many edible beans, including broad beans and soybeans, contain **oligosaccharides** (particularly **Raffinose** and **Stachyose**) a type of sugar molecule also found in cabbage. An anti- oligosaccharide enzyme is necessary to properly digest these sugar molecules. As a normal human digestive tract does not contain any anti-oligosaccharide enzymes, consumed oligosaccharides are typically digested by bacteria in the large intestine. This digestion process produces flatulence- causing gases as a by product. Because of their nutrient content, beans and pea are the only foods that appear in two food groups. Meat and Beans and Vegetable's.

Green beans are a nutritious vegetable that are rich in vitamins, minerals, and phyto-nutrients such as vitamins A and C, calcium, iron, manganese, beta-carotene, and protein. Green beans provide significant cardiovascular benefits due to their **omega-3 (alpha-linolenic acid)** content. They also contain ant-inflammatory compounds which make them highly beneficial for individuals who suffer with auto-immune disorders such as fibromyalgia, arthritis, COPD, chronic fatigue syndrome, irritable bowl syndrome, chronic sinusitis, bursitis, Raynaud's syndrome and lupus. They are also known to help prevent type 2 diabetes. Green beans are an excellent source of dietary fibre and can aid the digestive tract by promoting regular peristaltic action and aid in the removal toxic, cancer-causing substances in the digestive tract. They contain a wide variety of carotenoids such as lutein and neoxanthin and flavonoids such asquercetin and procyanidins which make them excellent for eye health and for preventing disease. Green beans can be snacked on raw, added to salads or soups, or steamed. Consider trying fresh green beans drizzled with olive oil, seasoned with your favorite spices, and roasted in the oven for 30 minutes for a healthy alternative to french fries.

Beans are in the Meat and Beans group, because they are a good source of protein and iron. Even better, beans provide a low fat, saturated fat-free, and cholesterol free source of protein.

Beans are listed in the Vegetable's group because they are a plant based food that provides fibre, folate, potassium and anti-oxidants. Researchers in Canada have found that proteins from a common garden pea could be used as a food additive or new dietary supplement to fight high blood pressure and chronic kidney disease (CKD). Researchers where of the view that the study is the first to report that a natural food product can relieve symptoms of CKD. "In people with high blood pressure, our protein could potentially delay or prevent the onset of kidney damage. In people who already have kidney disease, our protein may help them maintain normal blood pressure levels so they can live longer (Anonymous, 2009)."

Bean sprouts are sprinkled on salads or eaten as between meals and snacks. Sprouts are a great way of getting some nutrients. They are rich in vitamin C and a handful of fresh sprouts can provide you with three quarters of your daily vitamin-C requirement. In addition they also contain vitamin A, B and E as well as calcium, iron and potassium. Being high in protein, **beans** should be a regular feature in your diet. They are also very lean. Beans also provide the fatty acids your nails need to stay strong and prevent splitting. Excellent for the heart, ½ a cup of beans added in our diet (in the form of any dish or soup), helps in lowering cholesterol levels by 8 per cent. You should try black, kidney, or pinto beans; each one of them supplies about one-third of your day's fibre needs.

Health Benefits

Protects your heart: Studies have shown that the soluble fibre present in the beans is helpful in reducing the bad **cholesterol** levels. In addition, it also reduces inflammation and blood pressure.

Speeds up blood clotting: The vitamin K present in green beans helps speed up the healing process in your body. For example, if you are injured, vitamin K can help stimulate the clotting of blood which in turn stops excessive bleeding. Making beans a part of your daily diet means more intake of vitamin K. Vitamin K helps absorb calcium which in turn can help in prevention of bone density loss and osteoporosis?

Premenstrual Syndrome: Green beans are packed with manganese which helps alleviate PMS symptoms such as irritability and mood swings. In addition, eating beans also helps in relieving symptoms of osteoarthritis and osteoporosis.

Delay aging: Want to delay aging, avoid wrinkles and fine lines? Add green beans to your daily diet. Green beans are a great source of vitamin A and equally great for your skin. So, eat more beans and get rid of that dull skin and age spots now!

Rich dietary fibre: Green beans are rich in fibre and if you are suffering from problems such as constipation or irritable bowel syndrome, this vegetable is a must in your diet. Green beans also help in regulating your blood sugar levels and so if you are diabetic have some every day.

Low fat: If you are trying to lose weight or maintain it, green beans are ideal as they are low in fat. Add green beans to your diet as it gives you all the essential nutrients without any extra calories. You can make a side dish to go with your rotis or add them to your curries and soups.

For pregnant women: Green beans is said to be good for pregnant women as it helps in the healthy development of the baby's heart. Studies show that green beans also help protect babies from conditions such as asthma.

Beans have less sugar and more fibre and protein than any other vegetable. Hence they play an important role in bringing down the risk of diabetes while simultaneously controlling the body's cholesterol level.

Beans are good to excellent sources of eight important nutrients that many people do not get enough of their diet.

Beans are naturally cholesterol and saturated fat free.

Research suggests that beans can play a role in promoting health and preventing diseases, including reducing cancer risk, improving cardiovascular health, controlling blood sugar and managing weight.

Beans contain a lot of fibre that help reduce the absorption of fat in the body.

Beans also contain high amounts of protein that is beneficial in building body muscle. Body muscle cells are known to burn calories at much higher rate.

While beans are often associated with the gastrointestinal disturbances therefore, need to be cooked thoroughly because our digestive tracks are not adapted to breaking down some proteins that are contained in certain beans.

Some of the best kinds of beans to eat are:

Navy beans

Kidney beans

White beans

Lima beans

Green Beans Nutrition Facts

Tender, flexible green beans are delight of vegetarian lovers for their wholesome nutritional qualities. They are unripe or immature pods obtained from the bean plant belonging to common *fabaceae* family and known scientifically as *Phaseolus vulgaris.*

Fresh green beans are very low in calories (31 kcal per 100 g of raw beans) and contain no saturated fat; but are very good source of vitamins, minerals, and plant derived micronutrients.

They are very rich source of dietary fibre (9 per cent per100g RDA) which acts as bulk laxative that helps to protect the mucous membrane of the colon by decreasing its exposure time to toxic substances as well as by binding to cancer causing chemicals in the colon. Dietary fibre has also been shown to reduce blood cholesterol levels by decreasing re-absorption of cholesterol binding bile acids in the colon.

Green beans contain excellent levels of vitamin A, and many health promoting flavonoid poly phenolic antioxidants such as lutein, zeaxanthin and β-carotene in good amounts. These compounds help act as protective scavengers against oxygen-derived free radicals and reactive oxygen species (ROS) that play a role in aging and various disease process.

Zea-xanthin, an important dietary carotenoid in the beans, selectively absorbed into the retinal macula lutea in the eyes where it thought to provide antioxidant and protective UV light filtering functions. It is, therefore, green beans offer some protection in preventing *age related macular disease* (ARMD) in the elderly.

Fresh snap beans are good source of folates. 100 g fresh beans provide 37 µg or 9 per cent of folates. Folate along with vitamin B-12 is one of the essential components of DNA synthesis and cell division. Good folate diet when given during preconception periods and during pregnancy helps prevent from neural-tube defects in the offspring.

They also contain good amounts of vitamin-B6 (pyridoxine), thiamin (vitamin B-1), and vitamin-C. Consumption of foods rich in vitamin C helps body develop resistance against infectious agents and scavenge harmful oxygen free radicals.

In addition, beans contain healthy amounts of minerals like **iron**, calcium, magnesium, manganese, and potassium, which are very essential for body metabolism. Manganese is a co-factor for the anti oxidant enzyme *superoxide* *dismutase,* which is a very powerful free radical scavenger. Potassium is an important component of cell and body fluids that helps controlling heart rate and blood pressure.

Fresh beans contain vitamin A, B-complex vitamins, calcium and potassium.

Green beans are diuretic and may be used to treat diabetes.

A fresh bean should snap crisply and feels velvety to the touch.

Caution

Green beans contain oxalic acid, a naturally occurring substance found in some vegetables, which, may crystallize as oxalate stones in the urinary tract in some people. It is, therefore, people with known oxalate urinary tract stones are advised against eating vegetables belong to brassica and fabaceae family. Adequate intake of water is therefore advised to maintain normal urine output to minimize the stone risk.

PEA (*Pisum sativum*)

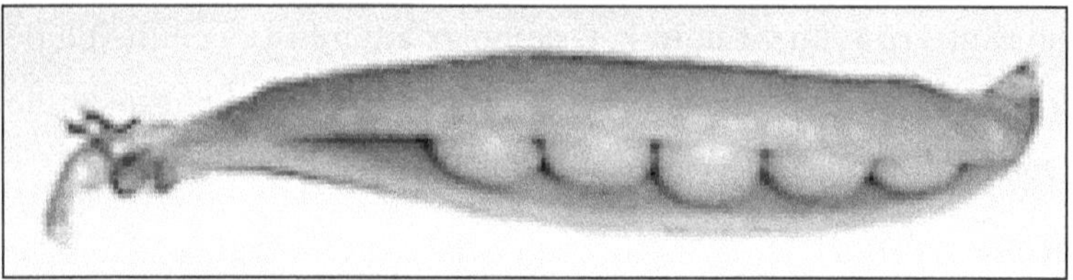

Green Pea Pod

Fresh peas are sweet and juicy and one of the most digestible and non-gassy legumes. They contain 78 per cent water, with only traces of fat, are low in sodium and provide a good source of iron and vitamin A, C, B1, B2 and niacin. Peas have a diuretic and mildly laxative effect and are recommended for strengthening digestion, reducing water retention and helping to promote elimination. They are popular as a summer vegetable and also make a nutritious addition to salads, soups and casseroles. Pea tendrils are also edible. They are tender top shoots of young pea plants, have similar pea taste and favored in cooking as well in salads.

Health Benefits

Peas are one of the most nutritious leguminous vegetable, rich in health benefiting phyto-nutrients, minerals, vitamins and **anti-oxidants**.

Peas are relatively low in calories when compared with beans, and cowpeas. 100 g of green peas provide only 81 calories, contain good amount of soluble and insoluble fibre but contains no cholesterol.

Fresh pea pods are excellent source of folic acid. 100 g provides 65 mcg or 16 per cent of recommended daily levels of folates. Folates are B-complex vitamins required for DNA synthesis inside the cell. Well established research studies suggest that adequate folate rich foods in expectant mothers would help prevent neural tube defects in the newborn babies.

Fresh green peas are very good in ascorbic acid (vitamin C). Contain 40 mcg/100 g or 67 per cent of daily requirement of vitamin C. Vitamin C is a powerful natural water-soluble anti-oxidant. Vegetables rich in this vitamin helps body develop resistance against infectious agents and scavenge harmful, pro-inflammatory free radicals from the body.

Peas contain phytosterols especiallyβ-sitosterol. Studies suggest that vegetables like legumes, fruits and cereals rich in plant sterols help lower cholesterol levels in the body.

Garden peas are also good in vitamin K. 100 g of fresh leaves contain about 24.8 mcg or about 21 per cent of daily requirement of vitamin K-1 (phylloquinone). Vitamin K has found to have potential role in bone mass building function by promoting osteo-trophic activity in the bone. It also has established role in Alzheimer's disease patients by limiting neuronal damage in the brain.

Fresh green peas also contain adequate amounts of anti-oxidants flavonoids such as carotenes, lutein and zeaxanthin as well as vitamin-A. Vitamin A is essential nutrient which is required for maintaining healthy mucus membranes and skin and is also essential for vision. Consumption of natural fruits rich in flavonoids helps to protect from lung and oral cavity cancers.

In addition to folates, peas are also good in many other essential B-complex vitamins such as pantothenic acid, niacin, thiamin, and pyridoxine. Furthermore, they are rich source of many minerals such as calcium, iron, copper, zinc and manganese.

CARROT (*Daucus carota*)

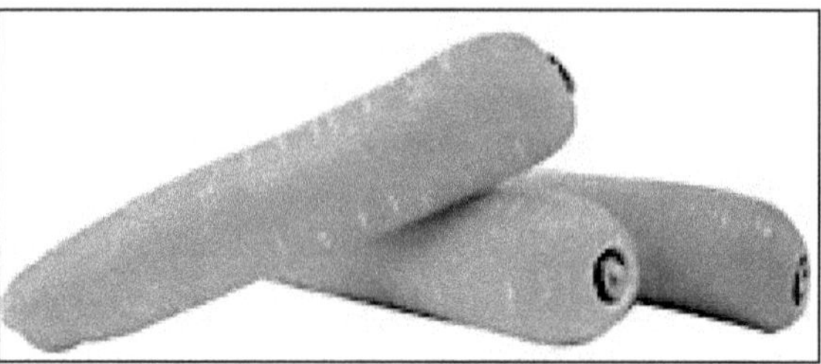

Orange and Cylindrical Type Carrots

Naturally sweet, delicious and crunchy, carrots are healthy additions you can make to the vegetable list in your diet. Indeed these root vegetables comes with wholesome health benefiting compounds such as beta-carotenes, vitamin A, minerals and anti-oxidants in ample amount. Carrots are full of nourishing properties for eyes, hair and skin. They are the best source of pro-vitamin and A carotenes. Known for improving night vision, carrots are rich in Vitamin C, Vitamin A which helps in regulating healthy blood sugar levels - good for the heart. Carrots are considered a goldmine of natural vitamins and nutrients. Among other raw vegetable juices, carrot juice is a sure shot winner. Carrot juice is known for its excellent cleansing effects and, in addition, it was considered a good remedy for constipation and fatigue. As we all know, no other vegetable contains as much beta-carotene as carrots. In our body, beta-carotene is converted into Vitamin A, which assists in improving our eyesight and the functioning of our immune system, strengthening our bones and teeth, preventing possible problems with the functioning of the thyroid gland. Carrot juice has very good anti -inflammatory, anti-cancerous and anti-aging properties. Carotene is a known anti-oxidant, which can prevent degenerative processes in the cells and has great anti-aging effects. Also, drinking raw carrot juice can be helpful for lowering the symptoms of stresses and normalizing the function of your digestive system.

Carrot is a good of source beta-carotene which our body converts into vitamin A and minerals such as potassium and manganese. It contains some amount of sodium, iron, zinc, copper and calcium. It contains about 87 per cent of water, rich in mineral salts and Vitamin (B, C, D, E). Raw carrots are an excellent source of vitamin A and potassium. Even though the colour of original carrots is orange it grows in other colours including white, yellow, red or purple. Carrot juice reduces the risk of many different types of cancer including skin and breast cancer.

Carrot juice is immensely beneficial. It is diuretic in action and helps in case of scanty urination and menstruation. Its juice eliminates unwanted uric acid from the blood, is thus helpful to gout patients. Carrots contain a high level of carotenoids and antioxidants. Psoralins and coumarins found in umbelliferous vegetables such as celery, parsley and parsnip bind to hepatic enzymes and inhibit mutagenicity in the experimental animals. Carotenoids - α- carotene, β- carotene, lycopene, lutein and zeaxanthin, present in carrots, tomatoes and corn have been shown to be anti carcinogenic in nature. Micro nutrients such selenium, and zinc repair the damage done by mutations likewise thiamine and vitamin- B_6 inhibit human platelet aggregation. Tannins and phenolic compounds found in vegetables are antimutagenic in humans [Hunang and Toshihiko, 1994). In **carrots** and other yellow or brown- fleshed vegetables such as pumpkin, amaranthus, capsicum, sweet potato, watermelon and the like, the colour is imparted by **carotenoids**, of which only vitamins B and C are water soluble and vitamins A, D, E and K are fat soluble. While as the orange-coloured carrots contains a high concentration of **beta-carotene**, which is a substance that is converted to vitamin- A in the human body. A half cup serving of cooked carrots contain four times the recommended daily intake of vitamin A in the form of protective **beta-carotene.** Beta-carotene is also a powerful antioxidant effective in fighting against some forms of cancer, especially lung cancer and protect against stroke and heart disease. Carrots are also very good source of dietary fibre, vitamin C, vitamin K, foliate, vitamin B6, pantothenic acid, iron and manganese. But to obtain maximum benefit it is best to eat them raw. Carrot juice is a excellent source of beta-carotene, a vitamin A precursor. They are also excellent sources of vitamin B1 and vitamin B2 and of the minerals potassium, sodium, silicon and anti-cancer nutrients, including **phthalide** and **glutathione** (antioxidants). A new Super-A carrots, intensely orange with sweet taste, 40 per cent richer in beta-carotene, an oxidant the body converts into vitamin and boosts health were developed by growers in Shropshire- London (Anonymous, 2009).

Carrots are excellent source of anti-oxidant compounds and the richest vegetable source of the provitamin A carotenes. Carrot's oxidant compounds help protect against cardiovascular disease and cancer and also promote good vision, especially night vision, inhibit cataracts and treat indigestion. Research demonstrated that high carotenoid diets are associated with a reduced risk of heart disease. After beta carotene is converted to vitamin A in the liver, it travels to the retina where it is transformed into **rhodopsin** – a purple pigment that is necessary for night vision. Plus beta carotene's powerful anti-oxidant actions help provide protection against muscular degeneration and the development of senile cataracts, the leading cause of blindness in the elderly. Carrots are by for one of the richest source of carotenenoids- just one cup provides **16, 679** IUs of beta carotene and **3, 432** REs (retinol equivalent), or roughly **686.3 per cent** the RDA for vitamin A. A high carotene intake has been linked with a 20 per cent decrease in post menopausal breast cancer and upto 50 per cent decrease in the incidence of cancers of bladder, cervix, prostate, colon, larynx and oesophagus. Research has suggested that physiological level's as well as dietary intake of carotenoids may be inversely associated with insulin resistance and high blood sugar levels. According to Acharya Balkrishna (2008), carrot enhances body heat and promotes digestion. It cures bleeding diathesis, piles, sprue and irritable bowel syndrome and cough. It also cures acidity and gas formation. While according to Raj Ninghantu, itcures

flatulence. It is anthelmintic and cures burning sensation and pain. It quenches thirst and cures pitta.

Carrots also contain a phyto-nutrient called **falcarinol** that may be responsible for the recognized epidemiological association between frequently eating carrots and a reduced risk of cancers. **Falcarinol** provides protection against colon cancer a study published in the Journal of Agricultural and Food Chemistry. However, massive over consumption of carrot's, can cause hyper- carotenemia, a condition in which the skin turns orange (although hyper-carotenemia is not itself dangerous, unlike over dose of vitamin A, which can cause liver damage). Researchers at the University of Newcastlefound that "boiled- before-cut" carrots contain 25 per cent more of anti-cancer compound **falcarinol** than those that were chopped up first. And the sugars which give the carrots its distinctively sweet flavour were also found in higher concentrations in the carrot that had been cooked whole, so the vegetable tasted better as well being healthier (Jeremy Laurance, 200p)

Carrots are nutritional heroes; they store a gold mine of nutrients. No other vegetable contains as many carotene as carrots, which our body coverts into vitamin A. For maintaining optimum health and weight loss carrots are ideal. ½ cup carrot peeled and boiled have carbs 4, fibre 2gms, energy 80kj. Cooked carrots are an excellent source of vitamin A, it contains Vitamin B6. Copper, Folic acid. The high level of beta-carotene is very important and gives carrots their distinctive orange color. In order to assimilate the greatest quantity of the nutrients present in carrots, it is important to chew them well. Choose bright colour carrots, smooth ones and medium size carrots as these are fresh. Carrots also contain minerals and vitamins like thiamin, niacin, vitamin B6, folate and manganese. Carrots are also a good source of fibre, vitamin A, vitamin C, vitamin K and potassium

Benefit's of Carrot in our Body

Carrots are considered superfoods for fertility and help boost the quality and quantity of sperm. But there are far more health benefits of eating carrots.

Ageing: Carrots are considered anti-ageing foods as they are rich in beta-carotene, an antioxidant that fights free radicals.

Boost beauty: Carrots are rich in antioxidants and vitamin A, both responsible for healthy skin, hair and nails.

Fight cardiovascular diseases: Carrots consists of beta-carotene, alpha-carotene and lutein are antioxidants that fight cholesterol. Carrots are also fibrous; hence it helps soak out bad cholesterol.

Dental health: Consumption of carrots improves the dental health by clearing plaque. Biting on carrots increases the production of saliva, hence balances the acid level which fight cavity-causing bacteria.

Improves liver health: Carrots are considered detox food, as it cleanses the liver. Carrots help reduce bile and fat storage in the liver.

Infection: Raw carrot or boiled are applied on cuts and wounds as an antiseptic

Carrot greens are high in vitamin K and pectin which may have cholesterol lowering properties.

Phytonutrients in carrot called polyacetylenes include compounds such as falcarinol and falcarindiol have been found to have anti-inflammatory property, keep blood cells from clumping together and inhibit the growth of cancer cells.

A single medium carrot delivers almost twice the DV of vitamin A, which is associated with improving the vision.

While carrots are good for our eyes, they also promote healthy skin, delay ageing and even prevent cancer. Initially used as medicine, carrots are commonly found in orange colour. But they are also available in yellow, red, white and purple hues.

Several studies show that certain properties in carrots help reduce the risk of lung, breast and colon cancer.

The beta-carotene in carrots is converted to vitamin A in the liver, which is then transformed in your retina, to rhodopsin – a pigment that improves night vision. Beta-carotene is also known to protect against macular degeneration and senile cataracts.

Carrots help slow down the ageing of cells and also work as an antioxidant.

The vitamin A present in carrots protects your skin from sun damage and even prevents premature pigmentation, wrinkling, dry skin, acne and uneven skin tone.

You can also use boiled or mashed carrots on cuts and bruises to prevent infection.

Boosting immunity (especially among older people), strengthening bones and teeth, preventing possible problems with the functioning of the thyroid gland.

Reducing photosensitivity (beta carotene protects the skin from sun damage).

Helping to heal minor wounds and injuries.

Reducing the risk of heart disease and high blood pressure.

Cleansing the liver when consumed regularly, can help the liver excrete fats and bile

Fighting bronchitis.

Fighting infection (vitamin A helps cell membranes healthy, making them stronger against disease causing micro-organisms.

Carrots boast of alpha-carotene, beta-carotene, and beta-cryptoxanthin. Eat them raw in the form of salad.

Improves muscle, flesh and skin health.

Its juice has very good anti-inflammatory, anti- cancer and anti- ageing properties.

Carrot juice can be helpful for lowering the symptoms of stress.

Helping fight enema.

Reducing acne

Improving eye health *i.e.* help night vision, inhibit cataracts.

Useful in treating infections of lung, digestive system and urinary tract.

Diet of raw or cooked carrots can improve skin appearance and calcium absorption.

Sweet and succulent carrot are notably rich in anti-oxidants, vitamins and dietary fibre; however, they provide only 41 calories per 100 g, negligible amount of fat and no cholesterol.

They are exceptionally rich source of **carotenes and vitamin-A**. Studies have found that flavonoid compounds in carrots help protect from skin, lung and oral cavity cancers.

Carotenes are converted in to vitamin A in the liver. Beta-carotene is the major carotene that is present in these roots. Beta carotene is one of the powerful natural anti-oxidant helps protect body from harmful fee radical injury. In addition, it also has all the functions of vitamin A such as vision, reproduction (sperm production), maintenance of epithelial integrity, growth and development.

Carrots are rich in **poly-acetylene anti-oxidant falcarinol**. Falcarinol in carrots may help fight against cancers by destroying pre-cancerous cells in the tumors.

Fresh roots are also good in vitamin C, which is s water soluble anti-oxidant. It helps ody maintain healthy connective tissue, teeth and gum. Its anti-oxidant property helps body protect from diseases and cancers by scavenging harmful free radicals.

This root vegetable is especially contain good amounts of many B-complex group of vitamins such as folic acid, vitamin B-6 (pyridoxine), thiamin, pantothenic acid, *etc.* that acts as co-factors to enzymes during substrate metabolism in the body.

It also has healthy levels of minerals like copper, calcium, potassium, manganese and phosphorus. Potassium is an important component of cell and body fluids that helps controlling heart rate and blood pressure by countering effects of sodium. Manganese is used by the body as a co-factor for the antioxidant enzyme, *superoxide dismutase.*

Caution

It is recommended that the intake of carrot juice should be limited to no more than four cups a day since over- consumption can cause Yellowing of the skin, a condition known as Xanthosis.

RADDISH (*Raphanus sativus*)

Radish-Scarlet Red

Radishes come in different forms varying in size, colour and duration of required cultivation time. Radishes can be broadly categorized into four main types-summers, fall, winter, and spring while growers classify them by shapes, colors, and sizes, such as black or white colored radishes, with round or elongated roots. Top greens can also be used as food. Radish are suggested as an alternative treatment for a variety of ailments including whooping cough, cancer, coughs, gastric discomfort, liver problems, constipation, dyspepsia, gallbladder problems, arthritis, gall stone, kidney stones and intestinal parasites. Radish are rich in ascorbic acid, folic acid and potassium. They are a good source of vitamin B6, riboflavin, magnesium, copper and calcium. One cup of sliced radish bulbs provide approximately 20 calories or less coming largely from charbohydrates, making radishes relative to their size, a very filling food for their caloric value.

Radishes and their juice are an old home remedy for cough, rheumatism and gall bladder problems. The leaves of the plant, which are usually discarded, are very nutritious, containing almost 10 times as much vitamin C as the roots. In Yemenite folk medicine, radishes are used to eliminate kidney stones. Half a cup of fresh radish juice each morning on an empty stomach usually dissolves even stubborn stones, passing then out of the body in the urine.

A close look reveals that it is largely made up of water (more than 90 per cent) yet contains as much as potassium as banana's and about half the ascorbic acid of oranges. It is an excellent source of vitamin C and K. It has a excellent levels of copper, manganese, potassium. It is a good source of calcium, magnesium, iron, phosphorus, zinc and sodium. Radish is low in saturated fat and very low in cholesterol but very good source of dietary fibre. It has anti-bacterial and anti-viral properties. It is beneficial for cough, respiratory problems, digestive disorders, asthama, bronchitis and liver and gall bladder troubles. According to Ayurved (Traditional System of Medicine developed over millennium by sages and seers of ancient India) radish promotes digestion, is anthelmintic, cures digestive disorders, piles, and all kinds of swelling. Good for heart, hiccoughs, dermatoses, cholera and in secondary amenorrhoea. It also cures tuberculosis, phthisis, bronchial asthama, eye problems, abdominal pain and kapha and pitta disorders. It also cures blood disorders and enhances appetite. Radishes have antibacterial and anti-fungal properties. They are a member of the cabbage family. Radishes contain vitamin C, potassium and other trace minerals

Health Benefits

It is an astringent and diuretic and is used to promote bile flow.

Radish is low in saturated fat and cholesterol.

High in dietary fibre, vitamin C, folate, vitamin K, riboflavin, vitamin B6, calcium, magnesium, copper and manganese.

The nutritional value of radish makes it ideal for maintaining optimum health and weight loss, by increasing your metabolism. It is low in calories and has a low glycaemic index (GI), which helps in easy digestion of food and thus helping in weight loss.

Radish helps in preventing Urinary Disorders. It is a diuretic *i.e.* increases production of urine. It is helpful in the treatment of dysuria or painful urination.

It is beneficial as a carminative for relieving gastric discomforts, and also serves as a laxative and stimulant. They are high in dietary fibres that make them ideal for people who have dietary issues. The water content is high in radish and thus eating it, can also make you full for a long time.

Avoid including too many radishes in your diet if you are interested in weight gain.

Radishes are very low calorie root vegetables, contains only 16 calories per 100 g. However they are very good source of anti-oxidants, minerals, vitamins and dietary fibre.

Fresh Raddishes are rich in vitamin C; provide about 15 mg or 25 per cent of DRI of vitamin C per 100 g. Vitamin C is a powerful water soluble anti-oxidant required by the body for synthesis of collagen. Vitamin C helps body scavenge harmful free radicals, prevention from cancers, inflammation and helps boost immunity.

In addition, they contain adequate levels of folates, vitamin B-6, riboflavin, thiamin and minerals such as iron, magnesium, copper, calcium and anthocyanins that acts as a detoxifier for your body. The anti-oxidants and isocyanates kills the cancerous cells and stops their multiplication. Thus its juice is recommended for cancer prevention.

They contain many phytochemicals like indoles which are detoxifying agents and zeaxanthin, lutein and beta carotene which are flavonoid antioxidants.

Radishes are very low calorie root vegetables; contains only 16 calories per 100 g. However they are very good source of anti-oxidants, minerals, vitamins and dietary fibre.

Fresh Raddishes are rich in vitamin C; provide about 15 mg or 25 per cent of DRI of vitamin C per 100 g. Vitamin C is a powerful water soluble anti-oxidant required by the body for synthesis of collagen. Vitamin C helps body scavenge harmful free radicals, prevention from cancers, inflammation and helps boost immunity.

In addition, they contain adequate levels of folates, vitamin B-6, riboflavin, thiamin and minerals such as iron, magnesium, copper and calcium.

They contain many phytochemicals like indoles which are detoxifying agents and zeaxanthin, lutein and beta carotene which are flavonoid antioxidants.

Vital minerals like zinc and phosphorus make your skin look naturally glowing. Also radishes are rich in water content that keeps your body hydrated, which in turn makes your skin healthy and youthful. As radishes containe 95 per cent watert, helps in moisturizing the skin well. Being rich in vitamin A, C and K, it keeps various skin disorders at bay and eating radish helps in better cell production and repairs.

Eating radish can lower the body temperature and relieve inflammation caused by fever. It also has anti-bacterial properties that heal your body quickly.

Because of its potassium, it lowers the blood pressure levels, as it relaxes the blood levels, hences increasing the blood flow.

According to Ayurved (Traditional System of Medicine developed over millennium by sages and seers of ancient India) radish promotes digestion, is anthelmintic, cures digestive disorders, piles, and all kinds of swelling. Good for heart, hiccoughs, dermatoses, cholera and in secondary amenorrhoea. It also cures tuberculosis, phthisis, bronchial asthama, eye problems, abdominal pain and kapha and pitta disorders. It also cures blood disorders and enhances appetite.

Caution: Radishes are not recommended for people with gastro- intestinal inflammations or ulcers. Radishes may contain goitrogens, a plant based compounds found in cruciferous and brassica family vegetables like cauliflower, broccoli *etc.* Goitrogens may cause swelling of thyroid gland and should be avoided in individuals with thyroid dysfunction. However, they may be used liberally in healthy persons.

TURNIP (*Brassica campestris*)

Turnip-PTWG

Although this bulbous root that is widely eaten; it is its top fresh greens that are more nutritious; several times richer in vitamins, minerals, and anti-oxidants. The root is rich in vitamins A, B and C and minerals such as sulphur, calcium, potassium, sodium and phosphorus. It is also rich in fibre and contains far fewer calories than potato. A serving of 100g has only 22 calories. Turnip greens supply many times the nutritional content of the root. They are excellent sources of vitamins and minerals. Turnips detroxify the body and alkalize the blood, promoting sweating, releasing mucous and improving the appetite. They are also generally beneficial in conditions such as indigestion, diabetes and jaundice. Turnip greens have traditionally been used in Asia in the treatment of lung congestion, bronchitis, asthma and sinus problems. Although raw turnips have a somewhat pungent smell, this is neutralized in cooking. Although the leafy greens of the turnip are also edible, we mostly consume its bulbous roots with a purple top. This vegetable also abounds in precious nutrients. It has a high content of sulforaphane, a compound with a bitter taste, which gives turnip a cancer fighting power. It has a high amount of fibre that provides multiple health benefits. It aids in weight loss by keeping you full. It keeps blood sugar stable, promotes a healthy digestive tract and prevents inflammation in the colon.

Health Benefits

The dietary nitrates in turnip reduce blood pressure and improves endothelial dysfunction.

Turnip also contains a lot of vitamin C. Eating two medium turnips meets vitamin C needs for the entire day.

Apart from boosting the immune system in the fight against colds and flue, vitamin C also keeps eyes healthy by protecting them against UV light damage.

Turnips are very low calorie root vegetables; contains only 28 calories per 100 g. However, they are very good source of anti-oxidants, minerals, vitamins and dietary fibre.

Fresh roots are indeed one of the vegetables rich in vitamin C; provide about 21mg or 35 per cent of DRA of vitamin C per 100 g. Vitamin-C is a powerful water-soluble anti-oxidant required by the body for synthesis of collagen. It also helps body scavenge harmful free radicals, prevents from cancers, inflammation, and helps boost immunity.

Turnip greens are the storehouse of many vital nutrients, in fact several times than the roots. The greens are very rich in antioxidants like vitamin A, vitamin C, arytenoidsxanthenes and lute in. In addition, the greens are excellent source of vitamin K.

Top greens are also very good source of B-complex group of vitamins such as foliates, riboflavin, pyridoxine, pantothenic acid and thiamin.

Fresh greens are also excellent sources of important minerals like calcium, copper, iron and manganese.

Caution

Turnips and top greens are generally very safe including in pregnant women. However, the roots and top greens contain small amounts oxalic acid (0.21 g per 100 g), a naturally occurring substance found in some vegetables belonging to brassica family which may crystallize as oxalate stones in the urinary tract in some people. It is therefore, those with known oxalate urinary tract stones may have to avoid eating them. Adequate intake of water is therefore advised to maintain normal urine output in these individuals to minimize the stone risk.

HORSE RADISH (*Armoracia rusticana*)

Horse Radish Roots - Long Coarse Tapering Roots

Horseradish is a long, tapering root used as a condiment in the kitchens. The root has strong, hot, and sharp flavor, which can be only be described after experiencing its unique taste! The plant is a small perennial herbgrown as annual field crop for its thick, rough, fleshy roots in many parts of Europe, America, and Asia.

Valued for the pungent effect of its long, white root, horse radish is put to good use as a flavouring and as a medicinal agent, especially in Japan. It contains mustard oil (**allylisothiocyanate),** which in fact an irritant. It alleviates colds, snotty noses and sinusitis by promoting nasal discharge and releasing phlegm from lungs.

Health Benefits

Horseradish is low in calories and fat; but contains good amount of dietary fibre, vitamins, minerals, and anti-oxidants. The active principles in the root found to have anti-inflammatory, *diuretic* (increase urine output), and nerve soothing effects.

The root contains many volatile phyto-chemical compounds, which give its much-famed pungent character. Some of the major constituents in the root are *allyl isothiocyanate, 3-butenyl isothiocyanate, 2-propenylglucosinlate* (*sinigrin*), *2-pentyl isothiocyanate,* and *phenylethyl isothiocyanate.* It has been found that these compounds have anti-oxidant as well as de-toxification functions.

It is a potent gastric stimulant; increases appetite, and aids in digestion. The volatile phyto-chemical compounds in the root stimulate salivary, gastric, and intestinal glands to secrete digestive enzymes, thereby facilitate digestion.

Horseradish has good amounts of **vitamin-C** which is a powerful water soluble anti-oxidant.

The root spice has some of vital minerals in moderation like sodium, potassium, manganese, iron, copper, zinc, and magnesium.

In addition, the root has small amounts of essential vitamins such as **folate**, vitamin B-6 (pyridoxine), riboflavin, niacin, and pantothenic acid

Caution

Horseradish can cause irritation to skin, mucus membranes, and eyes. This is due to release of allyl sulphide gas (**allyl-isothiocyanate**) while chopping, crushing, or grating the root. Disruption of cell wall activates enzyme myrosinase which when reacts with **glucosinolates** to form **allyl isothiocyanates**. **Lemon**citrus or vinegar stops this reaction and stabilizes the flavor. Its effect can be minimized by using blender/mixer in well-ventilated place and wearing protective gloves and mask.

Beet (*Beta vulgaris*)

Beets with Green Tops

Beets are highly nutritious and "cardiovascular health" friendly root vegetables. Certain unique pigment antioxidants present in root as well as top greens have found to offer protection against coronary artery disease and stroke, lower cholesterol levels in the body and have anti-aging effects. This colourful tuber is a great source of folate also known as vitamin B9. This vitamin promotes functioning of the liver and the nervous system, which usually gets congested under chronic stress.

Romans used beet root as a treatment for fevers and constipations, amongst other ailments, considered its juice as an aphrodisiac. The root is a rich source of natural sugar. It also contains potassium, calcium, chlorine, iodine, iron, copper and a lot of vitamin A and B. The iron content in the beet regenerates and reactivates the red blood corpuscles. This blood forming quality helps in cases of anemia. Although it is normally the root part of the plant that is eaten, in fact its leaves are a very good source of beta carotene. Its high sodium and potassium content acts as a good solvent for excess calcium in the body thus, preventing deposits of calcium forming in the joints. This complaint is common as the process of ageing begins. The root is also valuable in the treatment of hypertension and heart trouble. The glamorous root performs certain mundane functions too. It relieves the habitually constipated ones, cleanses the liver particularly during jaundice. Being alkaline in nature, the root is useful in combating acidosis. For those who have jumpy nerves, beat it with beet. It is so soothing that schools of thought believe that beet roots actually calm the nerves, particularly at menopause or even during pre-menstrual tensions. The red colour in beet is decided by slightly water soluble xanthine pigments called **betalains**. High ratio of betacyanin to betaxanthine leads to violet, medium to red and orange tuber colours and these ratios are controlled by a triallelic system at the R Locus in the chromosome with incomplete dominance. Beet root is excellent for those in their late middle age. **Betanins-** obtained from the roots are used industrially as red food colourants, *e.g.* to improve the colour of tomato paste, sauces, desserts, jams and jellies, ice creams and break fast cereals.

So make beetroot available for a large part of the year a regular part of one's meal. Researchers from the William Harvey Research Institute at Barts and the London School of Medicine have discovered that drinking just 500 ml of beetroot juice a day can significantly reduce BP in one hour. This decrease in BP resulted from the chemical formation of **nitrite** from the dietary nitrate in the juice. However, the human body in itself is incapable of changing nitrate into nitrite. It is the bacteria on the tongue or the saliva that converts nitrate in the juice into nitrite. This nitrite-containing saliva is swallowed and in the acid environment of the stomach is either converted into nitric oxide or re-enters the circulation as nitrite (Sinha, 2008). Beet juice cleanses the blood and strengthens the gall bladder and liver. It is a rich source of the mineral boron, which plays an important role in the production of human sex hormones. Today the beetroot is still championed as a universal panacea. Beetroot not only provides a multitude of health benefits, the vibrant vegetable is also a great source of long lasting energy. First of all, beetroot has a high sugar content which helps to provide an instant energy boost. It is also high in many energy-boosting nutrients including magnesium, iron, **vitamin C** and nitrate.

Health Benefits

There is a great number of amazing health benefits that beetroot has. Cleaning your fatty liver, retrieving your eyesight and preventing colon obstructions, are only some of its useful benefits.

This amazing plant contains tryptophan and betaine, substances that can calm anxiety and support your cardiovascular system and at the same time it will boost your energy levels to maximum. Consuming beets can also prevent free radical damage because it contains several antioxidant compounds.

Beets are an amazing blood purifier and are rich in iron and produce the disease-fighting white blood cells. They also stimulate red blood cells and improve the supply of oxygen to the cells. Beets prevent cancer and heart disease, and the detoxifying properties make them good for your organs. Beets are also high in fibre and nourishing for digestive health. It has been used in the treatment of leukemia for a long time, due to the amino acid betaine which has anti- cancer properties. Beet roots are also effective against illnesses caused by oxidative stress. The fibres in beets can reduce the cholesterol levels up to nearly 50 percent. Beets protect the elasticity of the blood vessels and balance the blood pressure levels.

Helps in detoxification: Beetroot is reckoned to be a great purifier. It detoxifies your body by pulling the toxins into the colon from where they can be evacuated. Some studies suggest that beetroot juice might also stimulate red blood cell production and build stamina.

Low in fat and calories: Although it has a high sugar content, it is low in calories and almost fat free. Since it is loaded with fibre it makes you full on lower calories. This makes it a nutritious option for those looking to keep their weight down.

Heart health: Studies have shown that the high content of nitrates in beetroot produces a gas called nitric oxide. This gas helps to relax and dilate your blood vessels which improves blood flow and lowers blood pressure.

Rich in antioxidants: Betanin, the pigment which gives beetroot its colour, is a potent antioxidant. Along with another class of antioxidants called polyphenols, these are getting more attention in the scientific community. According to the American Journal of Clinical Nutrition, antioxidants reduce the oxidation of bad cholesterol, protect the artery walls and guard against heart disease and stroke.

Folate, Fibre, Vitamin C and other minerals: Most people think that diabetics should avoid beetroot since its sweet. It is sadly misunderstood. Beetroot is a great source of fibre and minerals like iron, potassium and manganese which are essential for good health and in combination with other foods it can deliver a lot. Vitamin C boosts immunity, folate is essential for normal tissue growth and fibre helps in smooth digestive functions. It is particularly high in protein and iron than most other roots and tubers. Raw beets are an excellent source of **folates**; contains about 109 mcg/100 g (**Provides 27 per cent of RDA**). However, extensive cooking may significantly depletes its level in food. Folates are necessary for DNA synthesis in the cells. When given during peri-conception period folates can prevent neural tube defects in the baby.

Hair care: Beetroot is actually one of the best home remedies to fight the flakes and an itchy scalp. You can boil some beets in water and use the concentrated liquid to massage on the scalp. Alternatively, you can mix some beetroot juice, vinegar and ginger juice and apply to the scalp. Keep this for 20 minutes and rinse.

The nutrients in beetroots are heat sensitive. With the rise in cooking time and temperature, the antioxidant content decreases. Beetroot is rich in Vitamin C which is a water soluble vitamin that can be destroyed on cooking. Not only this, it also loses more than 25 per cent of its folate when cooked. It is best to mildly steam or bake it at lower temperatures.

Garden beets are very low in calories (contain only 45 kcal/100 g), and contain only small amount of fat. Its nutrition benefits come particularly from fibre, vitamins, minerals, and unique plant derived anti-oxidants. Beet greens (tops) are an excellent source of carotenoids, flavonoid anti-oxidants, and vitamin A; contain these compounds several times more than that of in the roots. Vitamin A is required maintaining healthy mucus membranes and skin and is essential for vision. Consumption of natural vegetables rich in flavonoids helps to protect from lung and oral cavity cancers.

Beet Greens/Root

1. Beet greens contain notable amounts of calcium, iron, magnesium and phosphorus
2. They also contain vitamins A, B-complex and C

Beet roots are high in carbohydrate levels and should therefore be used sparingly

Lowers Your Blood Pressure Levels: Beetroot is an extraordinary wellspring of nitrates, which when eaten, is changed over to nitrites and a gas called nitric oxides. Both these parts help to broaden the arteries and lower pulse. Specialists likewise observed that having pretty much 500 grams of beetroot consistently lessens a man's circulatory strain in around six hours.

Reduces "Bad" Cholesterol And Prevent Plaque Formation: Beetroot contains soluble fibres, flavanoids and betacyanin. Betacyanin is the ingredient that gives beetroot its purplish-red shading and is additionally effective cell reinforcement. It aides diminish the oxidation of LDL cholesterol and does not permit it to store on the walls of the artery. This secures the heart from potential heart attacks and stroke, lessening the requirement for prescription.

Good For Pregnant Women And Unborn Child: Another astounding benefit of the root is that it has supply of folic acid. Folic acid is crucial for pregnant mums and unborn children on the grounds that it is a key part for the best possible arrangement of the unborn kid's spinal cord, and can secure the child from conditions, for example, spine bifida (is an congenital issue where the kid's spinal cord does not frame totally and appears as though it has been separated into two at the base). Beetroot additionally offers mums-to-be that additional energy boost presupposed amid pregnancy.

Beats Osteoporosis: Beetroot is rich in mineral silica, an essential par for the body to utilize calcium effectively. Since calcium makes up our bones and teeth, having a glass of beetroot juice a day could help keep conditions, for example, osteoporosis and weak bone sickness under control.

Keeps Diabetes Under Check: Individuals experiencing diabetes can satisfy their sweet desire by adding a little beetroot in their eating regimen. Being a medium glycaemic record vegetable (means it discharges sugars gradually into the blood), it helps in keeping up your glucose levels low while satisfying your sugar needing. Also, this vegetable is low in calories and fat-free making it an ideal vegetable for diabetics.

Treats Anemia: It is a typical myth that beetroot replaces lost blood and is in this manner great to treat weakness. While this may sound a touch unbelievable to numerous, there is a truth covered up in the myth. Beetroot contains a considerable measure of iron. Iron aides in the arrangement of haemagglutinin, which is a part of the blood that helps transport oxygen and supplements to different parts of the body. It is the iron substance and not the colour that helps treat weakness.

Helps Relieve Fatigue: A study introduced at the American Diabetics association's gathering expressed that beetroot helps support a man's vitality. They said that because of its nitrate content it helped enlarge the arteries in this way helping in the best possible transportation of oxygen to different parts of the body, expanding a man's vitality. Another hypothesis was that in light of the fact that the root is a rich wellspring of iron, it helps in enhancing a man's stamina.

Improves Sexual Health And Stamina: Otherwise called 'natural Viagra', beetroot has been usually utilized as a part of various prehistoric traditions to support one's sexual wellbeing. Since the vegetable is a rich wellspring of nitrates it helps discharge nitric oxide into the body, enlarging the veins, and expanding blood stream to the genitals – a component that pharmaceuticals like Viagra look to replicate. Another variable is that beetroot contains a great deal of boron, a chemical ingredient that is crucial for the generation of the human sex hormone.

Protects you from Cancer: The betacyanin content in beetroot has another vital capacity. In a study done at the Howard University, Washington DC, it was found that betacyanin helped decelerate the development of tumors by 12.5 per cent in patients with breast and prostate cancer. This impact does not just aid in the analysis and treatment of tumors however it likewise helps malignancy survivors stay cancer- free.

Beats Constipation: Due to its high dissolvable fibre content, beetroot goes about as an incredible laxative. It helps in balancing your bowel movement. It likewise washes down the colon and flushes out the unsafe poisons from the stomach.

Boosts Brain Power: A study done at the University of Exeter, UK, demonstrated that drinking beetroot juice could build a man's stamina by 16 percent, in view of its nitrate content. Known to grow oxygen uptake by the body, the study additionally found that on account of this one component, it could likewise help in the best possible working of the brain and beat the onset of dementia. It has additionally been seen that nitrate when changed over to nitrite helps in the better transmission of neural driving forces, improving the cerebrum work.

Pregnant women should also regularly consume beet roots, as they contain folic acid, which protects the new born from several different diseases.

Beet roots also stimulate the gall bladder and the liver. When combined with carrots, beet roots become an effective remedy against kidney disease.

This vegetable is also helpful in case of bone pain, toothaches and head aches.

The root is rich source of phytochemical compound ***Glycine betaine.*** Betaine has the property of lowering *homocysteine* levels in the blood. Homocysteine, one of highly toxic metabolite, promotes platelet clot as well as atherosclerotic-plaque formation, which is otherwise can be harmful to blood vessels. High levels of homocystiene in the blood results in the development of coronary heart disease (CHD), stroke and peripheral vascular diseases.

It contains significant amounts of **vitamin-C**, one of the powerful natural antioxidant, which helps body scavenge deleterious free radicals one of the reasons for cancers development.

The root is also rich source of niacin (vit. B-3), pantothenic acid (vit. B-5), pyridoxine (vit. B-6) and carotenoids, and minerals such as iron, manganese, and magnesium.

In addition, the root indeed has very good levels of potassium. 100 g fresh root has 325 mg of potassium or 7 per cent of daily requirements. Potassium lowers heart rate and regulates metabolism inside the cells by countering detrimental effects of sodium.

1. It is used for purifying the blood.
2. Improves circulation.
3. Promotes menstruation.
4. Stimulates the bowels.
5. It helps calm nervousness and is of benefit in vascular congestion.
6. Recommended for intestinal cleansing of parasites

Caution

Beeturia is a harmless condition of passing red or pink colour urine after eating beets and its top greens. The condition can be found in around 10-15 per cent of the population who are genetically unable to break down betacyanin pigment.

Beet greens contain *oxalic acid*, a naturally occurring substance found in some vegetables, which may crystallize as oxalate stones in the urinary tract in some people. It is therefore; in individuals with known oxalate urinary tract stones are advised to avoid eating excess greens.

Green beets are rich in oxalic acid but, if eaten in excess, they can interfere with calcium metabolism and promote the formation of calcium metabolism and promote the formation of calcium – Oxalate kidney stones.

CELERY (*Apium graveolens* Dulce)

Leaf-Celery

It provides iron, calcium, sodium, potassium, phosphorus and traces of other minerals, as well as vitamin A, B1, B2, C as well as silicon. It is specially rich in potassium and sodium, four fresh stalks of celery provide 341 mg potassium and 126mg sodium and their juice makes a great electrolyte replacement drink. It has been recommended for rheumatism, arthritis, and nervousness and sciatica to liver disorders. Iranian boil the aromatic seed oil and inhale the vapour as a cure for headache (Levey, 1966). In Egypt, wild celery seed is used as a diuretic, digestive aid, and emmemagogue, a substance that induces menses in non-pregnant women (Levey, 1966). Celery juice contains the anti-cancer nutrients phthalid and polyacetylene (antioxidants). Being rich in potassium and sodium helps in lowering blood pressure. Celery is valuable in weight loss diets, where it provides low calorie fibre bulk. Bergapten in the seeds can increase photosensitivity, so the use of seed of essential oil externally in bright sun shine should be avoided. The oil and large doses of seeds should be avoided during pregnancy. They can act as a uterine stimulant. May help prevent certain cancers, regular blood pressure and reduce cholesterol. It contains vitamin C and several other active compounds that promote health, including phalides, which may help lower cholesterol and coumarins that may be useful in cancer prevention.

Celery is an excellent source of vitamin C, a vitamin that helps to support the immune system. Vitamin C rich food like celery help reduce cold symptoms or severity of cold systems. Celery contains compounds called **pthalides**, which can help relax the muscles around arteries and allow those vessels to dilate. With more space inside the arteries, the blood can flow at a lower pressure. Pthalides also reduce stress hormones, one of whose effects is to cause blood vessels to constrict. Celery has a reputation among some persons being a high-sodium vegetable and blood pressure reduction is usually associated with low- sodium foods. Celery contains compounds called **coumarins** that help prevent free radicals from damaging cells, thus decreasing the mutations that increase the potential for cells to become cancerous. Coumarins also enhance the activity of certain white blood cells, immune defenders that target and eliminate potentially harmful cells, including cancer cells. Coumarin compounds in the celery appear to be useful in toning the heart and blood vessels. These compounds can also be useful in cases of migraine. One of them, 3-n-butyl phthalide, was found to significantly lower blood pressure. In addition, compounds in celery called **acetylenics** have shown to stop the growth of tumor cells. The stalks of the plant, which are rich in iron, magnesium and carotene, are usually eaten raw in salads. Leaves are thought to stimulate the appetite and also increase urination and bring on menstruation. Celery juice has been used to alleviate oedema, treat rheumatism and clear skin problems. In addition to rich source of vitamin C and fibre, it contains negative calories *i.e.* you burn caloreies to digest them and this property of celery is very beneficial for those who want to lose weight.

According to Acharya Balkrishna (2008), it suppresses Kapha and Vata, elevates pitta and is a pain reliever, but gives heartburn. It stimulates heart, increases urine flow, but it is slightly aphrodisiac and excites the womb. It is beneficial in curing hiccoughs, vomiting, annul pain and whooping cough. Since it treats digestive system it is one of the main medicines for stomach disorders. It is also effective in curing piles and stones in kidney or gallbladder.

Celery may not be the food of choice that you dream of when thinking of sex, but it is a great source food for sexual stimulation. It contains **andro-sterone**, an odourless hormone released through male perspiration and turns women on. Celery is best eaten raw, wash and cut some and much away in bed just before you get down and dirty. Celery is the best vegetable source of naturally occurring sodium. It is high in potassium. The high water content in celery makes it ideal for vegetable juicing. As an easy way to reduce grains in your diet, spread peanut butter on celery rather than bread.

You may have heard people say that celery has negative calories, which means eating celery burns off more calo ries than your body absorbs after eating it. While celery is a very low-calorie food and an excellent choice when you're trying to achieve or maintain a healthy weight, it isn't necessarily a 'negative calorie' food.

Calories in Celery

Celery provides about 6 calories in each medium-sized stalk, according to the US Department of Agriculture National Nutrient Database. The same portion of celery contains about 1 gram of total carbohydrates -including 0.6 grams of fibre. Fibre isn't fully digested or absorbed by your body, according to the University of Massachusetts. However, the negligible number of calories provided by protein, sugar, and fat in celery are absorbed and used by your body.

Thermic Effect of Food

The number of calories your body burns by eating and digesting food, called the thermic effect of food, is about 10 per cent of your body's total daily energy expenditure. Therefore, if you're expending and eating 2, 000 calories in a day. Your body burns about 200 of those calories eating a digesting food. Therefore, if you're eating a medium-sized celery stalk, your body will burn almost 1 calorie eating and digesting that stalk.

Bottom Line

While celery is a very low-calorie food, it likely doesn't provide you with negative calories. While it's theoretically possible for negative-calorie foods to exist, there are no reputable scientific studies that prove certain foods cause "negative calorie" effects. In fact, protein is the macronutrient that causes your body to burn the most calories, according to a review.

Health Benefits

You can munch on the raw stalks, or load them up with dip or peanut butter, make juice out of them or chop them up and add them to your soups or stir fries. Here are some of the health benefits of eating celery.

Enables weight loss: A long stalk of celery has only 10 calories, so, skip the calorie-laden junk food and grab some celery. Celery also curbs your cravings for rich, sugary foods.

Lowers stress levels: Celery contains magnesium and essential oils that help calm you down. Eating celery before you sleep will ensure that you get a good night's sleep.

Contains healthy salt: Celery contains plenty of sodium, but the natural and healthy kind. People with high blood pressure are advised to consume less table salt, but can safely consume the salt in celery.

Prevents acidity: Celery contains properties that balance the pH levels in your body, thereby preventing acidity.

Improves vision: Celery is rich in Vitamin A, which sharpens your vision and takes care of your eyes. Eating celery regularly helps prevent the deterioration of vision that usually occurs with age.

Powers you up: Celery juice is the perfect drink to recharge your system after a workout. It replaces your electrolytes and nourishes you with its rich minerals.

Reduces cholesterol: A compound called butylphthalide is responsible for the unique flavour and aroma of celery. This same compound also lowers your LDL cholesterol levels, the bad kind of cholesterol. Eating just two stalks of celery every day can reduce your LDL levels by 7 points.

Improves kidney function: Celery helps to eliminate the toxins from your body, thereby improving your kidney function and preventing the formation of kidney stones.

Improves digestion: Celery consists mainly of water and dietary fibre, both of which are extremely good for your digestive system. It has diuretic and cleansing properties that prevent constipation.

Contains pheromones: Celery contains two pheromones called androstenone and androstenol that can boost your sex life Chewing on the celery stalk releases these pheromones.

Lowers blood pressure: Eating celery raw can significantly lower your blood pressure. A compound called phthalides in celery also improves your circulation.

Fights cancer: Celery contains a flavonoid called luteolin that actively blocks the growth of cancer cells. Studies have shown that eating celery regularly lowers your risk of developing pancreatic, colon and breast cancer especially.

Cools you Down

Celery juice is a great drink for hot weather, since it helps lower your body temperature. Try to drink a glass of celery juice every few hours during the summer months.

One of the very low calorie herbal plant, celery leaves contain only 16 cal per 100 g weight and lots of non-soluble fibre which when combined with other weight loss regimens may help to reduce body weight and blood cholesterol levels.

Celery is a functional food. Its leaves are rich source of flavonoid antioxidants such as *zeaxanthin, lutein* and *beta- carotene,* which have anti-oxidant, cancer protective, and immune-boosting functions.

It is also good source of vitamin-A. Vitamin-A and beta-carotene are natural flavonoid antioxidants. Vitamin A is also required for maintaining healthy mucus membranes and skin, and for vision. Consumption of natural foods rich in flavonoids helps body to protect from lung and oral cavity cancers.

The herb is also rich in many vital vitamins including folic acid (provides 9 per cent of RDA), riboflavin, niacin and vitamin-C, which are essential for optimum metabolism.

Fresh celery is an excellent source of **vitamin-K**, provides about 25 per cent of DRI. Vitamin-K help increase bone mass by promoting osteotrophic activity in the bones. It also has established role in Alzheimer's disease patients by limiting neuronal damage in the brain.

The herb is very good source of minerals like potassium, sodium, calcium, manganese, and magnesium. Potassium is an important component of cell and body fluids that helps control heart rate and blood pressure.

Its leaves and seeds contain many essential volatile oils that include terpenes, mostly **limonene** (75 to 80 per cent), and the sesquiterpenes like β-selinene (10 per cent) and humulene; but its characteristic fragrance is due to chemical compounds known as *phthalides* (butylphthalid and its dihydro derivate sedanenolid) in them.

Essential oil obtained from extraction of celery plant has been used in soothing remedies for nervousness, osteoarthritis, and gouty-arthritis conditions. In addition, its seeds, and root has diuretic (removes excess water from body through urine), galactogogue (help breast milk secretion), stimulant, and tonic properties.

Root has been used in the preparation of soups and sauces.

Medicinal Uses

Wild celery has been used in complementary medicines to reduce blood pressure, to relieve indigestion and as an anti-inflammatory agent. It also used as a diuretic to remove excess water from the body.

The essential oils in seeds, herb, and root have use as carminative, emmenagogue, galactogogue (help breast milk secretion), nervous system ailments such as headache and nervous irritability.

The herb has also been claimed useful in rheumatism and gouty conditions.

Caution

The herb especially wild celery can cause severe anaphylactic reactions in some sensitive individuals. It is also should not be eaten by pregnant women. People on diuretic medications and anticoagulant medications should use this herb sparingly.

It is also very high in soluble as well as insoluble fibre contents. Eating recipes with too much fibres may cause stomach-pain, indigestion and bloating, and sometimes complicates existing constipation condition.

PARSLEY (*Petroselinum crispum*)

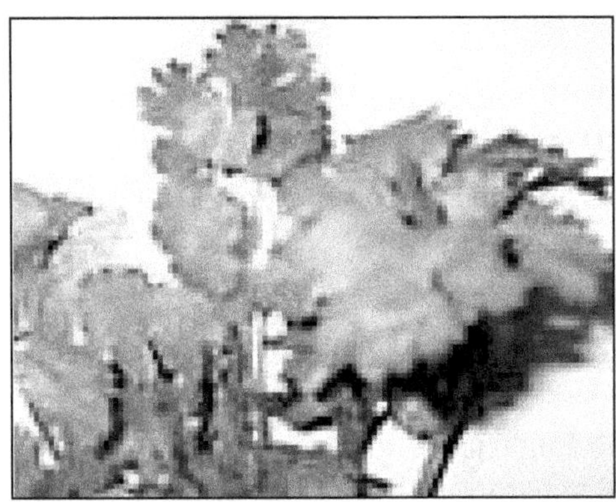

Parsley–Flat Leaves

Wonderfully nutritious parsley is a popular culinary as well as medicinal herb, which is recognized as one of the functional food for its unique anti-oxidants and disease preventing properties. Is a very popular biennial herb with aromatic leaves that are used as a culinary flavouring. The leaves are either eaten fresh in salads or steeped for tea. Due to its significant therapeutic benefits, parsley was first used as herbal medicine and later as food. Parsley is a wonderful aid to digestion. It is known to have laxative effects, relieves flatulence, reduces stomach scrams and stimulates appetite. It has been known to have the ability to shrink small blood vessels and is helpful in treating piles. It is also known to have liver protection properties. A recent animal study reported that degenerative changes observed in the liver cells of diabetic rates were significantly reduced or disappeared on treatment with parsley. Its anti septic properties make it useful in cystitis (inflammation of urinary bladder). Parsley has a tonic effect on reproductive system and is sometimes used during labour. It is known to be effective emmenagogue: a substance that induces or hastens menstrual flow and helps to regulate the menstrual cycles. It appears to increase diuresis by inhibiting the Na+\K+ ATPase pump in the kidney, thereby enhancing sodium and water excretion while increasing potassium re-absorption. It is also valued as an aquaretic. When crushed and rubbed on the skin, parsley can reduce itching in mosquito bites. When chewed, parsley can freshen bad breath. Folk medicine has a tradition of using parsley to treat health conditions. Tea made from leaves or roots (of parsley) can be used to treat jaundice, coughs, menstrual problems, rheumatism, kidney stones and urinary infection. Juice from parsley sooths conjunctivitis and other eye inflammations.

Both seeds and dried roots of parsley are used as spices. The essential oil, distilled mainly from seeds has a warm, spicy, herbaceous scent. The parsley leaf oil is also used extensively for garnishing and seasoning. Parsley leaves are used as a flavouring and as garnish either dried or fresh, in soups, meat dishes, poultry, fish and as stuffing. Parsley is an excellent diuretic and useful for kidney disorders and the elimination of stones and gravel. Both the fresh juice and a tea made from the leaves or seeds can be used to relieve water retention (oedema) and strengthen digestion. Parsley also stimulates delayed menstruation and promotes menstrual flow.

Health Benefits

The herb contains no cholesterol; but is rich in anti-oxidants, vitamins, minerals and dietary fibre which helps control blood cholesterol levels, prevents constipation, protects body from free radicals mediated injury and from cancers.

Parsley contains many health benefiting essential volatile oils that include *myristicin, limonene, eugenol, and alpha-thujene.*

The essential oil, **Eugenol,** present in this herb has been in therapeutic use in dentistry as a local anesthetic and anti-septic agent for teeth and gum diseases. *Eugenol* has also been found to reduce blood sugar levels in diabetics, however, further detailed studies required to establish its role.

Parsley is rich in poly-phenolic flavonoid anti-oxidants including *apiin, apigenin, crisoeriol, and luteolin*; and has been rated as one of the plant source with **highest anti-oxidant** activities.

The herb is a good source of minerals like potassium, calcium, manganese, iron, and magnesium. Potassium in an important component of cell and body fluids that helps control heart rate and blood pressure by countering the effects of sodium. Iron is essential for heme production inside red blood cells. Manganese is used by the body as a co-factor for the antioxidant enzyme, *superoxide dismutase.*

It is also rich in many antioxidant vitamins including **vitamin-A**, beta-carotene, **vitamin-C**, vitamin-E, zea-xanthin, lutein, and cryptoxanthins. The herb is also an excellent source of vitamin-K and folates. **Zea-xanthin** helps prevent age related macular degeneration (ARMD) in the retina of the eye in the old age population through its anti-oxidant and ultra-violet light filtering functions.

Fresh herb leaves are also rich in many essential vitamins such as pantothenic acid (vitamin B-5), riboflavin (vitamin B-2), niacin (vitamin B-3), pyridoxine (vitamin B-6) and thiamin (vitamin B-1). These vitamins are essential during carbohydrates, fat and protein metabolism by acting as co-enzymes.

It is probably the richest of the entire herb source for **vitamin K;** provides 1640 mcg or 1366 per cent of recommended daily intake. Vitamin K has been found to have potential role in bone health by promoting osteotrophic activity in the bones. It has also established role in the treatment of *Alzheimer's disease* patients by limiting neuronal damage in the brain.

Medicinal Uses

The leaves, stems and roots of this herb plant have antiseptic and carminative properties.

Eugenol *also has been found to reduce blood sugar levels in diabetics, however, further detailed studies required to establish its role.*

The extraction from the herb has been found to have diuretic effects.

Parsley is useful as a digestive aid.

It helps to purify the blood and stimulate the bowels.

Parsley is an anti-carcinogen.

It contains three times as much vitamin C as oranges, and twice as much iron as spinach.

Parsley contains vitamin A and is a good source of copper and manganese.

For a natural breath freshener, try a sprig of parsley!

PARSNIP (*Pastinaca sativa*)

Parsnips are closely related to carrots family of vegetables; grown for their sweet, succulent underground taproots. The root vegetable has similar appearance and growth characteristics as other apiaceae family members like carrots, parsley, celery, cumin, dill etc. A sweet culinary root vegetable and mildly diuretic, it lubricates the intestines and is beneficial for soothing stomach and intestinal irritations. It helps to clean the liver and gall bladder and is used in soups or teas. It promotes perspiration and is beneficial for coughs, cold, headaches and arthritis. Its fleshy, stout roots appear like that of carrots, but are white or cream in color and sweeter than that of carrots. Good winter frost is essential for better crop, as frost converts much of the starch to sugars and helps develop long, firm parsnips. The roots are generally harvested when they reach about six to ten inches long, by pulling the entire plant with its root (uprooting) as in carrots.

Health Benefits

Generally, parsnip contains more sugar than **carrots, radish, turnips**. In general, it has calories (100 g provide 75 calories) comparable to that of some fruits **like banana, grapes** *etc.* Nonetheless, its sweet, juicy root is rich in several health-benefiting phyto-nutrients, vitamins, minerals, and fibre.

It is one of the excellent sources of soluble and insoluble **dietary fibre**. 100 g root provides 4.9 mg or 13 per cent of fibre. Adequate fibre in the diet helps reduce blood cholesterol levels, obesity and constipation conditions.

As in carrots and other members of apiaceae family vegetables, parsnip too contains many **poly-acetylene** anti-oxidants such as *falcarinol, falcarindiol, panaxydiol,* and *methyl-falcarindiol.*

Several research studies from scientists at University of Newcastle at Tyne found that these compounds have anti-inflammatory, anti-fungal, and anti-cancer function and offer protection from colon cancer and acute lymphoblastic leukemia (ALL).

Fresh roots are also good in vitamin C; provide about 17 mg or 28 per cent of RDA. Vitamin C is a powerful water-soluble anti-oxidant, easily available to us from natural sources. It helps body maintain healthy connective tissue, teeth, and gum. Its anti-oxidant property helps protect from diseases and cancers by scavenging harmful free radicals from the body.

Further, the root is rich in many B-complex groups of vitamins such as **folic acid,** vitamin B-6 (pyridoxine), thiamin, and pantothenic acid as well as vitamin K and vitamin E.

In addition, it also has healthy levels of minerals like iron, calcium, copper, potassium, manganese and phosphorus. Potassium is an important component of cell and body fluids that helps controlling heart rate and blood pressure by countering effects of sodium.

Caution

Parsnip plant and its parts may cause hypersensitivity reactions like contact dermatitis and oral allergy syndrome (OAS) when handled in some sensitive individuals. The reaction symptoms may include rash and skin lesions. Some of common OAS symptoms may include itching or burning sensation in the lips, mouth, and throat. In severe cases swelling of the lips, tongue, and redness in eyes, and breathing difficulty may be observed. Individuals with known history of allergy to birch category pollen agents like walnuts, fig, carrots, parsley *etc.* may develop cross-sensitivity to parsnip and should be avoided.

Only the root part of the plant should be eaten. Parsnip leaves are poisonous.

PURSLANE (*Portulaca oleracea*)

Purslane is widely grown in many Asian and European regions as staple leafy vegetable. Its leaves appear thick, contain mucilaginous substance, and have a slightly sour and salty taste. Leaves and tender stems have slightly sour and salty taste. In addition to succulent stems and leaves, its yellow flower buds are also edible. It contain more **Omega- 3 fatty acids** (**alpha- linolenic acid** in particular) than any other leafy vegetable plant. This soft, succulent **Purslane** has more omega-3 fatty acids than in some of fish oils. If you are a vegetarian and want to avoid all forms of animal products, then here is the answer! Go for this healthy dark green leafy vegetable and soon you will forget fish! Purslane has 0.1 mg per gram of EPA and this is extra ordinary amount of EPA for land based vegetable source. EPA is an **Omega- 3 fatty acid** normally found mostly in fish and algae. It also contains vitamins, mainly vitamin A, vitamin C and some vitamin B and carotenoids as well as dietary minerals, such as magnesium, calcium, potassium and iron. Also present are two types of betalain alkaloid pigments, the reddish **betacyanins** (visible in the coloration of the stems) and the yellow **betaxanthins** (noticeable in the flowers and in the light yellowish cast of the leaves). Both of these pigment types are potent anti -oxidants and have been found to have anti-mutagenic properties in laboratory studies. 100 grams of fresh purslane leaves (about 1 cup) contains 300-400mg of **alpha-linolenic acid.** One cup of cooked leaves contain 90mg of calcium, 561mg of potassium and more than 2000 IU's of vitamin A.

This weed is also called pig weed and it is certainly better feed for pigs than the GMO grains used by most farmers for pig and other livestock feed. Amazingly, the purslane weed contains more omega 3 fatty acids than most fish oils. Furthermore, it has one of the highest levels of vitamin A among the leafy green vegetables. Vitamin A is excellent for boosting the health of the eyes and it is also a marvelous protector from several types of cancer.

Powerful Anti-oxidant

Purslane is loaded with two types of betalain alkaloid pigments called yellow beta-xanthins and reddish beta-cyanis. Both of these pigments are stong anti-mutagens and antioxidants. This weed is also rich in vitamin C, vitamin B-complex including niacin, riboflavin and niacin, carotenoids and trace minerals such as calcium, magnesium and iron.

It purslane offers many nutritional benefits and it should be considered as a booster of the health instead of a simple weed, especially because we spend a lot of money on health supplements.

Health Benefits

This wonderful green leafy vegetable is very low in calories (just 16 kcal/100g) and fats; but is rich in dietary fibre, vitamins, and minerals.

Fresh leaves contain surprisingly more Omega-3 fatty acids (α-linolenic acid) than any other leafy vegetable plant. 100 grams of fresh purslane leaves provides about 350 mg of α-linolenic acid. Research studies shows that consumption of foods rich in ω-3 fatty acids may reduce the risk of coronary heart disease, stroke, and help prevent development of ADHD, autism, and other developmental differences in children.

It is an excellent source of Vitamin A, which is one of the highest among green leafy vegetables. Vitamin A is a known powerful natural antioxidant and is essential for vision. This vitamin is also required to maintain healthy mucus membranes and skin. Consumption of natural vegetables and fruits rich in vitamin A is known to help to protect from lung and oral cavity cancers.

Purslane is also a rich source of vitamin C, and some B-complex vitamins like riboflavin, niacin, pyridoxine and carotenoids, as well as dietary minerals, such as iron, magnesium, calcium, potassium, and manganese.

Also present in purslane are two types of betalain alkaloid pigments, the redish *beta-cyanins* and the yellow *beta-xanthins*. Both of these pigment types are potent anti-oxidants and have been found to have anti-mutagenic properties in laboratory studies. [Proc. West. Pharmacol. Soc. 45: 101-103 (2002)]

Caution: Purslane contains oxalic acid, a naturally occurring substance found in some vegetables, which may crystallize as oxalate stones in the urinary tract in some people. 100 g fresh leaves contain 1.31 g of oxalic acid, more than in **spinach** (0.97 g/100 g) and cassava (1.26 g/100 g). It is therefore, people with known oxalate urinary tract stones are advised to avoid eating purslane and certain vegetables belonging to amaranthaceae and brassica family. Adequate intake of water is therefore advised to maintain normal urine output.

CHICORY (*Cichoium intybus*)

Chicory is used as a vegetable, most often found in a green salad, as a coffee alternative, by roasting and grinding its roots, or as a medicinal remedy. As a supplement, chicory contains cardioactive, probiotic and sedative properties that may be helpful in the treatment of cardiovascular disease, gastrointestinal issues such as diarrhea, constipation, dyspepsia and cancer. The leaves of the chicory plant can be eaten raw in salads or cooked like greens. The roots are usually roasted, dried and ground into coffee. Chicory contains many nutrients, including vitamins A, B6, C, E, and K, as well as zinc, folate, calcium, magnesium, potassium, iron and manganese. Chicory can be found in infusions or extracts; it is used as a supplement that may have laxative, diuretic, cardioprotective and anti-inflammatory properties.

Especially the flower- It is variously used as a tonic and appetite stimulant and as a treatment for gallstone, gastroenteritis, sinus problems and cuts and bruises (Howard, 1987). Root chicory contains oils similar to those found in plants in the related Tanacetum which includes Tansy and is likewise effective at eliminating intestinal worms. All parts of the plant contain these volatile oils, with the majority of the toxic compounds concentrated in the plant root. Chicory is well known for its toxicity to internal parasites. Studies indicate that ingestion of chicory by farm animals result in reduction of worm burdens which has prompted its wide spread use as a forage supplement. It has a bitter spicy taste, which mellows when it is grilled or roasted. It can also be used to add colour and zest to salads. The young leaves can be used in salads and the root can also be boiled and eaten like a vegetable. It is also grown for cattle feedin Europe.

Infusion or decoction of root or flowers can stimulate the appetite, aid digestion, promote bile secretion and help to relieve the pain and discomfort caused by gallstones. Dried ground chicory is used as a coffee substitute, either on its own or combined with cereals in cereal coffee. Chicory contains inulin, which helps diabetics regulate their blood sugar levels. It is closely related to lettuce and dandelion but is a member of the sunflower family. It may be cleansing to the liver and gallbladder. Chicory is beneficial for digestion, the circulatory system and the blood. Its leaves are a good source of calcium, vitamin A and potassium.

Chicory is also known to contain a number of nutritional and health benefits. It is also high in Vitamin A, calcium and potassium. It also contains iron, niacin, phosphorus, and other important vitamins and minerals. Chicory has been found to contain a substance called inulin, which is very beneficial to diabetics, as it helps to regulate blood sugar levels. As chicory is also utilized as a sweetener, with a taste comparable to sugar, it makes a great sugar alternative for diabetics as well. Other health benefits of chicory include its effectiveness at fighting intestinal worms and parasites, aiding in digestion and colon cleansing, and helping to cleanse the blood, liver and circulatory system of harmful toxins.

Many other health benefits have been associated with chicory, or, more specifically, chicory root, when used as a tea. Chicory tea can be made by simmering a tablespoon of chicory root in a cup of water for approximately 20-30 minutes. It is recommended to drink a cup of chicory tea every day in order to truly benefit from the health and nutritional properties inherent in the chicory root. Some herbal tea companies also sell a chicory root tea, or an herbal tea with chicory as an ingredient. Other ailments that have utilized the chicory root tea as a treatment include cramps, sore throats, and tuberculosis, among others.

GLOBEARTICHOKE (*Cyanara scolymus*)

Artichoke Bud

Artichoke globe measures about 6-10 cm in diameter and weigh about 150 g. Fuzzy, immature florets in the centre of the bud constitute "choke". These are inedible in older, larger flowers. Edible portion of the buds consists primarily of the fleshy lower portions of the involucres bracts (triangular scales) and the base, known as the "heart". The flower heads are commonly eaten as vegetables but extracts of leaves and roots were once used to treat liver disorders, jaundice and atherosclerosis. Artichokes provide a number of vitamins and minerals and one of their key active ingredients, **Cynarin** was shown by clinicalk studies to stimulate bile flow from gall bladder, thus mobilizing fatty stores and reducing cholesterol. Other studies found that its extracts can help digestive complaints such as poor appetite, nausea and abdominal discomfort.

The flower heads are considered useful in the dietary of diabetics. The leaves are bitter and considered to be useful in dropsy and rheumatism. Dried and fresh leaves and/or stems of cynara are used to increase bile production. **Cynarin,** an active constituent in cynara causes an increase in bile flow. Globe artichoke has (cynara) has a powerful effect on the production of bile and fat indigestion enzymes, stimulating liver functions and lowering cholesterol levels. Seemingly this latter action is due to increased bile production and reduced absorption of cholesterol in the intestine, with less cholesterol being synthesized in the liver and more being eliminated. Artichoke can also be made in to herbal tea, artichoke tea is produced as a commercial product in the Dalat region of Vietnam. Other studies found that artichoke extracts can help digestive complaints such as poor appetite, nausea and abdominal discomfort.

Health Benefits

Artichoke supports the liver. The substance **cynarin** gives artichoke its detoxifying qualities. Artichokes' B vitamins increase mental alertness and strengthen your immunity.

Artichoke is low in calories and fat, but is a rich source of dietary fibre; provides 5.4 g per 100 g, about 14 per cent of RDA. Dietary fibre helps control constipation conditions, decrease bad or "LDL" cholesterol levels by binding to it in the intestines and helps prevent colon cancer risks by preventing toxic compounds in the food from absorption.

Scientific studies have shown that bitter principles, **cynarin** and **sesquiterpene-lactones** in artichoke extraction have overall cholesterol reduction action in the body by inhibiting its synthesis and increasing its excretion in the bile.

Fresh artichoke is an excellent source of vitamin **folic acid**; provides about 68 mcg per 100 g (17 per cent of recommended daily allowance). Folic acid acts as a co-factor for enzymes involved in the synthesis of DNA. Scientific studies have proven that adequate levels of folates in the diet during pre-conception period and during early pregnancy help prevent from neural tube defects in the newborn baby.

It is also rich in B-complex group of vitamins such as niacin, vitamin B-6 (pyridoxine), thiamin, and pantothenic acid that are essential for optimum cellular metabolic functions.

Fresh globes also contains good amounts of anti-oxidant vitamin, **vitamin-C**. Provides about 20 per cent of recommended levels per 100 g. Regular consumption of foods rich in vitamin C helps body develop resistance against infectious agents and scavenge harmful, pro-inflammatory free radicals from the body.

It is one of the vegetable sources for vitamin K; provides about 12 per cent of DRI. Vitamin K has potential role bone health by promoting osteotrophic (bone formation) activity. Adequate vitamin-K levels in the diet helps limiting neuronal damage in the brain; thus, has established role in the treatment of patients suffering from Alzheimer's disease.

It is also good source of anti-oxidants such as **silymarin, caffeic acid and ferulic acid,** which help body protect from harmful free-radical agents.

It is also rich source of minerals like copper, calcium, potassium, iron, manganese and phosphorus. **Potassium** is an important component of cell and body fluids that helps controlling heart rate and blood pressure by countering effects of sodium. Manganese is used by the body as a co-factor for the antioxidant enzyme, *superoxide dismutase*. **Copper** is required in the production of red blood cells. **Iron** is required for red blood cell formation.

In addition, it also contains adequate levels of anti-oxidant flavonoid compounds like carotene-beta, lutein and zea-xanthin levels.

Caution

This veggie is very well tolerated in general population. However, its leaves are thought to exacerbate gall stone disease.

CUCUMBER (*Cucumis sativus*)

Fresh Cucumber Ready for Sale

It has a very cooling effect on body. It is rich in potassium, due to which it helps combat fatigue and muscle weakness. Its juice cleanses the kidneys, lowers high blood pressure and improves skin problems. Cucumber is a natural diuretic which is used in the fitness world by body builders and people trying to reduce their weight. It is also said to help digestion and constipation. Although less nutritious than most fruit and vegetables, the fresh cucumber is still a very good source of vitamin C and the mineral molybdenum. It is also a good source of vitamin A, potassium, manganese, folate, dietary fibre and magnesium. What you may not know is that this crisp, refreshing fruit also contains compounds called **sterols**, which have been shown to lower cholesterol in animals. The heaviest concentration of **sterols** is in the skin of the cucumber, so you should not remove the peel before eating. Many people use cucumbers on the skin as a beauty aid as well.

By taking a glass of cucumber juice every day, the skin tone gets improved and the skin starts glowing. It also contains innumerable foliate, besides Vitamin A and C. Cucumber has **diuretic** properties, which can help to eliminate water from the body and it also contains an enzyme that splits protein and cleanses the intestines. Its juice is beneficial to internal inflammations such as stomach and kidney inflammations and sore throats.

Cucumbers are best eaten with skin, which is rich in chlorophyll and silicon. Externally, a blend of juiced cucumber with equal parts of glycerine and rose water makes a soothing lotion for chapped hands and lips. These are excellent as anti-ageing food and improve tired and dehydrated skin. The high water content and high amounts of silica present helps to get a smooth and glowing skin. This wonderful, low calorie vegetable indeed has more nutrients to offer than just water and electrolytes. Cucumbers contain most of the vitamins you need every day. Just one cucumber contains vitamin B1, vitamin B2, vitamin B3, vitamin B5, vitamin B6, folic acid, vitamin C, calcium, iron, magnesium, phosphorus, potassium, and zinc.

Best muscle building foods: Cucumber contains silica which is a component of your connective tissue. It's a cheap muscle building food.

Nutritive Value

Cucumber	Partion	Carbohydrate	Fibre	Fat	Energy
Raw (un-peeled)	5 slices	(g)	(g)	(g)	(kj)
	45g	3	0.5	0	15

There is 1 calorie in 1 slice of cucumber (1 stick, 10cm long)

Cucumber is:

Low in saturated fat, cholesterol and sodium.

High in vitamin C, vitamin K, potassium and vitamin A, pantothenic acid, magnesium, phosphorus and manganese.

Maintaining good general health and losing weight.

But do not eat too much cucumbers if you are interested in gaining weight.

Health Benefits

- Hydrates 95 per cent water
- Aids in digestion
- Flushes out toxins
- Reduces cholesterol
- Regulates body temperature
- Promotes hair growth
- Regulates blood alkalinity
- Promotes Joint health
- Relieves sun burns
- Sooths diseased gums
- Dissolves kidney and bladder stones
- Heal stomach ulcers
- Builds connective tissues
- Relieves headaches
- Promotes healthy skin.

Cucumbers are a highly alkalinizing and hydrating food that are rich in nutrients such as vitamins A, C, K, magnesium, silicon, and potassium. Cucumbers are also packed with antioxidants and enzymes such as erepsin which helps to digest proteins and destroy parasites and tapeworms. The high chlorophyll and lignan content in the cucumber skin makes it a great anti-cancer food and can be particularly helpful in reducing the risk of estrogen related cancers such as breast, uterus, prostate, and ovarian cancer. The high fibre content of cucumbers makes it an excellent remedy for constipation by adding bulk and hydration directly to the colon. Cucumbers are also one of the best natural diuretics around, aiding in the excretion of wastes through the kidneys and helping to dissolve uric acid accumulations such as kidney and bladder stones. They have wonderful anti-inflammatory benefits which can significantly benefit autoimmune and neurological disorders. It can also help to diminish swelling and puffiness underneath the eyes when applied externally. Cucumbers also benefit teeth and gums as the fibre and nutrients help to massage the gums and remove bad bacteria from the teeth. Their high silica content promotes strong and healthy hair and nails which has earned them the reputation for centuries as being a "beautifying" food. Fresh cucumber juice has the ability to cleanse and detox the entire body as well as help to alleviate digestive problems such as gastritis, acidity, heartburn, indigestion, and ulcers. It is also an ideal way to properly hydrate the body since it is contains beneficial electrolytes that have the ability to bring nutrients and hydration deep into the cells and tissues making it far more effective than water alone.

Cucumbers are high in water and low in calories, so they are excellent for weight loss. Moreover, they are also rich in fibre and thus help digestion and prevent constipation.

Fight Inflammation: Cucumber in general has cooling properties and hence can be utilized for reducing the effects of inflammation. It can be effectively used to give relief from acne, pimples and other skin inflammations.

Antioxidant properties: Cucumber is laced with antioxidants like vitamin C and beta carotene in addition to antioxidant flavonoids like quercetin, apigenin, luteolin, and kaempferol, which are beneficial for the body. Kaempferol can help fight cancer and reduce the risk of chronic disease.

Rehydrates body: Cucumbers are constituted by 95 per cent of water. They are an excellent source of hydration of the body while also helping to flush out toxins from the body. A cucumber whether eaten in slices or in a freshly cut salad will provide instant water content to the body and make you refreshed. Also due to its high water content, it is a low calorie food.

Fights Bad Breath: Cucumber is very helpful for those with bad breath. A slice of cucumber when pressed against the roof of your mouth with your tongue for about 30 seconds will eliminate the bacteria that cause bad breath.

Control Stress: Cucumbers are a rich source of vitamin B, along with vitamin B1, B5 and B7. These vitamins are known to be effective in controlling feelings of stress and anxiety. These vitamins boost immunity and energize the body.

Aids in Digestion: Cucumber is mostly made up of water and fibre. These are excellent to help your system digest better. Even Cucumber skins contain insoluble fibre which helps food to move quickly through your digestion tract.

Promote joint health: Silicon in cucumbers strengthens the joint and the connective tissues, if mixed with carrots, it treate gout and arthritis pain by reducing uric acid.

Eliminate toxins: Cucumbers are loaded with water which eliminates the waste from the body and even dissolves kidney stones.

Fight Diabetes: Lowers cholesterol and controls blood pressure, fibre, magnesium and potassium, all of which are found in high levels in cucumbers.

Fight against cancer: They are high in three lignans *viz.* Secoisolariciresinol, Lariciresinol and Pinoresinol, which prevent various cancer types such as breast, ovary and uterus.

Apart from the above benefits, consumption of cucumber is also known to be good for your skin and hair. It rehydrates the skin, cleansing it thoroughly and eliminating the outbreak of pimples and acne. It's also beneficial for the hair, as it gives it a shiny and lustrous look.

Cucumber can be eaten in a variety of ways. They can be sliced, diced, juiced or added in a salad. It is recommended that the skin of the cucumber should also be consumed for better health benefits.

It is one of the very low calories vegetable; provides just 15 calories per 100 g. It contains no saturated fats or cholesterol. Cucumber peel is a good source of dietary fibre that helps reduce constipation, and offers some protection against colon cancers by eliminating toxic compounds from the gut.

It is a very good source of potassium, an important intracellular electrolyte. Potassium is a heart friendly electrolyte; helps reduce blood pressure and heart rates by countering effects of sodium.

It contains unique anti-oxidants in good ratios such as **β-carotene** and α-carotene, vitamin-C, vitamin-A, ***zea-xanthin*** and ***lutein.*** These compounds help act as protective scavengers against oxygen-derived free radicals and reactive oxygen species (ROS) that play a role in aging and various disease processes.

Cucumbers have mild diuretic property probably due to their high water and potassium content, which helps in checking weight gain and high blood pressure.

They are surprisingly have high amount of **vitamin K,** provides about 17 μg of this vitamin per 100 g. Vitamin-K has been found to have potential role in bone strength by promoting osteotrophic (bone mass building) activity. It also has established role in the treatment of ***Alzheimer's*** *disease* patients by limiting neuronal damage in their brain.

Besides for food, this vegetable is used in beauty treatments and beauty products for acne cleansing, the face, *etc.*

This vegetable is able to cleanse your intestines, as well as the entire digestive tract and boost your metabolism.

ASH GOURD (Petha) (*Benincasa hispida*)

The health benefits of ash-gourd have long been known. The ash-gourd is mentioned in ancient Ayurvedic texts like *Charaka Samhita* and *Ashtanga Hridaya Samhita* for its many nutritional and medicinal properties. Ash-gourd is loaded with nutrients. It's an excellent source of vitamin B1 (thiamine), a good source of vitamin B3 (niacin), and vitamin C. It is also rich in many minerals like calcium. Its high potassium content makes this a good vegetable for maintaining a healthy blood pressure. Containing almost 96 per cent water, this gourd is a dietitian's delight. Be sure to include this vegetable in your weight-loss diet.

In Ayurveda and other traditional eastern medicine, ash-gourd is used as a general tonic for its restorative properties. It is also used as brain food - to treat mental illnesses and nervous disorders such as epilepsy, paranoia, and insanity in **Ayurveda.** Ash-gourd is alkaline in nature and hence has a cooling and neutralizing effect on stomach acids and as such used effectively for treating digestive ailments like hyperacidity, dyspepsia, and **ulcers.** Ash-gourd juice is a popular home remedy for peptic ulcers. Ash-gourd juice is also used to treat diabetes. Ash-gourd is also useful in treating respiratory disorders like asthma, blood-related diseases, and urinary diseases like **kidney stones.**

Every part of this fruit is useful. Ash-gourd leaves are rubbed on bruises to heal them, while the seeds are used for expelling intestinal worms. The ash made from burning the rind and seeds are mixed with coconut oil and used to promote hair growth and to treat **dandruff.**

Inexpensive and versatile, ash-gourd is a healthful vegetable that should definitely be a part of any nutritious diet.

Traditional Vaidyasalas in Kerala use a specific ash gourd variety called Vaidya Kumbalam or Nei Kumbalam for preparing ayurvedic medicine called Kushmand rasayana. It is used as a tonic and health rejuvenator for treating diabetes, fever and also as a laxative.

Health Benefits

The dilate juice of ash gourd is useful in the treatment of peptic ulcer. The juice squeezed out of grated out of ash gourd with equal amount of water added to it, should be taken daily in the morning on empty stomach. No food should be consumed for 2 to 3 hours after wards. This also relieves inflammation in the alimentary canal.

A delicious sweet prepared from pulp of the fruit by boiling its pieces in water and adding sugar syrup to it is used as a medicine to increase weight, in tuberculosis, weakness of the heart, heat in the body, thinness of semen and anemia.

Its peel and seeds, when boiled in coconut oil are useful in hair growth; prevent dandruff and dryness of the scalp.

Ash gourd acts as a blood coagulant. From ancient times, its fresh juice mixed with a teaspoon of goose berry or lime juice is used as a specific medicine to stop profuse bleeding from lungs and presence of blood cells in the urine.

Due to its diuretic action, it increases the output of urine and washes out waste products from the body.

Shelled seeds of ash gourd are anabolic- that promotes tissue growth- especially when taken with coconut milk, they expel tape worm and other worms from the intestines.

The fruit is used to treat summer fevers. In **Ayurveda**, the fruit is used to treat **epilepsy**, lung diseases, asthama, coughs, urine retention and internal **hemorrhage.** The rind is used to treat **diabetes** and the seeds to expel tapeworms. The juice of the fruit is effective in cases of mercury poisoning and snakebites. Being extremely low in **calories,** the ash gourd is used to treat **obesity.** A delicious sweet made by boiling the pulp in sugar syrup is used to treat general debility, to increase weight after sudden **weight loss,** weakness of the heart and **anemia.**

It contains a lot of calcium, phosphorus, iron, thiamine, riboflavin, niacin and vitamin C.

It promotes metabolism and prevents sugar from getting converted into fat. Hence it is good to cure obesity as it is also very low in calories.

Cough, cold, asthma, bronchitis, influenza, fever, sinusitis can be controlled without any side effects using ash gourd.

It is also good for bleeding gums. Regular gargling with the juice of ash gourd helps to protect teeth and gums.

Severe **asthma** can be controlled by regular se of ash gourd. It is also useful to control thyroid related problems. It also helps to fight mouth cancer.

Being alkaline in nature, it acts as a good antacid. It maintains the chemistry between the alkaline and the internal acids thus maintaining the PH of the body. Hence it keeps the digestive function in place and reduces acidity, constipation and tones the digestive system.

The unripe ash gourd is cooked as a vegetable. The ripe fruit is largely used for making sweet meat. The leaves and buds are also cooked.

BOTTLE GOURD (*Lagenaria siceraria*)

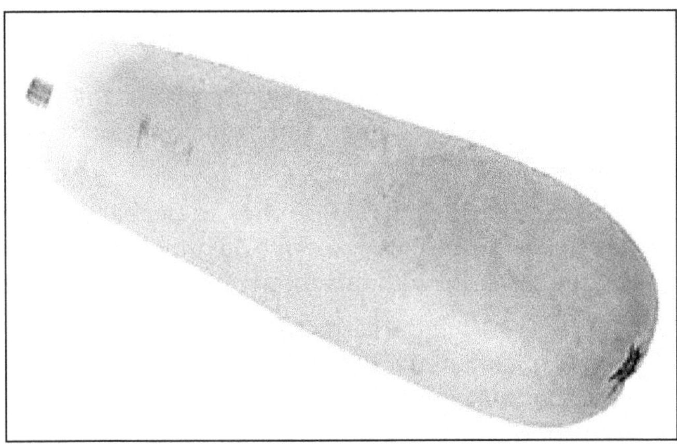

Bottle gourd (*Lagenaria siceraria*): Various medicinal uses of the leaves, fruits and seeds have been recorded from various countries for example as a pectoral, as an anthelmintic, a purgative and even as a headache remedy. The pulp is good for overcoming constipation, cough and night blindness and as antidote against certain poisons. A decoction made from the leaf is very good medicine for curing jaundice. It is good for people suffering from biliousness and indigestion. Its juice is used in the treatment of stomach acidity, indigestion, ulcers, epilepsy and other nervous diseases. According to **Ayurveda** Bottle gourd (lauki) can help lose weight. Cooked bottle gourd is a diuretic and good antidote to urinary disorders. Its juice is prescribed for stomach acidity, indigestion and ulcers. It is rich in dietary fibre with very low content of fat and cholesterol. Approximately 96 per cent of bottle gourd is water. Apart from the iron content it is also rich in vitamin B and vitamin C, it also contains sodium, potassium, essential minerals and trace elements. It is a suitable vegetable for light, low-cal diets as well as for children, people with digestive problems, diabetics and convalescents. It is a suitable vegetable for light, low-cal diets as well as for children, people with digestive problems, diabetics and convalescents. The vegetable is very effective in treating constipation and other digestive disorders.

Bottle Gourd Nutrition

Bottle gourd is rich in dietary fibre with very low content of fat and **cholesterol**. Approximately 96 per cent of bottle gourd is water and is therefore a great thirst quencher. Apart from the iron content it is also rich in vitaminB and vitamin C. it also contains sodium, potassium, essential minerals and trace elements.

High in sodium and potassium, bottle gourd is an excellent vegetable for hypertension patients. It is a suitable vegetable for light, low-cal diets as well as for children, people with digestive problems, diabetics and convalescents.

High in water content, calabash gourd is cooling, calming, diuretic and easy to digest when cooked. The vegetable is very effective in treating constipation and other digestive disorders. It helps in treating urinary disorders, since it serves as an alkaline mixture for treating burning sensation in urinary passage, caused by high acidity of urine.

Bottle gourd is recommended by **ayurvedic** physicians for balancing the liver function, when the liver is inflamed and cannot efficiently process food for maximum nutrition and assimilation.

Drinking a glass of bottle gourd juice once in the early morning helps in treating grey hair. Mixing the juice with sesame oil provides an effective medicine for insomnia. Massage the scalp with this preparation every night.

Bottle gourd juice is very helpful in treating insanity, epilepsy, stomach acidity, indigestion, ulcers and other nervous diseases. The bitter variety of calabash gourd is considered as a cardiac tonic, an antidote for poisoning and a tonic for alleviating bronchitis, cough, asthma, and biligenic affections.

A glass of bottle gourd juice, mixed with a little salt, helps in preventing excessive loss of sodium from the body. It quenches thirst, prevents fatigue and keeps one refreshed during summers. The juice extracted from the leaves of bottle gourd is beneficial in jaundice.

Bottle Gourd Calories

100 gm of bottle gourd contains 15 calories.0.1gm fat in 100gm edible portion. It is comprised of 96 per cent of water. Low in Saturated Fat, and Cholesterol, High in Dietary Fibre, Vitamin C, Riboflavin, Zinc, Thiamin, Iron, Magnesium and Manganese.

1. The cooked vegetable is cooling, diuretic, sedative and antibilious and gives a feeling of relaxation after eating it.
2. It is valuable for in urinary disorders. A glass of fresh juice prepared by grating the whole fruit, mixed with teaspoonful of lime juice if taken daily, is best treatment for burning sensation in the urinary passage due to high acidity of urine. As it serves as an alkaline mixture.
3. Its juice is also a valuable medicine for excessive thirst due to severe diarrhea, diabetes and excessive use of fatty foods. A glassful of plain juice with a pinch of salt if taken daily will overcome this condition. Its daily use (juice) during summer prevents excessive loss of sodium, quenches thirst thus helps prevent fatigue.
4. Mixture of fresh juice along with sesame oil acts as an effective medicine for insomnia. It should be massaged over scalp every night.
5. Bottle gourd juice is used in the treatment of insanity, epilepsy, stomach acidity, indigestion, ulcers and other nervous diseases. It helps constipation.

Famous 'YOGI' Swami Ram Dev Ji Maharaj, who is a household name today in his daily sermons, advises use of its juice along with **Pranayama (**meditation**),** for the treatment of various diseases viz obesity, increased cholesterol, hypertension and other heart ailments. **Ayurveda** says lauki can help lose weight. Cooked lauki is a diuretic and good antidote to urinary disorders. The juice of the fruit is prescribed for stomach acidity, indigestion and ulcers.

Dr Charu Dua, chief nutritionist at the Pushpanjali Crosslay Hospital, said, It's of vital importance that people check the lauki before drinking its juice. If bitter, it carries a highly toxic compound called **Tetracyclic Triterpenoid Cucurbitacins** that can cause serious side-effects and even death. However, as a vegetable it is healthy because it contains a lot **of water**. This makes it very low in calories.

BITTER GOURD (*Morordica charantia*)

Bitter Gourd Hanging Down in Plants

Bitter melon is the immature pod vegetable, popular in many Asian countries. This widely grown as edible pod is, in fact, among the most bitter of all culinary vegetables. Bitter gourd has excellent medicinal virtues. The fresh juice and the extract of the unripe fruit hasbeen found in various studies to have a blood sugar lowering effect. **It**is a herb that regulates blood sugar levels and keeps body functions operating normally. It is **anti-pyretic** tonic, appetizing, stomachic, anti-bilious and laxative. The fruit is said to wormicidal and a cure for stomach disorder. It is beneficial for diabetic patients and persons suffering from **arthritis, rhematism** and **asthama.** Its leaves are known to act as **galactogogs**, while the roots are astringent. Its juice from its leaves is used against lysmenorrhea, as well as for external eruptions, burns or boils. A powder prepared from the plants is good for treating ulcers. The seeds of better gourd contain "**Gourdin**"- which activates the inactive insulin present in blood and rejuvenates the pancrease, thus being beneficial to patients suffering with diabetes. Seeds also contain vaccine that trigger symptoms of **favism** in susceptible individuals. It contains **Gurmarin**, a polypeptide considered to be similar to bovine insulin, which has been shown in experimental studies to achieve a positive sugar regulating effect by suppressing the neural response to sweet stimuli. Bitter gourd contains a unique phyto-constituent that has been confirmed to have a **hypoglycemic** effect called **charantin**. There is also another insulin like compound (momordica), a polypeptide (Protein) which reduces blood sugar levels, in much the same way as insulin, when injected into insulin- dependent diabetics.

An anti diabetes drug gourdin extracted from the seeds of better gourd, showed lowering of blood pressure (Pushpa Khanna, 2007). It also helps in purifying the blood. It contains four very promising bioactive compounds. These compounds activate a pectin called AMPK, which is well known for regulating fuel metabolism and enabling glucose uptake, processes which are impaired in diabetes. Drinking, 50 – 60ml of juice has also shown positive results in clinical trials.

Bitter gourd contains a **lectin** that has insulin like activity. The insulin like bioactivity of this **lectin** is due to its linking together 2 insulin receptors. This lectin lowers blood glucose concentrations by acting on peripheral tissues and also, similar to insulin's effects in brain, suppresses the appetite. This lectin is likely to a major contributor to the hypoglycemic effect that develop after eating bitter gourd and it is why it is recommended as a way of managing adult onset diabetes. According to Kaiydey Nighuntu- it is bitter and Vata Karak. It promotes digestive functions, is purgative, bitter and cold. It is Aphrodiastic, and light pittakarak. It cures cough, anaemia, infection of wounds, burning sensation, dermatoses and is antipyretic. Whileas, according to Bhava Prakash- it is cool, purifies excretory system, purgative, slightly bitter and does not cause vata. It cures fever, blood disorders, cough, jaundice, diabetes, and is **anthelmintic**. Its fragrant volatile oil contains **carotene**, **glucosides**, **saponin** and **mamoridisine** alkaloid.

Nutrition Value of Karela Juice

Karela is an excellent source of all essential nutrients. The amazing health benefits of karela are all attributed by its excellent nutrient content. Be it the vitamins or the minerals, karela has it all. Regular consumption of karela juice fulfills the deficiency of major nutrients that we require on a daily basis. It is a rich store of all water-soluble vitamins like vitamin C, B1, B2, and B3. It also contains minerals like zinc, alkaloids, manganese, and folic acid. The vitamins and minerals in bitter guard are much higher when compared to other green vegetables that are popularly available.

Bitter Gourd is Beneficial

Controls Diabetes: Bitter gourd helps control diabetes due to its blood glucose-lowering effect. It has three active substances with anti-diabetic properties, including charantin, vicine and polypeptide-p. It helps increase pancreatic insulin secretion and prevents insulin resistance, making it beneficial for both Type 1 and Type 2 diabetics.

Purifies Blodo: Bitter gourd also works as a natural blood purifier. Impure blood can cause symptoms like constant headaches, allergies, fatigue and weakened immunity. Bitter gourdhelps in cleansing or detoxifying impurities in the blood. This helps in the treatment of blood disorders like blood boils and itching due to toxemia. It even imparts a "glowing" effect on the skin and keeps the skin free from problems like acne, psoriasis and eczema.

Supports Liver Health: Bitter gourd has hepatic properties, making it highly beneficial for your liver health. It promotes the filtering functions of your liver, which helps cleanse the blood of impurities, and aids in removing harmful toxins from the liver. A properly functioning liver reduces the risk of obesity, cardiovascular disease, chronic fatigue, headaches, digestive problems, jaundice and other health issues. Drink a cup of bitter melon juice at least once a day to keep your liver healthy.

Lowers Cholesterol: This healthy vegetable also helps lower your level of low-density lipoproteins (LDL or the 'bad' cholesterol). Moreover, its phytonutrients and antioxidants help increase your level of high-density lipoproteins (HDL or the 'good' cholesterol) and improves your LDL/HDL ratio. A low bad cholesterol level automatically reduces the risk of heart attacks, heart disease and strokes.

Boosts Immunity: Regular intake of bitter melon can also help reinforce your immune system and improve the body's ability to fight disease. Its antioxidant property works as a powerful defense mechanism against illness. It even protects the body from free radical damage that interferes with the immune system.

Protects against Cancer: Another health benefit of bitter melon comes from its anti-cancer properties. The antioxidants in it help fight free-radical damage that can cause numerous types of cancer. It is helpful in preventing or reducing the risk of cervical, prostate and breast cancers. Researchers analyzed the anti-tumor activity of bitter melon in a 2010 study published in the Pharmaceutical Research journal. They concluded that bitter-melon-related products induce cell cycle arrest and apoptosis without affecting normal cell growth. This helps in the treatment of different types of cancer. To help ward off cancer, include bitter gourd in your diet or drink its juice on a regular basis.

It is used as a folk medicine for diabetes, several research proved that it contains a hypoglycaemic insulin like principle, designed as **plant insulin** which has been found highly beneficial in lowering the blood and urine sugar levels.

It is highly beneficial in the treatment of blood disorders like blood boils, scabies, itching, psoriasis, ringworm and other fungal diseases.

The juice of fresh leaves is valuable in piles, for the treatment of alcoholism and is effective medicine in early stages of cholera and other types of diarrhea during summer.

Regular consumption of bitter gourd juice has been proven to improve energy and stamina level. Even sleeping patterns have been shown to be improved\ stabilized

The high beta-carotene and other properties in bitter gourd make it one of the finest vegetable-fruit that help alleviate eye problems and improving eyesight.

Bitter gourd juice may be beneficial in the treatment of a hangover for its intoxication properties. It also helps cleanse and repair and nourish liver problems due to alcohol consumption.

It juice can help build your immune system and increase your body's resistance against infection.

It hastens healing, when the paste of its roots are applied over the piles.

Regular consumption of its juice has also shown to improve psoriasis condition and other fungal infections like ring worm and athletes feet.

Toxemia – It cleanses the blood from toxins. If two teaspoonful of its juice are sipped daily, it helps cleanse the liver. Also helps in ridding jaundice for the same reason. Precaution should be taken by the pregnant women in taking too much of its juice, as it may stimulate the uterus that may lead to preterm labour.

Very low in calories provides only 17 cal per 100g. The pods are rich in phytonutrients like dietary fibre, minerals, vitamins and anti-oxidants.

Bitter melon notably contains phyto-nutrient, polypeptide-P; a plant insulin known to lower blood sugar levels. In addition it also contain hypoglycemic agent called charantin. *Charantin* increases glucose uptake and glycogen synthesis in the cells of liver, muscle and adipose tissue. Together, these compounds are thought to be responsible for reduction of blood sugar levels in the treatment of type-2 diabetes.

Fresh pods are an excellent source of folates; contains about 72 mcg/100g (Provides 18 per cent of RDA). Folate helps reduce incidence of neural tube defects in pregnant mothers when taken during early pregnancy.

Fresh bitter melon is an excellent source of vitamin-C (100g of raw pod provides about 140 per cent of RDI). Vitamin-C, one of the powerful natural antioxidant, helps body scavenge deleterious free radicals one of the reasons for cancers development.

The vegetable also an excellent source of health benefiting flavonoids such as b-carotene, a- carotene, lutein, zeaxanthins. It also contains good amount of vitamin A. These compounds help act as protective scavengers against oxygen-derived free radicals and reactive oxygen species (ROS) that play a role in aging, cancers and various disease processes.

Bitter melon stimulates digestion and peristalsis which can be helpful in relieving indigestion and constipation problems.

The vegetable is also good source of Niacin (vitamin B-3), Pantothenic acid (vit. B-5), Pyridoxine (vit. B-6) and minerals such as iron, zinc, potassium, manganese and magnesium.

Early laboratory tests suggest that compounds in bitter melon might be effective for treating HIV infection.

Caution

Bitter gourd may contain alkaloid substances like quinine and morodicine, resins and saponic glycosides which may be intolerable by some people. The bitterness and toxicity may be reduced somewhat by parboiling or soaking in salt water for up to 10minutes. Toxicity symptoms may include excessive salivation, facial redness, dimness of vision, stomach pain, nausea, vomiting, diarrhea, muscular weakness).

PUMPKIN (*Cucurbita* spp.)

Mature Pumpkin

Pumpkin is rich in vital anti-oxidants and vitamins. This humble vegetable is very low in calories yet good source of vitamin A, flavonoid poly-phenolic antioxidants like leutin, xanthins and carotenes. In structure, the pulp feature golden-yellow to orange color depending up on the poly-phenolic pigments in it. The fruit has hollow center, with numerous small, off-white colored seeds interspersed in the net like structure. Pumpkin seeds are great source of protein, minerals, vitamins and **omega-3 fatty acids.** It offers vitamins, minerals, fibre and all important antioxidants that our body needs to stay healthy and young. Pumpkin seed oil (PSO) has high Tocopherol (TC) and unsaturated fatty acids (UFA) making it well suited for improving human nutrition. PSO has been implicated in preventing prostate growth, retarding hypertension, mitigating hypercholesterolemia and arthritis, improved bladder compliance, alleviating diabetes by promoting hypoglycemic activity and reduced gastric, breast, lung and colorectal cancer. The high unsaturated and tocoperol content could potentially improve human nutrition. In a study, Kyung-Soo Hahm, Yoonk yung Park and colleagues extracted proteins from pumpkin rinds to see if the proteins inhibit the growth of microbes, including *Candida albicans* (*C. albicans*). That fungus causes vaginal yeast infection, diaper rash infection in infants and other health problems. One protein had powerful effects in inhibiting the growth of *C. albicans*, in cell culture experiments, with no obvious toxic effects. The study suggests that the pumpkin protein could be developed into a natural medicine for fighting yeast infections in humans. The protein also blocked the growth of several fungi that attack important plant crops and could be useful as an agricultural fungicide, the research found. The study has been published in the current issue of ACS journal of Agricultural and Food Chemistry-a bi-weekly publication (Tribune, 29th oct.2009). Pumpkin makes for a great moisturizerand exfoliating masks depending upon the ingredients you use. The enzymes in it give you a younger look and natural **alpha- hydroxy acids** in pumpkin helps to remove cells to reveal younger, fresher skin. Infact, a lot of face products in the market use pumpkin as their main ingredient. Pumpkin nourishes tired or dry skin instantly giving you a healthy glow any time your skin needs a much- needed boost! Pumpkin peels can also help when you are having one of those dreaded break outs and acne problems. The nutrients can help soothe and reduce inflammation naturally, helping skin recover from acne –related damage (Anonymous, 2011). Pumpkin contains a compound (Alpha Linolenic Acid or ALA) that the body converts into a similar type of beneficial omega fatty acid. Pumpkin seeds have lots of vitamin E, a powerful antioxidant that helps slow down skin ageing, and those omega-3 fatty acids that keep your skin smooth and your hair shiny. Pumpkin seeds are packed with zinc. Studies also have shown them to reduce the appearance of acne and other skin problems.

Health and Nutrition Benefits of Eating Pumpkin

Pumpkinis a very low calorie vegetable (actually technically a fruit) that is packed with antioxidants such as vitamins A, C, E, beta-carotene, and zea-xanthin which can help prevent age-related macular disease. Pumpkin is also a rich source of B-vitamins such as folates, niacin, thiamin, and pantothenic acid and minerals like calcium, copper, potassium, and phosphorus. It is an excellent food for aiding weight loss and for reducing cholesterol. The high amount of phytonutrients in pumpkin have been shown to help prevent the risk of cancers, particularly mouth, lung, and colon cancer. Pumpkin also contains several anti-aging benefits and the antioxidants help to keep skin wrinkle-free and radiating a healthy glow. It also contains immune-boosting properties which can help the body stay strong and ward off common colds and flus that may be going around. Pumpkin is excellent way to replenish and restore the body after a workout as it is high in potassium which helps to restore the body's balance of electrolytes after exercise and keeps muscles functioning at their best. Pumpkin is a versatile food and can be eaten raw, steamed, baked, or roasted. Frozen or canned pumpkin is an alternative option to include this healthy vegetable into your diet when you are short on time

Rich in Fibre

Pumpkin is fairly high in fibre. One cup of cooked, mashed pumpkin contains 3 grams of fibre, or roughly 11 per cent of your recommended daily intake of fibre. Pumpkin seeds also contain somefibre, about 1.1 g in 28 gm of seeds. You can consume boiled, roasted or baked pumpkin. You can also use pumpkin as a major ingredient in soup, breads and pies. Pumpkin is a great home remedy for a constipated pet because of pumpkin's fibre content and great taste.

Excellent Source of Vitamins A, K and C

Vitamin A is a key nutrient for keeping our eyes healthy and our vision good. Moreover, it also helps promote bone growth, keeps the immune system strong and maintains a vigorous reproductive system. Half a cup of canned pumpkin has 953 mg of vitamin A and only 42 calories and that vitamin A comes in the form of beta carotene. Yet another reason why pumpkin is good for you is because it contains a lot of Vitamin K, about 40 per cent of the recommended daily dose of it! Vitamin K is the overlooked bone builder and heart protector. In fact, one serving of pumpkin contains almost 20 per cent of your recommended daily dose of Vitamin C which is needed for the growth and repair of tissues in all parts of your body.

Loaded with iron, potassium and magnesium: One small serving of pumpkin contains 250 milligrams of potassium and the same small serving of pumpkin also contains a good amount of iron. One cup has a little less than 10 per cent of your recommended daily allowance of iron. Moreover, unlike other sources of iron, pumpkin is fat-free.

Our body needs magnesium for maintaining normal muscle function and for boosting our immune systems, among other things. Pumpkin seeds provide us with enough magnesium to perform these functions well.

Pumpkin is very rich in carotenoids, which is known for keeping the immune system of an individual strong and healthy.

Beta-carotene, found in pumpkin, is a powerful antioxidant as well as an anti-inflammatory agent. It helps prevent build up of cholesterol on the arterial walls, thus reducing chances of strokes.

Being rich in **alpha-carotene**, pumpkin is believed to slow the process of aging and also prevent cataract formation.

Pumpkins have been known to reduce the risk of macular degeneration, a serious eye problem than usually results in blindness.

The high amount of fibre, present in a pumpkin, is good for the bowel health of an individual. Being loaded with potassium, pumpkin is associated with lowering the risk of hypertension.

The presence of zinc in pumpkins boosts the immune system and also improves the bone density.

It is one of the vegetables which is very low calories; provides just 26 cal per 100 g and contains no saturated fats or cholesterol; but is rich a source of dietary fibre, anti-oxidants, minerals, vitamins. Recommended by dieticians in cholesterol controlling and weight reduction programs.

Pumpkin is a storehouse of many anti-oxidant vitamins such as vitamin-A, vitamin-C and vitamin-E.

It is one of the vegetable in the cucurbitaceae family with highest levels of **vitamin-A**, which is a powerful natural anti-oxidant and is required by body for maintaining the integrity of skin and mucus membranes. It is also an essential vitamin for vision. Research studies suggest that natural foods rich in vitamin A helps body protect against lung and oral cavity cancers.

It is also an excellent source of many natural poly-phenolic flavonoid compounds such as α and β carotenes, cryptoxanthin, lutein and zeaxanthin. Carotenes convert into vitamin A inside the body.

Zea-xanthin is a natural anti-oxidant which has UV (ultra-violet) rays filtering actions in the macula lutea in reitina of the eyes. Thus, it helps protect from "age related macular disease" (ARMD) in the elderly.

Rich in B-complex group of vitamins like folates, niacin, vitamin B-6 (pyridoxine), thiamin and pantothenic acid.

Not only is this vegetabe low in calories but it is also high on fibre.

Since it is high on fibre, it helps in keeping apetite in check by slowing down overall digestion. The disease fighting nutrients in pumpkin like vitamins, potassium and magnessium help in improving your immunity too.

It is also rich source of minerals like copper, calcium, potassium and phosphorus.

Health and Nutrition Benefits of Pumpkin Seeds

Pumpkin seeds are good source of dietary fibre and mono-unsaturated fatty acids which are good for heart health. In addition, they are very good in protein, minerals and many health benefiting vitamins. For example 100 g of pumpkin seeds provide 559 cal, 30 g of protein, 110 per cent RDA of iron, 4987 mg of niacin (31 per cent RDA), selenium (17 per cent of RDA), zinc (71 per cent) etc. but no cholesterol.

The seeds are an excellent source of health promoting amino acid **tryptophan**. Tryptophan is converted to GABA in the brain.

They promote overall prostate health, apart from alleviating the problem of difficult urination that is associated with an enlarged prostate.

They comprise of L-tryptophan, a compound that has been found to be effective against depression.

They are believed to serve as a natural protector against osteoporosis.

They have been known to reduce inflammation, without causing the side effects of anti-inflammatory drugs.

Studies have revealed that they help prevent calcium oxalate kidney stone formation.

Being rich in phytosterols, they have been associated with reducing the levels of LDL cholesterol.

Pumpkin seeds are high in zinc, which is good for the prostate and building the immune system.

They also contain fatty acids that kill parasites.

Raw pumpkin seeds contain essential fatty acids and beneficial proteins.

For maximum nutritional benefits, seeds should be eaten raw.

Roasted seeds contain damaged fat that can lead to plaque in the arteries.

It prevents heart attacks, strokes and cardic arrest.

It improves insulin regulation and prevents diabetic complications.

It improves better sleep at night, hence having them before bed time can be helpful.

It eliminates tape worms and other harmful parasites from the body.

It boosts your metabolism.

SQUASH (*Cucurbita morschata*)

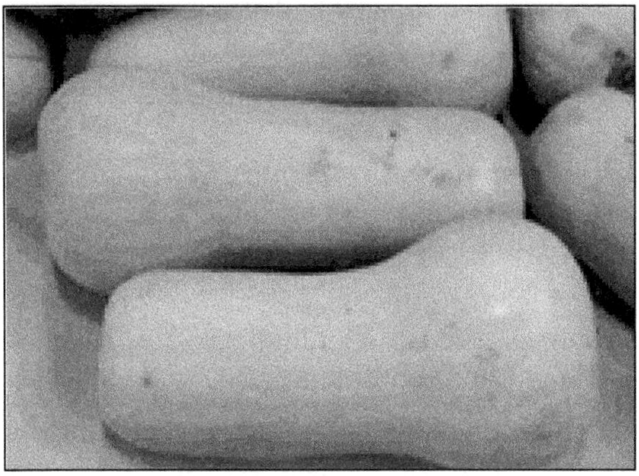

Ripe Squash Butternut

Squash Butternutis the most popular among winter squash varieties. Oftentimes, the squash is recognized as large pear shaped golden-yellow pumpkin fruit, which is put for sale in the markets. Butternuts are annual long trailing vines. They usually cultivated in warmer climates of South and Central American regions for their edible fruits, flowers, as well as seeds.

The butternut plant is monoecious as in **pumpkins**, and feature different male and female flowers that require honeybees for effective pollination. Butternut, in-fact, is the most common among winter squash fruits.

Externally, the squash is better described as large sized fruit featuring thick neck attached to a pear shaped bottom and has smooth, ribbed skin. However, the fruit varies widely in its shape and size; with individual fruit may weigh up to 15 kg. Interiorly, its flesh is yellow to orange in color. Cross-section of lower bulb part feature central hollow cavity containing mesh-like mucilaginous fibres interspersed with large, flat, elliptical seeds similar to pepita (pumpkin seeds). The fruit's unique golden yellow color comes from yellow-orange pigments in their skin and pulp.

Butternut squash seeds are used as nutritious snack food since they contain 35-40 per cent oil and 30 per cent protein. In Argentina, the fruit is also used to feed livestock.

Health Benefits Squash (Butternut)

Butternut squash contains many vital poly-phenolic anti-oxidants and vitamins. Similar to other cucurbitaceae members, it is very low in calories; provides just 45 cal per 100 g. It contains no saturated fats or cholesterol; but is rich source of dietary fibre and phyto-nutrients. Squash is one of the common vegetable that is often recommended by dieticians in the cholesterol controlling and weight reduction programs.

It has more **vitamin A** than that in pumpkin. At 10630 IU per 100 g, it is perhaps the single vegetable source in the cucurbitaceae family with highest levels of **vitamin-A**, providing about **354 per cent of RDA**. Vitamin A is a powerful natural anti-oxidant and is required by body for maintaining the integrity of skin and mucus membranes. It is also an essential vitamin for vision. Research studies suggest that natural foods rich in vitamin A helps body protect against lung and oral cavity cancers.

Furthermore, butternut squash has plentiful of natural poly-phenolic flavonoid compounds like α and β-carotenes, cryptoxanthin-β, and lutein. These compounds convert to vitamin A inside the body and deliver same protective functions of vitamin A on the body.

It is rich in B-complex group of vitamins like folates, riboflavin, niacin, vitamin B-6 (pyridoxine), thiamin, and pantothenic acid.

It has similar mineral profile as pumpkin, containing adequate levels of minerals like iron, zinc, copper, calcium, potassium, and phosphorus.

Butternut squash seeds are good source of dietary fibre and mono-unsaturated fatty acids that are good for heart health. In addition, they are very good in protein, minerals, and numerous health benefiting vitamins. The seeds are excellent source of health promoting amino acid **tryptophan**. Tryptophan converts to health benefiting GABA neuro-chemical in the brain.

SUMMER SQUASH

Summer squash are believed to be originated in Central America and Mexico. Different cultivars of summer squash are grown throughout the United States during the warm, frost free season. Almost all the members of the squash family feature tender skin and flesh, small edible seeds and high moisture content. Like other members of the summer squash group, the zucchini plant has the bush habit rather than the vine spread of the winter squashes. Its fruits are ready for harvesting about 40-50 days after seed implantation.

Health Benefits

One of the very low calories vegetable that is used during weight reduction and cholesterol control programs. Zucchinis provide only 17 calories per 100 g. Contains no saturated fats or cholesterol. Its peel is good source of dietary fibre that helps reduce constipation and offers some protection against colon cancers.

Courgette is relatively moderate source of folates, consists of 24 mcg or 6 per cent of RDA per 100 g. Folates are important in cell division and DNA synthesis. When taken adequately before pregnancy, it can help prevent neural tube defects in the fetus.

It is a very good source of potassium, an important intra-cellular electrolyte. Potassium is a heart friendly electrolyte; helps reduce blood pressure and heart rates by countering effects of sodium.

Fresh fruits are rich in vitamin A; provide about 200 IU per 100 g.

Furthermore, zucchinis, especially golden skin variety are rich in flavonoid poly-phenolic antioxidants such as carotenes, lutein and zeaxanthin. These compounds help scavenge harmful oxygen-derived free radicals and reactive oxygen species (ROS) from the body that play a role in aging and various disease process.

Fresh fruit is good source of anti-oxidant vitamin-C. Provide about 17.9 mcg or 30 per cent of RDA per 100g.

In addition, they are also good in B-complex group of vitamins like thiamin, pyridoxine, riboflavin and minerals like iron, manganese, phosphorus, zinc and potassium. Potassium in an important component of cell and body fluids, helps controlling heart rate and blood pressure.

Zucchini and other summer squash varieties contain vitamins A and C.

They also contain potassium and calcium.

The flavor of zucchini is best when it is less than six inches long.

Zucchinis can grow as large as baseball bats but have little flavor when they reach this size.

MUSKMELON (*Cucusmis melo*)

In general, melons feature round or oblong shape, measure 4.5- 6.5 inches in diameter and weigh 450 – 850 gm. Internally, flesh range from orange-yellow to salmon color, has soft consistency and juicy texture with a sweet, musky aroma that emanates best in the ripe fruits. At the center, there is a hollow cavity filled with small off-white color seeds encased in a web of mucilaginous netting.

It is mainly a dessert fruit which is very wholesome and nutritious. Muskmelon (cantaloupe) are a good source of potassium, vitamin A and folate. The potassium is helpful in preventing kidney stones and cantaloupes in general are a useful laxative. It is low in calories and very nutritious. The hardy green cover protects the rich red juicy flesh inside which has the properties of a powerful anti-oxidant. Consuming about a coup full of musk melon juice provides you healthy antioxidants that will make you immune to fight against the infections. The potassium present in the fruits helps maintain the blood pressure, regulates heart beat, prevents muscle cramping, maintain electrolyte balance in ectracellular fluid, helps reduce water retention, providing sufficient oxygen to the brain so that it can function normally and also helps in preventing stroke. Musk melons have good sedative properties and help those suffering from **insomnia**. It acts as a mild anti-depressant. Its juice is helpful in curing lack of appetite, acidity, ulcer, weight loss and urinary tract infection, reduces the heat in the body and is also helpful in reducing heat related disorders. Its high water content helps you stay hydrated during the hot season. Having this yellow-orange coloured succulent fruit early in the morning is also good for your digestive system. A great source of beta-carotene, folic acid, potassium, vitamins C and A, muskmelon not only helps you stay healthy, but is also great for your skin and hair. Here's why you should include this fruit in your diet.

Beneficial to be consumed during the following conditions:

1. Lack of appetite
2. Weight loss
3. Urinary tract infections
4. Constipation, acidity and ulcers

Health Benefits

Strengthens immune system: A rich source of Vitamin C, muskmelon helps boost your immunity by stimulating the White Blood Cells in the body.

Prevents cancer: According to studies, the high carotenoid content in this fruit can help to prevent cancer, especially that of lungs.

Good for the heart: The potassium content in the fruit helps control blood pressure and prevents hypertension. It also reduces the risk of heart diseases. Containing adenosine that has blood thinning properties, muskmelon also prevents blood clotting in the cardiovascular system.

Helps cure kidney diseases: A good diuretic, the fruit helps cure kidney disease. Having it with lemon early morning is helpful.

Good for vision: Rich in vitamin A and beta carotenes, studies have proven that regular consumption of foods like muskmelon can reduce the risk of developing cataracts by 40 per cent.

Helps treat insomnia: With effective laxative properties – a special compound that calms anxieties – muskmelon soothes your nervous system, getting rid of your sleeping disorders.

Treats menstrual problems: The vitamin C helps regulate menstrual flow, thus relieving cramps in women. Have this fruit daily during the painful time of the month.

Cures Toothache: The rind of this fruit is effective in curing toothache. Peel the fruit's skin and add it to boiling water, till it is cooked. Once the mixture cools, strain and use it as a mouth rinse.

Prevents nausea: Having muskmelon tea, prepared from the fruit's root, is an effective diuretic and helps reducevomiting sensation

Helps lose weight: The next time you scoop out the seeds of this fruit, don't throw them away. High in fibre, consuming these seeds will help in weight loss.

Anti-ageing agent: The fruit helps combat free radicals in the body that damage the skin and are responsible for causing premature ageing.

Rejuvenates skin: The collagen content, along with the protein compounds in this fruit, helps in your skin cell regrowth, providing a beautiful glow to your skin.

Cures skin problems: If you suffer from skin problems like eczema, make sure to have muskmelon juice regularly. You can also apply this juice as lotion to treat skin problems or as first aid for burns and abrasions.

Fights hair loss: **Insonitol**, a form of Vitamin B, required for your hair growth, is found in most citrus fruits as well as muskmelon.

Acts as natural conditioner: Mash the fruit in a cup and massage your hair with its pulp after shampooing. Wash off after 10 minutes.

Wonderfully delicious with rich flavour, muskmelons are very low in calories and fats; but rich in numerous health promoting poly-phenolic plant derived substances, vitamins and minerals that are required for optimum health.

The fruit is an excellent source of Vitamin A, (provides about 112 per cent of recommended daily levels) *one of the highest among fruits.* Vitamin A is a powerful antioxidant and is essential for vision. It is also required for maintaining healthy mucus membranes and skin. Consumption of natural fruits rich in vitamin A is known to help to protect from lung and oral cavity cancers.

It is also rich in antioxidant flavonoids such as ***beta carotene****, lutein, zeaxanthin* and *cryptoxanthin.* These **antioxidants** have the ability to help protect cells and other structures in the body from oxygen free radicals and hence; offer protection against colon, prostate, breast, endometrial, lung, and pancreatic cancers.

Zeaxanthin, an important dietary carotenoid, selectively absorbed into the retinal macula lutea in the eye where it is thought to provide **antioxidant** and protective light-filtering functions; thus it protects eyes from "Age related macular degeneration" (ARMD) disease in the elderly.

It is a good source of potassium. Potassium in an important component of cell and body fluids and helps control heart rate and blood pressure; thus offers protection against stroke and coronary heart diseases.

The fruit also contain good amounts of B-complex vitamins, such as niacin, pantothenic acid and **vitamin C** and minerals like manganese. Consumption of foods rich in vitamin C helps body develop resistance against infectious agents and scavenge harmful oxygen free radicals. Manganese is used by the body as a co-factor for the antioxidant enzyme superoxide dismutase. Commercially, muskmelons are being used to extract an enzyme known as *superoxide dismutase* (SOD) which is essential for maintaining strong antioxidant defenses in the human body.

Other Properties

It reduces heat in the body to a great extent.

Relieves tiredness

Increases appetite

Increases kapha more

Reduces vata and pitta

It is very good to digest

Sweet in taste

Unctuous in quality

Cold in potency

Diuretic and aphrodisiac

Laxative and strengthening

It does not contain cholesterol.

It counteracts the effect of cigrette smoke on your lungs by restoring the vitamin A level in the body.

Rich in vitamin A and betacarotene, which are important vision nutrients.

It helps maintain healthy skin because of vitamin A presence.

Rich in folic acid which is essential for pregnant women.

It helps to create healthy fetuses and can even prevent cervical cancer and **osteoporosis.**

WATERMELON (*Citrullus lanatus* Thunb)

Watermelon has everything you need to beat the scorching summer heat. Wonderfully delicious, Coolest thirst-quenching melons are the great source of much needed water (92 per cent) and **electrolytes** to beat the summer temperatures. It is not only cooling, but is also a natural diuretic. It is very alkaline in nature and has a soothing effect on the stomach and prevents biliousness. It is the thirst-quencher and is beneficial for those suffering from kidney problems and urinary tract infections. Pumpkin has folic acid, vitamin A, and vitamin C. Bitter gourd has polypeptide P (an insulin – like compound that can replace insulin).

They are indeed the best thirst quenchers, but are rich in anti-oxidants as well. Juicy watermelons are indeed packed with some of the best antioxidants in nature. Research has shown that they are good for cardiovascular diseases, **colon cancer** and **diabetes** too. They are rich in Vitamin A, B and C. What's more, they are low in calories and very nutritious. It is an ideal health drink because it is 92 per cent water, and does not contain any fat or cholesterol. It also fills you up fast because of its water content. So have it along with your barbeque or grills as watermelons will prevent you from overeating.

A team of US researchers led by Dr. Bhimangouda Bhimu Patil, Director of Texas A and M's Fruit and Vegetable Improvement, Center in College Station says that watermelon has ingredients that deliver **Vigra**-like effects to the body's blood vessels and may even increase libido. The study further says that the list of its very important healthful benefits grow longer with each study. Many fruits and vegetables contain nourishing ingredients known as phyto-nutrients, naturally occurring compounds that are bio-active, or able to react with the human body to trigger healthy reactions. In watermelon, these include lycopene –which is typically found in tomatoes; beta carotene; and the rising star among its phyto-nutrients- **citrulline-** an amino acid, whose beneficial functions include the ability to relax the blood vessels, much like **Viagra** and other impotence drugs do. Scientists say that when watermelon is consumed, citrulline is converted to Arginine, a complex amino acid through certain enzymes, that boosts nitric oxide, which relaxes the blood vessels in the same way as drugs that treate **erectile dysfunction**, thus increase the libido. Arginine is an amino acid that helps the heart and circulation system and maintains a good immune system (Chidanand, 2008). Juice from the rejected water melons is a potential source of biofuel as it can be fermented into ethanol says a study.

Watermelon is the richest edible natural source of **L-citrulline**, which is closely related to **L- arginine,** the amino acid required for the formation of nitric acid oxide essential to the regulation of vascular tone and healthy blood pressure. Once in the body, the L-citrulline is converted in to L-arginine. Simply consuming L-arginine as a dietary supplement is not an option for many hypertensive adults, because it can cause nausea, gastrointestinal tract discomfort and diarrhea. In addition to the vascular benefits of citrulline, watermelon provides vitamin A, B6, C fibre, potassium and lycopene, a powerful antioxidant. Watermelon may help to reduce serum glucose levels. It is both watermelon and its extract is the best source for L-citrulline. Individuals with increased blood pressure and arterial stiffness-especially those who are older and those with chronic diseases such as type 2 diabetes-would benefit from L-citrulline in either the synthetic or natural (watermelon) form The optimum appears to be 4-5 grams a day (Figueroa and Arjmandi, 2010). It contains about 6 per cent sugar by weight, the rest being mostly water. As with many other fruits, it is a source of vitamin C. It is not significant source of other vitamins or minerals unless one eats several kilograms per day. It contains a significant amount of **citrulline** and after consumption of several kilograms an elevated concentration is measured in the blood plasma, this could be mistaken for **citrullinaemia** or other urea cycle disorders. It is a traditional food plant in Africa, this fruit has potential to improve nutrition, boost food security, foster rural development and support sustainable land care. It is also packed with a giant dose of glutathione, which helps boost our immune system. They are also a key source of lycopene - the cancer fighting oxidant. Other nutrients found in watermelon are vitamin C and Potassium.

Watermelon juice is the best medicine for those suffering from blood pressure because the potassium and magnesium controls the blood pressure in the body. One of the greatest quality in this fruit is it contains 92 per cent of alkaline water, which is not there in any fruit. Adding watermelon to your weight loss diet can help you lose weight. Watermelon is a water-heavy fruit that is extra filling. Watermelon has 50 calories per cup. Many people are concerned with watermelon's high glycemic index ranking of 72, but watermelon has a glycemic load score of 4 out of 40. Watermelon will not dramatically effect your blood sugar because it has very few carbohydrates per serving.

This food is very low in Saturated Fat, Cholesterol and Sodium. It is also a good source of Potassium, and a very good source of Vitamin A and Vitamin C, but large portion of the calories in this food come from sugars. The juicy fruit is 92 per cent water, eight per cent sugar, a great source of vitamin C and only 71 calories per serving.

They are indeed the best thirst quenchers, **no doubt,** but did you know that they are rich in anti-oxidants as well? Juicy watermelons are indeed packed with some of the best antioxidants in nature. Research has shown that they are good for cardiovascular diseases, **colon cancer** and **diabetes** too. They are rich in Vitamin A, B and C. What's more, they are low in calories and very nutritious. It is an ideal health drink because it is 92 per cent water, and does not contain any fat or cholesterol.

Health Benefits

Watermelon is an excellent fruit that effectively hydrates, detoxifies, and cleanses the entire body on a cellular level. It is rich in vitamins A and C as well as lycopene, beta-carotene, lutein, and zeaxanthin which are excellent for providing protection from lung, mouth, pancreatic, breast, prostate, endometrial, and colon cancer. Watermelon is also known to significantly reduce inflammation, help flush out edema, aid in weight loss, and alleviate depression. Watermelon can also boost the immune system as well as strengthen vision. Watermelon is not nearly as high in sugar as most people think as it has half the sugar than an apple. Watermelon is loaded with antioxidants that have the ability to neutralize free radical molecules and aid in the prevention of chronic illnesses. The rind of the watermelon is equally beneficial as it is one of the highest organic sodium foods in nature and one of the best sources of chlorophyll and can be juiced for a delicious and healing drink. And, if you are lucky enough to get a watermelon with black seeds, even better! Crunch those seeds up too, they have an amazing effect on the nervous system, aiding in relaxing the body and lowering blood pressure and contain helpful amounts of iron, zinc, and protein.

Rich in electrolytes and water content, melons are nature's gift to beat tropical summer thirst.

Watermelons are very low in calories (just 30 cal per 100 g) and fats yet very rich source of numerous health promoting phyto-nutrients and anti-oxidants that are essential for optimum health.

Watermelon is an excellent source of Vitamin-A, which is a powerful natural anti-oxidant. It is essential for vision and immunity.

It is also rich in **anti-oxidant flavonoids** like *lycopene*, **beta-carotene,** *lutein, zeaxanthin and cryptoxanthin.* These antioxidants are found to be protective against colon, prostate, breast, endometrial, lung, and pancreatic cancers. Phyto-chemicals present in watermelon like **Lycopene** and **carotenoids** have the ability to help protect cells and other structures in the body from oxygen free radicals. Studies have also shown that *lycopene* protects skin damage from UV rays and from prostate cancer.

Watermelon fruit is a good source of potassium; Potassium is an important component of cell and body fluids that helps controlling heart rate and blood pressure; thus offers protection against stroke and coronary heart diseases.

It also contains good amount of vitamin-B6 (pyridoxine), thiamin (vitamin B-1), vitamin-C and manganese. Consumption of foods rich in vitamin-C helps body develop resistance against infectious agents and scavenge harmful oxygen free radicals.

It is 92 per cent water by weight. And it is also mildly diuretic, meaning that it helps increase urine flow but does not strain the kidneys (unlike alcohol and caffeine). It not only detox the kidney, it actually helps the liver process ammonia (waste product from protein digestion), which again, eases kidney strain while getting rid of excess fluids and sodium retension.

Watermelon is one of the most alkaline forming fruit on the planet. Eating alkaline forming fruits, help reduce your risk of developing disease and illness caused by a meat, egg and diary products (aka. Acid forming diet. With more alkalinity, you reduce your risk of developing cancer later in life.

Watermelon with red flesh is a significant source of l**ycopene** and a very concentrated source of the carotenoid. Lycoene helps in reducing colon, prostate, breast, endometrial and lung cancers. Lycopene present improves blood flow via vasodilation (relaxation of blood pressure).

Choosing to regularly eat lycopene-rich fruits, such as watermelon, *and* drink green tea may greatly reduce a man's risk of developing prostate cancer, suggests research published the *Asia Pacific Journal of Clinical Nutrition*

Watermelon is an excellent source of **vitamin C** and a very good source of **vitamin A,** notably through its concentration of beta-carotene. Pink watermelon is also a source of the potent carotenoid antioxidant, lycopene. These powerful antioxidants travel through the body neutralizing free radicals.

Watermelon is rich in the B vitamins necessary for energy production. Our food ranking system also qualified watermelon as a very good source of vitamin B6 and a good source of vitamin B1, magnesium, and potassium.

This sweet, crunchy, cooling fruit is exceptionally high in **citrulline**, an amino acid our bodies use to make another amino acid, **arginine,** which is used in the urea cycle to remove ammonia from the body, and by the cells lining our blood vessels to make nitric oxide. Nitric oxide not only relaxes blood vessels, lowering high blood pressure, it is the compound whose production is enhanced by Viagra to prevent **erectile dysfunction. Arginine** has been shown to improve insulin sensitivity in obese type 2 diabetic patients with insulin resistance. Am J Physiol Endocrinol Metab. 2006 Nov; 291 (5). Citrulline relaxes blood vessels much like Viagra without any side effects.

Watermelons may relieve muscle soreness. Eating watermelon could be effective means of alleviating the pain associated with muscle soreness

Watermelon helps eliminate free radicles, that can oxidize cholesterol and increase **flammation.**

Citrulline in watermelon relaxes and helps lower high blood pressure, which in turn help ease strain on the kidneys, improves their health.

Watermelon can help prevent cardiovascular disease: It is effective at preventing pre-hypertention (which can lead to cardiovascular disease). The amino acid L-citrulline and L-arginine from watermelon extract-**their aortic blood pressure lowered.**

Extra nitric oxide can also help treat angina, high blood pressure and other cardiovascular problems, the study noted.

Reduces high blood pressure and boosts immune system.

Reduces heart disease risk and blood sugar level.

Has high water content (92 per cent) cleans kidney and fighets cancer.

Highest alkalizing fruit and strengthens bones.

It contains a lot of vitamin C, which protects from free-dadiocal DNA damage and lycopene helps reduce inflammation and neutralize free radicals.

Watermelon Seeds

A rich source of iron, potassium, and vitamins, these seeds are great for your health, skin and hair. So, chew on the watermelon seeds or usetheir oil (Anonymous, 2014).

Good for your heart: A rich source of magnesium, the seeds help in maintaining normal heart functioning, blood pressure and support metabolic processes. They can also treat cardiovascular diseases and hypertension.

Control diabetes: Boiling a handful of watermelon seeds in water and drinking this liquid every day like tea is said to control your blood sugar levels.

Moisturise skin: Containing unsaturated fatty acids, the seeds help keeping your skin moisturised.

Delay ageing: Due to their antioxidant properties, eating these seeds can make your skin look younger and healthier.

Clear acne: Take a swab of cotton and apply watermelon seed oil on your face to remove the dirt and sebum from your skin as well as clean the pores. Suitable for all skin types, this oil also helps treat skin infections.

Strengthen your hair: Due to a high content of protein and essential amino acids, eating watermelon seeds can strengthen your hair. Also, having roasted watermelon seeds can add shine to your hair as they contain copper that produces melanin, a pigment that provides colour to your hair.

Treat itchy scalp: Since watermelon seed oil is light in texture, it gets absorbed easily without clogging the pores of your scalp. It acts as a moisturiser for your dry and flaky scalp.

Prevent hair breakage: Providing your hair with essential fatty acids, consuming the seeds regularly can help prevent breakage.

THYME (*Thymus vulgarius*)

Thyme is one of the sweet smelling herb, popular among culinary plants which has been used traditionally for common winter ailments. Thyme herb is packed with numerous health benefiting phyto-nutrients (plant derived compounds), minerals and vitamins that are essential for wellbeing. It has been in use extensively for the high **thymol**, **pheno**l and **carvo**l content in the thyme oil. The flowering thyme tops contain an essential oil consisting primarily of thymol and carvacrol, along with tannins, bitter compounds Saponins and organic acids. Thymes best use medicinally is as an antiseptic, but it also has expectorant, anti spasmodic, and deodorant properties. It aids in digestion and as such is excellent when combined with fatty meats that often causes gastrointestinal problems. It is also a popular culinary herb. The essential oil of common thyme (*Thymus vulgaris*) is made up of 20 -55 per cent **thymol.** Thymol, an antiseptic is the main active ingredient in Listerine mouth wash. Before the advent of modern antiseptics, it was used to medicate bandages. It has also been shown to be effective against the fungus that commonly infects toe nails. Medicinally thyme is used for respiratory infections in the form of a Tincture, tisane, salve, syrup or by steam inhalation. Because it is antiseptic, thyme boiled in water and cooled is very effective against inflammation of the throat when gargled 3 times a day. The inflammation will normally disappear in 2 – 5 days. Other infections and wounds can be dripped with thyme that has been boiled in water and cooled.

Thyme is a good tonic for stomach and nerves. Its tea relieves flatulence, promotes appetite, strengthens digestion, loosens phlegm and increases prespiration. Infusions have a calming effect, relax muscle spasms and alleviate exhaustion. Extracts and infusions are also used to treat bronchitis, laryngitis and coughs. Thyme vinager was used for centuries to relieve headaches. Essential oil of thyme makes a good antiseptic mouthwash, and is also used externally for warts and for relaxing bath (Michael Shron, 2009).

Health Benefits

Thyme contains many active principles that are found to have disease preventing and health promoting properties.

Thyme herb contains **thymol,** one of the important essential oils, which scientifically have been found to have antiseptic, anti-fungal characteristics. The other volatile oils in thyme include *carvacolo, borneol* and *geraniol.*

Thyme contains many flavonoid **Phenolic antioxidants** like *zeaxanthin, lutein, pigenin, naringenin, luteolin,* and *thymonin*. Fresh thyme herb has one of the highest antioxidant levels among herbs, a total ORAC (Oxygen Radical Absorbance Capacity).

Thyme is packed with minerals and vitamins that are essential for optimum health. Its leaves are one of the richest sources of **potassium**, **iron**, calcium, **manganese**, magnesium, and selenium. Potassium is an important component of cell and body fluids that helps controlling heart rate and blood pressure. Manganese is used by the body as a co-factor for the antioxidant enzyme, *superoxide dismutase.* Iron is required for red blood cell formation.

The herb is also a rich source of many important vitamins such as B-complex vitamins, beta carotene, vitamin A, vitamin K, vitamin E, vitamin C and folic acid.

Thyme provides 0.35 mg of **vitamin B-6** or pyridoxine; furnishing about 27 per cent of daily recommended intake. Pyridoxine keeps up GABA (beneficial neurotransmitter in the brain) levels in the brain, which has stress buster function.

Vitamin C helps body develop resistance against infectious agents and scavenge harmful, pro-inflammatory free radicals.

Vitamin A is a fat soluble vitamin and antioxidant that is required maintaining healthy mucus membranes and skin and is also essential for vision. Consumption of natural foods rich in flavonoids like vitamin A and beta-carotene helps protect from lung and oral cavity cancers.

Another spice that can be used as a preservative is thyme, for it too has antioxidant properties and can be used to prevent not only food from decaying but cosmetic products as well. It is most effective when used in conjunction with other preservatives.

Caution

Too much thyme can over stimulate the thyroid and cause poisoning symptoms.

CORIANDER (*Coriandrum sativum*)

Coriander Plant with Leaves

What we call coriander leaves are known as **Cilantro** in US and Canada, Chinese parsley, Mexican parsley and Kindza. **It** has been used as a folk medicine for the relief of anxiety and insomnia in Iranian folk medicine. Experiments on mice support its use as an anxiolytic. It was considered an aphrodisiac by ancient Greeks, Chinese and Egyptians and was believed to increase a man's sexual prowess even in the afterlife. Coriander seeds are used in traditional Indian medicine as a diuretic by boiling equal amounts of coriander seeds and cumin seeds, then cooling and consuming the resulting liquid. In holistic and some traditional medicine, it is used as a carminative and for general digestive aid. Additionally, coriander juice (mixed with either turmeric powder or mint juice) is used by some as a treatment for **acne**, applied to the face like toner. It is stimulant, aromatic and carminative. The powdered fruit, fluid extracted and oil are chiefly used for medicinally as flavouring to disguise the taste of active purgatives and correct their griping tendencies. It is an ingredient of the following compound preparations of the pharmacopceia confection, syrup and tincture of senna and tincture of syrup of Rhubrab and enters also into compounds with angelica gentian, jalap, quassia and lavender. As a corrigent it is considered superior to other aromatics. If used too freely the seeds become nacrotic. Coriander water was formerly much esteemed as a carminative for windy colic. In general it is anthelmintic and appetite stimulant. **I**t cures tridosas, swelling and oedema. It relieves pain and also thirst. It is a natural purgative and improves digestion. It excites liver and is rakta pitta shamak. It is beneficial for heart and cures cough, urine infections and related disorders. It is anti-pyretic and promotes physical strength. It reduces sperm count (Acharya Balkrishna, 2008). Coriander seeds, used as spice, are round to oval in shape, yellowish brown in color with vertical ridges and have flavour that is aromatic, sweet and citrus, but also slightly peppery. This popular culinary herb is quite similar to **dill**in the sense that both its leaves and seeds can be used as a seasoning condiment.

It has a long history as a traditional medicine with cholesterol lowering and blood sugar lowering effects. Coriander seed powder is an essential ingredient in Indian cuisine but for therapeutic benefits, a teaspoon or two of coriander seeds soaked overnight and consumed the next morning seems to be useful in heart disease and diabetes, The seeds are rich in two main compounds viz **Linalon** and **Decanoic acid.** It has a long history as a traditional medicine with cholesterol-lowering and blood suger- lowering effects.

Health Benefits

Toxic Metal Cleansing: Coriander is known for its heavy metal detoxification benefits. It has been found to remove heavy metals from the body (taken in via consumption of non-organic foods, using conventional water supplies in your home, eating fish, using deodorants, smoking cigarettes and/or being around second-hand smoke, cooking foods in aluminum cookware or aluminum foil, taking over-the-counter drugs like antacids, using vaccines, or if you have metal fillings in your teeth). Heavy metals are connected with many serious health problems including cancer, heart disease, brain deterioration, emotional problems, kidney disease, lung disease and weak bones. The chemical compounds in cilantro bind to toxic metals and loosen them from the tissues.

Cardiovascular Protection: All herbs are wonderful for cardiovascular support. The organic acids found in cilantro have been found to help lower bad cholesterol (LDL) and raise good cholesterol (HDL). It also helps dissolve cholesterol build up in the arteries to help those suffering from atherosclerosis and heart disease.

Anti-Diabetic Activity: In parts of Europe, cilantro has traditionally been referred to as an "anti-diabetic" plant, due to its wonderful ability to help lower cholesterol and lower blood sugar levels (it packs in a whopping 521 mg of potassium per 100 grams!). Individuals suffering from diabetes have problems regulating and keeping their blood sugar stable. Consuming coriander and keeping sodium intake to a minimum, could help eliminate this effect.

Antioxidant Activity: Coriander contains an antioxidant called quercetin (among others), which attributes to this herbs amazing antioxidant activity. Antioxidants help protect your cells against oxidative stress caused by free radicals (a primary cause of the aging process). Protecting yourself against free radicals with antioxidants is the most effective way to reduce developing chronic disease later in life. The antioxidant activity in cilantro helps protect against most degenerative diseases including cancer, heart disease, diabetes, arthritis, macular degeneration, Alzheimer's disease and many more.

Anti-anxiety Effects: Coriander has muscle relaxing qualities and may act as a mild sedative. It acts on calming the nerves, helping to relieve anxiety and reducing the harmful effects of stress. Sipping on some cilantro, cucumber and celery juice after a long day is an excellent way to get in B vitamins to help calm the mind and help you de-stress.

Improves Sleep Quality: Similar to the effects it has on reducing anxiety, cilantro is a calming herb, so it also helps improve the quality of your sleep.

Blood-sugar Lowering Effects: Coriander's health benefits also go past helping lower LDL cholesterol, but can also reduce hypertension by lowering blood pressure! Each 100-g serving of raw cilantro leaves provides over 521 mg potassium and only 46 mg sodium (naturally occurring cell salts). Consuming a high-potassium, low-sodium diet may help you control your blood pressure. Try juicing cilantro or putting it in salads!

Anti-bacterial and Anti-fungal Activity: Coriander acts as a natural antiseptic and anti-fungal agent for the skin and disorders like dermatitis and eczema. This herb also has wonderful anti-bacterial properties that can be used to help improve oral health. The antimicrobial substances in cilantro help prevent and cure small pox too!

Natural Internal Deodorant: The large amount of chlorophyll in coriander is a great way to detox the body from the inside out. Flushing out toxins from the liver, kidney and digestive tract help remove excess bacteria from the body that would normally sweat out and accumulate in your armpits and feet. Bacteria dislike chlorophyll (due to the high oxygen content), and thus, this herb helps your body deodorize and smell great (http://livelovefruit.com/benefits-of-cilantro/).

Adding 1 or 2 teaspoons of corriander juice to fresh butter milk is idealfor treating nausea and dysentery.

It is beneficial for women especially those suffering from heavy menstrual flow. Six grams of corriander seeds shouldbe boiled in 500ml water and after adding sugar, it can be consumed while warm.

It is known to lower blood sugar by stimulating the secretion of insulin. It has also been called as an anti-diabetic plant.

It helps the heart by lowering bad cholesteroland raising good cholesterol.

The herb is also great for the skin. Its juice mixed withturmeric powder can help treat pimples and black heads (Anonymous, 2012).

Coriander herb contains no cholesterol; but is rich in anti-oxidants and dietary fibre which help reduce LDL or "bad cholesterol" while increasing HDL or "good cholesterol" levels.

The leaves and seeds contain many essential volatile oils such as *borneol, linalool, cineole, cymene, terpineol, dipentene, phellandrene, pinene* and *terpinolene*.

The leaves of the coriander plant have anti-septic and carminative properties.

The leaves and stem tips are also rich in numerous anti-oxidant polyphenolic flavonoids such as *quercetin, kaempferol, rhamnetin* and *epigenin.*

The herb is a good source of minerals like potassium, calcium, manganese, iron, and magnesium

It is also rich in many vital vitamins including folic-acid, riboflavin, niacin, vitamin-A, beta carotene, vitamin-C that are essential for optimum health. Cilantro leaves provides 30 per cent of daily recommended levels of vitamin-C.

It provides vitamin-A, which is an important fat soluble vitamin and anti-oxidant, is also required for maintaining healthy mucus membranes and skin and is also essential for vision. Consumption of natural foods rich in vitamin-A and flavonoids helps body protect from lung and oral cavity cancers.

Coriander is one of the richest *herbal sources for* ***vitamin K.*** Vitamin-K has potential role in bone mass building by promoting osteotrophic activity in the bones. It also has established role in the treatment of Alzheimer's disease patients by limiting neuronal damage in their brain.

The coriander seeds oil have found application in many traditional medicines as analgesic, aphrodisiac, anti-spasmodic, deodorant, digestive, carminative, fungicidal, lipolytic (weight loss), stimulant and stomachic.

The herb contains many phytochemical compounds; phenolic flavonoid antioxidants like quercitin and essential oils have found application in many traditional medicines as analgesic, aphrodisiac, antispasmodic, carminative, depurative, deodorant, digestive, carminative, fungicidal, lipolytic, stimulant and stomachic.

Coriander seeds contain many compounds that are known to have anti-oxidant, disease preventing, and health promoting properties.

The seeds are excellent source of minerals likeiron, copper, calcium, potassium, manganese, zinc and magnesium

Unlike other dry spice seeds that are lack in vitamin C, coriander seeds contain ample amount of this anti-oxidant vitamin.

Furthermore, the seeds indeed are storehouse for many vital B-complex vitamins like thiamin, riboflavin, and niacin.

The characteristic aromatic flavour of coriander seeds comes from the many fatty acids and essential volatile oils. Some important fatty acids in the dried seeds include *petroselinic acid, linoleic acid (omega-6), oleic acid,* and *palmitic acid.* In addition, the seeds contain essential oils such as *linalool (68 per cent), a-pinene (10 per cent), geraniol, camphene, terpine etc.* Together, these active principles are responsible for digestive, carminative, and anti-flatulent properties of the seeds.

In addition, dietary fibres bind to bile salts (produced from cholesterol) and decrease their re-absorption in colon, thus help lower serum LDL cholesterol levels.

Together with flavonoid anti-oxidants, fibre composition of coriander helps protect the colon mucus membrane from cancers.

A very good food for digestive system, coriander promotes liver functions and bowel movements.

Coriander is good for diabetes patients. It can stimulate the insulin secretion and lower the blood sugar levels.

Vitamin K in it is good for the treatment of Alzheimer's disease.

The fat soluble vitamin and antioxidant- Vitamin A, protects from lung and cavity cancers.

Coriander contains anti-inflammatory properties. This is why it is good against inflammatory diseases such as arthritis.

Coriander's anti-septic properties help to cure mouth ulcer.

Coriander is good for the eyes. Antioxidants in coriander prevent eye diseases. It's a good remedy in the treatment of conjunctivitis.

Coriander seeds are especially good for the menstrual flow.

It's a very good herb to promote the nervous system. It can stimulate the memory.

Coriander helps those suffering from anaemia. Coriander contains high amounts of iron, which is essential for curing anemia.

It is diuretic, improves digestion and stimulates the function of liver.

It regulates the level of the blood sugar and is efficient in eliminating the excess fat from the body.

CLOVE [*Syzygium aromaticum* (L.)]

Some of the spices are found to be useful for heart health, include coriander seeds, turmeric, black pepper, cinnamon, fenugreek seeds, black cumin seeds, ginger and garlic because of the presence of bioactive components

Dried Cloves

Cloves are a very common spice, which is widely used in Indian cooking. It is renowned for its rich aroma and flavour. This tiny little spice has many curative properties. Traditionally cloves are used as a table spice and mixed with chilies, cinnamon, turmeric and other spices in the preparation of curry powder. It is hot and pungent. It is beneficial for eyes and promotes digestive functions. It enhances taste and cures Kapha, Pitta and blood disorders. It also quenches thirst and cures vomiting and flatulence caused by indigestion or accumulation of gas. It is beneficial in curing breathing problems, caught and tuberculosis. Its oil promotes digestive power, cures cough and vomiting in case of pregnant women (Acharya Balkrishna, 2008) It a volatile oil contains euginol, emitil and caryophyllene. Both cinnamon and cloves are used extensively in Indian cooking and have been found to improve the function of insulin and to lower glucose, total cholesterol, LDL and triglycerides in people with type 2 diabetes. Cloves boost your energy level. They also have one of the best aromatherapy scents that improve your sexual bevahour. Clove oil is used in the manufacture of perfumes, soaps, bath salts and as a flavouring agent in medicine and dentistry. In Chinese medicine, it is used for vomiting, indigestion and other related problems. If you are feeling stressed then, boil some water with basil leaves, mint leaves and clove. Then you can use this water for black tea. Add a little bit of honey to it. This will help you ease tension. You can also use cloves as a mosquito repellent. Clove can effectively help to ward off mosquitoes. Just add cloves to your diet in the form of curries, soups and main dishes to gain maximum benefits. They are also used to flavour paan.

Cloves are a highly prized medicinal spice that have been used for centuries in treating digestive and respiratory ailments. Cloves contain good amounts of vitamins A, C, K, and B-complex as well as minerals such as manganese, iron, selenium, potassium, and magnesium. They also contain powerful antiseptic, antiviral, anti-inflammatory and anesthetic properties making them tremendously useful in helping to heal a wide variety of illnesses and health conditions. Cloves are particularly beneficial for the digestive tract and are great for indigestion, gas, constipation, bloating, nausea, and countering the effects of heavy, rich food. They are excellent for relieving muscle spasms, headaches, and nerve pain. They are also often used to disinfect gums, teeth, kidneys, liver, skin, and bronchi. Clove oil contains **eugenol** which is a powerful anesthetic and natural pain reliever and is commonly used to help relieve toothaches and to numb gums in dentistry. Clove oil is also beneficial for the circulatory system and is a potent platelet inhibitor which prevents blood clots. Clove oil is also excellent for athletes foot and for healing cuts, bruises, burns, rashes, and psoriasis. Essential oil of Clove is an effective decongestant and should be used in a vaporizer, humidifier, or aromatherapy machine to help disinfect the air and to help benefit respiratory conditions such as sinusitis, tuberculosis, bronchitis, asthma, colds and coughs. Cloves are often combined with other herbs to create seasonings such as Curry Powder and Garam Masala in India, Chinese Five Spice in China, and Worcester Sauce in Great Britain. They are also the ideal addition to deserts, fruit salads, smoothies, and savory dishes alike. Clove tea is helpful for strengthening the immune system and detoxifying the body. Steep 2 tsp of whole cloves in two cups of hot water for at least 10 minutes, sweeten with honey if desired. Cloves can be found whole or powdered in you local supermarket or health food store. Capsules, extract, tincture, tea, and topical oils and creams can all be found online or at your local health food store.

Clove oil is used in the manufacture of perfumes, soaps, bath salts and as a flavouring agent in medicine and dentistry.

The good: This food is very low in Cholesterol. It is also a good source of Vitamin E (**Alpha Tocopherol**), Calcium and Iron, and a very good source of Dietary Fibre, Vitamin C, Vitamin K, Magnesium and Manganese.

Health Benefits

Cloves help stimulate sluggish circulation and thereby promote digestion and metabolism.

Cloves also have great purification and protection properties in them.

Consuming a couple of cloves can help to cleanse your body from harmful toxins and microbes. This will help to boost your health. Cloves also help to protect the damage caused to your body's cells by microbes, and help to stimulate energy in every part of your body.

The presence of eugenol, in cloves gives it a lovely aroma and helps in numbing dental pain and clearing out bacteria in the mouth.

Cloves also have great anti-inflammatory properties. Use cloves to treat inflammation and infections.

Cloves also aid with cardiac and diabetes health.

Due to its purification properties, cloves help in purifying your blood.

Tooth ache- Cloves can also help give you quick relief from toothache. Just place a piece of clove on the affected area, and gain instant relief The use of a clove in toothache decreases pain. It also helps to decrease infection due to its antiseptic properties. Clove oil, applied to a cavity in a decayed tooth, also relieves tooth ache.

Cloves promote enzymatic flow and boost digestive functioning. They are used in various forms of gastric irritability and dyspepsia. Licking the powder of fried cloves mixed with honey is effective in controlling vomiting. The anesthetic action of clove helps in dealing with stomach pain and stops vomiting.

Coughs- Chewing a clove with a crystal of common salt eases expectoration, relieves the irritation in the throat. Chewing a burn clove is also effective medicine for cough.

Digestive disorders - If you are troubled with digestion problems, chew on a couple of cloves. Cloves promote enzymatic flow and boost digestive functioning. They are used in various forms of gastric irritability and dyspepsia. Licking the powder of fried cloves mixed with honey is effective in controlling vomiting. The anesthetic action of clove helps in dealing with stomach pain and stops vomiting.

Cloves, or lavang as they are known in Hindi, have been used for thousands of years in Indian and Chinese medicines as a natural preservative. Containing high amounts of phenolic compounds, which have antioxidant properties, they keep food from going bad by preventing the growth of fungus and bacteria.

Sucking a piece of clove after meals helps in reducing acidity problem.

MINT (*Mintha spicata*)

Mint Leaves

It is one of the oldest and most popular **herbs** that is grown around the world. There are many different varieties of mint, each having its own subtle flavour and aroma. This herb is used in a range of dishes from stuffing to fruit salads. Mint is an essential ingredient in many Indian and Middle Eastern cuisine and is popularly mixed with natural plain yogurt to make a 'raita' or brewed with tea to make the famous Indian 'Pudina Chai'. In Thai cooking, it is added to soups and to some highly-spiced curries. Mint grown in Asia is much more strongly flavored than most European mints, with a sweet, cool aftertaste. Peppermint has been one of the popular herbs known since antiquity for its distinctive aroma and medicinal value. The herb has a characteristic refreshing cool breeze sensation when eaten on taste buds, palate and throat, and on nasal olfaction glands when inhaled. This unique quality of mint is due **menthol,** an essential oil in it. Mint sooths the digestive tract, provides relief from stomach aches and is known to have cancer-prevention properties.

Mint as Minta Spacata is a plant that has been long used in diverse cultures, such as India, Middle East and Europe. Mint has a sweet flavour, with a cooling after-sensation. Both, fresh and dried mint are used in preparing a large number of recipes, including curries, soups, chutneys, salads, juices, and ice creams.

The herb contains **menthol** and is popular as both a flavouring agent and a tea. It is also well known for its carminative effects – that is, it helps to relieve flatulence and indigestion. Since menthol is also an appetite stimulant, digestive and sedative, peppermint can be used to prevent or alleviate cramps, insomnia and vomiting. It is also claimed to be an aphrodisiac. In addition, peppermint has antiviral properties; its tannins have been found to suppress the activity of flu virus and inhibit Herpes simplex. Peppermint oil is used to relieve the symptoms of irritable bowel syndrome (IBS). Mint contains a host of medicinal properties. To relieve headache, migraine, cold and stomache upsets, chew on some fresh mint leaves. Also regular consumption of mint helps in flushing out toxins from the body.

Recent study has shown that spearmint tea may be used as a treatment of fevers, headaches, digestive diseases and for mild hirsutism in women. Its anti -androgenic properties reduce the level of free testosterone in the blood, while leaving total testosterone and DHEA unaffected. A major compound of oil is **R-carvone**; pure R-carvone is sufficient to produce a smell which people identify as a spearmint smell. The fresh flowering herb on distillation yields 0.25 to 0.50 per cent of volatile oil, known as spearmint oil. The characteristic constituent of the oil is **1-carvone**. The herb is considered stimulant, carminative, diuretic, restorative, stimulant, anti-emetic, anti-spasmodic and stomachic (Duke and Ayensu, 1985). A sweetened infusion of the herb is given as a remedy for infantile troubles, vomiting in pregnancy and hysteria. The leaves are used in fevers and bronchitis. The oil is a counter irritant. The leaves should be harvested when the plant is just coming it to flower and can be dried for later use. The stems are macerated and used as a poultice on bruises. The essential oil in the leaves is antiseptic, though it is tonic in large doses. Both the essential oil and the stems are used in the folk remedies for cancer. A poultice prepared from the leaves is said to remedy tumours (Duke and Ayensu, 1985).

Peppermint has been one of the popular herbs known since antiquity for its distinctive aroma and medicinal value. The herb has a characteristic refreshing cool breeze sensation when eaten on taste buds, palate and throat, and on nasal olfaction glands when inhaled. This unique quality of mint is due **menthol,** an essential oil in it. Peppermint is not just an ingredient in flavoured chewing gums. The herb is a hybrid of two plants, water mint and spearmint and is beneficial both cosmetically as well as medically. When extracted as an oil, it is one of the many essential oils and is used extensively in perfumes and colognes. Peppermint oil is both rejuvenating and refreshing. It not only strengthens the immune system but also eases mental exhaustion. The oil can be used to give you a healthy scalp and healthy hair. Add a few drops of peppermint oil to your daily shampoo and it can help reduce that dandruff. Another benefit is it allows taming frizzy hair and brings about a shine to it. Peppermint oil can also bring about hair growth. The oil usage is not just a boon to the hair but also the skin and lips. Not only does it help control oil secretion of the skin but also clears it and brightens tired skin. The oil prevents acne and helps open pores. Peppermint oil helps heal cracked lips, and this is the reason why this oil is an active ingredient in chopsticks and lip glosses. The oil has a soothing effect (because of menthol that's present in it), and so can be used on wind-chapped or sun-burnt lip as it will give a cooling and refreshing effect on the skin. Please note that peppermint oil is really strong and when used, it should be diluted with a carrier oil, else it can cause side effects.

Because of its active compounds contained, mint has sedative, disinfectant and cicatrizing properties. It can be successfully used in gastro-intestinal disorders; it helps the liver and calms indigestion. It contains menthol, menthone, menthofuran, a-pinene, limonene, cardinene, acetic aldehide, isovaleriana, vitamin C and antibiotic substances.

Peppermint has greenish-purple lance-shaped leaves while the rounder leaves of spearmint are more of a grayish green color. And this peppermint can do wonders for you in the health department (Chaturvedi, 2014).

Health Benefits

Soothe your tummy with peppermint: Trials have repeatedly shown the ability of peppermint oil to relieve symptoms of irritable bowel syndrome, including indigestion, dyspepsia, and colonic muscle spasms. These healing properties of peppermint are apparently related to its smooth muscle relaxing ability. Once the smooth muscles surrounding the intestine are relaxed, there is less chance of spasm and the indigestion that can accompany it. The menthol contained in peppermint may be a key reason for this bowel-comforting effect.

A potential anti-cancer agent: Perillyl alcohol is a phytonutrient called a **monoterpene**, and it is plentiful in peppermint oil. In animal studies, this phytonutrient has been shown to stop the growth of pancreatic, mammary, and liver tumors. It has also been shown to protect against cancer formation in the colon, skin, and lungs.

An anti-microbial oil: Esssential oil of peppermint also stops the growth of many different bacteria. These bacteria include Helicobacter pylori, Salmonella enteritidis, Escherichia coli O157:H7, and methicillin-resistant Staphylococcus aureus (MRSA). It is found to inhibit the growth of certain types of fungus as well.

Breathe easier with peppermint: Peppermint contains the substance rosmarinic acid, which has several actions that are beneficial in asthma. In addition to its antioxidant abilities to neutralize free radicals, rosmarinic acid has been shown to block the production of pro-inflammatory chemicals, such as leukotrienes. It also encourages cells to make substances called prostacyclins that keep the airways open for easy breathing. Extracts of peppermint have also been shown to help relieve the nasal symptoms of allergic rhinitis (colds related to allergy).

A rich source of traditional nutrients: Peppermint is a good source of manganese, copper, and vitamin C. Vitamin C seems to play a role in decreasing colorectal cancer risk. It is the main water-soluble antioxidant in the body is needed to decrease levels of free radicals that can cause damage to cells. Some studies have shown a link between increased vitamin C intake and a decreased risk for colon cancer, possibly by as much as 40 per cent, while other studies have shown that vitamin C intake can help to decrease the incidence of colon tumours.

Mint contains numerous chemical compounds that are known to have anti-oxidant, disease preventing and health promoting properties.

It contains no cholesterol; but is rich in anti-oxidants and dietary fibre which helps to control blood cholesterol and blood pressure levels.

It contains many essential volatile oils like *menthol, menthone, menthol acetate* that act on cold-sensitive receptors in the skin, mouth and throat, the property which is responsible for the well known cooling sensation that it provokes when inhaled, eaten, or applied to the skin.

The essential oil, **menthol,** also has analgesic (pain-killer), local anesthetic and counter-irritant properties.

Essential oils in the peppermint relax smooth muscles in the intestinal wall and sphincters by blocking calcium channels at cellular level. This property of mint has been applied as an anti-spasmodic agent in the treatment of "irritable bowel syndrome" or IBS and other colic pain disorders.

It is also rich in many antioxidant vitamins including vitamin A, beta carotene, vitamin-C and vitamin E. The leaves of mint also contain many important B-complex vitamins like folates, riboflavin and pyridoxine (vitamin B-6) ; and the herb is also an excellent source of vitamin-K.

It is soothing the digestive tract and if you are having stomach ache then it can be of great help.

Drinking herbal mint tea reduces irritated bowel syndromes, cleanses the stomach and also clear up skin disorders such as acne.

Mint acts as a cooling sensation to the skin and helps in dealing with skin irritations.

Mint helps in eliminating toxins from the body.

Crushed mint leaves helps in whitening teeth and combat bad breath.

Mint is a very good cleanser for the blood.

Medicinal Uses

As mentioned above, the essential oils in the peppermint act on cold-sensitive receptors in the skin, mouth and throat, the property which is responsible for the well known cooling sensation that it provokes when inhaled, eaten, or applied to the skin. This property of mint can be applicable in the preparation of cough/cold reliving remedies like syrups, lozenges and nose inhaler.

Peppermint oil has analgesic, local anesthetic and counter-irritant properties and has been used in the preparation of topical muscle relaxants and analgesics.

It is also being used in oral hygiene products and bad-breath remedies like mouthwash, toothpaste, mouth and tongue-spray, and more generally as a food flavor agent; *e.g.* in chewing-gum, candy.

The essential oils in the peppermint can relax smooth muscles in the intestinal wall and sphincters by blocking calcium channels in them. This property of mint has been applied in treating irritable bowel syndrome or IBS and as an anti-spasmodic agent.

Caution

Individuals with *gastro-esophageal reflex disease* (GRD) are advised to limit peppermint in their diet since compounds in mint leaves relaxes smooth muscles in the esophageal wall and sphincters by blocking calcium channels in them which can aggravate their reflux condition.

FENNEL (*Foeniculum vulgare*) Saunf

Bulb Fennel

Fennel is considered both a vegetable and a herb due to its wide ranging nutritional and healing benefits. Fennel is rich in folic acid, vitamin C, magnesium, cobalt, iron, and essentials oils that contain powerful anti-bacterial and anti-fungal properties. It is excellent for indigestion and is commonly used as a natural antacid in order to help reduce acidity and inflammation in the digestive tract and to facilitate proper absorption and assimilation of nutrients from food. Fennel has potent anti-flatulent and carminative properties which means it is able to prevent and stop the formation of gas in the stomach and intestines. It is also known to be highly beneficial for sinus congestion, bronchitis, renal colic, anemia, hypertension, macular degeneration, constipation, bloating, diarrhea, and irritable bowel syndrome. Fennel is used to help protect against both cardiovascular disease and cancer. It contains an important anti-inflammatory phytonutrient called **anethole** that blocks both inflammation and carcinogenesis, which is the mutation of regular cells into cancerous cells. Fennel also has the ability to ease and regulate menstruation by regulating hormonal action properly in the body. Fennel is used amongst nursing women to help stimulate consistent milk flow for their babies. It is also known to help strengthen hair, prevent hair loss, relax the body, and sharpen the memory. Fennel seeds can be chewed after a meal to aid in digestion and to remove bad breath. Fennel seeds can also be made into an effective medicinal tea by steeping the seeds in hot water for 10 minutes or more. Fresh fennel juice can also be used topically to swollen or inflamed eyes to reduce irritation, swelling, and fatigue. Fresh fennel has a crunchy, slightly sweet licorice flavour and is a wonderful addition to fresh salads, smoothies, soups, stir-fry, potatoes, and other vegetable dishes. Its succulent enlarged bulb imparts special "anise like" sweet flavour to the recipes. Bulb fennel is cultivated as vegetable for its beautiful, squatted stems in many regions of the country. Infusions of seeds and roots relieve flatulence, strengthen digestion, help suppress appetite and as a result aid weight loss. Fennel is also effective in treating colics and ulcers. The seeds and leaves are used to flavour dishes and the strems are used as a vegetable. Fennel contains an estrogen like substance (estirol) that turns out libido. Fennel contains the antioxidant flavonoid quercetin. This herb is anticarcinogenic and can be useful for cancer patients undergoing chemotherapy or radiation. Fennel can be useful for indigestion and spasms of the digestive tract. It also helps expel phlegm from the lungs. Armed with phyto-nutrients and antioxidants, cancer cells have nothing but to accept defeat when the spice is fennel. Anethole- a major constituent of fennel resists and restricts the adhesive and invasive activities of cancer cells. It suppresses the enzymatic regulated activities behind cancer cell multiplication. A tomato -fennel soup with garlic or fresh salads with fennel bulbs make for an ideal entree prior to an elaborate course meal. Roasted ennel with parmesan can be another star pick

Health Benefits of Fennel Bulb

Fennel bulb is a versatile vegetable, used since ancient times for its nutritional and medicinal properties. This winter season has some noteworthy essential oils, flavonoid anti-oxidants, minerals, and vitamins that have known health benefits.

Bulb fennel is one of very low calorie vegetables. 100 g bulb provides just 31 calories. Further, it contains generous amounts of fibre (3.1 g/100 g or 8 per cent of RDI), very little fat and zero cholesterol.

Fresh bulbs give sweet anise-like flavor. Much of it is due to high concentration of aromatic essential oils like *anethole, estragole, and fenchone* (*fenchyl acetate*). Anethole has been found to have anti-fungal and anti-bacterial properties.

The bulbs have moderate amounts of minerals and vitamins that are essential for optimum health. Their juicy fronds indeed contain several vital vitamins such as pantothenic acid, pyridoxine (vitamin B-6), folic acid, niacin, riboflavin, and thiamin in small but healthy proportions. 100 g fresh bulbs provide 27 μg of folates. Folic acid is essential for DNA synthesis and cell division. Their adequate levels in the diet during pregnancy can help prevent neural tube defects in the newborn babies.

In addition, fennel bulb contain average amount of water-soluble vitamin, vitamin-C. 100 g of fresh bulbs provide 12 mg or 20 per cent of vitamin C. Vitamin C helps body develop resistance against infectious agents and scavenge harmful, pro-inflammatory free radicals. Further, it has small amounts of vitamin A.

The bulbs have very good levels of heart-friendly electrolyte potassium. 100 g provides 414 mg or 9 per cent of daily-recommended levels. It is an important electrolyte inside the cell. Potassium helps reduce blood pressure and rate of heartbeats by countering effects of sodium. Fennel also contains small amounts of minerals such as copper, iron, calcium, magnesium, manganese, zinc, and selenium.

LOTUS (*Nelumbo nucifera*)

Crunchy, delicate flavored lotus root is an under-water edible rhizome of lotus plant. Since centuries, the lotus rhizome has held high esteem in some countries, especially in China and Japan. Almost all the parts of the plant: root, young flower stalks, and seeds are being used in the cuisine.

Lotus root is grown as annual root vegetable crop in customized ponds. The entire plant is used in medicine. The sacred water lotus has been used in the Orient as a medicinal herb for well-over 1500 years. The leaf juice is used in the treatment of diarrhea and is decocted with liquorice (*Glycyrrhiza* spp.) for the treatment of sun-stroke. A decoction of the flowers is used in the treatment of premature ejaculation. The flowers are recommended as a cardiac tonic. A decoction of the floral receptacle is used in the treatment of abdominal cramps, bloody discharges *etc.* The flower stalk is used in treating bleeding gastric ulcers, excessive menstruation, post partum hermorrhage. The stamens are astringent and used in treating urinary frequency, premature ejaculation, hemolysis, epistasis and uterine bleeding. A decoction of the fruit is used in the treatment of agitation fever, heart complaints *etc.* The seed is used in the treatment of poor digestion, enteritis, chronic diarrhea, insomnia, palpitation *etc.* The plumule and radicle are used to treat thirst in high febrile disease, hypertension, insomania and restlessness. The root starch is used in the treatment of diarrhea, dysentery *etc.*, a paste is applied to ring worm and other skin aliments. It is also taken internally in the treatment of hemorrhages, excessive menstruation and nose bleeds. The roots are harvested in autum or winter and dried for later use. The root nodes are used in the treatment of nasal bleeding, hermoptysic, hermaturia and functional bleeding of uterus. The plant has a folk history in the treatment of cancer, modern research has isolated certain compounds from the plant that shows anti-cancer activities. The leaves, which have anti-pyretic and refrigerant properties, are used against symptoms of summer-heat, such as head ache, respiratory congestion, chronic thirst and dark scanty urine. The peduncle relieves stomachaches, clams restless fetus and controls leucorrhea.

Health Benefits of Lotus Root

Lotus root is one of the moderate calorie root vegetables, composed of several health benefiting phyto-nutrients, minerals, and vitamins.

Lotus rhizome is very good source dietary of fibres.

Lotus root is one of the excellent sources of vitamin C

It contains moderate levels of some of valuable B-complex group of vitamins such as pyridoxine, folates, niacin, riboflavin, pantothenic acid, and thiamin.

The root provides some important minerals like copper, iron, zinc, magnesium, and manganese

The crunchy, sweet yet delicate flavour of root lotus is because of its optimum electrolyte balance. It has agreeable ratio of sodium to potassium at the value 1:4. While sodium gives the sweet taste to the root, potassium acts to counter negative effects of sodium by regulating heart rate and blood pressure.

The lotus seeds maintain energy balance and have a sedative effect on the central nervous system so as to stimulate calmess and relaxation, therefore ideal for individuals suffering from insomnia or restlessness.

The seeds have a mild effect on dilating the blood vessels, thereby causing a drop in blood pressure. Certain anti-aging enzymes in these seeds aid in repairing damaged proteins, reducing oxidative stress and hence delay ageing.

Caution

Lotus root may harbour parasites like *Fasciolopsis buski*, which is a trematode that commonly infests in aquatic plants like lotus, water caltrop (*Trapa natans*), Chinese water-chestnut (*Eleocharis dulcis*) *etc.* The symptoms may include stomach pain, vomiting, diarrhea, fever, and intestinal obstruction. Thorough washing and cooking in the steam destroys the larvae.

ROSEMARY (*Rosmarinus officinalis*)

Rosemary Plant with Narrow Leaves

A beautiful, fragrant ever green herb, it has a variety of herbal uses. An infusion of the leaves and flowering tops is a sedative that can be used to relieve flatulence and headaches, promote perspiration, increase bile flow and stimulate menstruation. Rosemary, especially its young leaves, contain **carnosic acid**, a powerful antioxiodant. Recent research identified rosemary as one of the first plants that can improve memory by inhibiting **acetylcholinesterase** (a brain chemical which switches off the connection between nerve cells). Rosemary leaves are used as a culinary seasoning and to make essential oils for aromatherapy. They are also used externally mainly in champoos and other hair preparations. Apart from culinary and medicinal purpose rosemary shoots, flowers and leaves are used in ceremonies such as weddings and festivals for decorating banquet halls as incense to ward off bad influences.

Rich in nutrients including iron, calcium, vitamins and antioxidants, rosemary in various forms –rosemary extract, rosemary oil and rosemarinic acid –is effective for many skin types and is found in skin care cleansers, soaps, face masks, toners and creams. This herb is used widely in aromatherapy as its stimulant properties refresh, rejuvenate and help you feel energised. Here's how this fragrant herb can benefit your skin (Anonymous, 2013).

Aromatherapy-uses rosemary essential oil as stress release agent, stimulating the nervous system without the dopey feeing and as a massage oil to relieve rigid and tired muscles. Having the properties of analgesic, rosemary is considred a solid pain reliever and therefore is good for clearing the head and treating migraines.

Health Benefits

Prevents ageing of skin: Since it contains naturally powerful antioxidants, rosemary oil strengthens the capillaries and helps to slow the effects of ageing on skin. It also tightens sagging and loose skin to help it look firmer and more elastic. Additionally, rosemary stimulates biological activity and cell growth to reduce fine lines and wrinkles.

Treats acne: With its powerful disinfectant and antibiotic properties, the herb can help to treat acne and oily skin conditions. When massaged into the face, rosemary oil can help lightens dark spots and blemishes on the skin resulting in an improved skin complexion.

Prevents infection: The anti-inflammatory properties of rosemary extract help to reduce swelling and puffiness of the skin. It also helps to heal burns and soothe the skin. This herb's medicinal qualities makesit a powerful cure for chronic skin conditions including dermatitis, eczema and psoriasis.

Rosemary leaves contain certain phyto-chemical compounds that are known to have disease preventing and health promoting properties.

The herb parts especially flower tops contain phenolic anti-oxidant ***rosmarinic acid*** as well as numerous health benefiting volatile essential oils such as *cineol, camphene, borneol, bornyl acetate, α-pinene etc.* These compounds are known to have rubefacient (counter irritant), anti-inflammatory, anti-allergic, anti-fungal and anti-septic properties.

Rosemary leaves provide just 131 calories per 100 g and contains no cholesterol.

The herb is exceptionally rich in many B-complex group of vitamin, such as **folic acid**, pantothenic acid, **pyridoxine**, and riboflavin.

It is one of the herbs contain high levels of folates; providing about 109 mcg per 100 g. Folates are important in DNA synthesis and when given during peri-conception period can help prevent neural tube defects in the newborn babies.

Rosemary herb contains very good amounts of **vitamin A**. Vitamin A is known to have antioxidant properties and is essential for vision. It is also required for maintaining healthy mucus membranes and skin. Consumption of natural foods rich in vitamin A is known to help body protect from lung and oral cavity cancers.

Fresh rosemary leaves are good source of antioxidant vitamin; vitamin-C containing about 22 mg per 100 g, about 37 per cent of RDA. It is required for the collagen synthesis in the body. Collagen is the main structural protein in the body required for maintaining the integrity of blood vessels, skin, organs, and bones.

Rosemary herb parts, whether fresh or dried, are rich source of minerals like potassium, calcium, iron, manganese, copper, and magnesium. Potassium is an important component of cell and body fluids, which helps control heart rate and blood pressure. This herb is an excellent source of **iron**, contains 6.65 mg/100 g of fresh leaves.

It increases circulation, can help relieve excess muscle tension.

Can help control dandruff and boost shine.

Several compounds (essential oils) in rosemary help alleviate cold symptoms and loosen chest congestion.

Rosemary has strong effects on the adrenal and other glands of the body.

It has anti-parasitical, anti-fungal and anti- bacterial properties.

It increases the ability to memorize key facts and concepts and increases test performances in students.

It helps to boost recall and mental clarity by working directly on the central nervous system.

Fatigue and depression brought on by weariness or feeling "run down" are relieved by the stimulating and energizing effects of rosemary essential oil.

Rosemary extract is made by distilling the leaves of the herb rosemary. A powerful preservative, it contains carnosic and rosmaranic acid, which are **antioxidants** that prevent decay. In fact, rosemary contains over 20 other antioxidants, which have been found to be much more effective and long lasting than the **antioxidants** in other foods.

Medicinal Uses

Rosmarinic acid, a natural polyphenolic antioxidant found in rosemary, has been found to have anti-bacterial, anti-inflammatory, and anti-oxidant functions. **Sage, peppermint, oregano, thyme** also contain appreciable levels of rosmarinic acid.

Rosemary oil, distilled from the flowering tops, contains volatile essential oil such as camphene, cineol, borneol, bornyl acetate and other esters. These compounds are known to have tonic, astringent, diaphoretic, and stimulant properties.

Its herbal oil is also being used externally as a rubefacient to soothe painful ailments in gout, rheumatism and neuralgic conditions.

Rosemary herb extractions when applied over scalp have stimulating function on the hair-bulbs and help preventing premature baldness. It forms an effectual remedy for the prevention of scurf and dandruff.

Rosemary tea is a good remedy for removing nervous headache, colds, and depression.

Caution

Rosemary herb might cause abortion in pregnant woman when eaten in large amounts. In some cases, rosemary oil products may cause allergic skin reactions. In toxic doses, rosemary has been found to cause kidney dysfunction, and might exacerbate existing neurological conditions like epilepsy, neuroses *etc.* Rosemary is an emmenagogue herb, encouraging blood flow in the uterus and promoting menstrual discharge and therefore pregnant women should not use it.

CURRY LEAF (*Murraya koenigii* (L.) Sprengel

Curry leaf is found almost throughout India up to an altitude of 1500 mtrs. It is much cultivated for its aromatic leaves. The leaves are highly valued for their distinctive flavour, making them very popular in the cuisines of South and West India and Sri Lanka. They are also used as a herbal tonic in Ayurveda. It is therefore, a good idea to include curry leaves in your diet. They can be used both raw or cooked. They can be ground and also used as a paste in curries. Incorporating curry leaves into your daily diet can help you lose weight. The leaves flush out fats and toxins, reducing fat deposits that are stored in the body as well as reducing bad cholesterol levels. If you are over weight, incorporate eight to ten curry leaves into your diet daily. Chop them finely and mix them into a drink or sprinkle them over a meal.

Curry leaf is a good source of vitamin A, calcium, and folic acid. Its richness in vitamin A and anti-oxidant may help explain its use in preventing early development of cataract. Being a fairly good source of folic acid, the leaves can also help in absorption of iron, other proposed benefits include boost in circulation and anti –inflammation. It is also anti-diabetic, anti-oxidant, anti-microbial, hepatoprotective, hypo-cholestrolemic and delays premature greying. With its anti-inflammatory benefits, it is used in treating bruises and skin eruptions. It can also be used as a sedative and a hair tonic. Its mildly properties aid digestion too. The studies have shown it to have anti-fungal activities, which explains why it is used to fight bad breath and gum diseases. Some practioners of herbal medicine advise consuming a few leaves in the morning, while others recommend therapeutic doses as juice. However, it should not exceed more than 15grams. Curry leaf is an essential ingredient in Indian cooking especially in South India. However, it's customary for most of us to simply remove and throw the leaf from our food and not consume it. Curry leaf has many medicinal properties. It stimulates digestive enzymes and helps break down food more easily. Curry leaves contain 2.5 per cent oil, alpha-selinene, beta-bisabolene, beta-cadinene, beta-caryophyllene, beta-elemene, beta-gurjenene, beta-phellandrene, beta-thujene and beta-transocimene. These ingredients give curry leaves it typical aroma. The leaf is used in South India as a natural flavouring agent in various curries. Volatile oil is used as a fixative for soap perfume. The leaves, bark and root of the plant are used in the indigenous medicine as a tonic, stimulant, carminative and stomachic.

Health Benefits

Boosts heart health: Its anti-oxidative properties helps to regulate the level of LDL or bad cholesterol in blood and increases HDL or good cholesterol level which is vital for good heart health.

Battles infection: The anti-microbial, anti-oxidant and anti-inflammatory properties of curry leafs help the body to fight against various diseases.

Boosts immunity: Curry leaves are loaded with several vitamins and are a good source of iron and calcium. It helps built immunity and protect you and your family from diseases.

Helps keep anaemia at bay: Curry leaves are a rich source of iron and folic acid. Folic acid is mainly responsible for carrying and helping the body absorb iron, and since kadi patta is a rich source of both the compounds it is your one-stop natural remedy to beat anaemia.

Fights diabetes: Not only does kadi patta help lower the blood sugar levels, but also keeps in check for a few days after the administration of curry leaves. Curry leaves help your blood sugar levels by affecting the insulin activity of the body and reduces ones blood sugar levels. Also the type and amount of fibre contained within the leaves play a significant role in lowering blood sugar levels. So, if you suffer from diabetes, kadi patta is the best natural method to keeping your blood sugar levels in check (Sinha, 2014).

Improves digestion: Curry leaves is known to help improve digestion and alter the way your body absorbs fat, thereby helping you lose weight. Since weight gain is one of the leading causes of diabetes, kadi patta treats the problem right at the root. Curry leaves are good for indigestion. It stimulates the release of digestive enzymes which help in breaking down the food inside your stomach.

Lowers cholesterol: Many research shows that curry leaves have properties that can help in lowering one's blood cholesterol levels. Packed with antioxidants, curry leaves prevent the oxidation of cholesterol that forms LDL cholesterol (bad cholesterol). This in turn helps in increasing the amount of good cholesterol (HDL) and protects your body from conditions like heart disease and atherosclerosis.

Improves eyesight: Curry leaves are rich source of vitamin A which is beneficial for good vision and eye health.

Regulates blood sugar: Curry leaves are rich in fibre content. Dietary fibre helps to slow down digestion which ensures proper utilization of insulin and keeps the blood sugar level in check.

Prevents greying of hair: Kadi patta has always been known to help in preventing greying of the hair. It is also very effective in treating damaged hair, adding bounce to limp hair, strengthening the shaft of thin hair, hair fall and treats dandruff. The best part about this benefit is that you can either choose to eat the curry leaves to help with your hair woes or apply it to your scalp as a remedy. Heat oil and add some curry leaves. When the curry leaves turns dark black remove the oil from heat and let it cool down. Store in a glass container. Apply this oil on your head and leave it overnight. This helps prevent premature aging of grey hair and also nourishes the hair roots. This oil also acts as a very good stimulant and helps in hair growth.

Curry leaf has many medicinal properties: It stimulates digestive enzymes and helps break down food more easily (have a glass of butter milk mixed with a little hing (asafoetida) with a few curry leaves thrown after meal for good digestion).

A good remedy for nausea and indigestion: Extract juice of curry leaves, squeeze a lime and add a pinch of sugar.

Chew a few leaves every day to lose weight.

Curry leaves are also known to improve eye sight, so make sure you do not throw away the leaves while eating. It is also believed to prevent cataract.

Curry leaves are also good for hair growth and colour: If you do not like its raw taste, you can buy the curry leaf powder widely available in the market and have it with dosa or hot rice.

Alternatively, you can also add a few curry leaves to your hair oil and boil it for a few minutes. Apply this to hair tonic which will keep your hair healthy.

Using curry leaves with food strengthens the functions of stomach and promotes its action. It is good for indigestion problem (Reshmi, 2012).

It also helps to control Diarrhea, and upset stomach problems.

It also helps prevent premature hair graying.

Mix lime juice, honey or sugar crystals to the juice of curry leaf and drink it. This prevents nausea and dry vomiting.

Mix Curry leaf with finely chopped ginger. Mix it with rice and take it early in the morning instead of water which will reduce dizziness, stomach disorders and constipation.

Grind Curry leaf, black pepper corns, dried ginger, little cumin seeds and salt to a fine powder. Mix it with rice, make two small balls and eat it.

SAGE (*Salvia officinalis*)

Sage-Dark Green Leaves

Traditionally used to improve mental acuity, new research now suggests that sage may restore mental functions and improve memory, Sage tea is astringent, sedative and expels gas; it clears the respiratory tract, makes a good gargle for sore throats and helps overcome colds. Sage is useful for night sweats as it reduces sweating. It also reduces milk flow in nursing mothers prior to weaning, prevents the formation of kidney stones by dissolving resides of uric acid and regularizes menstruation. An infusion of sage can be applied to the scalp to reduce dandruff.

The sage herb is greatly beneficial for its astringent, stimulant, nervine, diuretic, expectorant, memory boosting, anxiolytic, anti-inflammatory, anti microbial, antioxidant effects. It will benefit the swellings, the cuts, sprain, mouth ulcers, sore throat, dyspepsia, abdominal troubles, menopausal symptoms and joint pain.

Health Benefits

Sage herb parts have many notable plant derived chemical compounds, essential oils, minerals, vitamins that are known to have disease preventing and health promoting properties.

The primary biologically active component of common sage appears to be its essential oil which contains mainly ketone α- and β-thujone. In addition, sage leaf contains numerous other substances including cineol, borneol, tannic acid; bitter substances with cornsole and cornsolic acid; fumaric, chlorogenic, caffeic and nicotinic acids; nicotinamide; flavones; flavone glycosides; and estrogenic substances. These compounds are known to have counter-irritent, rubefacient, anti-inflammatory, anti-allergic, anti-fungal and anti-septic properties.

Thujone is GABA and Serotonin (5-HT3) receptor antagonist. It enhances concentration, attention span and quickens the senses; hence sage infusion has long been recognised as "thinker's tea". Its effects help deal with grief and depression.

Thee lobe sage (*S. triloba*) has flavone called **salvigenin**. Research studies found that vascular relaxant effect of salvigenin may provide benefits in the cardiovascular diseases.

This herb is exceptionally very rich source of many B-complex groups of vitamins, such as **folic acid**, thiamin, **pyridoxine** and riboflavin with many vitamins several times more than recommended levels.

The herb contains very good amounts of **vitamin A** and beta carotene levels. 100 g dry ground herb provides 5900 IU; about 196 per cent of RDA. Vitamin A is a powerful natural antioxidant and is essential for vision. It is also required for maintaining healthy mucus membranes and skin. Consumption of natural foods rich in vitamin A known to helps body protect from lung and oral cavity cancers.

Fresh sage leaves are good source of antioxidant vitamin; vitamin-C. Vitamin C helps in the synthesis of structural proteins like collagen. Adequate levels in the body help maintain integrity of blood vessels, skin, organs, and bones. Regular consumption of foods rich in vitamin-C helps body protect from **scurvy**; develop resistance against infectious agents (boosts immunity) and scavenge harmful, pro-inflammatory free radicals from the body.

Sage herb parts, whether fresh or dried, are rich source of minerals like potassium, zinc, calcium, iron, manganese, copper, and magnesium. Potassium is an important component of cell and body fluids which helps control heart rate and blood pressure. Manganese is used by the body as a co-factor for the antioxidant enzyme *superoxide dismutase.*

The fresh leaves or the tender branches are used as tooth brush as margossatree is used very commonly in rural India, as herbal tooth brush cum paste. It strengthens gums and cleans the oral cavity and teeth. Herbal tooth powders contain this herb.

Dipping a cloth in its hot tea works like a compress. It can be applied to forehead to relieve tension, headache. A compress applied to abdominal areas can help in cramps, flatulence and general intestinal and digestive discomfort. Sprains also respond to this compress. This hot compress can be used in itching and swelling after insect bites.

For purpose of fermentation or poultice, the bags of its powder are used.

The oil is used in preparations for joints pain cream or liniments.

Sage is another spice with antioxidant and antibacterial properties. It is used in several cultures as a natural preservative, keeping meats and cheeses from going bad. Especially used to preserve sausages, it has a strong flavour and is used sparingly, to keep it from overpowering the food.

Medicinal Uses

The essential oil obtained from sage has been found to have acetylcholinesterase (Ach) enzyme inhibition activities. This help rise Ach levels in the brain. Ach improves concentration and may play a role in the treatment methods for memory loss associated with the disease like Alzheimer's.

Sage oil, distilled from the flowering tops, contains volatile essential oil such as camphene, cineol, borneol, bornyl acetate and other esters. These compounds are known to have tonic, astringent, diaphoretic and stimulant properties.

Sage herb oil is also being used externally as a rubefacient to soothen painful ailments like muscle stiffness, rheumatism and neuralgic conditions.

Used as blended massage oil or in the aromatic therapy sage oil helps with nervousness, anxiety, headaches, stress and fatigue.

Caution

Sage herb when used in large amounts causes nervous irritation, convulsions and death. Hence its use is prohibited in known epileptic conditions. The herb should not be used in pregnancy as chemical compounds like **thujone** in it may cause uterine stimulation resulting in abortion.

WATER CRESS (*Nasturtium officinalis*)

Water Cress Leaves

Water cress is a green leafy vegetable that is very rich in vitamins A, C, K and beta carotene and minerals such as calcium, magnesium, copper, and potassium. It contains high levels of iodine which is beneficial for both the thyroid and immune system. Watercress also has antibiotic properties similar to those in the onion family which also gives it a spicy flavour and kick. It has the ability to help in breaking up kidney or bladder stones and is one of the best foods for purifying the blood and aiding in removing mucus from the body. Watercress is especially good for helping to prevent osteoporosis, anemia, cardiovascular diseases, muscular and skeletal problems, eye degeneration, and memory issues. It contains a special compound called **Gluconasturtiin** which is believed to help prevent colon, breast, and prostate cancers. Watercress is excellent freshly juiced with celery, cucumber, and apple. It is also delicious in salads, soups, or stews, and can be steamed like spinach. Watercress can also be made into an herbal tea by pouring boiling water over a handful of greens and allowing to steep for 15 minutes or more. Fresh watercress can also be crushed and applied to the skin to help relieve eczema, psoriasis, acne, and other skin irritations and infections (http://bit.ly/1eLD8Aa).

It is rich in vitamins and minerals, especially vitamin A and C, zinc and iron. The plant is also rich in potassium, calcium, phosphorus and iron, with good quantities of iodine, sodium and magnesium.

Although the healing powers of water cress have long been known, recent scientific studies have shown that it can inhibit the growth of some cancerous tumours. New research has revealed fresh watercress contains high levels of PEITC **(phenethyl isothiocyanate),** which neutralizes a dangerous carcinogen in tobacco called NNK and one study indicated that the consumption of 40g of fresh watercress three times a day for three days will help to protect smokers from lung cancer. For more lasting protection, this process can be repeated once a month.

A tea prepared from the leaves can strengthen digestion, increase urination and cleanse the respiratory system by releasing phlegm and mucus. Watercress is also recommended for the treatment of catarrh, anemia, weak digestion and gout.

If you're suffering from swollen breathing passages in the lung, coughs, or bronchitis, watercress may be just the herb for you. It acts as an anti-inflammatory, helping reduce inflammation and clear breathing passages. It helps to clear heat, and lubricate the lungs, and has even been found to be an effective chemopreventative agent against lung cancer induced by tobacco-specific lung carcinogens.

This powerful little herb purifies the blood, and stimulates the lymphatic system as well, contributing to overall health and wellness.

Health Benefits

In addition to its anti-cancer benefits, this precious leafy vegetable has high amounts of folic acid, iodine, calcium, iron, manganese and vitamin A, B6, C and potassium.

It is also known as a diuretic, digestive aid and expectorant.

This rich flavored green leafy vegetable is store house of many phytonutrients that have health promotional and disease prevention properties.

One of the very low calorie green leafy vegetables (11 kcal per 100 g raw leaves) and very low in fats; recommended in cholesterol controlling and weight reduction programs.

Cress leaves and stems contain gluconasturtiin, a glucosinolate compound that gives peppery flavor. Research studies suggest that the hydrolysis product of gluconasturtiin, 2-phenethyl isothiocyanate (PEITC), is believed to be cancer preventing by inhibition of phase I enzymes (mono-oxygenases and cytochrome P450s).

Fresh cress has more concentration of ascorbic acid (vitamin C) than some of fruits and vegetables. 100 g of leaves provide 47 mg or 72 per cent of RDA of vitamin C. As an anti-oxidant, vitamin C helps to quench free radicals and reactive oxygen species (ROS) through its reduction potential properties. Lab studies suggests that regular consumption of foods rich in vitamin C helps maintain normal connective tissue, prevent iron deficiency, and also helps body develop resistance against infectious agents by boosting immunity.

It is one of the excellent vegetable sources for **vitamin-K**; 100 g provides over 200 per cent of daily recommended intake. Vitamin K has potential role bone health by promoting osteotrophic (bone formation and strengthening) activity. Adequate vitamin-K levels in the diet helps limiting neuronal damage in the brain; thus, has established role in the treatment of patients suffering from **Alzheimer's disease.**

Cress is also excellent source of **vitamin-A** and flavonoids anti-oxidants like β **carotene**, lutein and zeaxanthin.

It is also rich in B-complex group of vitamins such as riboflavin, niacin, vitamin B-6 (pyridoxine), thiamin and pantothenic acid that are essential for optimum cellular metabolic functions.

It is also rich source of minerals like copper, calcium, potassium, magnesium, manganese and phosphorus. Potassium in an important component of cell and body fluids that helps controlling heart rate and blood pressure by countering effects of sodium. Manganese is used by the body as a co-factor for the antioxidant enzyme *superoxide dismutase*. Calcium is required as bone/teeth mineral and in the regulation of heart and skeletal muscle activity.

Regular inclusion of cress in the diet is found to prevent osteoporosis, anemia, vitamin A deficiency and believed to protect from cardiovascular diseases and colon and prostate cancers.

Caution

Excessive or prolonged use of watercress may cause kidney problems and it should not be used in pregnancy. Since, nowadays, wild water cress grows mainly in polluted water, it may contain various pollutants as well as the deadly liver fluke. Therefore, wild watercress is unsafe for gathering and only water cress grown commercially is filtered, shallow, graveled beds should be consumed.

DANDELION (*Taraxacum officinale*)

Dandelion with Golden Yellow Flower

The whole plant is highly nutritious and can be used as a tonic, diuretic and mild laxative. When prepared as an infusion, it can be used to stimulate bile formation and is also said to relieve the symptoms of jaundice and gall stones. It is useful for relief of water retention and to cleanse the body of poisons. The fresh, young leaves which are best used before the flower forms, can be added to salads or used for juice.

It is a natural diuretic that increases urine production by promoting the excretion of salts and water from the kidney. Dandelion may be used for a wide range of conditions requiring mild diuretic treatment, such as poor digestion, liver disorder and high blood pressure. It is a source of potassium, a nutrient often lost through the use of other natural and synthetic diuretics. The roots are stout, fusiform and fleshy, dark brown externally and white pulp inside somewhat yam-like. It contains bitter milky latex; more concentrated than in stems and leaves. Roots are generally dug when the plant turns second year of life. Generally, roots are harvested in summer for medicinal purposes or autumn for drying and grinding for coffee.

Dandelion greens are some of the best to help strengthen the bones and teeth, and help detox the liver. Dandelions provide the body with 535 per cent of the recommended daily value of vitamin K, which means that it is one of the most important plant-based sources to help strengthen bones. Vitamin K is needed for improving bone density – individuals who are low in vitamin K, often suffer from higher risk of fracture. It keeps bone mineralization in balance and helps raise osteocalcin, which controls the building of bone.

Dandelions also help stimulate digestion and bile flow, and are considered a diuretic, helping to lower blood pressure and relieve premenstrual fluid retention.

Researchers have also found that dandelions help promote healthy lipid profiles, and suppress fat accumulation in the liver.

Dandelion Greens

Dandelion is beneficial to digestion and is an antiviral that may be useful in the treatment of AIDS and herpes

It may also be useful in treating jaundice, cirrhosis, edema due to high blood pressure, gout, eczema and acne

Dandelion is also used to treat and prevent breast and lung tumors and premenstrual bloating

Dandelion greens are high in vitamin A in the form of antioxidant carotenoid and vitamin C

They also contain calcium and potassium

Dandelion root contains inulin, which lowers blood sugar in diabetics

Health Benefits

Fresh dandelion greens, flower tops, and roots contain valuable constituents that are known to have anti-oxidant, disease preventing, and health promoting properties.

Fresh leaves are very low in calories; providing just 45 calories per 100 g. It is also good source of dietary fibre (provide about 9 per cent of RDA per 100 g). In addition, its latex is a good laxative. These active principles in the herb help reduce weight and control cholesterol levels in the blood.

Dandelion root as well as other plant parts contains bitter crystalline compounds **Taraxacin**, and an acrid resin, **Taraxacerin**. Further, the root also contains *inulin* (not insulin) and *levulin*. Together, these compounds are responsible for various therapeutic properties of the herb.

Fresh dandelion herb provides one of the highest source of vitamin-A among culinary herbs. Vitamin A is an important fat-soluble vitamin and anti-oxidant, required for maintaining healthy mucus membranes and skin and vision.

Its leaves are packed with numerous health benefiting flavonoids such as *carotene-β, carotene-α, lutein, crypto-xanthin* and *zea-xanthn*. Consumption of natural foods rich in vitamin-A and flavonoids (carotenes) helps body protect from lung and oral cavity cancers. Zeaxanthin has photo-filtering functions and protects retina from UV rays.

The herb is good source of minerals like **potassium**, calcium, manganese, iron, and magnesium. Potassium is an important component of cell and body fluids, which helps regulate heart rate and blood pressure. Iron is essential for red blood cell production. Manganese is used by the body as a co-factor for the antioxidant enzyme, *superoxide dismutase.*

It is also rich in many vital vitamins including folic acid, riboflavin, pyridoxine, niacin, vitamin -E and vitamin-C that are essential for optimum health. Vitamin-C is a powerful natural antioxidant. Dandelion greens provide 58 per cent of daily-recommended levels of vitamin-C.

Dandelion is probably the *richest herbal sources of* **vitamin K; provides about 650 per cent of DRI.** Vitamin-K has potential role in bone mass building by promoting osteotrophic activity in the bones. It also has established role in the treatment of Alzheimer's disease patients by limiting neuronal damage in the brain.

Dandelion has traditionally been used internally for gall bladder and urinary disorders, gall stones, jaundice, cirrhosis, dyspepsia with constipation, edema associated with high blood pressure and heart diseases, chronic joint and skin complaint, gout, eczema and acne, soothing effect on bee stings, sores *etc.*

Skin care products to rejuvenate

Laxative properties.

Source of vitamin A and C, potassium, calcium, lethicin, iron, magnesium, niacin, phosphorus.

Cirrhosis- the herb increases bile production and cleanses the blood stream.

It is a natural diuretic

Arthritis - relieves the stiffness in the joints.

It also provides rest and sleep in those with fever.

Bile production by liver and urinary output from the kidney is increased with the use of the herb.

The leaves are particularly strong, being equivalent to frusemide, a drug used to treat hypertension.

It is powerful remedy, not only for hypertension but also for cardiac oedema, hepetogenic dropsy and water retention, due to stasis or congestion in the blood vessels serving the liver.

The diuretic effect of dandelion is helpful in the treatment of a number of other conditions, particularly chronic disorders like rheumatism, gout and eczema.

When the stomach is irritated and where active treatment would be injurious the decoction or extract of dandelion administration 3 or 4 times a day will often prove a valuable remedy.

It has a good effect in increasing the appetite and promoting digestion.

Medicinal Uses

Almost all the parts of dandelion herb found place in various traditional as well in modern medicine.

The principle compounds in the herb have laxative and diuretic functions.

The plant parts have been used as herbal remedy for liver and gall bladder complaints.

The herb is also a good tonic, appetite stimulant and is a good remedy for dyspeptic complaints.

The inside surface of the flower stems used as a smoothening agent for burns and stings (for example in stinging nettle allergy).

Caution

Although dandelion herb contains some bitter principles, it can be safely used in healthy individuals without any reservations. However, in patients on potassium sparing diuretic therapy, it may aggravate potassium toxicity. Dandelion herb can also induce allergic contact dermatitis in some sensitive individuals.

SWEET CORN (*Zea mays* var. *saccharata*)

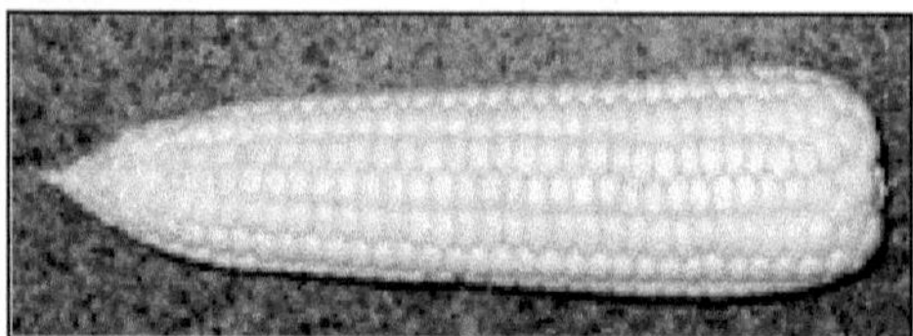

Sweet Corn

Sweet corn is a special maize variety in which its sweet kernels eaten as vegetable. In contrast to traditional field corn, sugar corn variety is harvested when the ears just reach milk stage and used fresh as the sugars in the kernels convert quickly to starch. Baby corns are very young, miniature ears harvested when they are very small. The kernels are in incipient stage and its core is sweet and tender enough to eat raw. Baby corns measure about 3-5 inches in length and weigh about 20-50 g.

It provides a good source of vitamin A, B1, B2, niacin and minerals such as iron, copper, phosphorus, magnesium, zinc, copper, iron and selenium. It has small amount of potassium. Corn silk, the fine tassel on the top of the corn cob, can be made into an infusion which soothes the urinary passages and act as a diuretic. This can be very beneficial in cases of kidney stones and cystitis, but to be effective several cups of the infusion should be drunk each day. Rich in vitamin B1, which is used in the metabolism of carbohydrates, vitamin B5, which helps with physiological functions, folate that helps generating new cells and vitamin C, which helps fight diseases.

Corn has vitamin B (Thiamin, Vitamin B6, Niacin, Riboflavin, and Folate). It has traces of vitamin A and vitamin E. Being rich in folate, corn helps the generation of new cells, especially important before and during pregnancy. Those suffering from anemia have shown positive effects after consuming corns. Regular consumption of corn, in moderate quantities, has been associated with better cardiovascular health

Corn is high in fibre, which helps you fight digestive problems like constipation. Fibre also helps lower cholesterol levels, reduces the risk of colon cancer and is also useful in helping to lower blood sugar levels in diabetics. Folic acid present in corn is known to prevent neural- tube birth defects. Expert recommend corn for people who suffer from anemia. Therefore, corn should be a must in your diet.

Corn silk is the strands found under the green colour outer covering of the corn. As kids, you must have been fascinated about it since they resemble a dolls hair. Many people just throw it away because we don't know that it has so many uses and it can benefit us so much.

Health Benefits

Here are some of the amazing benefits of corn silk:

Prevents Kidney Stones: Corn silk has been used to prevent kidneystone since ancient times. Kidney stones are formed by the accumulation of small crystals in the kidney. It helps in the proper flow of urine that prevents the accumulation of crystals.

However, it does not remove the stone already in the kidney.

Assists in blood clotting: It contains vitamin K which helps in blood clotting. Vitamin K makes sure that you don't loose excessive blood from your body when you get hurt hence helping it clot and preventing blood loss.

Controls blood sugar: In many recent studies corn silk has been proven as an agent that controls the blood sugar levels. It increases insulin levels and aids in the repair of damaged cells in the pancreas, where insulin is produced.

Control cholesterol: It also controls the cholesterol in the body. High cholesterol leads to several heart diseases. Hence consumption of outer silk of corn can help in keeping you away from the heart-related ailments.

Diuretic properties: It has diuretic properties which helps flush out excessive fluids and toxins from the body hence preventing the risk of cardiovascular disease and Urinary tract infecting (UTI).

At 86 calories per 100 g, sugar corn kernels are moderately high in calories on comparison to other vegetables. However, fresh kernels have much lower in calories than field corn and other cereals like wheat, rice *etc.* Their calorie mainly comes from more simple carbohydrates like glucose, sucrose than complex sugars like amylose and amylopectin as in cereals.

Sweet corn is gluten free cereal and may be used safely much like rice, quinoa *etc.*, in celiac disease individuals.

Corn features high quality phyto-nutrition profile comprising of dietary fibre, vitamins, and antioxidants in addition to moderate proportions of minerals. It is one of the finest source dietary fibres; 100 g kernels provide 2 g or 5 per cent of daily-requirement of dietary fibre. Together with slow digesting complex carbohydrates, moderate amounts of fibre in the food regulates gradual rise in blood sugar levels.

Yellow variety corn has significant levels of phenolic flavonoid pigment antioxidants such as β-*carotenes, and lutein, xanthins and cryptoxanthin* pigments along with vitamin A. 100 g fresh kernels provide 187 IU or 6 per cent of daily-requirement of vitamin A. Altogether, these compounds are required for maintaining healthy mucus membranes, skin and vision. Consumption of natural foods rich in flavonoids helps to protect from lung and oral cavity cancers.

Corn is a good source of phenolic flavonoid antioxidant, **ferulic acid**. Several research studies suggest that ferulic acid plays vital role in preventing cancers, aging, and inflammation in humans.

It also contains good levels of some of valuable B-complex group of vitamins such as thiamin, niacin, pantothenic acid, folates, riboflavin, and pyridoxine. Many of these vitamins functions as co-factors to enzymes in metabolism.

It contains healthy amounts of some important minerals like zinc, magnesium, copper, iron, and manganese.

Baby corn is a good source of folate and vitamin B. It also contains potassium, vitamin B6, riboflavin, **vitamin C** and fibre.

Good for people on a diet as it contains a good amount of nutrients, fibre and is low in calories.

Corn is an excellent source of dietary fibre and potassium and contains vitamin C, beta carotene and niacin (one of the B group vitamins which assist in the functioning of the digestive system, skin, and nerves. It is also important for the conversion of food to energy.)

Being rich in folate, corn helps the generation of new cells, especially important before and during pregnancy. Those suffering from anemia have shown positive effects after consuming corns. The Pantothenic acid present in corns helps with the physiological functions of the body.

Owing to the presence of thiamin, corns have been said to help in the metabolism of carbohydrates. Corn has been found to be helpful in treating kidney problems, including renal dysfunction.

Regular consumption of corn, in moderate quantities, has been associated with better cardiovascular health.

The beta-cryptoxanthin in corn makes it good for the health of the lungs and may even help prevent lung cancer.

SWISS CHARD (*Beta vulgaris*-Cicla group)

Chard plant features distinctly large dark green leaves with well-developed edible stalks. Generally, chard leaves are harvested at various stages of maturity. While whole plant with tender young leaves harvested for salad preparation; individual matured large sized leaves with slightly tougher stems picked up for sautéing and cooking. In North it is known as "Chauli Saag" and in Andhra "Thotakura". Swiss chard has amazing health benefits - it can regulate blood sugar levels, prevent various types of cancer, improve digestion, boosts the immune system, reduces fever and combats inflammation, among many others.

Swiss Chard with White Stalks

Swiss chard's many health benefits can be attributed to its impressive list of vitamins, nutrients and organic compounds, which include vitamin K, vitamin C, vitamin A and vitamin E, riboflavin and vitamin B6. Swiss chard is also packed with minerals, in fact, it includes a wealth of magnesium, manganese, potassium, iron, sodium and copper. It is also a good source of dietary fibre. Swiss chard has a significant amount of polyphenolic antioxidants, phytonutrients and enzymes that are unique and highly beneficial to your health. It is also rich source of minerals like copper, calcium, sodium, potassium, iron, manganese and phosphorus. Potassium is an important component of cell and body fluids that helps controlling heart rate and blood pressure by countering effects of sodium. Manganese is used by the body as a co-factor for the antioxidant enzyme, *superoxide dismutase*. Iron is required for cellular oxidation and red blood cell formation.

Health Benefits

It helps manage diabetes: Swiss chard has been found to regulate blood sugar levels in the body. One of the unique flavanoids found in Swiss chard is **syringic acid**, which inhibits the activity of a specific enzyme named alpha-glucosidase. This means that less carbs are broken down to simple sugars, which allows the blood sugar levels in the body to remain stable and prevent the plunges and peaks, known to be dangerous for diabetic patients, or those who are at a high risk of developing diabetes.

It improves bone health: Swiss chard contains a significant amount of calcium in its leaves, giving it a major boost for bone health. Alongside other minerals found in this vegetable, calcium helps stimulate bone growth and development, including magnesium and vitamin K.

It boosts your brain power: Swiss chard is a great source of both potassium and vitamin K, both of which are found in significant amounts in the brain and are integral parts of boosting cognitive development and abilities.

It boosts blood circulation: Both iron and copper are essential elements of red blood cells. Without these two minerals, people can develop anemia, which includes symptoms like weakness, fatigue, stomach disorders and lack of concentration. Ensuring a proper amount of red blood cells by eating such foods can increase your blood flow and oxygenation of essential organs within the body.

It improves blood pressure and heart health: Swiss chard contains anti-inflammatory and phytonutrient antioxidants, which, along with potassium, all contribute to reducing blood pressure and stress on the cardiovascular system. Some hypertension is due to pro-inflammatory enzymes within the body, which the organic compounds in Swiss chard are able to neutralize. Furthermore, Swiss chard protects those who eat it from a variety of conditions like atherosclerosis, heart attacks and strokes.

It promotes eye care: Swiss chard contains a substantial amount of beta-carotene, which has been linked to optimal eye health and a reduction in macular degeneration, glaucoma, night blindness and other vision-related conditions.

It promotes healthy hair: The organic compound biotin has directly been linked to healthy hair, the stimulation of follicles as well as luster and texture. Swiss chard also has significant amounts of biotin, among its many other beneficial compounds.

It can prevent cancer: As with many leafy, green vegetables, Swiss chard also contains anti-cancer properties due to the large amounts of antioxidants found in it. Antioxidants neutralize free radicals, which are a dangerous byproduct of cellular metabolism, causing healthy cells to become cancerous. It also contains significant amounts of vitamin E, C, zinc, lutein, zeaxanthin, kaempferol, beta-carotene and quercetin - many of which have been connected to preventing a wide array of cancers, specifically, colon cancer.

Swiss chard, like **spinach**, is the store-house of many phytonutrients that have health promotional and disease prevention properties.

Chard is very low in calories (19 kcal per 100 g fresh, raw leaves) and fats, recommended in cholesterol controlling and weight reduction programs.

Chard leaves are an excellent source of anti-oxidant vitamin, **vitamin-C.** Its fresh leaves provide about 33 per cent of recommended levels per 100 g. As an anti-oxidant, vitamin C helps to quench free radicals and reactive oxygen species (ROS) through its reduction potential properties. Research studies suggests that regular consumption of foods rich in vitamin C helps maintain normal connective tissue, prevent iron deficiency, and also helps body develop resistance against infectious agents by boosting immunity.

Chard is one of the excellent vegetable sources for **vitamin-K**; 100 g provides about 700 per cent of recommended intake. Vitamin K has potential role bone health by promoting osteotrophic (bone formation and strengthening) activity. Adequate vitamin-K levels in the diet helps limiting neuronal damage in the brain; thus, has established role in the treatment of patients suffering from **Alzheimer's disease.**

It is also rich source of omega-3 fatty acids; **vitamin-A** and flavonoids anti-oxidants like β **carotene**, α-carotene, lutein and zeaxanthin.

It is also rich in B-complex group of vitamins such as folates, niacin, vitamin B-6 (pyridoxine), thiamin and pantothenic acid that are essential for optimum cellular metabolic functions.

Regular inclusion of Swiss chard in the diet is found to prevent osteoporosis, iron deficiency anemia, vitamin A deficiency and believed to protect from cardiovascular diseases and colon and prostate cancers.

Caution

Because of its high vitamin K content, patients taking anti-coagulants such as warfarin are encouraged to avoid this food since it increases the vitamin K concentration in the blood, which is what the drugs are often attempting to lower. This effectively raises the effective dose of the drug and causes toxicity. Chard contains oxalic acid, a naturally occurring substance found in some vegetables which may crystallise as oxalate stones in the urinary tract in some people. It is, therefore, advisable to avoid eating chard in people with known oxalate urinary tract stones. Adequate intake of water is therefore advised to maintain normal urine output.

ALOE VERA [*Aloe vera* (L) Burn]

Aloe vera is a clear thin gelatinous material that comes from inside the aloe vera leaves. Aloe vera products are available in various forms like capsules, gel, juice and those that are applied directly to the skin. it has cooling effect and bitter in taste, It contains aloin that is responsible for its purgative action so it is known to relieve constipation. It regulates the peristaltic movements of intestines and promotes digestion, the liver and spleen function are stimulated by the use of this herb. Research work carried out over many years points conclusively to a toxic colon being the cause of a very wide range of illness, in establishing a regime of regular daily dose of aloe vera juice it may not only be found that the effects are gradual gentle and with no irritant or harmful side effects, but also blood circulation will be improved due to aloe Vera's ability to detoxify. Also being a natural healer, any internal ulcers or lesions will be soothed and healing will be enhanced. Aloe vera leave has some ingredients like Vitamin, Minerals, Amino acid, Polysaccharides, Enzymes, Plant steroids, Saponins, Lignin, Anthraquinones Salicylic acid which are necessary for human bodies.

Aloe Vera is one of the most miraculous findings of the modern medicine. Recently, scientific studies have discovered, confirmed, and even extended, many of the claims about the healing effects of this plant. Aloe Vera is confirmed as one of the most healing and most powerful medicinal plants in the world. Whether we use it externally or internally, this plant will only have a beneficial effect on our health

The Egyptians called it " the immortality plant" and the Indians called it the stick of Eden, " which should not come as a surprise taking into consideration that the Aloe vera has numerous health benefits. Aloe vera contains over 200 biologically active compounds, among which are vitamins, amino acids, enzymes, polysaccharids as well as minerals that promote better nutrient absorption.

Aloe Vera (barbadensis miller) contains over 20 minerals, all of which are essential to the human body. The human body requires 22 amino acids for good health – eight of which are called "essential" because the body cannot fabricate them. **Aloe Vera** contains all of these eight essential amino acids, and 11 of the 14 "secondary" amino acids. Aloe Vera has Vitamins A, B1, B2, B6, B12, C and E. Vitamins cannot be manufactured within the body, and some cannot be stored by the body, so it is necessary for the diet to sustain a continuous supply. Therefore to get a steady supply of these minerals, amino acids and vitamins - one should drink between two to four ounces of Aloe Vera Gel twice daily. Calcium, magnesium, chromium, selenium, sodium, iron copper and manganese are only some of the minerals found in the Alovera, when these minerals are combined they improve the health of the metabolic tract, which also makes them a good solution for losing weight.

The plant also contains vital enzymes like amylases and lipases which assist the process of sugar and fat decomposition, but also bradykinin which soothes inflammations.

The most popular among all beauty foods, this magic herb contains adequate amounts of amino acids (20 of them), calcium, sodium, magnesium, enzymes, vitamins, nitrogen and more. Capsules of aloe vora help among other things in blood circulation which in turn boosts the appearance of skin. The aloe vera juice contains many important vitamins and minerals, as well as β-carotene, enzymes, amino acids and complex carbohydrate mucopolysaccharide. These are responsible for its many healing actions, which include the soothing of inflammation of the digestive tract and the relief of constipation, flatulence and symptoms of irritable bowel syndrome (IBS). The gelatinous juice of the fresh leaves can be used externally. It is reputed to heal wounds and can also be rubbed on the skin to alleviate sunburns, wrinkles, skin irritations and minor cuts. Infusions are good for bathing wounds and eyes. The juice has a somewhat repellent taste so, when taken orally, it is usually blended with fruit juice to make it more palatable.

Aloe vera juice is excellent if you want younger-looking skin says Deepshikha and adds, "Have 30 ml aloe vera juice diluted with 100 ml water early morning on empty stomach. And after 20-30 minutes you can have your regular breakfast."Aloe vera works as Anti-septic, Antibacterial, and Anti inflammatory. It cures Eczema, Diabetes, Arthritis and Prevent infections. It also improves human immune system and digestive system.

But do you know the health benefits of drinking aloevera juice? Completely safe and versatile, aloe vera juice stems from the fact that it naturally contains many different nutrients: vitamins, minerals, amino acids and other trace elements.

Aloe vera's anti-ageing properties have been known for ages. Scientific evidence suggests that Aloe Vera juice can heal skin and reverse skin ageing. Skin ages because of UV light damage and loss of collagen in the skin. Collagen helps in keeping the skin firm and elastic. Ageing breaks down the collagen matrix in your skin which leads to wrinkles. Drinking aloe vera juice is known to reduce the wrinkle depth and improve skin elasticity due to increased collagen production (Anonymous, 2014). Aloe vera has a bitter taste which can be unpleasant in the raw state. The term gel refers to the inner leaf only, where as juice refers to a bitter substance found just under the skin of the leaf." Aloe- latex"

A aloe vera leaf is made up of:

Rind is the outer protective layer, where the production and synthesis of the nutrients occur's.

Sap –A layer of bitter fluid which helps protect the plant from animals and contains anthraquinones which has a bitter taste.

The mucilage gel -The inner part of the leaf that is filleted out to make Aloe vera gel which **contains polysaccharides.**

Health Benefits

It has amazing therapeutic potential; it is used internally as well as externally.

Skin cancer: Several studies discovered that the plant may prevent and treat melanoma. Research from the Belgrade's School of Medicine confirmed that the ingredients in Aloe vera inhibit proliferation of cells that morph into tumors.

Research discovered that Emodin contained in Aloe vera stops cell proliferation. They concluded that even though isolated Aloe vera compounds are protective. Aloe vera also contains acemannan, which naturally boosts the immune system. Gachon University of Medicine discovered that Aloe –emodin prevents the growth of human cancerous liver cells and caused cell death of tumor cells.

Helps digestion: Drinking aloe vera juice naturally allows the body to cleanse the digestive system. It encourages the bowels to move and helps with elimination if a person is constipated. And if you have diarrhea, it will help slow it down.

Increases energy levels: Our diets include many substances which can cause fatigue and exhaustion. Taken regularly, aloe vera juice ensures a greater feeling of well-being, allowing energy levels to increase and also helps maintain a healthy body weight.

Builds immunity: It is especially great for those who have chronic immune disorders like polysaccharides or fibromyalgia since the polysaccharides in aloe vera juice stimulate macrophages, the white blood cells that fight viruses.

Detoxifies: Aloe vera juice is a great natural aid to detox. With our stressful lives, the pollution around us and the junk foods we eat, we all need to cleanse our systems from time to time. Drinking aloe vera juice provides a fantastically rich cocktail of vitamins, minerals and trace elements to help our bodies deal with these stresses and strains every day.

Reduces inflammation: It improves joint flexibility and helps in the regeneration of body cells. It strengthens joint muscles which therefore reduces pain and inflammation in weakened or aged joints.

The Aloe Vera gel and juice contain more than 240 medicinal and nutritional components, among which we have the vitamins A, thiamin, riboflavin, niacin, B6, folic acid, B12, C, E, choline; and the minerals calcium, magnesium, potassium, iron; then amino acids, saponins, enzymes, *etc.*

It does wonders for the skin, such as helps heal burns, treats bites of insects, allergic rashes, sores, acne, eczema, and the list goes on. This extraordinary plant prevents the growth and spread of fungi and bacteria, and speeds up the regeneration process of cells. Therefore, Aloe successfully accelerates the wound healing process and solves the skin problems.

Helps treat diarrhea, colitis, irritable bowel syndrome and constipation

Regulates stomach acidity and digestion, and helps ease stomach pain and heartburn

Removes the toxic, harmful substances of the human organism and purifies the blood

Regulates the level of sugar in the blood

Boosts the metabolism and promotes weight loss

Improves pulmonary circulation

Speeds up the tissue healing process

Helps prevent vascular diseases and normalizes blood pressure levels

Makes the immune system stronger and more resistant to diseases

Prevents growth of tumor cells

Decreases swellings and inflammation

Relieves allergies and symptoms of asthma

Helps treat eye diseases, such as trachoma, cataract, keratitis, glaucoma, and so on.

Destroys bacteria, viruses, parasites and fungi

Helps when trying to quit smoking, drugs and alcohol

It also possesses antimicrobial and antifungal properties and as such helps in strengthening immune system and clearing the organism from toxins.

ENDIVE (*Cichorium endiva*)

Endive Flat Leaf Variety

Endive, commonly popular as **escarole**, is a green leafy vegetables with a hint of bitter flavor. However, this well known salad plant is much more than just a leafy green. Escarole is packed with numerous health benefiting plant nutrients such as vitamin C, vitamin A *etc.*

Health Benefits

Endive is a very low calorie leafy vegetable. 100 g fresh leaves provide just 17 calories but contribute about 8 per cent of RDA of fibre.

Current research studies suggest that high inulin and fibre content in escarole helps reduce high glucose and LDL cholesterol levels in diabetes and obese patients.

Endive is enriched with good amount Vitamin A and beta-carotenes. Both of these compounds are known to have antioxidant properties. Carotenes convert to vitamin A in the body. Also, vitamin A is required for maintaining healthy mucus membranes and skin. In addition, it is also essential vitamin for vision. Consumption of natural vegetables/greens rich in vitamin A helps to protect from lung and oral cavity cancers.

It contains good amounts of many essential B-complex groups of vitamins such as folic acid, pantothenic acid (vitamin B5), pyridoxine (vitamin B6) and thiamin (vitamin B1), niacin (B3). These vitamins are essential in the sense that body requires them from external sources to replenish and required for fat, protein and carbohydrates metabolism.

Escarole also good source of minerals like manganese, copper, iron, and potassium. Manganese is used as a co-factor for the antioxidant enzyme *superoxide dismutase.* Potassium is an important intracellular electrolyte helps counter the hypertension effects of sodium.

Caution

Although this greeny leaf vegetable contains high concentrations of bitter glycosides and inulin, no known side effects so far notified when eaten in moderate.

RHUBARB (*Rheum rhabarbarium*)

Fresh Rhubarbe

Stalks Rhubarbe Plant

Rhubarb is easy to grow and lives for many years (10-15 years) once established. The plant is usually propagated by dividing the old rhizomes (roots). Well grown plant feature broad heart shaped, dark green leaves with 12 to 18 inches long leaf petioles. It is these stalks, which are used, and their top greens discarded, as they are unfit for human consumption. Usually its stalks can be harvested from second year onwards after planting when the foliage spread and stalks reached sufficient girth of about one to two inches thick.

It is used mainly for its roots and rhizomes. It is an appetite stimulant and astringent, It is effective for both constipation and diarrhea, depending on the amounts used. Larger amounts can promote diarrhea while tiny amounts will have a constipative effect.

Health Benefits

Rhubarb is one of the least calories vegetable. 100 g fresh petioles provide just 21 calories. Nonetheless, it contains some vital phyto-nutrients such as dietary fibre, poly-phenolic anti-oxidants, minerals, and vitamins. Further, its petioles contain no saturated fats or cholesterol.

The stalks are rich in several B-complex vitamins such as folates, riboflavin, niacin, vitamin B-6 (pyridoxine), thiamin, and pantothenic acid.

Red color stalks contain more vitamin-A than green varieties. Furtehr, the stalks also contain small amounts of poly-phenolic flavonoid compounds like β-carotene, zeaxanthin, and lutein. These compounds convert to vitamin A inside the body and deliver same protective effects of vitamin A on the body. Vitamin A is a powerful natural anti-oxidant and is required by body for maintaining the integrity of skin and mucus membranes. It is also an essential vitamin for vision. Research studies suggest that natural foods rich in vitamin A helps body protect against lung and oral cavity cancers.

Like in other greens like **kale, spinach**; rhubarb stalks also provide good amounts of **vitamin-K**; 100 g of fresh stalks provide 29.3 µg or about 24 per cent of daily recommended intake. Vitamin K has potential role in bone health by promoting osteotrophic (bone formation and strengthening) activity. Adequate vitamin-K levels in the diet helps limiting neuronal damage in the brain; thus, has established role in the treatment of **Alzheimer's disease.**

Its stalks also contain healthy levels of minerals like iron, copper, calcium, potassium, and phosphorus. However, most of them may not absorb into the body as they are chelated by oxalic acid into insoluble complexes and excreted out.

Caution

Top green part of rhubarb leaf (blade) contains unusually high amounts of oxalic acid, a naturally occurring substance found in some vegetables. 100 g of leaves contain about 0.59 - 0.72 mg of oxalates. Lowest published lethal dose (LDLo) of oxalate in humans is 600 mg/kg. Oxalate can cause severe symptoms even at much lower concentrations than this on the human body. Symptoms may include burning in the eyes, mouth, and throat; skin edema, difficulty breathing. In severe cases it can result in kidney failure, convulsions, coma, and death. The leaves are high in oxalic acid, which can bind calcium and promote calcium crystals in people susceptible to kidney stones.

ARUGULA (*Eruca sativa*)

Arugula, also known as salad rocket, is a nutritious leafy green vegetable of Mediterranean origin. It belongs to the *brassicaceae* family like **mustard greens, cauliflower, kale** *etc.* and has scientific name *Eruca sativa.* Rocket-salad is a low growing annual herb features **dandelion** like succulent, elongated, lobular leaves with green veins. Young plant features plain light green color leaves which appear somewhat identical to that of spinach. Young plant has mildly sweet, less peppery leaves.

Health Benefits

Like other greens, arugula is one of very low calorie vegetable. 100 g of fresh leaves provides just 25 calories. Nonetheless it has many vital phytochemicals, anti-oxidants, vitamins, and minerals that can immensely benefit health.

Rocket salad is rich source of certain phytochemicals such as *indoles, thiocyanates, sulforaphane* and *isothiocyanates*. Together they have been found to counter carcinogenic effects of estrogen and thus help benefit against prostate, breast, cervical, colon, ovarian cancers by virtue of their cancer cell growth inhibition, cytotoxic effects on cancer cells.

In addition, di-indolyl-methane (DIM), a lipid soluble metabolite of indole has immune modulator, anti-bacterial and anti-viral properties (by potentiating Interferon-Gamma receptors and production). DIM has currently been found application in the treatment of recurring respiratory papillomatosis caused by the Human Papilloma Virus (HPV) and is in Phase III clinical trials for cervical dysplasia.

Rocket is very good source of folates. 100 g of fresh greens contain 97 mcg or 24 per cent of folic acid. When given around conception period it helps prevent neural tube defects in the newborns.

Like kale, salad rocket is an excellent source of vitamin A. 100 g fresh leaves contain 1424 mcg of beta carotene and 2373 IU of vitamin A. Beta carotenes converts into vitamin A in the body. Studies found that vitamin A and flavonoid compounds in green leafy vegetables help protect from skin, lung and oral cavity cancers.

This vegetable also rich in B-complex group of vitamins such as thiamin, riboflavin, niacin, vitamin B-6 (pyridoxine), and pantothenic acid those are essential for optimum cellular enzymatic and metabolic functions.

Fresh rocket leaves contain good levels of vitamin C. Vitamin C is a powerful, natural anti-oxidant. Foods rich in vitamin C helps body protect from scurvy disease; develop resistance against infectious agents (boosts immunity) and scavenge harmful, pro-inflammatory free radicals from the body.

Salad rocket is one of the excellent vegetable sources for vitamin-K; 100 g provides about 90 per cent of recommended intake. Vitamin K has potential role bone health by promoting osteotrophic (bone formation and strengthening) activity. Adequate vitamin-K levels in the diet helps limiting neuronal damage in the brain; thus, has established role in the treatment of patients suffering from *Alzheimer's disease.*

Arugula is good in minerals especially copper and iron. In addition it has small amounts of some other essential minerals and electrolytes such as calcium, iron, potassium, manganese and phosphorus.

YAMS (*Dioscorea* spp.)

Yams Tubers

The plant is a perennial vine cultivated for its large, edible, underground tuber, which can grow up to 120 pounds in weight and up to 2 meters in length. Yam is similar in appearance **to sweet potato;** but not at all related to it. Important differences: yams are monocotyledons, larger in size, features thick, rough, dark brown to pink skin depending up on cultivar type whereas sweet potatoes (*Ipomoea batatas*) are dicotyledonous, relatively smaller in size and possess very thin peel. Unlike sweet potatoes which can be eaten raw, yams should not be eaten raw since they contain many naturally occurring plant toxins including dioscorin, diosgenin and tri-terpenes. They must be peeled and cooked in order to remove these bitter proteins.

Health Benefits

Yam is a good source of energy; 100 g provides 118 calories. It mainly composed of complex carbohydrates and soluble dietary fibre. Together, they raise blood sugar levels rather very slowly than simple sugars and therefore recommended as low glycemic index healthy food. In addition, dietary fibre helps reduce constipation, decrease bad or "LDL" cholesterol levels by binding to it in the intestines and prevent colon cancer risks by preventing toxic compounds in the food from adhering to colon mucosa.

The tuber is excellent source of B-complex group of vitamins. Provides adequate daily requirements of pyridoxine (viamin B6), thiamin (vitamin B1), riboflavin, folic acid, pantothenic acid and niacin. These vitamins mediate various metabolic functions in the body.

Fresh root also contains good amounts of anti-oxidant vitamin, vitamin-C. Provides about 29 per cent of recommended levels per 100 g. Vitamin C has important roles in anti-aging, immune function, wound healing, bone growth.

Also contains good amount of vitamin-A and beta carotene levels. Carotenes convert to vitamin A in the body. These compounds are strong antioxidants. Vitamin A has many functions like maintaining healthy mucus membranes and skin, night vision, growth and protection from lung and oral cavity cancers.

This tuber is indeed one of the vegetable rich sources of minerals like copper, calcium, potassium, iron, manganese and phosphorus. 100 g provides about 816 mg of Potassium. Potassium is an important component of cell and body fluids which helps controlling heart rate and blood pressure by countering hypertensive effects of sodium. Copper is required in the production of red blood cells. Manganese is used by the body as a co-factor for the antioxidant enzyme *superoxide dismutase.* Iron is required for red blood cell formation.

Medicinal Uses

The mucilaginous tuber milk contains **allantoin,** a cell-proliferant that speeds the healing process when applied externally to ulcers, boils and abscesses. Its decoction is also used to stimulate appetite and to relieve bronchial irritation, cough *etc.*

Caution

Yams of African species must be cooked to be safely eaten, because various natural toxin substances such as dioscorine can cause illness if consumed.

BASIL (*Ocimum basilicum*)

Basil with Pink Flowers and Leaves

The king of herbs, **basil herb** is one of the oldest and popular herbal plant rich in many notable health benefiting phyto-nutrients. This highly prized plant is revered as "holy herb" in India. It is also known as Tulsi, is known to possess antioxidant properties and boosts metabolism. It is great for skin because it contains vitamin C, calcium, phosphorus and carotein. The leaves have appetite- stimulating properties and are generally used as a culinary flavouring. Infusions of leaves can be used to relieve flatulence, fermentation, stomach cramps and constipation and the plant is also said to relieve nausea. To relieve stomach ache or an upset stomach, chew fresh basil leaves. Basil is a medicinal plant which has a lot of compounds called metabolites, meaning the leaves of a basic basil plant, like any other plant, make a lot of stuff.

Health Benefits

Basil leaves contain many chemical compounds that are known to have disease preventing and health promoting properties.

Basil herb contains many *polyphenolic flavonoids* like **orientin** and **vicenin**.

Basil leaves contains many health benefiting essential oils such as *eugenol, citronellol, linalool, citral, limonene* and *terpineol.* These compounds are known to have anti-inflammatory and anti-bacterial properties.

The herbs parts are very low in calories and contain no cholesterol, but are very rich source of many essential nutrients, minerals, and vitamins that are essential for optimum health.

Basil herb contains exceptionally high levels of *beta-carotene*, ***vitamin A***, *cryptoxanthin, lutein* and *zea-xanthin*. These compounds help act as protective scavengers against oxygen-derived free radicals and reactive oxygen species (ROS) that play a role in aging and various disease process.

Zeaxanthin, a yellow flavonoid carotenoid compound, is selectively absorbed into the retinal macula lutea where it found to filter harmful UV rays from reaching retina. Herbs, fruits, and vegetables that are rich in this compound help to protect from *age related macular disease* (AMRD), especially in the elderly.

Vitamin A is known to have antioxidant properties and is essential for vision. It is also required for maintaining healthy mucus membranes and skin. Consumption of natural foods rich in vitamin-A has been found to help body protect from lung and oral cavity cancers.

Vitamin K in basil is essential for many coagulant factors in the blood and plays vital role in the bone strengthening function by helping mineralization process in the bones.

Basil herb contains good amount of minerals like potassium, manganese, copper, and magnesium. Potassium is an important component of cell and body fluids, which helps control heart rate and blood pressure. Manganese is used by the body as a co-factor for the antioxidant enzyme, *superoxide dismutase.*

Basil leaves are an excellent source of **iron**, contains 3.17 mg/100 g of fresh leaves (about 26 per cent of RDA). Iron, being a component of hemoglobin inside the red blood cells, determines the oxygen carrying capacity of the blood.

Medicinal Uses

Basil leaves contains many health benefiting essential oils such as eugenol, citronellol, linalool, citral, limonene and terpineol. These compounds are known to have anti-inflammatory and anti-bacterial properties.

An important essential oil, **eugenol** has been found to have anti-inflammatory function by acting against the enzyme *cycloxygenase* (COX), which mediates inflammatory cascade in the body. This enzyme-inhibiting effect of the eugenol in basil makes it an important remedy for symptomatic relief in individuals with inflammatory health problems like rheumatoid arthritis, osteoarthritis, and inflammatory bowel conditions.

Oil of basil herb has also been found to have anti-infective functions by inhibiting many pathogenic bacteria like *Staphylococcus, Enterococci, shigella* and *Pseudomonas.*

Basil tea (basil water-brewed) helps relieve nausea and is thought to have mild anti-septic functions.

SAFFRON (*Crocus sativus*)

Saffron Plant with Flower and Stigma

It is both an expensive aromatic spice and a medicinal herb with several benefits. Saffron is also believed to strengthen the appetite, soothe the alimentary canal, increase bile flow, clear liver stagnancy, help menopausal difficulties and relieve phlegm. In small doses, saffron has been used to treat coughs, bloated stomach, colic and insomnia and it is sometimes used in herb liqueurs as an appetite stimulant. A natural carotenoid dicarboxylic acid called 'Crocetin' is the primary cancer fighting element that saffron contains. It not only inhibits the progression of the disease but also decreases the size of the tumour by half, guarantying a complete goodbye to cancer. Though it is the most expansive spice in the world for it is derived from around 250, 000 flower stigmas (saffron crocus) that make just about half a kilo, a few saffron threads come loaded with benefits you won`t regret paying. Saffron threads can be used in various ways.

Health Benefits

Saffron contains many chemical compounds known to have anti-oxidant, disease preventing and health promoting properties.

The flower stigma are composed of many essential volatile oils but the most important being **safranal**, which gives saffron its distinct hay-like flavor. Other volatile oils in saffron are cineole, phenethenol, pinene, borneol, geraniol, limonene, p-cymene, linalool, terpinen-4 oil, *etc.*

It has many non-volatile active components; the most important of them is **α-crocin**, a carotenoid compound, which gives the stigmas their characteristic golden yellow color. It also contains other carotenoids including zeaxanthin, lycopene, α- and β-carotenes. These are important antioxidants that helps protect body from oxidant-induced stress, cancers, infections and acts as immune modulators.

The active components in saffron have many therapeutic applications in many traditional medicines as antiseptic, antidepressant, anti-oxidant, digestive, anti-convulsant.

Is a good source of minerals like copper, potassium, calcium, manganese, iron, selenium, zinc and magnesium.

It is also rich in many vital vitamins including vitamin A, folic acid, riboflavin, niacin, **vitamin-C** that are essential for optimum health.

Medicinal Uses

The active components present in saffron have many therapeutic applications in many traditional medicines since long time ago as anti-spasmodic, carminative, diaphoretic.

Safranal, a volatile oil found has antioxidant, cytotoxicity towards cancer cells, anticonvulsant and antidepressant properties.

Alfa-**crocin,** a carotenoid compound, which gives the spice its characteristic golden yellow color, has anti-oxidant, anti-depressant, and anti-cancer properties.

Caution: Saffron contains a poison that can damage the kidney and nerves and 10g can be a fatal dose forhumans. It should therefore, be used only in small amounts. High doses of saffron can cat as uterine stimulant and in severe cases can cause miscarriage. Therefore, pregnant woman may be advised to avoid it in their dishes.

CARAWAY (*Carum carvi*)

Caraway Seeds

This biennial herbaceous plant blooms once in every two years to produce creamy flowers in umbels. It grows to about 30cm in height bearing small feathery leaves. The seeds which appear like **cumin** are crescent shaped and dark brown with up to five vertical ribs. The seeds are harvested during early hours of the day to avoid spilling on the ground. The cut plants are then staked till they dry and then the seeds are threshed. Stimulate the appetite, relieve flatulence and improve digestion. It can also promote the onset of menstruation and alleviate utrine cramps. To prepare an infusion, use one teaspoon of crushed seeds boiled in half a cup of water.

Health Benefits of Caraway Seeds

In addition to its use as medicinal values, caraway indeed has many health benefiting nutrients, minerals, vitamins and anti-oxidants.

Seds are rich source of **dietary fibre**. They increase bulk of the food and helps prevent constipation by speeding up movement of food through the gut.

Fibre also binds to toxins in the food and helps protect the colon mucus membrane from cancers. In addition, dietary fibres bind to bile salts (produced from cholesterol) and decrease their re-absorption in colon, thus help lower serum LDL cholesterol levels.

Caraway has many essential oils and the principle volatile compounds are *carvone, limonene, carveol, pinen, cumuninic aldehyde, furfurol,* and *thujone*. These active principles in the caraway seeds are known to have antioxidant, digestive, carminative and anti-flatulent properties.

It has many health benefiting flavonoid anti-oxidnats such as *lutein, carotene, crypto-xanthin* and zeaxanthin. These compounds act as an powerful anti-oxidants by removing harmful free radicals from the body thus protect from cancers, infection, aging and degenerative neurological diseases.

Caraway spice is an excellent source of minerals like **iron, copper, calcium, potassium, manganese, selenium, zinc** and **magnesium**.

The seeds indeed are storehouse for many vital vitamins. Vitamin A, vitamin E, vitamin C as well as many B-complex vitamins like thiamin, pyridoxine, riboflavin and niacin particularly are concentrated in the caraway seeds.

Caraway seeds are an excellent source of macro and micronutrients and play a significant role in preventing infections, aiding in digestion and maintaining oxidative balance through their antioxidant role.

Medicinal Uses

Caraway water is sometimes used in treating flatulence and indigestion in traditional medicines, especially used to relive infantile colic.

It is also used in pharmaceuticals as flavoring agent in mouth-wash and gargle preparations.

Caraway extraction is used as rubefacient (to soothe muscle sores), clear the cold, as a remedy in bronchitis and ***irritable bowel syndrome*** in many traditional medicines.

They are also a rich source of fibre which helps in the reduction of cholesterol levels.

BLACK PEPPER (*Piper nigrum* L.)

Popular black pepper often referred as "king of spice" is a popular spice known to the world since ancient times. The peppercorn plant is native to tropical evergreen rain forest of South Indian state, **Kerala**, from where it spread to rest of the world. The Pepper fruit, also known as *peppercorn*, is actually a berry obtained from this plant. And the **peppercorns** are collected as whole berries or as ground black pepper. Peppers have been in use since ancient times for its anti-inflammatory, carminative, anti-flatulent properties. Improves digestion and promotes intestinal health, prevents gas formation, has diuretic (promotes urination) properties, has impressive antioxidant and antibacterial effect. It stimulates liver and increases salivation. Cures bronchitis and bronchial asthama. It is a pain reliever, anthelmintic, purifies urine and is beneficial in erectile dysfunction, amenorrhoea, skin disorders, fever and dermatoses (Acharya Balkrishna, 2008). It is also good for eyes. The important alkaloids that cause bitterness in black pepper are **piperin, piperidin, piperetine** and **chavicine.** Technically, the pepper fruit is a drupe, measuring about 5 mm in diameter, containing single large seed at its center.

Black pepper has more healthy properties than most people know about. Black pepper is a spice that has the potential to make food more beneficial when used in various recipes and as a table spice. In Kerala, many start their day with a cup of black coffee with a pinch of black pepper powder. The aroma of freshly ground black pepper is difficult to ignore.

Research studies have shown that black pepper has quite a handful of healthy properties.

It has the ability to enhance the function of the digestive tract. Black pepper is a great way to combat this. More than just a spice, it has been shown to improve **digestion** and stimulate the secretion from the **taste buds.** This taste bud stimulation tells the stomach that it's supposed to increase its own digestive juices, namely hydrochloric acid. This, in turn breaks down protein in the stomach, improving the process of digestion. Black pepper is known to have a great amount of **antioxidant** properties. It also has benefits against bacterial growth, particularly in the intestinal trac9S (Sahu, 2015).

The research pinpoints piperine – the pungent-tasting substance that gives **black pepper** its characteristic taste, concluding that piperine also can block the formation of new **fat** cells. Soo-Jong Um, Ji-Cheon Jeong and colleagues describe previous studies indicating that piperine reduces fat levels in the bloodstream and has other beneficial **health** effects.

Black pepper and the black pepper plant, they noted, have been used for centuries in traditional Eastern medicine to treat gastrointestinal distress, pain, inflammation and other disorders.

Their laboratory studies and computer models found that piperine interferes with the activity of genes that control the formation of new fat cells. In doing so, piperine may also set off a metabolic chain reaction that helps keep fat in check in other ways. The group suggests that the finding may lead to wider use of piperine or black-pepper extracts in fighting **obesity** and related diseases.

So, just add freshly ground black pepper powder to your dish and see the difference.

Health Benefits

The chemical compounds present are known to have disease preventing and health promoting properties.

Peppercorns contain many health benefiting essential oils such as **piperine,** an amine alkaloid, which gives strong spicy pungent character to the pepper. It also contains numerous monoterpenes hydrocarbons such as *sabinene, pinene, terpenene, limonene, mercene etc.* that gives aromatic property to the pepper.

The above-mentioned active principles in the pepper may increase the motility of the gastro-intestinal tract as well as increase the digestion power by increasing gastro-intestinal enzyme secretions. It has also been found that piperine can increase absorption of selenium, B-complex vitamins, beta-carotene, as well as other nutrients in the food.

Black pepper corns contain good amount of minerals like potassium, calcium, zinc, manganese, iron, and magnesium.

They are also excellent source of many vital B-complex groups of vitamins such as Pyridoxine, riboflavin, thiamin and niacin.

Peppercorns are rich source of many anti-oxidant vitamins such as **vitamin-C** and **vitamin-A.**

Medicinal Uses

Peppers have been used therapeutically in dentistry as an antiseptic for tooth-decay and gum swellings.

Peppercorns are also being used in traditional medicines in treating flatulence and indigestion in traditional medicine.

Piperine, is a major active component in both black and white pepper (de-husked pepper), and has many physiological and drug like actions. Studies have shown that black pepper has cholesterol lowering properties. And may help in cardiac functions recovery after heart attacks.

Caution

Piperine can strengthen or modify the effects of numerous other medicines, particularly blood thinning agents. Therefore, it is important to seek advice from a qualified professional before using it in therapeutic dose.

AMARANTH (*Amaranthus* spp.)

Amaranth Plant with Dark Green Leaves

Amaranth's great nutritional qualities are it's high protein, particularly in the amino acid, Lysine, the building blocks for the synthesis of protein, which is low in the cereal grains. In fact, Amaranth has the highest lysine content of all the grains with Quinoa coming in a close second. About 3.5 oz of amaranth seeds provide 15 per cent of the recommended daily allowance of calcium, 76 per cent of the iron, and over 25 per cent of the folic acid recommended in diets today. Flour made from amaranth seeds would have been a critical food source in diets that lack protein rich foods. To make your whole wheat bread a complete protein, substitute about 25 per cent of your wheat flour with Amaranth flour. Amaranth, by itself, has a really nice amino acid blend. Just 150 grams of the grain is all that's required to supply an adult with 100 per cent of the daily requirement of protein. Amaranth is one of the highest grains in fibre content. This makes Amaranth an effective agent against cancer and heart disease. Amaranth is also the only grain that contains significant amounts of phytosterols which scientists are just now learning play a major part in the prevention of all kinds of diseases. Amaranth is also rich in many vitamins and minerals. The crimson flowers have an astringent property and infusion of these can be effective in the treatment of diarrhea and dysentery. Amaranth has a high nutritional value because of the high levels of essential micronutrients like carotene, vitamin C, iron and calcium. It is especially rich in lysine, and essential amino acid that is lacking in diets based on cereals and tubers. The protein found in young plants can be important for people without access to meat or other sources of protein. Amaranth seed contains more protein than other grains such as wheat, maize, rice, sorghum orbeans. It also contains high levels of minerals especially iron, phosphorous and magnesium more than what is found in animal products like milk and meat. It also has high levels of vitamin A, B, and E. The protein value of grain amaranth is highlighted when amaranth flour is mixed with other cereal grain flours on a ratio of 1:1 or 1:2. It is the flowers and leaves which are of value for medicinal purposes.

The fat content in amaranth seed is high (7 – 8 per cent) double other cereals. An utmost important constituent squalene is present with 4 – 6 per cent, more than four fold concentration compared with olive oil. Grain amaranth is highly recommended for infants because of its protein digestibility, absorption and retention by the baby's body system. These seeds are one of the most nutritious foods grown. Not only are they richer in protein than the major cereals, but the amino acid balance of their protein comes closer to nutritional perfection for the human diet than that in normal cereal grains.

Amaranth has been found to be having medicinal values, which can reduce or combat common diseases such as diabetes, hypertension, liver disease, hemorrhage, TB, HIV/AIDS, wound healing, skin disease among others. Amaranth seeds and biomass are rich in soluble and insoluble diet fibres important in prevention of coronary heart diseases of the colon. The compounds in amaranth can enhance human growth and development, improve general health, and strengthen immune responses to combat diseases. The food is low in saturated fat and sugar but very low in cholesterol and sodium. It is also a good source of iron, manganese, magnesium and phosphorus.

Part Used and Uses

Roots, leaves, flowers, fruits and seeds – Antidote for snake-bites.

Root – Antidote for scorpion-stings; Muscle spasticity; Dermatitis.

Root decoction – Haematemesis; Menorrhagia; Leucorrhoea; Heals boils and sores.

Leaf – Dysuria; Gonorrhoea; Urolithiasis; Hepatitis; Haemorrhoids; Stops epistaxis; Antidote for spider toxin.

CUMIN (*Cuminum cyminum* L.)

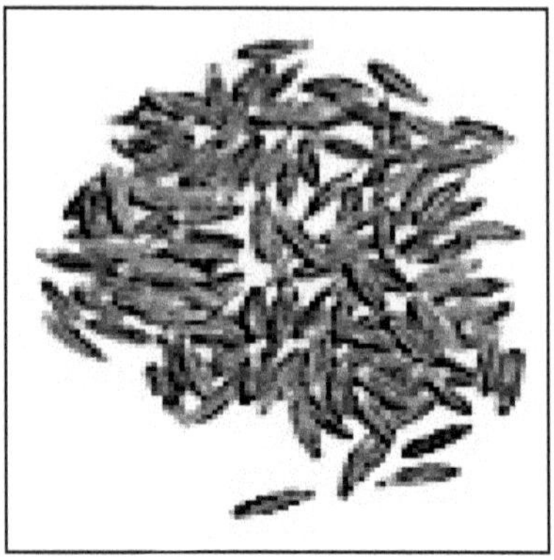

Cumin Seeds

A portent herb with **anti-oxidant** characteristics, cumin seeds contain a compound called **Thymoquinone** that checks proliferation of cells responsible for prostate cancer. It aids digestion and probably that is why we like chewing a handful of cumin seeds at the end of every meal. All the types of cumin seeds are dry, hot, carminative and pungent. They promote potency and digestive functions, cure pitta, promote intellect, purify uterus. They are anti-pyretic, digestive, enhance taste, cures cough and are beneficial for eyes. They promote longevity, relieve flatulence, cure vomiting and diarrhea (Acharya Balkrishna, 2008). The seeds and their essential oils are also used medicinally as an aid to stimulate gastric juices, increase appetite and relieve flatulence. Cumin is also said to increase milk secretion in nursing mothers. It bears small, gray-yellow colored, oblong shaped seeds with vertical ridges on its outer surface and single centrally placed seed that closely resemble **caraway seeds**in appearance.

Distinctive flavour and strong, warm aroma of cumin's is due to its essential oils. The main constituent and important aroma compound is **cuminaldehyde** (4-isopropylbenzaldehyde).

Health Benefits of Cumin Seeds

Research has shown that cumin power and cumin seeds can help jumpstart weight loss, decrease body fat and improve unhealthy cholesterol levels naturally. It helps improve digestion and metabolism level. The presence of thymol and other essential oils in cumin seeds stimulate the salivary glands thereby helping the digestion of food. Apart from this, they strengthen the sluggishness in our digestive system.

Common cold: Antiseptic properties of cumin can help fight flu, by boosting your immune system. A cup of water boiled with cumin seeds, ginger, basil leaves and honey, can give great relief.

Anemia: Cumin seeds contain a good amount of iron. Iron is an essential element for the formation of hemoglobin in the blood required for transport of oxygen. Anemia [lowered levels of hemoglobin in the blood] has always been a concern in **women**, children and adolescents. It's a good idea to include jeera in everyday preparations like parathas, curries, cookies, soups, rice, lentil preparations.

Digestion: The presence of Thymol and other essential oils in cumin seeds stimulate the salivary glands thereby helping in the digestion of food.

Constipation: Due to its higher fibre content, it boosts the activity of the gastrointestinal tract which in turn stimulates enzyme secretion. This is why its powder is commonly used as a natural laxative.

Regulates blood pressure and heart rate: Being high in potassium, a mineral that helps maintain the electrolyte balance in the body. This seed is an elixir for heart patients. This mineral not only helps in the regulation of cell production but also helps maintain your blood pressure and heart rate.

Fights cancer: This is because it contains an active compound known as Cuminaldehyde that helps in retarding the growth of Packed with zinc and potassium, it is an all round when it comes to improving your performance in the bed. Zinc is the important for sperm production and potassium maintains a healthy heart rate and blood pressure. This spice also helps deal with fertility issues and prevents conditions like erectile dysfunctional, tumors (kala zera).

Relieves astahma and cold: The potent anti-inflammatory, anti-bacterial and anti-fungal properties of cumin seed, makes it a great home ready for cold and asthama.

Helps prevent anaemia: It is rich in iron, which is the main component in the production of haemoglobin-a substance that carries and transports oxygen throughout the body.

Improves sexual health: Packed with zinc and potassium. It is all round when it comes to improving your performance in the bed. Zinc is important for sperm production and potassium maintains a healthy heart rate and blood pressure. This spice also helps deal with fertility issues and prevents condition like E. D (erectile dysfunctional).

Pregnant mothers: It helps in relieving constipation and improves digestion, greatly helps pregnant women deal with pregnancy symptoms like nausea and constipation.

Insomia: Certain high amounts of melatonin, which when consumed with banana increases the production of chemicals within the brain. This concoction helps beat insomnia and gives you a good night's sleep.

Enhances memory: Minerals like riboflavin, zeaxanthin, vitamin B6, niacin and many more, jeera is well known for its ability to maintain and restore memory and mental health.

Cumin helps control stomach pain, indigestion, diarrhoea, nausea and morning sickness.

Effective in stimulating menstrual cycle in women. Cumin can be used in the treatment of piles due to its fibre content, anti-fungal, laxative and carminative properties. So, you make it a point to add a dash of jeera powder in your diet (Deepika Sahu, 2012).

Seeds contain nmerous phyto-chemicals that are known to have **antioxidant, carminative** and anti-flatulent properties. The seeds are an excellent source of dietary fibre. Black Cumin Seeds: A study in 2009 revealed that black cumin seeds have a diversified effect on the lipid profile. It was also found to have a significant impact in lowering total and bad cholesterol (LDL). Its seeds contain many health benefiting essential oils such as *cuminaldehyde* (4-isopropylbenzaldehyde), pyrazines, 2-methoxy-3-sec-butylpyrazine, 2-ethoxy-3-isopropylpyrazine, and 2-methoxy-3-methylpyrazine.

The active principles in the cumin may increase the motility of the gastro-intestinal tract as well as increase the digestion power by increasing gastro-intestinal enzyme secretions.

This spice is an excellent source of minerals like iron, copper, calcium, potassium, manganese, selenium, zinc and magnesium.

It also contains very good amounts of B-complex vitamins such as thiamin, vitamin B-6, niacin, riboflavin, and other vital anti-oxidant vitamins like vitamin E, vitamin A and vitamin C.

The seeds are also rich source of many flavonoid phenolic anti-oxidants such as carotenes, zeaxanthin, and lutein. A study in 2009 revealed that black cumin seeds have a diversified effect on the lipid profile. It was also found to have a significant impact in lowering total and bad cholesterol (LDL).

Medicinal Uses

Its seeds are used to prepare decoction, which is sometimes used in treating flatulence and indigestion in traditional medicine.

The seeds are used in traditional medicines to stave off common cold.

CINNAMON (*Cinnamomum zeylanicum*)

Cinnamona "Quills" with Powder

It is sedative. In folk medicine it has been used in various conditions such as insomnia, menstrual cramps, flatulence and nausea. Cinnamon stabilizes blood sugar levels. A teaspoon of cinnamon a day has recently been reported to prevent or delay the onset of non- insulin dependent diabetes which develops in older age and can also reduce cholesterol. It is a natural food preservative and a source of iron and calcium. It is useful in reducing tumour growth and blocks the formation of new vessels in the human body. It is very low in saturated fat, cholesterol and sodium, but is a good source of vitamin K, iron, dietary fibre, calcium and manganese. Cinnamon is used in winter season as it is warm in nature. It cures common cold, cough and congestion of lungs, digestion troubles, toothaches, bad breath and diarrhea. It is used in the treatment of Type 11 diabetes and insulin resistance and reduces the risk of colon cancer.

Circulatory stimulant effects of cinnamon have been reported in several books on medicinal plants and Ayurveda. It helps in reduction of total and bad cholesterol (LDL). It also helps improve insulin resistance, thereby making it useful in diabetes management.

It is the brown bark of the cinnamon tree which is available in its dried tubular form known as a quill or as ground powder.

Traditionally, the inner bark is bruised with a brass rod, peeled and longincision are made in the bark. Its bark is then rolled by hand and allowed to dry.

It is the bark of the tree from where aromatic essential oil (makes up 0.5 per cent to 1 per cent of its composition) is extracted. Usually, the oil is processed by roughly pounding the bark, macerating it in seawater, and then quickly distilling the whole. The oil features golden-yellow colour, with the characteristic odor of cinnamon and a very hot aromatic taste. Some of the effective ways of including cinnamon in your diet are.

Start your day with a cup of cinnamon tea (in leaf or sachet)

Make your breakfast meal a super healthy one; just add this wonder spice to your morning oatmeal and you are going well.

A fruity delight comprising chopped apples, a few walnuts and your magic potion cinnamon

Honey and cinnamon inyour glass of milk before going to bed; no cancer nightmares assured.

Health Benefits

The active principles in the cinnamon spice are known to have anti-oxidant, anti-diabetic, anti-septic, local anesthetic, anti-inflammatory, rubefacient (warming and soothing), carminative and anti-flatulent properties.

Cinnamon hashighest anti-oxidant strength of all the food sources in nature.

The spice contains many health benefiting essential oils such as **eugenol**, a phenylpropanoids class of chemical compound, which gives pleasant, sweet aromatic fragrances. **Eugenol** has got local anesthetic and antiseptic properties, hence; useful in dental and gum treatment procedures.

Other important essential oils in cinnamon include *ethyl cinnamate, linalool, cinnamaldehyde, beta-caryophyllene*, and *methyl chavicol.*

Cinnamaldehyde in cinnamon-sticks has been found to have anti-clotting action, prevents clogging of platelets in the blood vessels, and thus helps prevent stroke and coronary artery disease.

The active principles in this spice may increase the motility of the intestinal tract as well as help increase the digestion power by increasing gastro-intestinal secretions.

This spicy bark is an excellent source of minerals like potassium, calcium, manganese, iron, zinc, agnesium and very good amounts of vitamin A, niacin, pantothenic acid, and pyridoxine.

The spice is also very good source of flavonoid phenolic anti-oxidants such as *carotenes, zeaxanthin, lutein* and *cryptoxanthins.*

A half teaspoon of cinnamon powder every day keeps cancer risk away.

It may help to lower cholesterol and reduce inflammation as it contains a myriad of compounds that have anti-oxidantal properties.

Cinnamon, or dalcheeni as it is known in Hindi, is an aromatic spice that is also used to preserve food. It has antioxidant properties, but like thyme it does not protect food from all the bacteria and microbes that can decay it; it is more organism specific, meaning it kills only certain organisms.

It helps in reduction of total and bad cholesterol (LDL) and increase in good cholesterol (HDL). It also helps improve insulin resistance, thereby making it useful in diabetes management.

Medicinal Uses

The essential oil, eugenol, has been in therapeutic use in dentistry as a local anesthetic and antiseptic for teeth and gum.

Eugenol also has been found to reduce blood sugar levels in diabetics,

The extraction from the sticks (decoction) is sometimes used in treating flatulence and indigestion in traditional medicine.

The spice is used in traditional medicines to stave off common cold and oxidant stress conditions.

Uncooked cinnamon spice can cause choking and respiratory distress. Excessive use of cinnamon stick may cause inflammation of taste buds, gum swelling, and mouth ulcers. Large quantities can cause difficulty breathing, dilate blood vessels, and cause sleepiness, depression, or even convulsions.

Half a teaspoon of cinnamon daily can lower your cholesterol.

It may also lower LDL 'bad' cholesterol, triglycerides (which are fatty acids in the blood) and total cholesterol.

Reduces blood sugar levels

As little as ½ teaspoon per day of cinnamoncan improve sensitivity and blood glucose.

In weight control it can improve insulin resistance

It decreases the risk of heart diseases.

It contains anti inflammatory compounds which can reduce pain, inflammation and stiff joints.

It helps in removing gas from the stomach and intestines and it also removes acidity and is referred to as a digestive tonic.

Acompound found in it prevents unwanted clumping of blood platelets.

If platelets clump together they can slow the blood flow.

It helps by removing nervous tension and memory loss.

Cinnamon raises the activity of the brain acting as a good brain tonic.

It increases one's alertness and concentration

When added to food, it prevents bacterial growth and food spoilage, making it a natural food preservative.

Reduces urinary tract infections

It is a diuretic and helps in secretion and discharge of urine and fats in the blood.

It is a blood thinning agent which acts to increase circulation

It helps reduce pain.

Ensures oxygen supply to the blood cells leading to higher metabolic activity.

Meaning it helps in weight loss.

I has anti-ageing properties.

It can be used to improve one‘s health and boost one‘s immune system.

Note: It is not recommended for pregnant women.

CARDAMOM (*Amomum subulatum*)

Black Cardamom Pods

Cardomom, is one spice, which is widely used in Indian cuisine, for its amazing taste. Besides, adding taste to your recipe, this spice is a medicine for various treatment like mouth ulcers, digestion etc. This is a thermogenic herb that increases metabolism and help burn body fat. Cardamom is considered one of the best digestive aids and is believed to soothe the digestive system and help the body process other foods more efficiently. Cardamom is known as a carminative, relieving flatulence, stimulating the stomach and aiding digestion. However, it is mainly used as a cooking spice or for flavouring drinks and medicines. These green wonders increase energy and relieve fatigue and help you rock your love making process.

Both varieties of cardamom feature three sided pods with a thin papery outer cover and small black seeds that are arranged in vertical rows. Elettaria pods are small and light green in color, while Amomum pods are larger and dark brown. The pods are being used as flavoring base in both food and drink, in cooking recipes and as well as in medicine. Black cardamom (*Amomum subulatum*) also known as *Nepal cardamom* is relatively big sized pod of same zingiberaceae family. The pod has dark brown rough outer coat, measure about 2-4 cm in length and 1-2 cm in diameter. The pods have camphor like intense flavor commonly used in spicy stews in sub Himalayan plains of India, Pakistan, Nepal as well as in China.

Health Benefits

Digestive disorders: This spice is great for boosting digestive health. It helps in stimulating the digestive system and reduces gas and bloating. It also helps in relieving acidity. Choose cardamom tea to treat headaches caused by indigestion. Just chew 2-3 cardamom pods for a few minutes.

Enhances appatite: Cardamom essential oil acts as an apatite stimultant, which once taken along with milk or any other break fast item can increase apatite.

Relieves inflammation: It is rich in anti-inflammatory properties that can lower pain in the joints. Not just this, if you have swelling in your gums, just put a cardamom in your mouth and it will help you in reducing it.

Bad breath: If you are troubled with bad breadth, then chew some cardamom. It will work as a mouth freshener and help get rid of bad breath. Green cardamom is loaded with anti-bacterial properties that can enhance oral health. A very simple remedy to treat this is by chewing a pod of a green cardamom after every meal.

Depression: If you are one of the victims of depression or anxiety, then powder some cardamom and boil it in hot water to prepare cardamom tea. The pleasing aroma will help you ease your depression and other mental problems.

Improves blood pressure: The potassium present in green cardamom is very helpful in lowering the blood pressure while stablising the heart rate.

Aid detoxification: Green cardamom has diuretic properties that can help you remove toxins and waste from your body. It also helps the body reduce natural anti-oxidants that can aid in detoxification process.

Prevents respiratory distress: Green cardamom is quite useful in respiratory diseases like cough, cold and asthma. It enhances the blood circulation in the lungs that can improve breathing and wheezing.

Reduces blood clots: Green cardamom can prevent the gathering of platelets that cause blood clot in the first place.

Releives hiccups: Green cardamom helps you in treating hiccups which is the result of involuntary contraction of diaphragm muscle.

Expectorant action: Cardamom helps to improve circulation to the lungs, so it is recommended for people suffering from asthma and bronchitis.

Impotency: It is a proven aphrodisiac and has a compound called Cineole, that can release nerve stimulant, so if you have a low libido boil half a teaspoon of cardamom power in a cup of milk and drink it warm in the night. I will prove you extra energy. Therefore, if your sex life is going down the drain, then use cardamom. Cardamom can be used to treat sexual dysfunctions like impotency and premature ejaculation. Besides, the above benefits, cardamom also has a cooling effect; stimulates the appetite and eases stomach cramps.

This exotic spice contains many plant derived chemical compounds that are known to have anti-oxidant, disease preventing and health promoting properties.

The spicy pods contain many essential volatile oils that include pinene, sabinene, myrcene, phellandrene, limonene, 1, 8-cineole, terpinene, p-cymene, terpinolene, linalool, linalyl acetate, terpinen-4-oil, a-terpineol, a-terpineol acetate, citronellol, nerol, geraniol, methyl eugenol, and trans-nerolidol.

The therapeutic properties of cardamom-oil have found application in many traditional medicines as antiseptic, antispasmodic, carminative, digestive, diuretic, expectorant, stimulant, stomachic and tonic.

Cardamom is a good source of minerals like potassium, calcium, magnesium, manganese and iron

The pods are rich in many vital vitamins including riboflavin, niacin, vitamin-C that are essential for optimum health.

Medicinal Uses

The therapeutic properties of cardamom oil have found application in many traditional medicines as antiseptic and local anesthetic, antioxidant and; health promoting and disease preventing roles.

MUSHROOMS

Mushrooms not only taste good but also a great source of healthy food. Mushrooms of button variety are rich in potassium, phosphorus, copper and iron. They are also a good source of thiamine (vitamin B_1) and riboflavin (vitamin B_2). As mushrooms mature, their caps open and expose their gills. It is best to avoid wide open caps and also to select mushrooms that are firm, not spongy. They are known to be beneficial in reducing blood fat levels and to have antibiotic properties; they are also claimed to have anti-tumour actions and to boost the immune system action against disease- producing micro-organisms by increasing white blood cell count. Mushrooms are easily digested and are recommended for any one suffering with digestive problems. Mushrooms are full of proteins, vitamins and minerals, amino acids, anti-biotic and anti-oxidants. Mushrooms have a significant amount of vitamin D. The amount varies according to its type or varieties. Shitake mushrooms are considered to be one of the best sources of vitamin D.

The health benefits of mushroom include the following:

The power of mushrooms comes from their ability to enhance the activity of natural killer T cells (NKT).

These NKTs attack and remove cells that are damaged or infected by a virus. Mushrooms are associated with decreasing most cancers and significantly reducing the risk of breast cancer in women. They prevent DNA damage, slow cancer or tumor growth, and prevent tumors from acquiring a blood supply.

B Vitamins are vital for turning food (carbohydrates) into fuel (glucose), which the body burns to produce energy. They also help the body metabolize fats and protein. Mushrooms contain loads of vitamin B_2 and vitamin B_3.

Mushrooms have zero cholesterol, fats and very low carbohydrates. The fibre and certain enzymes in them also help lower cholesterollevel. The high lean protein content in mushrooms helps burn cholesterol when they are digested.

Mushrooms can be an ideal low energy dietfor diabetics. They have no fats, no cholesterol, very low carbohydrates, high proteins, vitamins and minerals, a lot of water and fibre. Moreover, they contain natural insulin and enzymes which help breaking down of sugar or starch of the food.

Ergothioneine, a powerful anti-oxidant present in mushrooms is very effective in giving protection from free radicals as well as boosting up immunity. Mushrooms contain natural antibiotics (similar to penicillin, which itself is extracted from mushrooms) which check microbial and other fungal infections.

Mushrooms are the only vegetable and the second known source (after cod liver oil) to contain vitamin-D in edible form. They are rich in calcium (good for bones), iron (cures anemia), potassium (good for lowering blood pressure) and selenium. The best source of selenium is animal proteins. So, mushrooms can be the best choice for vegetarians.

Mushrooms stimulate the reproduction and activity of immune cells. In Japan they are often used to complement the chemotherapy, inorder to support the immune system.

SOYBEAN (*Vicia faba*)

Fava or broad beans are large, flattened light green pods usually eaten shelled for their delicious beans. Lima beans (*Phaseolus lunatus*) are large, plump, pale green pods with kidney shaped seeds. Soybeans have been found to be a storehouse of phytochemicals – beneficial antioxidant nutrients such as saponins, phytosterols and phenolic acids - which protect the body from free radical damage, reducing the risk of the degenerative diseases of ageing, such as heart attack, cancer and strokes. Two of the best anti-carcinogens in soya are some plant oestrogens, which prevent breast cancer and **genistein** which inhibits the growth of cancer cells. In fact, soy beans have a mild oestrogenic activity due to their **isoflavones,** also called phytoestrogens. Eating soya products can therefore be beneficial for women in HRT and can help offset some post menopausal symptoms. Soy beans have also been found to be beneficial in reducing cholesterol levels and preventing heart attack. A new Canadian University study found that three glasses of soya milk and one soya dessert consumed daily lowered the **'bad' LDL cholesterol** by 11 per cent and increased the **'good' HDL cholesterol** by 9 per cent in 70 per cent the subjects (Michael Sharon, 2009). Soybeans are ground to make soy flour, which comes in full fat and low fat op tions. It bursts with vitamins and minerals, and is also one of the best vegetarian sources of Omega-3 fatty acids. Soy protein is great for women post menopause and also for elderly women.

YAM BEAN

It is crisp with a slight sweetness like water chestnuts. There are many nutritional properties of merit in *Pachyrhizus tuberosus*; for example vitamin A, vitamin B complex, vitamin- C, phosphorus can be found. It is most frequently used cooked or fresh. Certain types of soya protein inhibits fat accumulation and reduces inflammation. Soybeans rich in **beta-conglycinins** limit lipid accumulation in fat cells by inhibiting an enzyme called fatty acid synthesis. Aspecific peptides (digested proteins) to do this mechanism has been identified. This exciting researchcould lead to the development of nutraceuticals to fight obesity (de Mejia-Food Science and Human Nutrition-AN, Tribune, 7th, April.2010)

Nutmeg: It is one of the most popular aphrodisiaces. Research proves that nutmeg has the same effect on mating behavour as vigra.

CASSIA

Its leaves or sap are used to treat fungal infection such as ringworm. They contain a fungicide, **chrysophanic acid.** Because of its antifungal properties, it is a common ingredient in soaps, shampoo's and lotions in Philippines. The effectiveness of this plant against skin diseases, it is also used to treat a wide range of ailments from stomach problems, fever, asthama to snake bite and venereal diseases (syphilis, gonorrhoea). Its leaves contain a chemical called **adenine** which has been documented as an effective platelet aggregating inhibitor (reduces sticky blood and arterial plaque).

SWEDE

Its fibre content stimulates regular bowel movements, while its high levels of potassium provide it with natural diuretic properties. In addition, it is very low in calories.

POINTED GOURD

The fruit is particularly recommended during convalescence. It is easily digested and is diuretic, laxative and cardiatonic. It is effective against bronchitis, high fever and nervousness.

WINTER SQUASH

It provides an outstanding source of alpha and beta carotenes (provitamin A) and contains good amounts of calcium, phosphorus and potassium. Squash can benefit the stomach, reduce inflammations and improve circulation, and squash seeds are reputed to be effective in destroying worms. Some healers recommend that a handful of seeds should be eaten daily for three weeks in order to eliminate parasitic worms.

SEAWEEDS

Seaweeds or marine algae, is a general descriptive name for sea vegetable. They are extremely rich in **iodine** and in other minerals, such as calcium, iron, and fluorine and has been known for centuries for their ability to promote health. They are especially beneficial to thyroid function, and can also lower cholesterol, alkalize the blood, remove radioactivity residues, help with weight loss and are important in bone mineralization and density. Seaweeds are used to treat conditions such as goiter, water retention (oedema) and swollen lymph glands. They come in several colours- brown, red, green, blue-green and yellow-green- each of which has its own individual properties in addition to the properties to all.

Rue (*Ruta graveolens*)

The aromatic leaves are high in **rutin,** the bioflavonoid that strengthens capillaries and blood vessels. An infusion of the leaves is diuretic, slightly increases blood pressure and has an abortifacient effect (induces abortion). In Yemenite folk medicine, rue is used to treat nervous breakdowns and to stimulate the onset of delayed menstruation.

Caution

Rue must not be taken during pregnancy.

Bathua (*Chenopodium album*)

It is also known as all good or goose foot. It is mainly eaten in the form of saag and in raita. It is rich in vitamin A, calcium, phosphorus and potassium.

The leaves and stems are edible and absolutely delicious, with an earthy, mineral- can be compared rich flavour that to spinach. This herb is one of the most nutritious wild foods you can eat. According to Ayurveda, the whole herb is generally used as medicine. It improves appetite. It is oleaginous, anthelmintic, laxative (at over dose), diuretic, aphordisiac and tonic. It is useful in treatment of biliousness, abdominal pains, eye diseases, throat infections, piles, blood disorders and troubles of heart and spleen. The leaves are antiscorbutic and yield ascaridole, which is used to treat round and hookworms. The leaves are sometimes dried and used for relieving some diseases. The powered leaves are used externally as an antiseptic and the leaf juice is used to relieve burns. A decoction of the leaves and stems, mixed with alcohol is used as a rub to relieve arthritis and rheumatism. The seeds of the herb are used to treat hepatic disorders and spleen enlargement. It contains eight essential amino acids, vitamins, minerals and fibre. It gives instant energy and keeps you active. It's easy to cook and can be cooked like rice in only 10 minutes.

Health Benefits

For kidney stone: it is very effective and useful in problem of kidney stone. Take tender leaves and branches of leaves and grind them to extract its juice and take 10-15 gm of it daily with or without water. This also reduces the tendency of stone formation.

For inter swelling: When bathua taken internally it helps to reduce internal swelling.

For external swelling: For external swelling cook leaves of bathua in steam and apply at place of swelling this help to reduce swelling.

For jaundice: Take bathua and giloy ras and mix them in equal proportion and take 25- 30 gm of it daily twice a day.

For irregular period: Take bathua seeds and sonth (dry ginger powder) make powder 15-20n gm boil in 400 gm of water when it reduces to 100 gm drain it and drink twice a day.

For curing infections after delivery: Take 10 gm of bathua, ajwain, methi and jaggery add dashmol and take for 10-15 days.

For anemia: Take 25-50 gm of bathua ras (juice extract from leaves after crushing) daily.

For urinary tract infections: Take 10 gm of bathua leaves ras add 50 ml of water and take with misri (candy sugar) take regularly.

For Blood purification: When bathua ras taken with 4-5 neem leaves ras it works as blood purifier.

Bathua finds a use in the treatment of some skin conditions, and the oil made from these leaves is used to treat hook worms.

It is rich in vitamin A, B-complex vitamins, vitamin C, calcium, potassium, phosphorus, trace minerals, iron and fibre

Caution

Please also note that you should eat bathua in moderation as consuming it more may result in diarrhea. In Ayurveda, it is said that pregnant women should not eat Bathua as it may result miscarriage.

OREGANO

Health Benefits

Hippocrates, the great Greek philosopher and scholar, used oregano for medicinal purposes. He used it as both an anesthetic and for digestive problems. The herb is endowed with several medicinal properties, the most prominent among them being its applications as an alexipharmic, appetiser and laxative. It is used in ayurveda to treat colds, influenza mild fevers, vomiting, diarrhea, jaundice, itchy skin, indigestion, stomachs upsets and painful menstruation conditions, reflecting anti-microbial (anti-bacterial) properties, strongly anti-septic, antispasmodic, carminative, cholagogue (help gall bladder secretion), diaphoretic (sweet production), expectorant, stimulant and mild tonic properties. It is also believed to calm nerves and is used to cure seasickness.

Oregano tea is a strong sedative and traditionally used to treat colds, bronchitis, asthma, fevers, and painful menstruation because of its antiseptic properties. Oregano can also be used as an antidote for venomous bites from snakes and insects. It is also used to treat skin disorders. Rosmaric acid reduces fluid build-up and even swelling and inflammation during an allergy attack, making it a natural and effective histamine reducing compound.

This spice can be used on meat and fish dishes as it resembles thyme in flavour and aroma. Oregano can be used in a number of recipes like soups, syrups, salad, dressings, cheese mixtures, seafood, omelets, sausages, ice cream, custards and all sorts of tomato based dishes and different Italian sauces. Fresh and dried oregano leaves have volatile oils possesses carminative, stomachic, diuretic, diaphoretic, antibacterial, antioxidant and anti-microbial properties and are shown to inhibit stubborn bacteria like pseudomonas aeruginosa and staphylococcus aureus. The phytonutrients, thymol, carvacrol and rosmarinic acid, present in oregano leaves acts like a strong antioxidant that fights free radicals in the body (Pinto, 2014).

Our experiences with oregano are usually limited to flavour, but in fact it can be used as a preservative as well. As an antioxidant, it has antibacterial and antifungal properties that can be used to preserve meats, but the disadvantage of oregano is that it could distort the flavour of the food.

Oregano's key chemical components are Acetate, Borneol, Bisabolene, Carvacrol, Caryophyllene, Cymene, Geranyl, Linalool, Linalyl Acetate, Pinene, Terpinene and Thymol. All of these chemicals are present in the leaves and in the oils, so they have both topical and internal benefits.

It has been thought to be an effective treatment for bacteria and parasite infestation in the colon and intestines. In fact, Mexican researchers have found it effective in combating giardia, an intestinal infection caused by a microscopic parasite.

Because of its anti-parasitical affect, its oil has been used in head lice treatments. Herbalists recommend it for the treatment of E-coli.

Oregano also has anti-inflammatory benefits. Some people rub the oil on inflamed joints and muscles. Topically, it can also be used as an antiseptic and anti-bacterial spread to relieve acne, cold sores, and minor cuts and scrapes.

It has been used in the treatment of allergies and even to regulate menstrual periods. Some cultures use it as a powerful pain killer. A few drops of the oil in juice consumed for 3-5 days may help clear up a sinus infection.

Oregano also has a large amount of antioxidants in its oil and leaves. It has 42 times the antioxidants as a medium sized apple, 30 times more than a white potato and 12 times more than an orange.

Oregano Cautions and Concerns

Not everyone should take oregano. Especially women who are pregnant should avoid digesting or absorbing oregano. It can weaken the lining of the embryonic sack. If you are allergic to mint, sage, basil or thyme, you might also be allergic to oregano. Though used in the aid of digestive problems, it can cause the opposite effect in some people.

Oregano oil may reduce the body's ability to absorb iron, which is another reason women and children in particular should not consume large amounts of oregano.

NONI

Noni (*Morinda citrifolia*) It is a small fruit, that grows as a shrub. It is popularly known as Indian mulberry, belonging to family Rubiaceae. There are several species of this plant, which are cultivated in the tropical regions of South East Asia and Australia.

Health Benefits

This round little fruit is a power house of nutrients and can prevent several diseases such as common cold, flu, diabetes, high blood pressure, aches, pains, burns, arthritis, inflammation, tumors, the effect of aging, parasitic, viral and bacterial infections. Dry power of Noni is a rich source of charbohydrates and dietary fibre. However juice is more nutricious as compared to the powdered form of this fruit. Its juice is rich in potassium, calcium vitamin A, vitamin B, vitamin C. The concentration of nutrients is much more in the pulp of this fruit as compared to other forms.

Reduces the risk of gout: Its juice can effectively reduce the concentration of uric acid crystals or particles in the blood. The consumption of nuni juice can help in treating this severe bone disease.

Reduces Arthritis: Not only does noni and its juice helps in reducing uric acid but at the same time its analgesic properties help in effectively reducing the pain of arthritis.

Boosts immunity: Regular consumption of nuni juice helps in increasing the overall strength of the body.

Fights aging: Loaded with selenium and vitamin, which help fight free radicals and retain skin elasticity. It makes the skin glowing and youth ful.

Helps in treating cancer

Boosts energy

QUINOA

Quinoa is a 100 per cent vegetarian reference protein - which means that all the protein present in it is absorbed by the body. The only other food that does this is egg white, a non vegetarian option. The flour can be made at home simply by mashing up the quinoa and using the powder for any dish.

Exotic superfoods

These exotic superfoods have been used in other parts of the world for their powerful healing punch successfully. Get to know some of these unusual superfoods and figure out how to implement them into your overall superfood regimen.

HIJIKI (*Sargassum fusiforme*)

It is a sea vegetable that looks like black angel hair pasta. Called the "beauty vegetable" in Japan for the shiny hair and beautiful skin it gives to those who indulge, this sea vegetable helps nourish the thyroid, promotes healthy skin and hair, and promotes a healthy immune system.

Health Benefits

Hormonal Health: Iodine plays a major role in the manipulation and balance of our hormones, as it directly interacts with the thyroid gland, one of the most important aspects of the entire endocrine system.

Digestive Health: Seaweed is the ultimate leafy green vegetable, and like so many other veggies, it is packed with dietary fibre, which ensures that your digestive process is smooth and healthy. Dietary fibre stimulates peristaltic motion to move food through the digestive tract, maximizing nutrient intake, and reducing constipation along the way.

Energy Levels: The iron content of hijiki is unprecedentedly high for a vegetable; some varieties that have been tested have up to 5 times more iron. This means that the prevention of anemia is very easy with hijiki as a part of your weekly diet. Proper levels of iron in your body increase your red blood cell count, which can increase oxygenation to your extremities and boost overall energy levels.

Bone Health: Hijiki contains far more calcium than milk, which is often considered one of the best dietary sources. Hijiki can definitely help keep you strong and active for many years.

Cholesterol and Diabetes: Hijiki is not only a low-calorie food that fills you up, due to the high level of fibre, but that same dietary fibre also eliminates excess cholesterol from the cardiovascular system and helps to balance glucose and insulin levels within the body. Optimizing the digestive system has many unforeseen benefits for the rest of the body, and sea vegetables can be a great place to take advantage of them!

Aids in digestion

Helps to prevent anemia

Boosts energy leveling in body

Helps to improve bone health

Aids in relaxation and stress release

Helps in eliminating excess cholesterol

Beneficial in maintaining hormonal balance

Helps to balance glucose and insulin levels.

WAKAME (*Undaria pinnatifida*)

Another great sea vegetable, it provides a salty taste that comes from a balance of sodium and other minerals from the sea. Wakame is a tender grayish green sea vegetable, and when you soak wakame, it expands many times its original size. Eat it raw as a snack, add it to soups and stir-fries, or roast it and sprinkle on salads and stews. Wakame becomes soft and melts in your mouth when cooked. What a great way to add minerals to your foods! Sea vegetables also give you a good dose of fibre to promote digestive health.

Health Benefits

Magnesium: This mineral is critical in the contraction and relaxation of muscles, function of certain enzymes in the body, production and transport of energy, and the production of protein.

Iodine: Iodine is needed for strong metabolism of cells - the process of converting food into energy. It also maintains the balance of the thyroid gland and is needed for the production of thyroid hormones.

Calcium: Wakame easily allows for the absorption of calcium into the human body. Each 100 grams of raw wakame contains 150 milligrams of calcium. Calcium is needed for strong healthy bones and the prevention of osteoporosis.

Iron: We need iron because it is essential for the production of red blood cells and the prevention of anemia.

Vitamins: Vitamins A, C, E, and K. These vitamins are all amazing for skin health and repair as well as immunology. Vitamin D. Promotes the absorption of calcium for healthy bones and enhances the nerve, muscle, and immune systems.

Riboflavin (Vitamin B2) : We need riboflavin to use the carbohydrates, fats, and proteins in the foods we eat. Riboflavin helps us use these nutrients for energy in our bodies for growth and is also necessary for red blood cell production. Riboflavin functions as an antioxidant and works in the body with other vitamins such as niacin, folate, and vitaminB6.

Folate: Helps the body make new cells and is especially important for pregnant women.

Lignans: Thought to play a role in preventing certain types of cancer, particularly breast cancer.

TARRAGON (*Artemisia dracunculus*)

It is a culinary and medicinal herb that is rich in vitamins A, C, and B-complex and minerals such as zinc, copper, iron, and magnesium. It is known to help stimulate the appetite, relieve flatulence and colic, balance the body's acidity, alleviate the pains of arthritis, rheumatism and gout, regulate menstruation, stop hiccups, prevent dyspepsia, and expel worms from the body. Tarragon contains poly-phenolic compounds that are known to help lower blood glucose levels. Tarragon is often used to help prevent strokes and heart attacks due to its ability to prevent clot formation inside narrow blood vessels in the heart and brain. It is also known to help prevent or slow down the oxidative process that forms cataracts and other degenerative diseases. Tarragon tea is an excellent tonic after a heavy meal as it is a mild, natural diuretic that helps the system flush out toxins produced by the digestion of heavy, rich protein based meals. Drinking tarragon tea before bed can also help to overcome insomnia and promote a restful and healing nights sleep. To make a tea, use two teaspoons of fresh or dried herb to two cups of water and add lemon and/or honey if desired. Fresh tarragon leaves act as a local anesthetic and can be applied to aching gums or teeth, cuts, and/or sores to help numb and relieve pain. Tarragon is a wonderful addition to fresh salads, guacamole, soups, stews, steamed or roasted vegetables, rice, and potatoes. Tarragon can be found online or at you local health food store in tincture, extract, capsule, cream, essential oil, or tea form. Fresh tarragon is generally available in the produce section at your local grocery store or farmers market.

Health Benefits

This herb is rich is rich in health benefiting phyto-nutrients that are indispensable for optimum health.

The main essential oils in tarragon are **estragole (methyl chavicol),** cineol, ocimene and phellandrene.

Traditionally, tarragon has been employed as a traditional remedy to stimulate appetite and alleviate anorexic symptoms.

Scientific studies suggest that poly-phenolic compounds in this herb help lower blood-sugar levels.

CLEAVERS (*Galium aparine*)

It is a medicinal wild herb that grows throughout the United States, Britain, Europe, Siberia, and the Himalayas. It is one of the most effective herbs for cleansing the lymphatic system. It is known to help move and dissolve lymphatic congestion, reduce swollen glands, ease upper respiratory congestion, and eliminate mucous from the body. Cleavers is also highly beneficial for removing toxic debris out of the blood and can help to tone and strengthen the entire circulatory system. It is also good for alleviating edema, bloating, and water retention. Cleavers is often used to reduce and eliminate lumps in the breast as well as reduce swelling and pain associated with urinary tract infections and cystitis. It is also known to help reduce swelling with enlarged prostates as well. Cleavers is excellent for the liver and can help to treat and prevent jaundice and/or any liver disorders. It also work as a tonic for the stomach and is a good remedy for ulcers and hemorrhoids. Cleavers contains anti-tumor compounds and is an effective natural treatment taken both internally and externally to help reduce the effects of cancer. Topically, cleavers can be used as a poultice, salve or cream to help reduce swollen lymph nodes and breast tissue as well as for skin irritations, abscesses, boils, burns, eczema, and psoriasis. Cleavers makes an excellent tea and is especially good when one is experiencing heavy mucus and congestion from a cold or flu. Use 2 tsp of dried herb to 1 cup of boiling water and let steep for at least 10 minutes, sweeten with raw honey if desired.

Health Benefits

Herbalists have long regarded cleavers as a valuable lymphatic tonic and diuretic. The lymph system is the body's mechanism to wash tissues of toxins, passing them back into the bloodstream to be cleansed by the liver and kidneys. This cleansing action makes cleavers useful in treating conditions like psoriasis and arthritis, which benefit from purifying the blood. Cleavers is a reliable diuretic used to help clean gravel and urinary stones and to treat urinary infections. [1]In cats, these actions make cleavers a safe long-term aid in the treatment of feline lower urinary tract disease (FLUTD), and the herb may also be useful for chronic low-grade kidney inflammation.2 In studies cleavers extract lowered blood pressure without slowing heart rate or having any health-threatening side effects.

GOLDEN-SEAL ROOT (*Hydrastis cadensis*)

It is a popular herb from North America that works as a powerful herbal antibiotic and immune system enhancer. Golden-seal is a good source of vitamins A, C, E, and B-complex and minerals such as calcium, iron and manganese. It contains potent anti-bacterial, anti-microbial, anti-fungal, and anti-inflammatory properties as well as alkaloids that are known to be an effective treatment for diarrhea and stomach problems that are caused by influenza or food poisoning. It is also very helpful for other digestive problems such as peptic ulcers, gastritis, dyspepsia, and colitis. Goldenseal helps to increase digestive enzymes and significantly enhances liver and spleen functions. It is an amazing infection fighter and is known to be particularly useful at treating sinus, respiratory, mouth, throat, bladder, yeast, and urinary tact infections. It is also an effective remedy for hemorrhoids, athletes foot, canker sores, and to help stop heavy menstrual bleeding. Goldenseal is an essential herb during cold and flu season as it can help to prevent and relieve the symptoms of colds, flu, fevers, bronchitis, heavy congestion, and even pneumonia. It is also helpful against hay fever, laryngitis, cystitis, hepatitis, and liver disease. Externally, a wash can be made with goldenseal extract to help treat conjunctivitis and inflamed eyelids. A mouthwash can also be made and used as a gargle for sore throats and gum infections and goldenseal creams/ointments are very effective for treating eczema, ringworm, boils, cuts, and rashes. The medicinal claims about goldenseal range from its powers as a cure for toenail fungus to its effectiveness against pancreatic cancer. The medicinal claims about goldenseal range from its powers as a cure for toenail fungus to its effectiveness against pancreatic cancer.

DONG QUAI

Dong Quai ***(Angelica sinensis)*** has been used for thousands of years as a medicinal herb and is prized for its ability to benefit both the female and male reproductive system. Dong Quai is considered to be the "premier hormone regulator" and has the ability to reduce estrogen levels if they are too high or increase estrogen levels if they are too low. It is often used as a natural infertility treatment and can also help improve sperm quality in men due to its **ferulic acid**. Dong Quai is high in vitamins and minerals such as vitamin B-12, folic acid, biotin, cobalt, and iron. It is highly beneficial for hypothyroidism, migraines, irritable moods, depression, low energy, heart palpitations, insomnia, hypertension, kidney disease, angina, arthritis, nerve pain, shingles, sciatica, fibromyalgia, and rheumatoid arthritis. Dong Quai is also highly beneficial for the cardiovascular system by helping to reduce blood pressure levels and blood sugar levels. It is also good for reducing anxiety, alleviating stress damage, calming the nervous system, and promoting overall relaxation in the body. Some people claim that dong quai can help you feel happier and less overwhelmed with the stresses of life. Dong Quai contains anti-spasmodic properties that can ease cramps and other symptoms of PMS. It is also known to help prevent hot flashes and other menopausal symptoms as well. Dong Quai is an excellent blood builder and can help to replenish and rebuild the blood after injury or surgery. It contains anti-aging properties and can help increase circulation, improve one's complexion, and aid in detoxification. Topically, dong quai is helpful for skin problems such as rosacea, hives, eczema, neurodermititis, and vitiligo.

Health Benefits

It is used primarily for health issues in women and has been termed "female ginseng." It reported to be a blood strengthener and has been used for cardiovascular conditions, inflammation, headache, infections, and nerve pain. It is also used to treat a wide range of conditions including menstrual disorders and other gynecological issues, as an analgesic in rheumatism, and in suppressing allergy symptoms

Caution should be used with those who have breast, uterine, or prostate cancer or women who are pregnant.

Chapter 4

Dark Green Leafy Vegetables

It is not possible to meet your nutritional needs without having leafy vegetables in your diet. They are available in abundance and are huge source of many vitamins and minerals your body needs to stay healthy, such as vitamins A, C and K, folate, iron and calcium. They are also great sources of fibre. They are also a vision protector and provides four essential minerals, *i.e.* calcium, magnesium, iron and potassium. Therefore, try to have them daily in your diet and darker the better. Research suggests that the nutrients found in dark green vegetables may prevent certain types of cancers and promote heart health. Green, especially leafy vegetables are to be eaten in the raw form, if you want to get the best of their nutritional properties. It is recommended that teenage girls eat 3 cups of dark green vegetables per week or about ½ a cup every day. Dark green vegetables are also high in fat -soluble vitamins such as vitamins A, K, D and E. These vitamins require a little bit of dietary fat in order for the body to absorb them. Therefore, when you eat dark green vegetables, make sure to add a teaspoon of dietary fat, such as butter, olive or canola oil, cheese or salad dressing to make sure your body absorbs all of the vitamins you eat. **While dark green, leafy vegetables are great for overall health, they are extremely beneficial for brain health. Be generous with your portions of spinach and broccoli because they are packed with antioxidants, folate, beta-carotene and vitamin C – all important to keep your brain in good health and improve memory**.

From kale, collards to spinach and Swiss chard, dark-green leafy vegetables probably are considered a "one stop shop" for all the best nutrients your body needs to fend off cancerous cells, *i.e.* fibre, vitamin B, phytochemicals, chlorophyll and more. It's time to add some greens to your diet

Dark green vegetables like broccoli and spinach have hair-healing powers because of Vitamins A and C, which secrete natural scalp oil. This works as a hair conditioner, providing essential oils that are important for hair growth and renewal. It is these natural oils that will always keep your hair shiny (Purvaja Sawant, 2013). Examples of dark green leafy vegetables:

Arugula has a peppery taste and is rich in vitamins A, C and calcium. It can be eaten raw in salads or added to stir-fry, soups and pasta sauces.

Broccoli has both soft florets and crunchy stalks and is rich in vitamins A, C, K, folate and fibre. Broccoli can be eaten raw or steamed, sautéed or added to a casserole.

Collard Greens have a mild flavour and are rich in vitamins A, C, K, folate, fibre and calcium. The best way to prepare them is to boil them briefly and then add to a soup or stir-fry. Collard greens can also be eaten as a side dish. Just add your favourite seasoning and enjoy.

Dandelion Greens have a bitter, tangy flavour and are rich in vitamin A and calcium. They are best when steamed or eaten raw in salad.

Kale has a slightly bitter, cabbage-like flavour and is rich in vitamins A, C and K. It is tasty when added to soups, stir- fries and sauces.

Mustard Greens have a peppery or spicy flavour and are rich in vitamin A, C, K, folate and calcium. They are delicious when eaten raw in salads or in stir-fries and soups.

Romaine Lettuce: Lettuce contains vitamin B6, C, E, folic acid and minerals like manganese and chromium. Romaine lettuce which is more dark green in colour in comparison to the Iceberg variety is said to be more nutritious. It a nutrient rich that is high in vitamins A, C, K and folate. If your hair is not having an optimum growth, it is good idea to consume lettuce as a regular part of your diet. It is best when eaten raw in salads, sand witches or wraps.

Spinach has a sweet flavour and is rich in vitamins A and K, folate and iron. Spinach tastes great when eaten raw in salads or steamed.

Swiss Chard tastes similar to spinach and is rich in vitamins A, C, K, potassium and iron. It is best stir-fried or eaten raw in salads.

Celery contains vitamin A and B, calcium, sodium, and iron. It is said to have negative calories and digesting them burns more calories. They are good for people who want to lose weight.

Popular Leafy Vegetables or Pot herbs (vegetables eaten as the cooked leaves)

Agati	Chard	Kale	Spinach beet
Amaranth	Colocasia	Kohlrabi greens	Nasturtium
Artichoke	Dandelion	Leek	Turnip green
Arugula	Endive	Lettuce	Water cress

Beet greens	Escarole	Mustard greens	Water spinach
Brussels sprout tops	Fenugreek	Purslane Romaine	Yarrow
Collard greens	Garden cress	Sorrel	
Cabbage	Iceberg lettuce	Spinach	

Health Benefits

Dark green leafy vegetables are, calorie for calorie, perhaps the most concentrated source of nutrition of any food. They are a rich source of minerals (including iron, calcium, potassium and magnesium) and vitamins, including vitamins K, C, E and many of the B vitamins. They also provide a variety of phyto nutrients including beta-carotene, lutein and zeaxanthin, which protect our cells from damage and our eyes from age-related problems, among many other effects. Dark green leaves even contain small amounts of **omega-3** fats. Dark greens *viz.* spinach is a good source of iron a nutrient that may help protect against the sleep robber known as restless legs syndrome.

Perhaps the star these nutrients is vitamin K. A cup of most cooked greens provide at least nine times the minimum recommended intake of vitamin K and even a couple of cups of dark salad greens usually provide the minimum all on their own. Recent research has provided evidence that this vitamin may be even more important than we once thought (the current minimum may not be optional) and many people do not get enough of it.

Vitamin K

- Regulates blood clotting.
- Vitamin K is a fat –soluble vitamin, so make sure to put dressing on your salad, or cook your greens with oil.
- Helps protect bones from osteoporosis
- May help prevent and possibly even reduce atherosclerosis by reducing calcium in arterial plaques.
- May be key regulator of inflammation and may help protect us from inflammatory diseases including arthritis
- May help prevent diabetes.
- Vitamin K is a fat –soluble vitamin, so make sure to put dressing on your salad, or cook your greens with oil.

Greens have very little carbohydrates in them, and the carbohydrates that are there are packed in layers of fibre, which make them very slow to digest. That is why, in general, greens have very little impact on blood glucose. In some systems greens are even treated as a "freebie" carb-wise (meaning the carbohydrate does not have to be counted at all). However, some greens contain substances called oxalates which may bind some percentage of the calcium in the greens.

Benefits	*Vegetarian Diet*	*Non Vegetarian Diet*
Detoxifies	Contains fibre *e.g.* bottle gourd, pumpkin, spinach, cabbage, which flushes toxins out of the body.	*e.g.* eggs, fish and mutton is a poor source of fibre.
Stronger Bones	Calcium extraction is rare amongst vegetarians.	Gorging on meat can lead to protein over load. This can tax our kidneys, interfere with the absorption of calcium and prompt the body to extract existing calcium from the bones
Carb deficiency	Abundant in root and tuber vegetables	A non-vegetarian diet is a poor source of carbohydrates, which can lead to ketosis – a condition where the body starts breaking down fat (instead of cabs) as a source of energy.
Easy digestion	Complex carbohydrates in vegetarian food are digested gradually providing a steady source of glucose.	Where as meats rich in fat and protein are difficult to digest.
Healthy skin	Eating beet root, tomato, pumpkin and bitter gourd can clear of blemishes.	Not possible
Weight management	Eating whole grains, legumes, vegetables, nuts and fruits lowers cholesterol levels, blood pressure and obesity.	Avoiding meat is the simplest way to reduce fat intake.
Easy on the teeth	Our molars are more suited for grinding grains and vegetables than tearing flesh.	Digestion begins with the sliva, which can only digest complex carhohydrates present in plant foods.
Phyto -nutrients	Diabetes, cancer, kidney disease, stroke and bone loss are partially preventable with a good intake of phytonutrients.	As these are present only in vegetarian diet, the non vegetarians are at a loss.

Bibliography

Acharya Balkrishna. (2008). Secrets of Indian Herbs, for good health. Divya Prakashan Patanjali Yogpeeth. Maharishi Dayanand Gram, Delhi- Hardwar Highway, Bahadarabad, Hardwar-249408, Uttarkhand, India.

Anonymous (2009). Turmeric can sooth bowls. Tribune, April 22nd 2009, p13

Anonymous (2009). Vegetable juice may help people lose weight. Tribune, 22nd April.p.12

Anonymous (2009). Health and Science. Garlic can treat cancer. Times of India. 9th April, 2009.

Anonymous (2009). Health Notes Proteins in garden peas may help fight high blood pressure, urinary diseases. Tribune, Wednes day, 03. 25th 2009 p.11

Anonymous (2009). Health Notes. Broccoli, cabbage, can help fight skin cancer, Tribune.04.03.09, p.12.

Anonymous (20011). Eating veggies and fruits cuts the risk of certain blood cancers. Times of India, Sep.28th. 20011, p18

Anonymous (20011). Pumpkin works magic on your skin. Times of India, IndiaTimes.comlife Style/beauty/pumpkin.

Aonymous (2012). Why men should eat berries. Times of India. 18th April 2012

Anonymous (2012). Sip a cherry juice to sleep betterTimes of India.10.05.2012.

Anonymous (2012). Viagra, move over, pomerganate's here. Daily Mirror.27th August, 2012.

Anonymous (2012). Eating asparagus may prevent hangover (study). Times of India, 27th Dec. 2012

Anonymous (2013). Eat Cauliflower to ward off cancer. Times of India: Health and Fitness. March, 28th 2013

Anonymous (2013). Foods that don't let you slim down. Daily Mirrow. Apr 22, 2013.

Anonymous (2013). Eating tree nuts can boost health and cut heart disease risk. Health Notes. The Tribune, April, 24th, 2013

Anonymous (2013).7 reasons you Must eat olives. Times of India. Health and Fitness. May 2, 2013,

Anonymous (2013). 5 Fruits for a glowing skin. Times of India. Health and Fitness. May 16, 2013.

Anonymous (2013). Figs are extremely nutritious. Times of India. Health and Fitness.28th, May 2013.

Anonymous (2013). Garlic: Wonder drug for healthy life. Times of India. Diet. Jun 3rd. 2013.

Anonymous (2013). Want to live longer? Turn vegan. Times of India. Diet. June 8th.2013

Anonymous (2013). Rosemary for a glowing, soft skin. Times of India. Beauty. Jul 11, 2013

Anonymous (2013). Kiwi fruit more nutritious than 27 other fruits. Times of India. Health and Fitness. 6th August 2013.

Anonymous (2013). Fruits, Vegetables may lower women's bladder cancer risk. Times of India. IANS 24th August 2013

Anonymous (2013). Eating broccoli can help prevent osteoarthritis. Times of India. ANI Science, Aug 28th, 2013.

Anonymous (2013). Must-eat foods that improve your mood. Times of India. Health Me Up. 29th August.2013

Anonymous (2013). Walnuts may prevent diabetes and heart disease. Tribune Spectrum, Health Capsules. 6th October, 2013

Anonymous (2013). Pomegranates help burn fat, increase blood flow. Times of India. Health and Fitness. 12th October 2013.

Anonymous (2014). Enjoy grapes for good health. Times of India. Health and Fitness. 9th March 2014

Anonymous (2014). Don't throw those watermelon seeds away. Times of India, Beauty. 28th March 2014

Anonymous (2014). Foods that fight wrinkles and ageing. Times of India. Health and Fitness. 7th March, 2014

Anonymous (2014). Aphrodisiac foods you must eat. Times of India, Health and Fitness. 28th June 2014

Anonymous (2014). Turmeric 'could help brain heal itself. Times of India.'Heather Saul, The Independent. Sep. 28th, 2014.

Anonymous (2014). Foodsfor yourhair. TimesofIndia. Health and Fitness. Oct.9th2014

Anonymous (2014). Foods that boot your immunity. Times of India. Health and Fitness.19th October 2014

Anonymous (2014). Natural remedies to cure fertility problem. Times of India. Health and Fitness.21st. December 2014.

Anonymous. (2015). How papaya, fenugreek leaves can control dengue. Times of India. Health and Fitness. August 30, 2015.

Anonymous. (2015). Beetroot juice can help you beat mountain sickness. Times of India. Health and Fitness.13th October 2015.

Anonymous (2016). Handful of nuts cuts cancer risk. Short cuts. Time of India. 6th December 2016.

Anonymous. (2017). How to battle arthritis. Daily Mirrow.p.25, June 24th.2017

Anand Holla (2012). Spers are important, care them. Mumbai Mirror.1st. August 2012

Charanjit Parmar (2012). Fruit facts; Fully ripe Bananas have anti cancer properties. Tribune, 25thMarch.p.6.

Chaturvedi Vinita (2014). Try the magic of peppermint for health. Times of India. Health and Fitness. May 5th 2014

Cummings, J. H; Beatty, E. R; Kingman, S. M, Bingham, S. A; Englyst. (1996). Digestion and Physiological properties of resistant starch in human large bowls. Br. J. Nutr. 75: 733-747.

Deepika Sahu (2012). Health benefits of Cumin. Times of India. Dec.30th 2012.

Duggal, R. K (1995). Native spices of alien plant. Daily Excelsior. 19th Nov. 1995, col 3.p.6

Englyst, H. N; Kingman, S. N; Cummings, J. H. (1992). Classification and measurement of nutritionally importan starch fractions. Eur. J. Lin. Nutr. 46: 533-550.

Gita Angara. (2007). Fighting Cancer? Eat some cabbage. V. Care foundation, 183,

Golden Jubilee Block, Tata Memorial Hospital, Parel, Mumbai.4300012, vol.14. No.2 June 2007 p.6

Gopalkrishnan, T. R (2007). Vegetable Crops. Horticultural Science. Series-4. Publishing Agency. Pitampora, New Dehli. CU Block; L. S. C Market p.66-75, 308.

Gupta, Sanjana (2014).14 Natural painkillers. Times of India. Diet and fitness. 23rd October 2014

Ipshita Mitra (2012) Eight Indian Spices that prevent cancer. Times of India. July 14th 2012.

Ipshita Mitra (2013). Top 10 foods that prevent breast cancer. Times of India. May 11th, 2013

Ishi Khosla (2013). Good Health The Sunday Tribune Spectrum, (Other Seeds) Chandigarh. 8th September 2013.

Howard Michael (1987). Traditional Folk Remedies (Remedies, 1987) p.120.

Johnson, E. J (2002). The role of carotenoids in human health. Nutr. Chin. Care. 5: 56-65 orlutein content in Sweet Potato leaves. Hort. Sci.41 (5) :1269-1271.

Kounteya Sinha (2008). Garlic effective in treating arsenic poisoning. Studies. Times of India, New Delhi. 28th January. P.7

Khosla, Ishi (2015). Spice up for a healthy heart. The Sunday Tribune Spectrum. Good Health. 9th August 2015.

Krutika Behrawala (2014), Have garlic during winters. Times oflndia. New Delhi. Health and Fitness. Jan 4, 2014,

Levy, M. T (1966). The medicinal formulary or Agradbin of Al- Kindi Madison Wi: Uni. of Wiscosin Press: 410.

Makhija, Pooja (2014). Fruit=Health? Not always. Times of India. Health and Fitness, Oct 16th 2014.

Maneka Gandhi. (2005). Meat may develop Alzheimer‘s disease. Daily Excelsior, May, 15th p.3 Sunday Magazine.

Manu Vipin (2012). Eatlychee to fight breast cancer. Times of India, August 4th, 2012

Methew, Meghana (2008). Broccoli may slow down effects of ageing says the study. Good Health. Times of India, April 1sth 2008, p.28.

Michael Sharon. (2009). Nutrients A-Z. A user‘s guid to foods, herbs, vitamins, minerals and supplements. Carlton Book Ltd. 20 Mortimer Street London. W1 T 3

Meghna Mukherjee. (2013). Beauty benefits of Walnut oil. Times of India. 10th Jan. 2013.

Moghul. Sobiya. N. (2013). 20 best muscle building foods. Times of India. Health Me 6th, Jun, 2013.

Moghul. Sobiya. N. (2013). Why you must eat eggplant. Times of India. Health Me Up 11th June 2013

Mukerjee, Anjali, (2014). This water is not good for wellbeing. The Sunday Tribune Spectrum, Fitness.24thAugust 2014.

Nandkarni, K. N (1954). Allium cepa linn and Allium sativum linn in the Indian Materia Media, 3rd ed (part 1) Puranik, M. V. Popular Book Depo. Mumbai.

Nandi, Partha (2020). The health benefits of Tamarind. Askdrnandi.com

Parmar, Chiranjit. (2012). Two Bananas give 90minutes' body fuel. Fruit Facts. The Sunday Tribune. Spectrum, 2012.

Parmar, Chiranjit. (2013). Litchi is a super fruit. Fruit Facts. Tribune, July, 14th 2013.

Pinto, Renita Tisha (2014). Health benefits of oregano. Times of India. Health Me Up. Jan 21, 2014

Pinto, Renita Tisha (2014). Fat releasing foods. Times of India. Health Me Up. October3rd 2014

Purvaja Sawant (2013).3 Best foods for your hair. Times of India. Health and Fitness. May 13th, 2013,

Purvaja Sawant (2014). Foods that fight acne. Times of India. Health and Fitness. April 4th 2014

Purvaja Sawant, (2013). Beauty benefits of papaya. Times of India. Health Benefits, Oct 11, 2013.

Renita Tisha Pinto (2013). Top 20 fruits for diabetics. Times of IndiaHealth Me Up, 18.03.2012

Reuters, K. N (2012) Health benefits of curry leaf. Times of India.1st. August, 2012

Roy, S. K and Chakraborti, A. K (1993). Vegetable's of Temperate climate. Commercial and Dietary importance in Encyclopedia of Food Science. Food Technology and Nutr. Academic Press London.p.4743-4743.

Romana D'Souza (2014). Herbs and spices for cancer prevention. Times of India. Healthy living.19thOctober 2014.

Sahu, Deepika (2015). Extraordinary health benefits of black pepper. Times of India. Health and Fitness. Diet Sep 30, 2015,

Setalvad, Naini. (2014). Time to go nutty. Good health. SundayTribune Spectrum, Chandi. 7th December 2014.

Sinha, Kounteya (2008). Beetroot juice helps lower blood pressure, says the study. Times of India New Delhi, Feb.8th. 2008, p.12

Sinah, Kunteya. (2013). Good news for 13 million Indian asthmatics. Ginger abundantly found, shown to ease asthma attack. Times of India. Science. May 20th, 2013,

Sinha Seema (2014). Health benefits of curry leaves. Times of India, Feb.23.2014

Sinha, Seema. (2014). Green papaya has many nutritional benefits. Times of India. Sep 13, 2014

Sobiya N. Moghul. (2013), Cancer: 12 foods to battle cancer. Health Me Up. Times of India, Apr 14, 2013,

Sobiya N. Moghul. (2013).10 Healthy reasons to eat apples. Times of India. Health Me Up. Jun 29th, 2013.

Sharma, Gargi. (2016). Why soaked Almonds are better than raw Almonds. Times of India. May, 9th 2016.

Shah, Bonny (2020). The many health benefits of red chillies. YTime of India 31st July 2020. (TATA Nutrikorner.com).

Sumitra Nair (2013). The right way to eat fruits. Times of India. Health and Fitness.10th, Jun 2013

Sunderagan, P (2005). Turmeric- a potential weapon against Alzheimer's. The Hindu. 25th April, 2005.

Takuya Sakurai. (2008). Biotechnol Biochemic. (Kyorin University, Japan.72 (2), 463- 476.

Thadani, Khusboo, (2017). Foods the boost brain power. Mumbay Mirrow, June 19th.2017.

Trina Remedios. (2013). About fertility: Foods for fertility. Times of India. Jan.9th.2013.

Trina Remedios (2013). Are you getting enough fruits ?Times of India. Health me up.2nd October 2012

Trina Remedios (2013). Health benefits of pineapple. Times of India. Health and Fitness. Health Me Up. Oct 17, 2013,

Tina Remedios (2014). Health benefits of pomegranate. Times of India. Health Me Up. August 31st.2014

Vatsyayan, R (2010). Use Onion, be healthy. TheTribune. December, 29th 2010, p12

Vatsyayan, R (2011). Ginger: The universal Medicine. The Tribune, Wednesday, August 3rd, p.14

Willcox, J. K; Catignani, R. L and Lazarus, S (2003). Tomatoes and Cardio-vascular health. Crit. Rev. Food Science. Nutr. 43: 1-18

Zeenia Baria (2013). Health benefits of pears. Times of India. Health and Fitness. 8th September. 2013.

INTERNET

http://www.timesofindia.indiatimes.com/life-style/beauty/Beauty-benefits-of-walnut oil/articleshow/17887375.cms

http://www.timesof india.indiatimes.com/life-style/health-fitnes/diet/benifits-ofcoriander/articleshow/5912718.cm

http://www.prokerala-com/health/medicininal-herbs/annonasquamosa-sugarapple.php.

http://www.times of india.indiantimes.com/life-style/health-fitness/diet/figs-are-extremely-nutritious/articleshow/7626531.cms

http://timesofindia.indiatimes.com/life-style/health-fitness/health/Health-benefits-of-pomegranate/articleshow/176645

http://timesofindia.indiatimes.com/life-style/health-fitness/diet/About-fertility-Foods-for- fertility/articleshow/17937114.

http://www.nutrition-and-yuy.com/fruit-nutrition.html

http://www.nutrition-and-yuy.com/vegetable-nutrition.html

http://www.nutrition-and-yuy.com/nuts - nutrition.html

http://www.nutrition-and-yuy.com/healthy-herbs.html

http://www.nutrition-and-yuy.com/apple-fruit.html

http://www.nutrition-and-yuy.com/apricot.html

http://www.nutrition-and-yuy.com/banana-fruit.html

http://www.nutrition-and-yuy.com/cherry-fruit.html

http://timesofindia.indiatimes.com/life-style/health-fitness

http://www.nutrition-and-yuy.com/coconut.html

http://www.nutrition-and-yuy.com/dates.html

http://www.livestrong.com/article/87197-nutritional-value-dragon-fruit

http://www.drgranny.com/2011/02/18/health-benefits-of-dragon-fruit/

http://www.healthmeup.com/photogallery-healthy-living/15-herbs and -spices for cancer prevention/32949114

http://www.theopenews.com/should-bananas-be-eaten-ripe-or-green-209:html

http://www.nutrition-and-yuy.com/fig-fruit.html

http://www.nutrition-and-yuy.com/guava.html

http://www.nutrition-and-yuy.com/orange-fruit.html

http://www.nutrition-and-yuy.com/grape fruit-html

http://www.nutrition-and-yuy.com/lemon.html

http://www.nutrition-and-yuy.com/papaya-fruit.html

http://www.nutrition-and-yuy.com/grapes.html

http://www.nutrition-and-yuy.com/pears.html

http://www.nutrition-and-yuy.com/peaches.html

http://www.nutrition-and-yuy.com/plums.html

http://www.nutrition-and-yuy.com/lychee.html

http://www.nutrition-and-yuy.com/mango-fruit.html

http://timesofindia.indiatimes.com/life/health-fitness

http://times of india.indiatimes.com/life-style/health-fitness

http://healthydietcent.com/apricot-effective-fighter-against-anemia-and-cancer

http://www.nutrition-and-yuy.com/loquat-fruit.html

http://www.nutrition-and-yuy.com/pomegranate.html

http://www.nutrition-and-yuy.com/pineapple.html

http://www.nutrition-and-yuy.com/jackfruit.html

http://www.nutrition-and-yuy.com/strawberries.html

http://www.nutrition-and-yuy.com/raspberry.html

http://www.nutrition-and-yuy.com/gooseberries.html

http://www.nutrition-and-yuy.com/almonds.html

http://timesofindia.indiatimes.com/life-style/health-fitness

http://choosehealthylife.com/this-will-happen-to-your-breasts-after-you-eat-pineapple/

http://www.nutrition-and-yuy.com/walnut.html

http://www.nutrition-and-yuy.com/avacados-fruit.html

http://www.nutrition-and-yuy.com/blackberries.html

http://www.nutrition-and-yuy.com/cranberries.html

http://www.nutrition-and-yuy.com/kiwi-fruit.html

http://timesofindia.indiatimes.com/life-style/health-fitness

http://www.liveandfeel.com/nutrition/fruit-vegetable-role.html

https://www.facebook.com/medicinalmedium/photos/pb.1467896254 56224-2207520000.1444334119.16874107

http://food.ndtv.com/opinions/why-soaked-almonds-are-better-than-raw-almonds-726909?

Index

A

B

C

W

Y

Z